Advances in Intelligent Systems and Computing

Volume 761

Series editor

Janusz Kacprzyk, Polish Academy of Sciences, Warsaw, Poland
e-mail: kacprzyk@ibspan.waw.pl

The series "Advances in Intelligent Systems and Computing" contains publications on theory, applications, and design methods of Intelligent Systems and Intelligent Computing. Virtually all disciplines such as engineering, natural sciences, computer and information science, ICT, economics, business, e-commerce, environment, healthcare, life science are covered. The list of topics spans all the areas of modern intelligent systems and computing such as: computational intelligence, soft computing including neural networks, fuzzy systems, evolutionary computing and the fusion of these paradigms, social intelligence, ambient intelligence, computational neuroscience, artificial life, virtual worlds and society, cognitive science and systems, Perception and Vision, DNA and immune based systems, self-organizing and adaptive systems, e-Learning and teaching, human-centered and human-centric computing, recommender systems, intelligent control, robotics and mechatronics including human-machine teaming, knowledge-based paradigms, learning paradigms, machine ethics, intelligent data analysis, knowledge management, intelligent agents, intelligent decision making and support, intelligent network security, trust management, interactive entertainment, Web intelligence and multimedia.

The publications within "Advances in Intelligent Systems and Computing" are primarily proceedings of important conferences, symposia and congresses. They cover significant recent developments in the field, both of a foundational and applicable character. An important characteristic feature of the series is the short publication time and world-wide distribution. This permits a rapid and broad dissemination of research results.

More information about this series at http://www.springer.com/series/11156

Wojciech Zamojski · Jacek Mazurkiewicz
Jarosław Sugier · Tomasz Walkowiak
Janusz Kacprzyk
Editors

Contemporary Complex Systems and Their Dependability

Proceedings of the Thirteenth International Conference on Dependability and Complex Systems DepCoS-RELCOMEX, July 2–6, 2018, Brunów, Poland

 Springer

Editors
Wojciech Zamojski
Department of Computer Engineering
Wrocław University of Technology
Wrocław
Poland

Jacek Mazurkiewicz
Department of Computer Engineering
Wrocław University of Technology
Wrocław
Poland

Jarosław Sugier
Department of Computer Engineering
Wrocław University of Technology
Wrocław
Poland

Tomasz Walkowiak
Department of Computer Engineering
Wrocław University of Technology
Wrocław
Poland

Janusz Kacprzyk
Polish Academy of Sciences
Systems Research Institute
Warsaw
Poland

ISSN 2194-5357   ISSN 2194-5365 (electronic)
Advances in Intelligent Systems and Computing
ISBN 978-3-319-91445-9   ISBN 978-3-319-91446-6 (eBook)
https://doi.org/10.1007/978-3-319-91446-6

Library of Congress Control Number: Applied for

Printed on acid-free paper

This Springer imprint is published by the registered company Springer International Publishing AG part of Springer Nature
The registered company address is: Gewerbestrasse 11, 6330 Cham, Switzerland

Preface

In this volume, we would like to present the reader with proceedings of the Thirteenth International Conference on Dependability and Complex Systems *DepCoS-RELCOMEX*, which took place in the Brunów Palace in Poland from 2 to 6 July 2018.

DepCoS-RELCOMEX is an annual conference series organized since 2006 at the Faculty of Electronics, Wrocław University of Science and Technology, initially by Institute of Computer Engineering, Control and Robotics (CECR) and now by Department of Computer Engineering. Its idea came from the heritage of the other two cycles of events: RELCOMEX (1977–1989) and Microcomputer School (1985–1995) which were organized by the Institute of Engineering Cybernetics (the previous name of CECR) under the leadership of Prof. Wojciech Zamojski, now also the DepCoS chairman. In this volume of "Advances in Intelligent Systems and Computing", we would like to include results of studies on selected problems of contemporary complex systems and their dependability. Proceedings of the previous DepCoS events were published (in historical order) by the IEEE Computer Society (2006–2009), by Wrocław University of Technology Publishing House (2010–2012) and recently by Springer in AISC volumes no. 97 (2011), 170 (2012), 224 (2013), 286 (2014), 365 (2015), 479 (2016) and 582 (2017).

Published by Springer Nature, one of the largest and most prestigious scientific publishers, the AISC series is one of the fastest growing book series in their programme. Its volumes are submitted for indexing in ISI Conference Proceedings Citation Index (now run by Clarivate), Ei Compendex, DBLP, SCOPUS, Google Scholar and SpringerLink, and many other indexing services around the world.

The selection of papers in these proceedings illustrates a broad variety of topics which are investigated in dependability analyses of today's complex systems. Dependability came naturally as a contemporary answer to new challenges in their reliability evaluation. This kind of systems cannot be interpreted only as (however complex and distributed) structures built on the base of technical resources (hardware), but their analysis must take into account a unique blend of interacting people (their needs and behaviours), networks (together with mobile properties, iCloud organization) and a large number of users dispersed geographically and producing an unimaginable number of applications (working online). A growing number of research methods apply the newest results of artificial intelligence (AI) and computational intelligence (CI). Today's complex systems are *really* complex and are applied in many different fields of contemporary life.

Dependability approach in theory and engineering of complex systems (not only computer systems and networks) is based on multidisciplinary approach to system theory, technology and maintenance of the systems working in real (and very often unfriendly) environment. Dependability concentrates on efficient realization of tasks, services and jobs by a system considered as a unity of technical, information and human assets, in contrast to "classical" reliability which is more restrained to the analysis of technical resources (components and structures built from them). This difference has shaped natural evolution in topical range of subsequent DepCoS conferences which can be seen over recent years.

The Programme Committee of the 13th International DepCoS-RELCOMEX Conference, its organizers and the editors of these proceedings would like to gratefully acknowledge participation of all reviewers who evaluated conference submissions and in this way helped to refine the contents of this volume. The list includes, in alphabetic order, Andrzej Białas, Ilona Bluemke, Eugene Brezhniev, Dejiu Chen, Frank Coolen, Mieczysław Drabowski, Francesco Flammini, Manuel Gil Perez, Zbigniew Gomółka, Zbigniew Huzar, Igor Kabashkin, Vyacheslav Kharchenko, Wojciech Kordecki, Alexey Lastovetsky, Jan Magott, István Majzik, Jacek Mazurkiewicz, Marek Młyńczak, Yiannis Papadopoulos, Krzysztof Sacha, Rafał Scherer, Mirosław Siergiejczyk, Robert Sobolewski, Janusz Sosnowski, Jarosław Sugier, Victor Toporkov, Tomasz Walkowiak, Bernd E. Wolfinger, Wojciech Zamojski and Wlodek Zuberek.

Expressing our thanks to all the authors who have chosen DepCoS as the publication platform of their research, we would like to stress our desire that their papers will help in further developments in design, analysis and engineering of dependability aspects of complex systems, creating a valuable source material for scientists, researchers, practitioners and students who work in these areas.

Wojciech Zamojski
Jacek Mazurkiewicz
Jarosław Sugier
Tomasz Walkowiak
Janusz Kacprzyk

Thirteenth International Conference on Dependability and Complex Systems DepCoS-RELCOMEX

Organized by Department of Computer Engineering, Wrocław University of Science and Technology, Brunów Palace, Poland, July 2–6, 2018

Programme Committee

Wojciech Zamojski (Chairman)	Wrocław University of Science and Technology, Poland
Ali Al-Dahoud	Al-Zaytoonah University, Amman, Jordan
Włodzimierz M. Barański	Wrocław University of Science and Technology, Poland
Andrzej Białas	Institute of Innovative Technologies EMAG, Katowice, Poland
Ilona Bluemke	Warsaw University of Technology, Poland
Eugene Brezhniev	National Aerospace University "KhAI", Kharkiv, Ukraine
Dariusz Caban	Wrocław University of Science and Technology, Poland
Frank Coolen	Durham University, UK
Mieczysław Drabowski	Cracow University of Technology, Poland
Chen De-Jiu	KTH Royal Institute of Technology, Stockholm, Sweden
Francesco Flammini	University of Naples "Federico II", Italy
Manuel Gill Perez	University of Murcia, Spain
Zbigniew Huzar	Wrocław University of Science and Technology, Poland

Igor Kabashkin   Transport and Telecommunication Institute, Riga,
                              Latvia
Janusz Kacprzyk   Polish Academy of Sciences, Warsaw, Poland
Vyacheslav S. Kharchenko   National Aerospace University "KhAI", Kharkiv,
                              Ukraine
Mieczysław M. Kokar   Northeastern University, Boston, USA
Krzysztof Kołowrocki   Gdynia Maritime University, Poland
Wojciech Kordecki   The Witelon State University of Applied Sciences
                              in Legnica, Poland
Leszek Kotulski   AGH University of Science and Technology,
                              Krakow, Poland
Henryk Krawczyk   Gdansk University of Technology, Poland
Alexey Lastovetsky   University College Dublin, Ireland
Jan Magott   Wrocław University of Science and Technology,
                              Poland
Istvan Majzik   Budapest University of Technology and
                              Economics, Hungary
Jacek Mazurkiewicz   Wrocław University of Science and Technology,
                              Poland
Marek Młyńczak   Wrocław University of Science and Technology,
                              Poland
Yiannis Papadopoulos   Hull University, UK
Ewaryst Rafajłowicz   Wrocław University of Science and Technology,
                              Poland
Krzysztof Sacha   Warsaw University of Technology, Poland
Elena Savenkova   Peoples' Friendship University of Russia,
                              Moscow, Russia
Rafał Scherer   Częstochowa University of Technology, Poland
Mirosław Siergiejczyk   Warsaw University of Technology, Poland
Czesław Smutnicki   Wrocław University of Science and Technology,
                              Poland
Robert Sobolewski   Bialystok University of Technology, Poland
Janusz Sosnowski   Warsaw University of Technology, Poland
Jarosław Sugier   Wrocław University of Science and Technology,
                              Poland
Victor Toporkov   Moscow Power Engineering Institute (Technical
                              University), Russia
Tomasz Walkowiak   Wrocław University of Science and Technology,
                              Poland
Max Walter   Siemens, Germany
Bernd E. Wolfinger   University of Hamburg, Germany
Min Xie   City University of Hong Kong, Hong Kong SAR,
                              China

| Irina Yatskiv | Transport and Telecommunication Institute, Riga, Latvia |
| Włodzimierz Zuberek | Memorial University, St. John's, Canada |

Organizing Committee

Chair

Wojciech Zamojski

Members

Włodzimierz M. Barański
Jacek Mazurkiewicz
Jarosław Sugier
Tomasz Walkowiak
Mirosława Nurek

Contents

Dynamic Time Warping Analysis for Security Purposes in Wireless Sensor Networks

Tomasz Andrysiak[✉] and Łukasz Saganowski

Institute of Telecommunications and Computer Science,
Faculty of Telecommunication, Information Technology and Electrical
Engineering, UTP University of Science and Technology,
Kaliskiego 7, 85-789 Bydgoszcz, Poland
{andrys,luksag}@utp.edu.pl

Abstract. The article presents network traffic anomaly detection method for the tested Wireless Sensor Networks (WSN) infrastructure with the use of Dynamic Time Warping (DTW) analysis. There are described essential aspects of threats to safety of correct operation of a WSN network with regard to possible attack types. A two-stage method of anomaly/attack detection was proposed. Stage One consisted in finding and eliminating of any outlying observations in the analyzed parameters of the WSN network traffic by means of Cook's distance. Data prepared in this manner were used in Stage Two for estimation of mean time series, which describe mean variability of the analyzed WSN's parameters. There was also proposed an effective method of profile creation for the examined network traffic parameters on the basis of Dynamic Time Warping analysis. In the suggested method, there were used relations between the estimated traffic model and its real variability in order to detect abnormal behavior, which is probably a consequence of an attack. The experiment results confirmed efficiency and effectiveness of the presented method.

Keywords: Wireless Sensor Networks · Anomaly detection
Dynamic Time Warping analysis · Outliers detection
Cook's distance · Network attacks

1 Introduction

Protection of a WSN network infrastructure against various types of attacks is currently a domain widely surveyed and developed. One of the possible solutions to this problem is detection and classification of abnormal behaviors reflected in the analyzed network traffic of WSN infrastructure. The advantage of this approach is that there is no need for upstream identification and memorizing patterns of potential attacks. Therefore, in the decision making process, it is only necessary to define what is and what is not normal behavior of the network traffic in order to detect the ongoing attack [1].

The classic sensor networks are being optimized for energy efficiency and low delay times, often disregarding safety requirements, which must be retained in critical infrastructures. Usually, there is a need to ensure reliability of the network's operation, resistance to the assault claims of the sent data, and fast enough response time of the network to alarming events. The wireless sensor networks, due to their arrangement of

© Springer International Publishing AG, part of Springer Nature 2019
W. Zamojski et al. (Eds.): DepCoS-RELCOMEX 2018, AISC 761, pp. 1–12, 2019.
https://doi.org/10.1007/978-3-319-91446-6_1

the elements in a particular area, are exposed to many threats. They can be random (e.g. a breakdown of the node), or they can be a result of a conscious and willful activity of intruders (like Denial-of-Service attacks). In case of WSN networks, whose main function is to provide the public telemetry, the risk of intentional actions of intruders significantly increases [2].

The Wireless Sensor Network used in the networks oriented onto safety must be able to deal with, among others, an attempt to impersonate the user, manipulate data, jam the system, flood or introduce latency in the network. The intruder may also try to damage physically the elements of WSN. Regardless of the human, there can also be other factors influencing the operation of a sensor network, such as environmental disruption, e.g. a strong radio signal or fire. The sensor networks, in their most applications, have limited resources of energy and computational abilities of the nodes in comparison to other cable and wireless IT networks. Due to significant differences in the required features of a sensor network, the solutions must be tailored in such a way that they are able to ensure a suitable level of security [3, 4].

Currently intensively examined and developed methods for detection of intrusion/ attacks are those utilizing phenomenon called anomaly in network traffic [5–7]. One of possible solutions is detection of abnormal behavior by means of statistical models, which describe the analyzed network traffic. The most often used are autoregressive models, heteroskedastic models or exponential smoothing models [8–10]. They allow for estimation of the characteristics of the analyzed network traffic. All the mentioned solutions require measure of similarity between the model of the network traffic and its real conduct. Usually, the measures which are used are based on Euclidean distance, but also there are more effective ones, like Dynamic Time Warping analysis [11]. In literature, there can also be found hybrid methods combining signal's decomposition elements (e.g. discrete wavelet transform), and then statistical models are estimated on such transformed signal [12].

In the present article we proposed anomaly detection system for WSN infrastructure. The solution which has been proposed is based on *(i)* Cook's distance for outliers detection and elimination, *(ii)* calculating mean time series of WSN traffic features value and *(iii)* Dynamic Time Warping (DTW) analysis to generation of network profiles. The process of anomaly detection consisted in a comparison between the parameters of normal behavior of WSN network estimated with the use of DTW measure and parameters of variability of the real analyzed WSN network traffic.

This paper is organized as follows. After the introduction, in Sect. 2, the overview of security risks in WSN infrastructure is presented. In Sect. 3, there are generally discussed the techniques and methods used in methodology of anomaly detection system. Then, in Sect. 4, the experimental results of the proposed solution are shown. Conclusions are given thereafter.

2 Security Risks in WSN Infrastructure

Designing and realization of the Wireless Sensor Networks requires serious input of technical means in order to ensure a suitable level of security of their operation in a given environment. The wireless networks are particularly exposed to a series of threats, such as terrorist attacks, eavesdropping, or impersonating the user. The attacks destabilizing the work of WSN are in many ways, unfortunately, quite easy to perform. Generating strong electromagnetic signal in the vicinity of network nodes can effectively disrupt or block transmission. However, in the case of gaining or acquisition of authorizing user data it is possible to impersonate his identity to falsify the data collected or to carry out new attacks from inside the network [1, 3].

The attacks on wireless sensor networks security can be divided into two main groups: passive attacks and active attacks. Passive attacks are any attempts of unauthorized access to data, or to the infrastructure of the wireless sensor network system, in which the perpetrator does not use the emission of signals that may interfere with or impede its correct operation. In turn, the active attacks are any steps related to unauthorized access by emitting some signals or by actions that can be detected [2, 4].

2.1 Passive Attacks on WSN Infrastructure

When performing a passive attack on a wireless sensor network, the attacker camouflages its presence and attempts to access data transmitted in the WSN via passive monitoring of the network. One of the passive attacks methods on the WSN is the interception of data transmitted between the nodes of the network. Due to the characteristics of the WSN radio medium in which the data transmission is performed in a transparent manner, the network is relatively susceptible to this kind of passive attack. In order to minimize the risk of such an event, broadcasting modules used in the WSN nodes have low-power transmitters – which makes the communication coverage minimized to a specific area. This brings the attacker relatively close to the WSN node, in order to make the eavesdropping of the radio baseband transmission effective. In order to protect against such an event, cryptographic mechanisms should be applied in the WSN, and where it is not required or possible – the monitoring of personnel moving in the area of the WSN operation should be introduced and the radio emissions from a particular area should be monitored [13].

Another passive method of an attack on WSN is the analysis of the traffic inside the network. In this case, the attacker's intention is not to know the content of data packets transmitted in the WSN, but to gain knowledge about the wireless sensor network topology. Due to the nature of the WSN a part of nodes is greatly charged with the transmission of information. Increased data transmission load on these nodes is also associated with retransmission of the information sent from the neighboring WSN nodes to the base station. Other nodes with a relatively high communication interface load are the nodes supervising the WSN clusters (the clusters are created for increasing the scalability of the WSN). The node collects data coming from only a given cluster and retransmits the data to the higher layers of the system. This means that gathering information based on the analysis of traffic in the WNS gives the intruder the knowledge about the critical nodes of a sensor network in terms of ensuring the accuracy of its work [14].

2.2   Active Attacks Within the Frame of WSN Infrastructure

In contrast to the passive methods of attacking the WSN described above, in the active forms of attack the attacker affects directly or indirectly the content of the information sent by the WSN. Attacks of this kind are more easily detected, as compared to passive attacks, because they directly affect the quality of the sensor network. One of the effects of the active attack can be, for example, the degradation of services in the WSN or, in extreme cases, the lack of access to certain services or even the complete loss of control over the network. Direct attacks on the WSN hardware infrastructure are particularly dangerous in the case of critical network infrastructures. These types of attacks are aimed at reducing the area monitored by the sensor network or at a total disabling of the given WSN [15, 16].

Manipulating the WSN nodes is aimed at diverting attention of the sensor network operator from the main attack, which may be spoofing or denial-of-service attack. The attacks on the WSN infrastructure using the high-energy short-term electromagnetic pulse are aimed at annihilation of a given sensor network (in the narrow sense of "annihilation") or of all electronic devices found in the destructive area of the electromagnetic pulse (in the broad sense of the word) [1].

Targeted attacks on the integrity or confidentiality of data are particularly dangerous because they give the attacker an unauthorized access to the network and to the data transmitted by the network. A Sybil Attack is a special form of spoofing and it consists in spoofing by broadcasting over a malicious node of many identifiers or on compromising a legal network node and the acquisition of its identity with the access to the WSN infrastructure. This type of attack is often carried out against the data correlation and aggregation systems. Sybil attacks can also be directed against the routing algorithms and algorithms determining the location of individual network nodes. It is also worth to remember that the number of identifications transmitted by a malicious node in relation to the Sybil Attack helps to hide the attack [13].

DoS attacks in sensor networks involve causing an overload of the affected WSN nodes, thereby preventing obtaining data from the affected nodes or preventing use of the services offered by the attacked sensor network. DoS attacks are targeted at all layers of the ISO/OSI network model [17].

An attack on the services carried out on the first layer of the ISO/OSI sensor network model consists in interfering the radio frequency band used by a given sensor network. DoS attacks in the data link layer come down to flooding the node with plenty of information, which dramatically increases the probability of packet collisions or it forces the attacked node to a continuous packets retransmission. The attacker can thus lead to the rapid depletion of the node energy resources. The transport layer is responsible for the compilation of connections between the network devices. Because of the large number of possible attacks on the transport layer protocols, as well as the number of protocols used for this purpose in the WSN, some examples of the types of attacks may be mentioned, such as spoofing acknowledgments, repeating acknowledgment, interfering acknowledgement, sequence number change or falsification of connection requests [18]. An example of an application-layer attack is a broadcast from, a malicious node with a smaller or a bigger output power than desired – this may hinder the correct localization of nodes [16].

3   Methodology of Anomaly Detection System

An idea for protection against new, unknown attacks may be a rather radical shift in concept of network infrastructure security. Instead of searching for attack signatures in the network traffic, it is better to define the profiles of allowed activities. Then, all the noticed deviations from those profiles can be treated as symptoms of new attacks/anomalies [19].

The profiles can be whatsoever, but in the systems of anomaly detection, an optimal solution seems to be the choice of such features for the profile, which would suggest symptoms of an unauthorized activity [20] when their values/tolerance is exceeded.

The strength of such an approach is protection against attacks unknown so far, or directed onto particular resources of network infrastructures, or simply constituting so called "zero-day exploits". Then, the anomaly detection systems, on the basis of defined network traffic profiles, can play an important role and turn out to be the only effective solution in such environments.

3.1   The Proposed Solution

We suggest a two-stage method of anomaly/attack detection. In the first step, IP packets are captured by the network sensor, then WSN traffic is extracted from IP packets. Next, we detect and eliminate any outlying observations in the analyzed WSN network traffic parameters by means of Cook's distance. The data prepared in such a way are used in Stage Two for estimation of mean time series, which describe mean variability of the analyzed WSN's parameters, and for creation of profiles for normal behavior of WSN network traffic (without attacks/ anomalies). In the profiling process, we use DTW analysis. In order to detect abnormal behavior, which is probably an aftermath of an attack, we analyze relations between the estimated traffic profile and its real variability on the basis of DTW measures.

The main hardware and software elements of proposed anomaly/attack detection method are presented in Fig. 1. WSN network consist of sensor network and WSN to IP gateway. Traffic from WSN network is packed into IP packets and it is analyzed by means of the proposed detection algorithm.

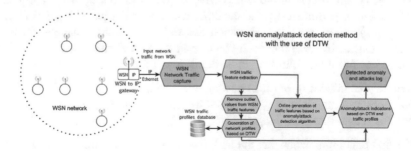

Fig. 1. Block scheme of the proposed anomaly/attack detection method.

3.2   Outliers Detection and Elimination Based on the Cook's Distance

In our approach, identification of outliers in the analyzed WSN traffic parameters is performed by means of a method using the Cook's Distance [21]. The essence of this method is estimation of the distance which states the level of data matching for two models: *(i)* a complete model, which includes all observations from the learning set, and *(ii)* a model built on a set of data, from which one *i* observation was omitted

$$D_i = \frac{\sum_{j=1}^{n} \left(\hat{Y}_j - \hat{Y}_{j(i)} \right)^2}{m \cdot MSE}, \tag{1}$$

where $\hat{Y}_j$ is the forecasted value of x variable for observations number j in the complete model, i.e. built on the whole learning set; $\hat{Y}_{j(i)}$ is the forecasted value of x variable for observations number j in the model built on the set in which the i - number observation was temporarily deactivated, MSE is the mean-model error, and m is the number of parameters used in the analyzed model.

For the Cook's distance D_i threshold value, above which the given observation should be treated as an outlier, in compliance with criterion (1), 1 is accepted, or alternatively

$$\frac{4}{n - m - 2}, \tag{2}$$

where n is the number of observations in the learning set.

Using the above formulated rules, we perform detection and elimination of outlying values for the analyzed WSN network traffic parameters in order to prepare them properly for the stage of profiles creating.

3.3   Generation of Network Profiles Based on Dynamic Time Warping Analysis

The main aim of use of DTW is to compare or determine the dissimilarity of two time series of real valued vectors $X \in (x_1, x_2, \ldots, x_N)$ and $Y \in (y_1, y_2, \ldots, y_M)$, where the lengths $N \in \mathbb{N}$ and $M \in \mathbb{N}$ of the consecutive sequence do not have to be equal.

This action is performed by the mean of use a warping pat $P \in [p_1 = (x_{n_1}, y_{m_1}), p_2 = (x_{n_2}, y_{m_2}), \ldots, p_L = (x_{n_L}, y_{m_L})]$, which determine the alignment between the two time series that is, by assigning each element in X to an element in Y. In DTW, this method has to fulfill the following three conditions [22]:

- Boundary condition: $p_1 = (1, 1)$ and $p_L = (N, M)$,
- Monotonicity condition: $n_1 \leq n_2 \leq \ldots \leq n_L$ and $m_1 \leq m_2 \leq \ldots \leq m_L$,
- Step size condition: $p_{l+1} - p_l \in \{(1, 0), (0, 1), (1, 1)\}$ for $l \in (1, 2, .., L)$.

The first condition claims that the path has to begin in the initial part of both time series, whereas the last path element has to include the end of both time series. The second condition claims that path must not return to an earlier element in any of the two time series. The last condition claims that the path must not skip an element in any of the two time series, hence, the path is only able to take one step in one or both time series.

The dissimilarity of each element (x_{n_l}, y_{m_l}) for $l \in (1, 2, \ldots, L)$ in the warping path P is defined by a local cost measure $c(x_{n_l}, y_{m_l})$. A low cost for (x_{n_l}, y_{m_l}) would appoint that this pair is similar, whereas a large cost defines a difference. The total cost $C_p(X, Y)$ of a path P can in such case be estimated as

$$C_p(X, Y) = \sum_{l=1}^{L} c(x_{n_l}, y_{m_l}). \tag{3}$$

The dynamic time warping distance $DTW(X, Y)$ between two time series X and Y can be estimated as the minimal total cost over all possible paths, which can be calculated as

$$DTW(X, Y) = \min\{(C_p(X, Y)) \mid p \text{ is an } (N, M) - warping\ path\}. \tag{4}$$

To estimate this minimal cost effectively, an accumulated cost matrix C of size $M \times N$ is presented in Eq. (4). Each element in this matrix is estimated recursively by means of adding the local cost to the minimal cost of all possible previous states, which can be formulated as

$$C_{n,m} \begin{cases} c(x_1, y_m) & n = 1, m = 1 \\ C_{1,m-1} + c(x_1, y_m) & n = 1, m > 1 \\ C_{n-1,1} + c(x_n, y_1) & n > 1, m = 1 \\ min(C_{n,1,m-1}, C_{n-1,m}, C_{n,m-1}) + c(x_n, y_m) & otherwise \end{cases} \tag{5}$$

The optimal path p^* consequently is to be calculated in reverse order from accumulated cost matrix C [23].

For proposed method we captured several time series for a given traffic feature from Table 1. In the next step we calculate mean time series for a given traffic feature. DTW is used in the context of variability and similarity calculation for a given traffic feature. We calculate distances by means of DWT between mean time series and every time series collected in the first step separately for every traffic feature. In the end, we selected two biggest distance values Δ_1 and Δ_2. We can understand Δ_1 and Δ_2 values as a distance measure between mean time series and two boundary time series (upper and lower). Presented operations are calculated for every traffic feature from Table 1. Additional information about profiles calculation are presented in Sect. 4.

4   Experimental Results

To check usefulness of the proposed anomaly/attack detection method, we used WSN network based on sensors with TinyOS [26] embedded operation system and WSN to IP gateway. A block scheme of the proposed method is presented in Fig. 1. A part of the testbed used for experiments (sensors and WSN to IP gateway) is presented in Fig. 2.

Fig. 2. A part of testbed used for experiments – WSN to IP gateway (on the left) with 7 example WSN sensors.

The main hardware and software elements of proposed anomaly/attack detection method are presented in Fig. 1. WSN network consists of sensor network and WSN to IP gateway. Traffic from WSN network is packed into IP packets and it is analyzed by the proposed detection algorithm. In the first step, IP packets are captured by a network sensor, then WSN traffic is extracted from IP packets. After traffic features extraction process, first we have to calculate profiles of normal (without attack/anomalies) WSN traffic network behavior.

WSN traffic features are presented in Table 1. Traffic features are represented as a time series (we collected traffic during 7 days' cycles). For every given traffic feature we collected several time series in order to calculate mean time series. Traffic is analyzed in one-minute windows (a window time can be set arbitrary). After that, we calculate for given analysis, window distance (by means of DTW) between the mean time series and remaining time series gathered during the traffic collecting process (the mean time series and the remaining time series represent the same traffic feature from Table 1). In the profile database we save two biggest distance values Δ_1 and Δ_2, which represent range of time series' similarity and mean time series for a given traffic feature.

Table 1. WSN traffic feature description

WSN Feature	WSN traffic feature description
WSNF1	QIRF: quality indicator of radio link (changes from: 0–127)
WSNF2	PFR: packet failure rate per time interval [%]
WSNF3	PPTM: packets per time interval value
WSNF4	RSSI: received signal strength indication for a given WSN sensor [dBm]
WSNF5	TTL: WSN packets TTL value
WSNF6	PSV: supply of WSN sensor in [V]
WSNF7	ST: WSN sensor temperature in [C]
WSNF8	SR: number of WSN sensor resets

Table 2. Results of detection rate DR[%] and FP[%] for three scenarios SC1-SC3 ("x" – character means that feature is not important for this anomaly/attack scenario)

WSN	DR[%]			WSN	FP[%]		
Feature	SC1	SC2	SC3	Feature	SC1	SC2	SC3
WSNF1	96.2	x	x	WSNF1	5.4	x	x
WSNF2	91.4	x	x	WSNF2	9.4	x	x
WSNF3	81.3	94.1	92.6	WSNF3	9.5	7.2	7.4
WSNF4	94.2	x	x	WSNF4	7.8	x	x
WSNF5	71.4	90.5	90.8	WSNF5	10.2	7.5	8.4
WSNF6	x	82.2	75.4	WSNF6	x	9.3	9.6
WSNF7	x	81.3	x	WSNF7	x	10.2	x
WSNF8	70.4	70.2	x	WSNF8	10.6	10.4	x

During normal work of the proposed anomaly/attack detection method, we extracted online traffic features from Table 1 and calculate distance for one-minute analysis window between the present time series values and the mean time series from profile database by means of DTW. When the online calculated distance values for a given analysis window exceeded values stored in the profile database, we indicate possible anomaly or attack.

In order to evaluate usability of the proposed method we prepared three anomaly or attack scenarios SC1-SC3. Testbed used in our experiments was partially artificial. We simulated real world WSN network conditions for example by spreading sensors on different rooms and different building levels and create additional obstacles between selected WSN sensors in order to make WSN network operation as close as possible to real world condition. In case of network attacks for Scenario 2 and Scenario 3 we added malicious sensors to WSN test network.

- SC1 - Scenario 1:

In the first scenario, an anomaly is realized by applying Radio Frequency Interferences RFI. In our case, we used ISM transmitter that operates on frequencies used by our WSN network. Distortion source may transmit packets that are recognized or not from the WSN protocol point of view used in our tested WSN network. Source of distortions should be placed near WSN/IP gateway or in close proximity to the group of sensors to achieve the best results. A different way to attack WSN hardware is to shield sensor or antennas of selected devices in WSN network.

The presented scenario has the biggest impact on QRIF (WSNF1), RSSI (WSNF4) and PFR (WSNF2) WSN traffic features. Detection rate and false positive values for SC1 are presented in Table 1. Indirectly EMI distortions can have an impact on, for example, SR (WSNF8), when the source of distortions is close to the sensor or WSN/IP gateway.

- SC2 – Scenario 2:

In the second scenario we simulate DoS attack by WSN flooding attack. The main aim of attack is to disturb packet exchanging by WSN sensor in test network. Attack is carried out by sending in broadcasts mode any type of packets that are recognized by sensors with time delay preventing sensors to process any usable packets. This type of attack has the biggest impact on data link and network features. Impact of flooding attack can be seen noticed in PPTM (WSNF3) and TTL (WSNF5) directly (see Table 1). Indirect effect of this attack can be noticed in PSV (WSNF6), ST (WSNF7) and SR (WSNF8) (impact is observed in longer time period). We can observe power supply (battery) degradation and hardware temperature increasing. More detailed results for SC2 are presented in Table 2.

- SC3 – Scenario 3:

Last anomaly/attack scenario requires adding malicious sensors to the WSN network. This type of attack is called Wormhole attack [25]. Malicious nodes create additional communication tunnel that disturbs communication between other sensors in WSN network. As a result, for example, routing protocol of network is disturbed and additional traffic is generated so usable bandwidth of WSN links are constrained.

Effect of this attack can be seen especially through PPTM (WSNF3) and TTL (WSNF5) WSN network traffic features. In longer time period, existence of larger amount of traffic will have impact on power supply PSV (WSNF6) of sensors that take part in communication process.

Based on experiments we can conclude that the examined traffic features can be divided into those connected to data link and network layer (WSNF1-WSNF5) which are important for communication process. Remaining traffic features from Table 1. WSNF6-WSNF8 can be classified to the WSN maintenance features where attacks or anomalies can be observed indirectly. The traffic features which are not important from a given scenario point of view, are not included in Table 2.

Taking into account all scenarios, we achieve detection rate up to 96.2%, while FP was 5.4%. All anomaly/attack detection systems have bigger values of FP because we try to recognize also unknown anomalies or attacks in contrary to IDS Intrusion Detection class systems, where mechanisms of attacks are known. FP values approximately up to 10% are acceptable for these anomaly/attack detection systems [24, 25].

5 Conclusion

The increasing number of new attacks, their global scope and level of complexity enforce dynamic development of network protection systems. In particular, maintaining adequate level of security and safety of resources and critical infrastructure realized as Wireless Sensor Networks is currently an intensively researched and developed domain.

WSN networks, due to their nature, are exposed to a great number of threats coming from both outside and inside of their own infrastructure. Thus, these networks require providing integrity and confidentiality of transmission, as well as protection of their nodes and data sent by their means. The most often implemented mechanisms, the aim of which is to provide safety, are the methods of detection and classification of abnormal behaviors reflected in the analyzed network traffic.

In this article, we presented a method of WSN network traffic parameter anomaly detection. To detect anomalies/attacks, there were used differences between the real network traffic and the estimated profile of this traffic, for the analyzed WSN network parameters, on the basis of DTW measures. For proper data preparation, any outlying observations were detected and eliminated from the analyzed WSN network traffic parameters by means of Cook's distance. Data prepared in this way were used for estimation of mean time series, which describe mean variability of the analyzed WSN's parameters, and for creation of profiles for normal WSN network traffic.

References

1. Dargie, W., Poellabauer, C.: Fundamentals of Wireless Sensor Networks. John Wiley & Sons Ltd, Chichester (2010)
2. Cayirci, E., Rong, C.: Security in Wireless Ad Hoc and Sensor Networks. John Wiley & Sons Ltd, Chichester (2009)
3. Butun, I., Morgera, S.D., Sankar, R.: A survey of intrusion detection systems in wireless sensor networks. IEEE Comm. Surv. Tutorials **16**(1), 266–282 (2014)
4. Goszczyński, T.: Problemy bezpieczeństwa w bezprzewodowych sieciach sensorowych (2011). http://www.par.pl/automatyka/bezpieczenstwo/413-problemy-bezpieczenstwa-w-bezprzewodowych-sieciach-sensorowych.html
5. Chondola, V., Banerjee, A., Kumar, V.: Anomaly detection: a survey. ACM Comput. Surv. **41**(3), 1–72 (2009)
6. Xie, M., Han, S., Tian, B., Parvin, S.: Anomaly detection in wireless sensor networks: a survey. J. Netw. Comput. Appl. **34**(4), 1302–1325 (2011)
7. Rodriguez, A., Mozos, M.: Improving network security through traffic log anomaly detection using time series analysis. In: Computational Intelligence in Security for Information Systems, pp. 125–133 (2010)
8. Andrysiak, T., Saganowski, Ł.: Network anomaly detection based on ARFIMA model. In: Image Processing & Communications, Challenges 6. Advances in Intelligent Systems and Computing, vol. 313, pp. 255–261. Springer (2015)
9. Andrysiak, T., Saganowski, Ł.: Network anomaly detection based on statistical models with long-memory dependence. In: Theory and Engineering of Complex Systems and Dependability. Advances in Intelligent Systems and Computing, vol. 365, pp. 1–10. Springer (2015)
10. Andrysiak, T., Saganowski, Ł., Maszewski, M.: Time series forecasting using Holt-Winters model applied to anomaly detection in network traffic. In: International Conference CISIS 2017. Advances in Intelligent Systems and Computing, pp. 567–576. Springer (2017)
11. Keogh, E., Pazzani, M.: Derivative dynamic time warping. In: First SIAM International Conference on Data Mining (SDM 2001), pp. 1–11 (2001)
12. Saganowski, Ł., Andrysiak, T., Kozik, R., Choraś, M.: DWT-based anomaly detection method for cyber security of wireless sensor networks. Secur. Commun. Networks **9**(15), 2911–2922 (2016)
13. Waraksa, M., Żurek, J.: Bezpieczeństwo transmisji danych w sieciach sensorowych, Zeszyty Naukowe Akademii Morskiej w Gdyni. Wybrane zagadnienia telekomunikacji, Gdynia, str., pp. 88–98 (2011)
14. Perrig, A., Stankovic, J., Wagner, D.: Security in wireless sensor networks. Commun. ACM **47**, 53–57 (2004)
15. Hu, Y.C., Perrig, A., Johnson, D.B.: Packet leashes: a defense against wormhole attacks in wireless networks. In: IEEE Infocom (2003)

16. Karapistoli, E., Economides, A.A.: Wireless sensor network security visualization. In: 4th International Congress on Ultra-Modern Telecommunications and Control Systems and Workshops (ICUMT), pp. 850–856 (2012)
17. Wood, A.D., Stnakovic, J.A.: Denial of service in sensor networks. IEEE Comput. **35**(10), 54–62 (2002)
18. Karlof, C., Wagner, D.: Secure routing in wireless sensor networks: attacks and countermeasures. Ad Hoc Sensor Networks **1**, 293–315 (2003)
19. Loo, C.E., Ng, M.Y., Leckie, C., Palaniswami, M.: Intrusion detection for sensor networks. Int. J. Distrib. Sensor Networks (2006)
20. Jackson, K.: Intrusion Detection Systems (IDS), Product Survey. Los Alamos National Library, LA-UR-99-3883 (1999)
21. Cook, R.D.: Detection of influential observations in linear regression. Technometrics **19**(1), 15–18 (1977)
22. Kruskall, J., Liberman, M.: The symmetric time warping problem: from continuous to discrete. In: Time Warps, String Edits and Macromolecules: The Theory and Practice of Sequence Comparison, pp. 125–161, Addison-Wesley Publishing Co., Reading (1983)
23. Keogh, E., Pazzani, M.: Scaling up dynamic time warping for datamining applications. In: Proceedings of the Sixth ACM SIGKDD International Conference on Knowledge Discovery and Data Mining, Boston, Massachusetts, pp. 285–289 (2000)
24. Cheng, P., Zhu, M.: Lightweight anomaly detection for wireless sensor networks. Int. J. Distrib. Sensor Networks, vol. 2015, Article ID 653232 (2015)
25. Ji, S., Chen, T., Zhong, S.: Wormhole attack detection algorithms in wireless network coding systems. IEEE Trans. Mob. Comput. **14**(3), 660–674 (2015)
26. TinyOS WSN operation system. https://github.com/tinyos/tinyos-main

Computerization of Operation Process in Municipal Transport

Karol Andrzejczak[1], Marek Młyńczak[2], and Jarosław Selech[3]([⊠])

[1] Faculty of Electrical Engineering,
Poznań University of Technology, Poznań, Poland
[2] Faculty of Mechanical Engineering,
Wrocław University of Science and Technology, Wrocław, Poland
[3] Faculty of Industrial Machines and Transport,
Poznań University of Technology, Poznań, Poland
jaroslaw.selech@put.poznan.pl

Abstract. Municipal transportation systems are complex, multi-object systems of required high reliability and safety level and on the other hand ensuring high level of quality and low cost of service. These requirements and limitations need reliable information for making reasonable decisions. Data stored in operational databases is getting more and more large and more complex for processing. Operators tend to use advanced information concerning daily usage, maintenance, reliability, costs, and many other aspects. Paper shows assumptions and framework of the information system oriented on municipal rail transportation system. There are shown and shortly described analytical methods supporting decision process and schematic structure of the system. The proposed information system links manufacturer and end-user (operator) for making decisions regarding vehicle improvement and decreasing operational costs. Database covers all available data derived from usage, maintenance, accidents and failures as well as natural environment. Novelty of the presented approach is linking data from various fields of operation (Man-Engineering-Environment) and sharing it among two opponents: manufacturer and operator. Processed information is used by manufacturer to optimize design and estimate warranty costs while transport operator uses Life Cycle Cost assessment to specify reasonable operation strategy.

Keywords: Information system · Municipal transport · Rail vehicle

1 Introduction

The paper presents the necessity and components of computerized system supporting decision making in operation of transportation system, particularly based on city trams. Presented Information Monitoring System (ISM) allows for acquisition, processing and analysis of safety, reliability and cost-related data in the process of technical objects operation like busses or city trams.

An idea is created according to advanced analyses and models including Life Cycle Costing (LCC) [11, 15] and Reliability, Availability, Maintainability and Safety (RAMS) [10] concepts compliant with the International Railway Industry Standard

© Springer International Publishing AG, part of Springer Nature 2019
W. Zamojski et al. (Eds.): DepCoS-RELCOMEX 2018, AISC 761, pp. 13–22, 2019.
https://doi.org/10.1007/978-3-319-91446-6_2

(IRIS) [8]. An implementation of the ISM system at the rolling stock supplier and in the operator practices makes possible monitoring malfunction-related data i.e. the RAMS indicators, cost related parameters, identification of the most cost-generating components and performance of multidimensional analyses on the acquired historical data of the rolling stock. This tool has been developed within a research project "Increase in the efficiency of operation of public transport following the implementation of the LCC and RAMS concept compliant with the IRIS standard based on an integrated information system" PBS3/B6/30/2015 [12].

2 Review of the Data Acquisition and Processing Systems

There exists a very large group of IT tools aiding the performance of joined reliability and cost-related analyses covering a range from initial estimation of reliability and cost indicators, complex simulations of the processes of maintenance and operation to advanced tools allowing an optimization of systems costs, safety and reliability. Packages of computer aided reliability and cost-related analyses have been developed by such companies as Systecon, Reliass, Relcon Scandpower, TDC, Isograph Ltd, Item Software, Relex Software, Prenscia (ReliaSoft), Ramentor, TDC, DNV or CAB Innov. In the commercial market, over 100 different IT packages can be distinguished from the discussed research field. They have a good position on the market in their relevant software segments and have similar complex approach to the performance of the analyses. They also provide access to many component reliability libraries compliant with international standards. These packages can be systematized according to two basic criteria of reliability-related research:

- methods of analysis (fault tree analysis, reliability block diagrams, Bayes methods, functional needs analysis, Markov analysis, Petri nets),
- the aim of the research (analysis of types of malfunctions and their consequences, preventive actions, malfunction forecasting, life cycle analysis, planning and optimization of maintenance).

The following IT tools are selected as the most significant for the research topic at hand have been described in detail below [11]. Majority of them is composed of many modules enabling a wide spectrum of analyses. These modules can operate independently but also in connection with other modules.

The **FMECA** method serves the purpose of performing analyses of types, consequences, and critical levels of unworthiness (fault). This is a tool that allows minimizing the time and simplifying the procedure of FMECA or FMEA report generation for process installations and new projects. The software allows modeling of a hierarchical system structure and a rapid development of the table of causes and effects of unworthiness (fault) with the use of a complex but ergonomic interface. The FMECA module allows gathering professional documentation for the analyses of the system operation safety [2, 11].

Fault Tree Analysis by Item Software is a module aiding the analysis of the tree of faults. SAPHIRE by a Canadian company Mitek and RiskSpectrum FT Professional by Relcon Scandpower have similar functionalities. That software has a multilevel graphic analytical environment integrated with the techniques of safety and reliability analysis allowing a quick construction of models of FTA fault trees and their analysis. All software has advanced algorithms for quantitative analysis and search for minimum ranges of fault in complex and large FTA models [2, 11].

This group includes **Reliability Block/Network Diagram** (Item Software), Reliability Block Diagram (Isograph), RBD Module (Item UK) and BlockSim (ReliaSoft).

These programs aid the analysis of systems presented in the form of a model of reliability structure i.e. block reliability diagram or network diagram.

BayesiaLab is an advanced Artificial Intelligence application with a complex graphic user interface that allows the researchers a comprehensive machine learning environment, knowledge modeling, diagnostic, analysis, simulation and optimization. BayesiaLab utilizes advanced learning algorithms, automatically generates structural models from the data, which makes it a highly specialized tool dedicated for specific purposes.

In the **TDC** Need software by a French company TDC, a functional analysis method was applied based on the European and French EN 1325-1, NF X50-151 standards and based on the APTE (R) method. This software is used as an aid in the process of design, value analysis, supplier advisory services as well as automatic generation of documents. TDC Structure is an application developed for assigning service functions to components of different technical solutions created by the designer. A functional block diagram (FBD) assists in the identification of internal technical functions (closed flow) resulting from a project.

Supercab Windchill Markov, CARE-RBD-Markov Markov Analysis and FaultTree+ Markov (Isograph) are all applications aiding the performance of time-dependent reliability analysis and worthiness based on the Markov chains and processes.

The GRIF – Workshop module is designed to model large, complex industrial systems using stochastic Petri nets with predicates and theorems.

RAM Commander (A.L.D.), MEADEP (SoHaR) RAM Studio, MIRIAM, FieldSim (ExproSoft) are software tools designed for the analysis and forecasting of reliability and maintenance, optimization of spare stocks, analysis of the fault tree and the event tree as well the evaluation of safety. The included reliability and safety modules include all generally known standards of reliability and approach to analysis of malfunctions of technical objects.

PRISM (RAC), **Windchill Prediction** (Relex) and **Reliability Workbench** – MIL-HDBK 217 (Isograph) are designed for the forecasting of reliability of technical objects. They include libraries and data related to the malfunction rate of electronic components according to the MIL-HDBK-217 standard (module MIL-217), electronic components according to other standards as well as mechanical components according to the NSWC6 standard (Mechanical module).

Software for **life cycle cost analysis** ensures the assessment of the costs of a technical object throughout all of the stages of its life cycle. This tool enables the analysis of the costs as a whole including the entire life cycle of an object i.e. from the concept stage, design, production, operation until the end of its life (scrapping, recycling). These analyses may include a variety of costs during the subsequent cycles such as the costs of the development of the design, production, warranty, repair and operation according to different scenarios [2].

The presented IT tools serve the purpose of computer aided reliability and cost analyses, worthiness calculations, maintainability and operational safety of technical systems. In majority of cases, there is a possibility of a simultaneous multi-aspect analysis of easy exchange of information among different modules of a given package as well as the export of data and results to external applications. The review does not include all programs available on the market but is limited to the most popular and most frequently utilized examples owing to their modular character, which ensures great application potential in all branches of the industry. A selection of the right tool depends on individual requirements, the scope and aim of the performed analysis as the potential of each of the programs may vary widely. The purchase price is also important (in majority of the discussed cases these are commercial tools that must be paid for). In some cases a free version can be installed with limited functionalities or a trial version available for a limited period of time.

3   Operation of Transport Systems

Contemporary transport systems are very efficient and, at the same time, complex and expensive, and each malfunction, fault and the resultant downtime results in a significant economic loss. It is expected high availability of the system with fault tolerant vehicles. For the purpose of roadworthiness, safety and environment protection, theory and practice of technical servicing are being upgraded, thus improving the existing operation strategies [1]. Modern information technologies provide many original solutions in the field of acquiring and data processing, thus facilitating cause-and-effect modeling, inference, failure forecasting and its causes searching [14].

Deterioration processes of technical objects force the need of supervising the changes in operation condition like natural, social and technical environment. The stage of operation and maintenance must be performed according to certain principles within an adopted operational scheme [13]. Besides, the process of operation alone requires aiding in the form of special measures that enable its functioning and are provided in a proper form and in a predetermined time. Given the limited reliability of a system and the related risk of disturbance of its operation, which may result in decreasing task performance, the structure of the system organization should cover a variety of cause-and-effect relations enabling the supervision and safety of the task performance [6]. The system of mass transport operation is composed not only of units of a fleet operator responsible for the operation and maintenance of vehicles but also other interested parties such as the manufacturer, the suppliers and the end-users [7]. Within the external environment, legal regulations and standards also exist (mandatory or self-imposed) along with competition depending on the conditions, under which the system operates. The mentioned above units are heavily intertwined with one another in terms of relations within the mass transport operation system and in terms of more distant relations with other units in the environment of this system. Figure 1 schematically presents the transport system in a relation with the involved parties.

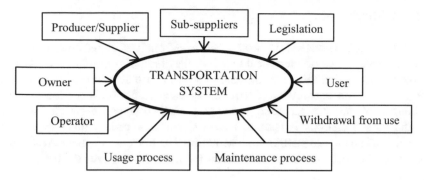

Fig. 1. Relations of a transport system with the interested environment

Irrespective of the number of both external and internal factors influencing the mass transport operation system, it should:

- ensure proper operation of vehicles used for the performance of transport tasks,
- enable proper flow of information within the system and proper exchange of information between the system and the environment [5],
- have clearly defined cost-generating factors and their impact on the costs of operation,
- be adapted to the requirements related to the transport services.

In order to reduce the costs of vehicle operation and maintenance, it is important to define the impact of the organization level of processes of keeping a given system operative (functional and task-ready) including [4, 9]:

- conditions of operation determining the relation between the occurrence of malfunctions and the operation,
- conditions of maintenance determining the relation between the maintenance operation and the process of modernization,
- external conditions (environment),
- reliability characteristics of a system.

Besides, the identification of a transport system in the analyzed field requires three following elements [9]:

- system composition – type and relations among subsystems and elements, quantity, type of malfunctions, required functions and performance, reliability characteristics,
- expected operational tasks,
- maintenance procedures and strategies,
- procedures in emergency situations impacting the decision-making processes.

4 ISM Structure

The aim of the RAMS analysis is the acquisition and analysis of data for the assessment of the malfunction rate of a technical object in the phase of operation. This analysis is to provide information related to the malfunctions and their causes, the maintenance works, hazards and consequences. This information is the basis for the experts for providing a cost analysis according to the LCC methodology. Before the information from the RAMS analysis is used in the LCC analysis, certain input data must be provided [3]. The process starting from acquiring data to be used in the RAMS analysis ending with the cost assessment (LCC) can be divided into three, tightly knit blocks (Fig. 2):

- data acquisition block,
- data analysis block,
- cost analysis block.

Data acquisition block is the basis of each RAMS analysis and covers passive and active acquisition of input data along with its initial processing. Data is acquired from the observation of the operation and diagnostic processes and it is related to the information on events related to malfunctions of technical objects and the costs of their elimination. This is a very important stage because no reliability model, even the most complex one, is capable of providing good results without proper input data.

RAMS analysis block pertains to the tool for the RAMS analysis. This block is the core of the ISM system. The input data is calculated and converted by appropriate algorithms. The main objective of the performed RAMS analyses of mass transport, particularly in relation to rail vehicles, is the reduction of the costs related to the operation and maintenance as well as improvement of quality and safety of passenger carriage. RAMS is composed of a set of tools for risk reduction (including economic risk) and safety management through reduction of hazards. The performed analyses will be compliant with the IRIS standard. The application of the RAMS analyses significantly contributes to a well-organized and managed system of maintenance of technical objects, particularly mass transport fleets.

Cost analysis block pertains to the throughout the life cycle of a technical object. It covers the stage of design, production, operation and maintenance, modernization and rebuilds of components, particularly the subcomponents of rail vehicles. Based on the results of the RAMS analysis, it is possible to generate economic indicators significant in terms of strategic planning in both production and maintenance. The RAMS methods are applied on the stage of design of new products and in the area of maintenance of objects already in operation. The data related to the costs of operation and maintenance makes possible identification of the most cost-generating factors and factors having impact on the environment and immediate surroundings.

The direct benefits of the implementation of the ISM system result from the integration of the RAMS and cost generation (operation and maintenance of technical objects) analysis in relation to urban transport (rail vehicles). The identification of the main cost carriers is performed according to the PN-EN 60300-3-3 standard [11]. The ISM procedures are related with the RAMS procedure and are developed

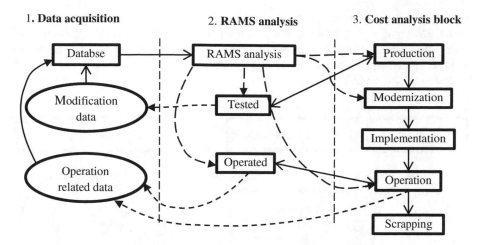

Fig. 2. RAMS analysis and cost analysis relation (based on [1])

according to the PN-EN 50126 IRIS standards [8, 10]. The scope of applications of the ISM package includes:

- identification of the most cost-generating components (vehicle components of the highest malfunction rate),
- optimization of the vehicle operation costs,
- limiting of losses due to transport system downtime,
- increased safety of passengers,
- reduction of negative impact on the environment,
- improved quality of service,
- adaptation of operational policy to the existing infrastructure.
- reliability and cost related comparative analysis of means of transport,
- adaptation of operational parameters of trams to the existing infrastructure.

The final objective of the performed research is the optimization of the investment capital expenditure by a fleet operator; hence the reduction of costs for the end users i.e. the passengers.

5   ISM System Characteristics in the Process of Object Operation

Today, the efficiency of almost any enterprise heavily depends on the reliable and measurable flow of constantly updated information. Hence, new IT tools are being constantly developed both diagnostic/advisory and record/decision-making, let alone such gathering data and assisting the management whose main carrier of information are usually properly prepared databases.

Strong competition results in a constant growth of efficiency of enterprises by finding new solutions and technologies, including modern IT systems. Fast technological access in the development of IT systems is a fundamental condition for the development of any business field. The reduction of time needed for the flow of information and convening it in an accurate way influences rapid achievement of decision-related and executive objectives [16]. The scheme of information flow through IT system is shown in Fig. 3.

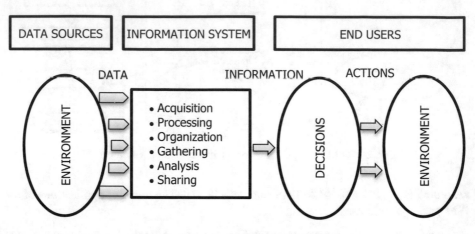

Fig. 3. System of information flow [17]

In the systemic approach it should be identified all elements of transportation system assuming that observed transportation vehicle is a core object in the chain Man-Engineering-Environment (MEE). Operation chain consists of direct operators or maintenance staff members, supporting systems (operation infrastructure) and environment (objects not involved directly in operation) (Table 1).

Table 1. Objects identification in operation system

Classification criteria	Operational objects		
	Supporting systems	Operation chain	Operation environment
Man	Supporting workers	Operator, maintenance staff	Outsider
Engineering	Supporting technical objects (infrastructure)	**Transportation Vehicle**	Technical object in surroundings
Environment	Work environment	Operation environment	Natural environment

The illustrative list of objects creating transportation vehicle operation system may contain: tram/bus, operator, other people like passengers, walking people, other vehicles in the street, other technical infrastructure (tram stops, traffic signals, rail, etc.), air (temperature, humidity, dustiness, composition) and others. All these elements may provide data and consequently information, relations, regressions or probability distributions giving criteria to decision undertaking.

Very often, IT systems are dedicated to a given branch of industry or even a single enterprise in order ensure efficient gathering of information and organization of data flow including the possibility of its monitoring and report preparation.

Monitoring is an organized and continuous method of long-term observations or observations and measurements of processes occurring in technical objects. If complex transport systems are monitored such as a fleet of mass transport vehicles, it should be characterized with proper organization and continuity. Continuity is to be understood as regular control-measurement actions with a systematic analysis of the obtained results.

6 Summary

The presented concept of operation of transport systems informatization using ISM allows combining several sources of data existing with different enterprises (supplier and operator of a fleet of vehicles). The exchange of data is performed:

- automatically–does not require work load of either of the enterprises on a daily basis,
- safely–data will not be accessed by third parties and it will be secured to be accessed only with a predetermined application,
- correctly–appropriate algorithms will validate the correctness of the data in terms of proper construction of the imported files and individual data,
- in a sustainable way–the data will be protected against erasure and the database will have a back-up copy.

The work based on the presented concept will translate into a reduction of the delivery time, as it will improve the flow of malfunction-related information. The results of analyses of malfunction-related data would enable evaluating the costs of operation, which provides the possibility of estimating and forecasting of the costs of the lifecycle of vehicles. The application of the ISM system will improve the organization within the business entity (supplier or fleet operator).

The ISM system will allow in the future for reduction of failure rate thanks to ad hoc (faster response to malfunctions), short-term (material and organizational planning) and long-term (design optimization) activities.

Besides, the described IT system is a dedicated tool for the performance of the RAMS reliability analysis. This analysis is required by the IRIS standard and obtaining this standard is an important factor in the development of a company through increase in sales of vehicles on a global level.

Acknowledgments. The research was carried out within the Applied Research Program PBS III/B6/30/2015 financed by National Centre for Research and Development (NCBiR) and the project of Poznan University of Technology - 04/43/DS PB/0096.

References

1. Bałuch, H.: The problem of rail tack monitoring and its place in GISRAIL. Problemy Kolejnictwa, zeszyt 140, Warszawa (2005)
2. Chybowski, L.: Comparison of complex software packages for multiaspect reliability analysis of marine technical systems. Szczecin: Akademia Morska w Szczecinie, Biblioteka Cyfrowa Świat Morskich Publikacji
3. Construction Equipment Management for Engineers, Estimators and Owners, D. Gransberg, C. Popescu, R. Ryan, Taylor & Francis Group (2006)
4. Dhillon, B.S.: Life Cycle Costing: Techniques, Models and Applications. Gordon and Breach Science Publishers, New York (1989)
5. Fricker, J.D., Whitford, R.K.: Fundamentals of Transportation Engineering. A Multimodal Systems Approach. Pearson Education, Inc., Upper Saddle River (2004)
6. Gill, A.: Optimization of the technical object maintenance system taking account of risk analysis results. Eksploatacja i Niezawodność – Maintenance and Reliability (2017)
7. Gramza, G.: Selected problems in the assessment of quality of urban mass transport. Urban Buses (2011)
8. IRIS—International Railway Industry Standard. Revision 02. Brussels: UNIFE (2009)
9. Młyńczak, M., Nowakowski, T., Restel, F., Werbińska-Wojciechowska, S.: Problems of reliability analysis of passenger transportation process. In: Proceedings of the European Safety and Reliability Conference. A. A. Balkema, Leiden (2004)
10. PN-EN 50126:2002 Railway applications – The specification and demonstration of Reliability, Availability, Maintainability and Safety (RAMS)
11. PN-EN 60300-3-3:2006 Reliability Management. part 3-3. Application Guide – life cycle cost assessment
12. Research project NCBiR PBS-246314 Increase in the efficiency of operation of public transport following the implementation of the LCC and RAMS concepts compliant with the IRIS standard based on an integrated IT system (2014–2016)
13. Sika, R., Hajkowski, J.: Synergy of modeling processes in the area of soft and hard modeling. In: Proceedings of the 8th International Conference on Manufacturing Science and Education, Trends in new industrial revolution, Sibiu, MATEC Web of Conference 121, 04009 (2017)
14. Skrzyński, E., Ochociński, K.: IT systems in railway infrastructure. Problems of Railway, vol. 176 (2017)
15. Csuzi, I.: The life cycle cost management (LCCM) aspects of maintenance in urban public transport (UPT) Systems, arguments for a guide of trams modernization. J. Sustain. Energy 6 (3), 102–111 (2015)
16. Trojanowska, J., Kolinski, A., Galusik, D., Varela, M.L.R., Machado, J.: A methodology of improvement of manufacturing productivity through increasing operational efficiency of the production process. In: Hamrol, A., Ciszak, O., Legutko, S., Jurczyk, M. (eds.) Advances in Manufacturing. Lecture Notes in Mechanical Engineering. Springer, Cham (2018)
17. Types of IT Systems. www.uci.agh.edu.pl/uczelnia/tad/.../07-Typy%20systemów% 20informacyjnych.ppt. Accessed 02 Apr 2018

Common Criteria IT Security Evaluation Methodology – An Ontological Approach

Andrzej Bialas[(⊠)] [iD]

Institute of Innovative Technologies EMAG,
Leopolda 31, 40-189 Katowice, Poland
Andrzej.Bialas@ibemag.pl

Abstract. The paper deals with the Common Criteria assurance methodology, particularly with the IT security evaluation process specified by the Common Criteria Evaluation Methodology (CEM). To better organize this very complex evaluation process the ontological approach is proposed. The previously developed ontology focused on the IT product development according to Common Criteria is extended by evaluation issues. Ontology classes, properties and individuals are elaborated to express the IT security evaluation according to CEM. The ontology use is exemplified on the vulnerability analysis of a simple firewall. The paper points out the need to extend this ontology to the full vulnerability analysis of different IT products and assurance levels. The readers should have basic knowledge about Common Criteria and the ontology development.

Keywords: Common Criteria · Vulnerability assessment · Ontology
Knowledge management · IT security evaluation · Security assurance

1 Introduction

Today's societies and economies are based on Information and Communication Technologies (ICT). It is necessary to apply trustworthy ICT products and systems to minimize the risk inherent to the use of ICT. One of the ways to get such trustworthiness is the rigorous development of ICT products, their security independent evaluation and certification. It allows to achieve security assurance for them. The Common Criteria approach [1], presented in the ISO/IEC 15408 standard, is the basic security assurance methodology. The assurance is measurable by EALs (Evaluation Assurance Levels) in the range EAL1 to EAL7. Currently there are over 2,000 IT products certified according to Common Criteria (CC) and over 170 registered protection profiles (technology independent sets of security requirements) [2]. Common Criteria is based on the paradigm that security assurance depends on the rigour applied to the security development, vulnerability assessment, testing, verification, documenting etc. and on the independent evaluation and certification of the IT product, called here Target of evaluation (TOE). The TOE embraces a set of software, firmware and/or hardware with possible documentation. The evaluation is based on the Common Criteria Evaluation Methodology (CEM) [3].

© Springer International Publishing AG, part of Springer Nature 2019
W. Zamojski et al. (Eds.): DepCoS-RELCOMEX 2018, AISC 761, pp. 23–34, 2019.
https://doi.org/10.1007/978-3-319-91446-6_3

The [1]/part 2 includes components to express security functional requirements (SFRs) used to model the IT product security behaviour. The [1]/part 3 includes components to express security assurance requirements (SARs) of this product. Both kinds of components are grouped by families, which, in turn, are grouped by classes. Components include elements. More information is available in [4–7] and in the author's publications [8–10].

The paper concerns the knowledge engineering approach to the IT security evaluation according to the Common Criteria Evaluation Methodology (CEM). In computer science and information science, "an ontology is a formal naming and definition of the types, properties, and interrelationships of the entities that really exist in a particular domain of discourse" [11]. Many disciplines, where "common understanding", "common taxonomy", "interoperability" or "reasoning" are important issues, have adopted ontologies recently. These domains are: web-based applications, medicine, public administration, biology, and information security. The IT security development/ evaluation domain has similar needs. Therefore the research on applying an ontology-based method in this domain may bring new advantages. The paper bases on the knowledge engineering principles and the Protégé v.5 tool elaborated at the Stanford Center for Biomedical Informatics Research [12, 13]. Extensive information how to build and use ontologies is available in [12].

The CC development and evaluation processes are complex, difficult and require good organization and computer support. The aim of the research presented in the paper is to work out an ontological model of the Common Criteria evaluation process allowing to better structure the knowledge related to this process and to improve the evaluation process. The research motivation is implied by the situation that due to the very complex and time-consuming certification process a lot of companies abstain from CC certification.

Section 2 presents the current state of research in the paper domain. Section 3 describes ontology concepts and relationships needed to express the IT security evaluation process. The ontology is exemplified on the firewall evaluation process. Conclusions summarize the research and present the planned works in this field.

2 Current State of Research

The review was focused on the applications of the knowledge engineering methodology in information security, especially in the Common Criteria domain.

The paper [14] features a very extensive literature survey focused on "the security assessment ontologies". The authors analyze 47 works, from 80 preliminarily identified, and conclude that:

- "Most of works on security ontologies aim to describe the Information Security domain (more generic), or other specific subdomains of security, but not specifically the Security Assessment domain";
- there is "… a lack of ontologies that consider the relation of "Information Security" and "Software Assessment" fields of research";

- there is "a lack of works that address the research issues: Reusing Knowledge; Automating Processes; Increasing Coverage of Assessment; Secure Sharing of Information; Defining Security Standards; Identifying Vulnerabilities; Measuring Security; Protecting Assets; Assessing, Verifying or Testing the Security".

The book [15] features a structured process for elicitation of threat analysis elements for a CC certification, a tool-supported identification of assets, assumptions and threats, and reasoning of Common Criteria threats based upon certain attacker types.

The paper [16] presents the results of research focused on the development of the CC ontology and the ontology-based tool supporting CC knowledge query, markup, review, and report functions to improve the understandability of CC and enhance the efficiency and effectiveness of the CC-certification process.

The paper [17] presents a CC Ontology tool which is based on an ontological representation of the Common Criteria components, to support the evaluator during the certification process. The tool supports the planning of an evaluation process, the review of relevant documents or making reports. The authors declare that the tool decreases the time and costs of certification.

The paper [18] discusses an ontological model of the CC functional components mapped to the security objectives with the use of a specialized tool. It concerns only this stage of the IT security development process.

The author's paper [9] discusses an IT security Development Ontology applied in the intelligent sensors domain. This ontology will be extended here to the IT security evaluation process. The paper [10] concerns the elaborated CCMODE Tools [6] supporting the evaluation evidences preparation, where ontologies are applied to elaborate a TOE security model injected into the security target (ST) document, to elaborate the knowledge base and to integrate systems components. An extensive survey of information security ontologies is included in the paper [19].

There is no common ontology-based methodology focused on the CC development and evaluation process.

3   Towards the Common Criteria Ontology

The Common Criteria Ontology (CCO) is developed as an extension of the IT Security Development Ontology (ITSDO) discussed in the author's previous publications, e.g. [9, 19, 20]. ITSDO complies with Common Criteria v3.1 and embraces concepts and relationships related to the IT security and TOE development processes. It is also based on the methodology [13] and tools [12]. ITSDO will be extended now by concepts and relationships related to the evaluation process. This way the CCO ontology will embrace all three CC processes, i.e.:

- the IT security development process of the IT product (TOE); as a result of security analyses, a security target document (ST) is prepared; the ST embraces the security problem definition (SPD), its solution by specifying security objectives (SO), security requirements and functions; security functional requirements (SFRs) derived from SOs; EAL-related security assurance requirements (SARs), determine how much assurance we can have in an IT product; the ST includes TOE security functions (TSF) meeting SFRs; TSFs are later implemented on the claimed EAL;

- the TOE development process output includes the IT product (TOE) and its documentation (evaluation evidences); they are transferred together with the TOE and ST to the security evaluation process;
- the security evaluation process carried out according to CEM in an independent laboratory, accredited by the given national certification body; it embraces the TOE, ST and evidences implied by the claimed EAL.

Figure 1 presents a general view on the basic elements of the developed Common Criteria Ontology (CCO): classes, properties, individuals and assertions. Certain elements belong to the earlier developed ITSDO ontology [9, 19, 20] representing:

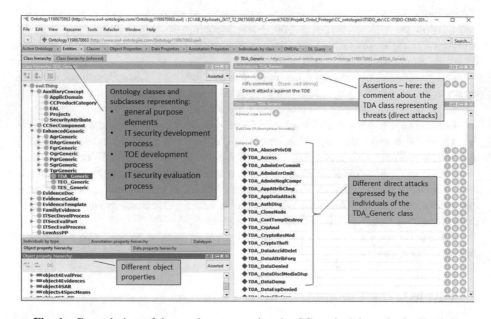

Fig. 1. General view of the ontology expressing the CC methodology (in the Protégé).

- IT security development process, e.g.: ITSecDevelProcess and TDA_-Generic classes, objects4SpecMeans property, TDA_Access individual of the TDA_Generic class,
- TOE development process, e.g.: EvidenceDoc and FamilyEvidence classes, object4SAR property,
- general purpose elements, e.g.: EAL, SecurityAttribute, Projects classes.

Figure 1 presents also newly developed elements related to the IT security evaluation process discussed in this paper, e.g.: ITSecEvalProcess, ITSecEvalPart classes, object4EvalProc property.

There is no strict partition between the elements belonging to these four groups, e.g. evaluation evidences elements, which are worked out during the IT security development (Security target) and the TOE development (EAL related evidences), are used in the IT security evaluation process.

3.1 Modelling the IT Security Evaluation Process

The IT security evaluation process [3] consists of four elements (the corresponding ontology classes are shown in Fig. 2):

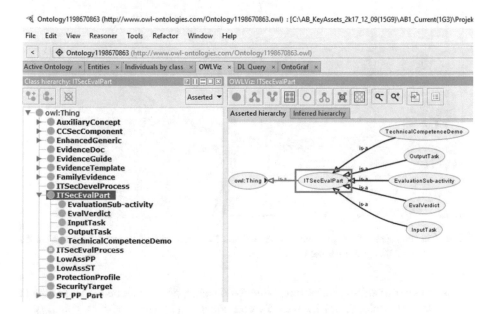

Fig. 2. IT security evaluation process visualized in the Protégé. The left panel shows the class hierarchy, the right – IT security evaluation process elements (the OWLViz plugin).

- one input task (InputTask class) responsible for the management of evaluation evidences delivered by developers for each SAR component included in the EAL;
- a certain number of evaluation sub-activities (EvaluationSub-activity) corresponding to the security assurance components embraced by the claimed EAL; for each component a verdict is assigned: Pass, Fail or Inconclusive;
- one output task (OutputTask) responsible for the report generation, i.e. Evaluation Technical Report (ETR) and Observation Reports (OR); ETR summarizes the evaluation process and includes a cumulative verdict; the verdict is Pass when all sub-activities embraced by the evaluation process are Pass;
- one demonstration of the technical competence task (TechnicalCompetenceDemo) presenting the operations required by the given national certification scheme.

The EvalVerdict class, discussed later, represents verdicts assigned during the evaluation, i.e. Pass, Fail, Inconclusive.

The evaluation sub-activities are identified with the SAR components. All SAR components have the same structure and the sub-activities corresponding to the components have the same structure as well (Fig. 3).

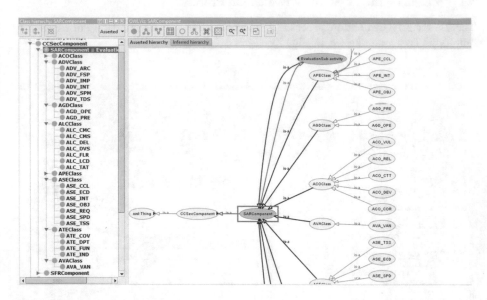

Fig. 3. Classes representing the Security Assurance Requirements – classes/subclasses on the left and the OWLViz graph on the right.

The number of sub-activities (components) depends on the claimed EAL for the IT product. For example, EAL4 includes 24 components ([1]/part 3, Tables 1 and 5):

- ASE_CCL.1 (Conformance claims), ASE_ECD.1 (Extended components definition), ASE_INT.1 (ST introduction), ASE_OBJ.2 (Security objectives), ASE_REQ.2 (Security requirements), ASE_SPD.1 (Security problem definition), ASE_TSS.1 (TOE summary specification); the TOE developer provides the security target (ST), which groups evaluation evidences for all these components; they concern the IT security development process;
- ADV_ARC.1 (security architecture), ADV_FSP.4 (complete functional spec.), ADV_IMP.1 (implementation), ADV_TDS.3 (basic modular design), AGD_OPE.1 (operational user guidance), AGD_PRE.1 (preparative procedures – installation, configuration), ALC_CMC.4, ALC_CMS.4 (configuration management system and coverage), ALC_DEL.1 (delivery procedure), ALC_DVS.1 (development security), ALC_LCD.1 (live cycle), ALC_TAT.1 (tools and techniques), ATE_COV.2, ATE_DPT.2 (test coverage and depth), ATE_FUN.1 (functional tests); the TOE developer provides evidences for each component; they concern the TOE development process;
- AVA_VAN.3 (focused vulnerability analysis), ATE_IND.2 (independent testing); these evidences are elaborated by the evaluator.

Each SAR component has an evaluation evidence assigned, provided by a developer. The corresponding evaluation sub-activity task (CEM) is focused on the assessment of this evidence. Let us consider the SAR component structure and its interpretation by CEM. It will be exemplified on the AVA_VAN.3 component (evaluation sub-activity) shown in Fig. 4.

Fig. 4. The ontological representation of the AVA_VAN.3 security assurance component in the Protégé.

The left panel shows the class hierarchy. The SARComponent ontology class (CC) is identified with the EvaluationSub-activity ontology class (CEM). It is the key link between CC and CEM. The SARComponent ontology class has subclasses: from ACOClass (composition) to AVAClass (vulnerability assessment), representing CC classes of SARs, which have families and the families group similar components. This way the structure of the CC catalogues of requirements was modelled within the CCO ontology. Please note that the term "class" is used in both domains: Common Criteria (grouping similar requirements) and knowledge engineering (grouping similar individuals).

The upper middle panel shows 5 individuals of the AVA_VAN class: AVA_VAN_1 to AVA_VAN_5 of the highlighted AVA_VAN ontology class. They are ontological representations of the SAR components: AVA_VAN.1 to AVA_VAN.5 [1]. The right panels presents the use and property assertions of one individual, i.e. AVA_VAN_3.

According to the CC methodology, each SAR component consists of three kinds of elements:

- D – evidence item which ought to be delivered by the developer, e.g. AVA_-VAN.3.1D expressed by the AVA_VAN_3.1D ontology class;
- C – contents and presentation of the item (AVA_VAN_3.1C ontology class),
- E – evaluator action element – how this item is verified by the evaluator; sub-activities are identified with E elements; for each sub-activity (component) one or more evaluator actions elements may exist, here the following four, expressed by the ontology classes: AVA_VAN_3.1E, AVA_VAN_3.2E, AVA_VAN_3.3E, AVA_VAN_3.4E.

The right lower panel shows all elements as the AVA_VAN_3 individual properties: hasDelem, hasCelem and hasEelem, e.g.: the AVA_VAN_3 hasDelem AVA_VAN_3.1.D, where the AVA_VAN_3 class is the property domain and the AVA_VAN_3.1.D class is the property range. This panel also presents the claimed EAL level (EAL_4), the provided evaluation evidences (AVA_VAN_EAL4) and the AVA_VAN.3 dependencies, i.e. components which must be also be considered during the AVA_VAN.3 evaluation, e.g. AGD_PRE_1.

All classes pointed by the hasDelem, hasCelem and hasEelem properties are shown in Fig. 5. Each of them has its own string-type data property whose values correspond to the AVA_VAN.3 elements description [1]/part 3, p. 186. These property values are expressed in English (denoted by @en).

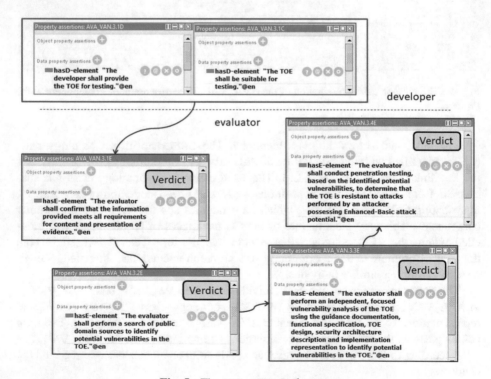

Fig. 5. The AVA_VAN_3 elements.

The most granular structure to which a verdict is assigned is the evaluator action element (E). `EvalVerdict` is assigned to an applicable E-element as a result of the corresponding CEM action and its work units (D, C elements). The `EvalVerdict` class has 3 individuals: `FAIL`, `INCONCLUSIVE`, `FAIL` – see Fig. 6.

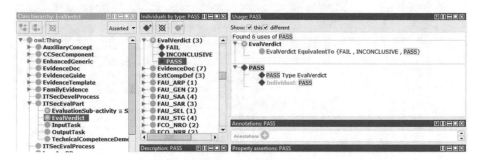

Fig. 6. The `EvalVerdict` class and its individuals.

The above defined classes and properties allow to express evaluation processes of different IT products (TOE).

3.2 Exemplifying the IT Security Evaluation Process

The evaluation process will be validated on a simple example related to the firewall project (MyFWL on the EAL4 assurance level). The ontological representation of the firewall development was presented in the papers [19, 20].

The `ITSecEvalProcess` ontology class (Fig. 7, Upper middle panel) represents evaluation processes of many IT products. The individual `ITSEP_MyFWL` concerns the firewall evaluation process expressed by object properties (right lower panel). The evaluation process embraces input-, output-, technical competencies demonstration tasks and the sub-activities corresponding to all SARs included in EAL4 (not all are shown).

The validation is restricted to the `AVA_VAN_3` evaluation sub-activity. Its object properties (Fig. 8) express the related `EAL4`, evaluation evidences as a whole (`AVA_VAN_EAL_4`), all dependencies and D, C, E elements.

The `AVA_VAN_3` sub-activity has also data properties (string-type), expressing different hints and references, e.g. to public vulnerability databases.

The E-elements have verdicts assigned during the evaluation (Fig. 5). The current ontology version stops on the D, C, E elements level. The additional level representing textual details concerning evaluation sub-activities, i.e. activities (related to the E-elements) and their evaluation work-units, can be added later. Instead of this, the evaluator is directed to the proper part of CEM. Similarly, all sub-activities can be performed producing the cumulative evaluation verdict.

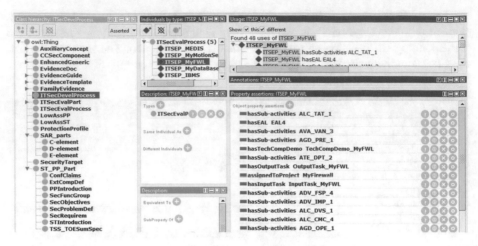

Fig. 7. IT security evaluation process of the MyFWL/EAL4 project.

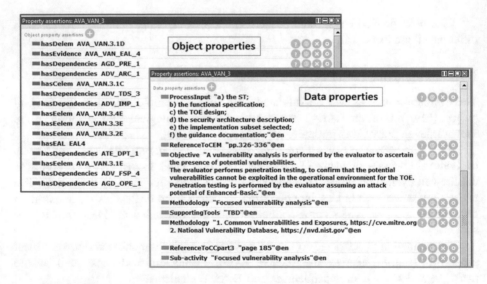

Fig. 8. IT security evaluation process – properties of the AVA_VAN_3 sub-activity.

3.3 The Proposed Ontology Extension to the Vulnerability Assessment Process

The vulnerability analysis is the key activity of evaluation. There is no strict routine to perform it. It depends on the IT product character (hardware/software/firmware) and the claimed EAL, which implies the considered attack potential [3]/Annex B.

Vulnerabilities to the relevant attacks should be analyzed. The E-elements of the AVA_VAN class components require (see Fig. 5) that the evidences should be checked, potential vulnerabilities searched, less or more restrictive vulnerability analysis done, and penetration tests performed to determine whether the TOE is resistant to attacks of the considered potential [21]. These tasks can be supported by the ontology as well, but it requires extra research on attack methods, vulnerabilities, standards, attack tools, etc. with respect to the IT product character. The objective of this research is to support the management of the vulnerability assessment for different products and EALs.

4 Conclusions

The paper presents multidisciplinary research and development works encompassing security engineering and knowledge engineering domains. The major contribution of the paper is to introduce the Common Criteria Ontology (CCO) expressing terms and relationships of the IT security development and evaluation processes compliant with the Common Criteria standard [1, 3]. CCO is an extension of the early elaborated IT Security Development Ontology (ITSDO) focused on the evaluation evidences preparation. The extension expresses how these evidences can be evaluated. The ontology testing, retrieving information is similar to those presented in the papers [9, 20].

Thanks to a knowledge engineering approach to the Common Criteria domain, the author's works aim at providing developers and evaluators with: design patterns, methodology, tools and related knowledge, which all help to elaborate and evaluate evidences. It is planned to extend the work to express vulnerability assessment.

The basic advantages and possibilities of the ontological approach allow better formalized and precise development and evaluation, tool support, project knowledge management and reusability. Currently the Protégé tool [12] is used but the ontological model can be also implemented in software, similarly to the CCMODE Tools [6].

Acknowledgement. "This work was conducted using the Protégé resource, which is supported by grant GM10331601 from the National Institute of General Medical Sciences of the United States National Institutes of Health." The paper results will be used in the R&D project focused on the CEM implementation.

References

1. Common Criteria for IT Security Evaluation, part 1–3, version 3.1 rev. 5 (2017). http://www.commoncriteriaportal.org/. Accessed 24 Jan 2018
2. CC Portal. http://www.commoncriteriaportal.org/. Accessed 10 Jan 2018
3. Common Methodology for Information Technology Security Evaluation, version 3.1 rev. 5, (2017). http://www.commoncriteriaportal.org/. Accessed 15 Jan 2018
4. Hermann, D.S.: Using the Common Criteria for IT Security Evaluation. CRC Press, Boca Raton (2003)
5. Higaki, W.H.: Successful Common Criteria Evaluation. A Practical Guide for Vendors. Wesley Hisao Higaki, Lexington (2011)
6. CCMODE: Common Criteria compliant, Modular, Open IT security Development Environment. http://www.commoncriteria.pl/. Accessed 24 Jan 2018
7. BSI. Guidelines for Developer Documentation according to Common Criteria, version 3.1 (2007)

8. Bialas, A.: Common criteria related security design patterns—validation on the intelligent sensor example designed for mine environment. Sensors **10**, 4456–4496 (2010)
9. Bialas, A.: Common criteria related security design patterns for intelligent sensors—knowledge engineering-based implementation. Sensors **11**, 8085–8114 (2011)
10. Bialas, A.: Computer-aided sensor development focused on security issues. Sensors **16**, 759 (2016)
11. Ontology. https://en.wikipedia.org/wiki/Ontology_(information_science). Accessed 11 Jan 2018
12. Protégé. https://protege.stanford.edu/. Accessed 3 Jan 2018
13. Musen, M.A.: The Protégé project: a look back and a look forward. AI Matters (Association of Computing Machinery Specific Interest Group in Artificial Intelligence) **1**(4) (2015). https://doi.org/10.1145/2557001.25757003
14. de Franco Rosa, F., Jino, M.: A survey of security assessment ontologies. In: Rocha, Á., Correia, A., Adeli, H., Reis, L., Costanzo, S. (eds.) Recent Advances in Information Systems and Technologies, WorldCIST 2017. Advances in Intelligent Systems and Computing, vol. 569. Springer, Cham (2017)
15. Beckers, B.: Pattern and Security Requirements: Engineering-Based Establishment of Security Standards. Springer, Cham (2015)
16. Chang S.-C., Fan C.-F.: Construction of an ontology-based common criteria review tool. In: Proceedings of the International Computer Symposium (ICS 2010). IEEE Xplore (2010)
17. Ekelhart, A., Fenz, S., Goluch, G., Weippl, E.: Ontological mapping of common criteria's security assurance requirements. In: Venter, H., Eloff, M., Labuschagne, L., Eloff, J., von Solms, R. (eds.) New Approaches for Security, Privacy and Trust in Complex Environments, pp. 85–95. Springer, Boston (2007)
18. Yavagal, D.S., Lee, S.W., Ahn, G.-J., Gandhi, R.A.: Common criteria requirements modeling and its uses for quality of information assurance. In: Proceedings of the 43rd Annual ACM Southeast Conference, Kennesaw, GA, USA, 18–20 March 2005, vol. 2, pp. 130–135 (2005)
19. Białas, A.: Ontology based model of the common criteria evaluation evidences. Theor. Appl. Inform. **25**(2), 69–92 (2013)
20. Białas, A.: Validation of the ontology based model of the common criteria evaluation evidences. Theor. Appl. Inform. **25**(3), 201–223 (2013)
21. Bialas, A.: Software support of the common criteria vulnerability assessment, In: Zamojski, W., et al. (eds.) Advances in Intelligent Systems and Computing, vol. 582, pp. 26–38. Springer, Cham (2017)

Availability Analysis of Transport Navigation System Under Imperfect Repair

Agnieszka Blokus-Roszkowska$^{(\boxtimes)}$

Gdynia Maritime University, Morska 81-87, 81-225 Gdynia, Poland
a.blokus-roszkowska@wn.am.gdynia.pl

Abstract. The results of the classical renewal theory are used for the availability analysis of multistate systems under the assumption of imperfect renovation. A repairable system with the time of renovation ignored is considered. The system can be repaired after exceeding its critical reliability state, assuming that after a certain time or after a certain number of repairs the system cannot be restored to the state of full ability. The expected values of the times until the successive times of exceeding the reliability critical state, and the expected values of the numbers of exceeding the system critical state at a certain time point, are estimated. The theoretical results are applied to the availability evaluation of an exemplary transport navigation system. The results are compared for a single master navigation system and a navigation system with a back-up system.

Keywords: Renewal process · Imperfect repair · Renewal function
Multistate system · Navigation system

1 Renewal and Availability of Multistate Systems

1.1 Introduction

Many real technical systems are aging systems with multistate components degrading with time. In such systems, due to aging, their components degrade ranging from the full ability state to complete failure and full inability [1–6]. Considering maintenance and renewal of multistate systems [7–10], the replacement of a system component by a new one is often assumed. For a multistate aging system it means that the component after repair is in the best "as good as new" reliability state, which in fact is not always possible. Moreover, even the perfect repair, due to conditions resulting from the state of the remaining components, may cause the component not to behave as new. In addition, its lifetimes in the reliability state subsets can be shorter than for a component in a new system. It is assumed that a system is repaired after exceeding its critical reliability state. Taking into account this assumption under imperfect repair, the renewal of the system means that the system is restored to a better reliability state than the critical state, one of the previous degraded state, but not to the best state. In this paper, the results of the classical renewal theory [7–9] are combined with the multistate approach to reliability and availability analysis [1–3, 11, 12] of repairable systems assuming imperfect renovation [13–15] of a system after exceeding its critical state.

© Springer International Publishing AG, part of Springer Nature 2019
W. Zamojski et al. (Eds.): DepCoS-RELCOMEX 2018, AISC 761, pp. 35–45, 2019.
https://doi.org/10.1007/978-3-319-91446-6_4

1.2   Renewal Stream and Renewal Process

It is assumed that all the components and the system under consideration have the reliability state set $\{0,1,\ldots, z\}$ $(z \geq 1)$, where state 0 is the worst and state z is the best. The state of the system and components degrades over time.

Similarly as in [4, 16], it is assumed that a system is repaired after exceeding its critical reliability state r $(r = 1,2,\ldots,z - 1)$ and that the time of its renovation can be ignored if the lifetime in the reliability state subset is not worse than that of the critical state. $T(r)$ denotes the system's lifetime in the reliability state subset $\{r, r + 1, \ldots, z\}$.

A random variable $T^{(N)}(r)$ $(r = 1, 2, \ldots, z - 1$ and $N = 1, 2, \ldots)$ describes the time between the moment of the N-1 system renovation and the Nth time that the system critical state is exceeded. Wherein $T^{(1)}(r)$ denotes the time between the commencement of the system operation and the moment of its first renovation. It is assumed that the random variables $T^{(1)}(r)$, $T^{(2)}(r), \ldots (r = 1, 2, \ldots, z - 1)$ are independent.

System renovation often means the operation that causes the system to return to its best reliability state z. However, in reality after a certain time or after a certain number of renewals, the return of a system to the best "as good as new" state is no longer possible. Thus, we assume that after N_1 number of system renovations to the best state z, the next system renovation will only be possible to state $z - 1$. Similarly, we assume that the system renewal from the $N_1 + 1$ number to the N_2 number takes place to state $z - 1$, so the number of such repairs is $N_2 - N_1$. Further, by the renewal of the system from the $N_2 + 1$ number to the N_3 number, we mean the operation that causes the system to return to the reliability state $z - 2$. Generally, we assume that the system renewal from the $N_x + 1$ number to the N_{x+1} number takes place to state $z - x$ $(x = 1,2,\ldots,z - r$ and $r = 1,2,\ldots,z - 1)$. From the number $N_{z-r} + 1$ to N_{z-r+1}, system renewal can only be carried out as a renewal to the critical reliability state r. After N_{z-r+1} number of renovations, another system renewal is no longer possible and after $N_{z-r+1} + 1$ time the system exceeds the reliability critical state r, the system remains unable to work.

The system renewal is characterized by a sequence of random variables [7–9]

$$S^{(N)}(r) = T^{(1)}(r) + T^{(2)}(r) + \ldots + T^{(N)}(r), \ r = 1, 2, \ldots, z - 1, \ N = 1, 2, \ldots. \quad (1)$$

Subsequently, a sequence of random variables $S^{(1)}(r)$, $S^{(2)}(r), \ldots (r = 1, 2, \ldots, z - 1)$ is a renewal stream. The random process $\{N(t, r), t \geq 0\}$ is a renewal process of a multistate system, where $N(t,r)$ is the number of systems exceeding the critical state r, i.e., the number of renewals of the system, up to the time point t.

1.3   Characteristics of a Renewal Stream with Imperfect Repair

In this Section, some basic characteristics of a repairable system with the time of renovation ignored are determined under assumptions presented in Sect. 1.2. Namely, the expected values of the times until the successive times that the reliability critical state is exceeded, and the expected values of the numbers of times that the reliability critical state is exceeded at a certain time point, are given.

If it is assumed that first N_1 system's renewals include operations that allow the system to return to the best state z, then $T^{(1)}(r)$, $T^{(2)}(r)$, ... $T^{(N_1+1)}(r)$ are independent random variables from the same distribution with expected value $\mu(r)$ and variance $\sigma(r)$.

The expected value of a random variable $S^{(\omega)}(r)$ ($\omega = 1, 2, \ldots, N_1, N_1 + 1$), representing the time until the ωth time the system exceeds the reliability critical state r, is [3, 4]

$$E[S^{(\omega)}(r)] \cong \omega \cdot \mu(r), \ \omega = 1, 2, \ldots, N_1, N_1 + 1, \ r = 1, 2, \ldots, z - 1. \tag{2}$$

Subsequently, the system renewal from the $N_1 + 1$ number to the N_2 number takes place to state $z - 1$. If a random variable $T^{(\omega)}(r)$ ($\omega = N_1 + 2, \ldots, N_2 + 1$) represents the time between the moment of the $\omega - 1$ system renovation and the ωth time that the system critical state is exceeded, then its expected value is given by

$$E[T^{(\omega)}(r)] = \mu(r) - \mu(z), \ \omega = N_1 + 2, \ldots, N_2 + 1, \ r = 1, 2, \ldots, z - 1. \tag{3}$$

Therefore, the expected value of a random variable $S^{(\omega)}(r)$ ($\omega = N_1 + 2, \ldots, N_2 + 1$) is

$$
\begin{aligned}
E[S^{(\omega)}(r)] &\cong (N_1 + 1)\mu(r) + (\omega - N_1 - 1)(\mu(r) - \mu(z)), \ \omega = N_1 + 2, \ldots, N_2 + 1, \ r \\
&= 1, 2, \ldots, z - 1,
\end{aligned}
\tag{4}
$$

More generally, random variables $T^{(\omega)}(r)$ ($\omega = N_x + 2, \ldots, N_{x+1} + 1, \ x = 1, 2, \ldots, z - r$) have the same distribution with expected values

$$
\begin{aligned}
E[T^{(\omega)}(r)] &= \mu(r) - \mu(z - x + 1), \\
\omega &= N_x + 2, \ldots, N_{x+1} + 1, \ x = 1, 2, \ldots, z - r, r = 1, 2, \ldots, z - 1,
\end{aligned}
\tag{5}
$$

and the expected values of random variables $S^{(\omega)}(r)$ are

$$
\begin{aligned}
E[S^{(\omega)}(r)] &\cong (N_1 + 1)\mu(r) + \sum_{j=1}^{x-1} (N_{j+1} - N_j)(\mu(r) - \mu(z - j + 1)) \\
&\quad + (\omega - N_x - 1)(\mu(r) - \mu(z - x + 1)), \\
\omega &= N_x + 2, \ldots, N_{x+1} + 1, \ x = 1, 2, \ldots, z - r, \ r = 1, 2, \ldots, z - 1.
\end{aligned}
\tag{6}
$$

As assumed before, the system cannot be repaired after $N_{z-r+1} + 1$ time that the system exceeds the reliability critical state r. Thus, time to complete system damage is

$$
\begin{aligned}
E[S^{(N_{z-r+1}+1)}(r)] &\cong (N_1 + 1)\mu(r) + \sum_{j=1}^{z-r} (N_{j+1} - N_j)(\mu(r) - \mu(z - j + 1)), \\
r &= 1, 2, \ldots, z - 1.
\end{aligned}
\tag{7}
$$

1.4   Renewal Function Under Imperfect Repair

The expected value of the renewal process is called the renewal function. As assumed in Sects. 1.2 and 1.3, first N_1 system's renewals cause the system to return to the best state z, and $T^{(1)}(r)$, $T^{(2)}(r)$,..., $T^{(N_1+1)}(r)$ are independent random variables from the same distribution with expected value $\mu(r)$ and variance $\sigma(r)$. In that case, the expected value of the number $N(t,r)$ of times the system exceeds the reliability critical state r up to the time point t, $0 \leq t \leq S^{(N_1+1)}(r)$, i.e. not larger than the time until the $N_1 + 1$ time the system exceeds the reliability critical state r, is given by [3, 4]

$$E[N(t,r)] \cong \frac{t}{\mu(r)}, \ 0 \leq t \leq S^{(N_1+1)}(r), \ r = 1,2,\ldots,z-1. \tag{8}$$

Similarly, the expected value of the number $N(t,r)$ of times the system exceeds the reliability critical state r up to the time point t, $S^{(N_1+1)}(r) < t \leq S^{(N_2+1)}(r)$, can be estimated from following formula

$$E[N(t,r)] \cong N_1 + 1 + \frac{t - (N_1+1)\mu(r)}{\mu(r) - \mu(z)}, \ S^{(N_1+1)}(r) < t \leq S^{(N_2+1)}(r),$$
$$r = 1,2,\ldots z-1. \tag{9}$$

Generalizing, if random variables $T^{(\omega)}(r)$ $(\omega = N_x+2,\ldots,N_{x+1}+1, x = 1,2,\ldots, z-r)$ are independent random variables from the same distribution with expected value (5), then expected value of the number $N(t,r)$ of times the system exceeds the reliability critical state r up to the time point t, $S^{(N_x+1)}(r) < t \leq S^{(N_{x+1}+1)}(r)$, $(x = 1,2,\ldots,z-r)$ is

$$E[N(t,r)] \cong N_x + 1 + \frac{t - (N_1+1)\mu(r) - \sum\limits_{j=1}^{x-1}(N_{j+1} - N_j)(\mu(r) - \mu(z-j+1))}{\mu(r) - \mu(z-x+1)}, \tag{10}$$
$$S^{(N_x+1)}(r) < t \leq S^{(N_{x+1}+1)}(r), \ x = 1,2,\ldots,z-r, \ r = 1,2,\ldots,z-1.$$

Taking into account obtained above results, the expected value of the number $N(t,r)$ of times the system exceeds the reliability critical state r up to the time point t, $t \geq 0$, can be determined using the following procedure:

$x = 1;$

$E[S^{(N_x+1)}(r)] \cong (N_1+1)\mu(r);$

```
if  (t ≤ E[S^(Nx+1)(r)])
   then  E[N(t,r)] ≅ t/μ(r);
else {
   do{
```

$x := x+1;$

$$E[S^{(N_x+1)}(r)] \cong (N_1+1)\mu(r) + \sum_{j=1}^{x-1}(N_{j+1}-N_j)(\mu(r)-\mu(z-j+1));$$

```
} while  (t > E[S^(Nx+1)(r)])  and  (x<z-r+1);
```

$$E[N(t,r)] \cong N_{x-1}+1+\frac{t-(N_1+1)\mu(r)-\sum_{j=1}^{x-2}(N_{j+1}-N_j)(\mu(r)-\mu(z-j+1))}{\mu(r)-\mu(z-x+2)};$$

```
if  (t > E[S^(Nz-r+1+1)(r)])
   then  E[N(t,r)] ≅ Nz-r+1 +1; }
```

2 Application

2.1 Master Navigation System

As an application, a navigation system [17, 18], consisting of four independently operating and connected in series systems: data collection system S_1, data processing system S_2, data presentation system S_3, and user interface S_4, is considered. It is assumed that a navigation system and those four systems are multistate systems. Following four reliability states have been distinguished: state "3" of system full ability, state "2" of system impendency over safety, state "1" of system unreliability, and state "0" of system full inability, [19].

Further, in reliability analysis of a navigation system, it is assumed that the four systems S_1, S_2, S_3 and S_4 forming the navigation system have exponential distribution. Their reliability functions are given by a vector [3, 4]

$$R_i(t,\cdot) = [1, R_i(t,1), R_i(t,2), R_i(t,3)], t \in <0,\infty), i = 1,2,3,4, \qquad (11)$$

with the coordinates

$$R_i(t,u) = \exp[-\lambda_i(u)t], i = 1,2,3,4, u = 1,2,3. \qquad (12)$$

The lifetimes in the safety states of systems S_1, S_2, S_3, S_4 are expressed in years and they have the exponential reliability functions (11) and (12) with the intensities of departure from the safety subsets, by the assumption, given by

$$\lambda_i(1) = 0.125, \ \lambda_i(2) = 0.167, \ \lambda_i(3) = 0.5, \ i = 1, 4, \tag{13}$$

$$\lambda_i(1) = 0.1, \ \lambda_i(2) = 0.125, \ \lambda_i(3) = 0.25, \ i = 2, 3. \tag{14}$$

Assuming the navigation system is a multistate series system, in case S_1, S_2, S_3, S_4 have exponential reliability functions (11) and (12) with the intensities of departure (13) and (14), the reliability function of the navigation system is given by the vector

$$\mathbf{R}_{S_M}(t, \cdot) = [1, \mathbf{R}_{S_M}(t, 1), \mathbf{R}_{S_M}(t, 2), \mathbf{R}_{S_M}(t, 3)], \ t \geq 0, \tag{15}$$

Where

$$\mathbf{R}_{S_M}(t, 1) = \exp[-0.45t], \ \mathbf{R}_{S_M}(t, 2) = \exp[-0.58t], \ \mathbf{R}_{S_M}(t, 3) = \exp[-1.5t]. \tag{16}$$

And afterwards, the mean lifetimes of the navigation system in the subsets $\{1,2,3\}$, $\{2,3\}$, $\{3\}$, counted in years are:

$$\mu_{S_M}(1) \cong 2.22, \ \mu_{S_M}(2) \cong 1.71, \ \mu_{S_M}(3) \cong 0.67. \tag{17}$$

and their mean values in the states 1, 2, 3, in years, respectively are:

$$\bar{\mu}_{S_M}(1) \cong 0.51, \ \bar{\mu}_{S_M}(2) \cong 1.04, \ \bar{\mu}_{S_M}(3) \cong 0.67. \tag{18}$$

2.2   Navigation System with a Back-up System

In a second case a navigation system consisting of a master system and an identical back-up system [17], connected in parallel, is considered. Both master system and back-up system includes data collection system, data processing system, data presentation system and user interface, similarly as in a first case. Thus, the navigation system is a series-parallel system with a reliability structure presented in Fig. 1.

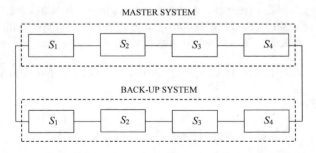

Fig. 1. The reliability structure scheme of a navigation system with a back-up system.

It is assumed that reliability functions of systems S_1, S_2, S_3, S_4 are exponential (11) and (12) and the intensities of departure from the safety subsets, as before, are given by (13) and (14). Then, the reliability function of such navigation system is a vector [3, 4]

$$R_{S_B}(t, \cdot) = [1, R_{S_B}(t, 1), R_{S_B}(t, 2), R_{S_B}(t, 3)], \ t \geq 0, \tag{19}$$

with the coordinates

$$R_{S_B}(t, 1) = 1 - [1 - \exp[-0.45t]]^2, \ R_{S_B}(t, 2) = 1 - [1 - \exp[-0.58t]]^2,$$
$$R_{sB}(t, 3) = 1 - [1 - \exp[-1.5t]]^2. \tag{20}$$

The mean lifetimes of the system in the subsets {1,2,3},{2,3},{3}, counted in years are:

$$\mu_{S_B}(1) \cong 3.33, \ \mu_{S_B}(2) \cong 2.57, \ \mu_{S_B}(3) \cong 1.00. \tag{21}$$

and the mean values of system lifetimes in the states 1, 2, 3, in years, respectively are:

$$\bar{\mu}_{S_B}(1) \cong 0.76, \ \bar{\mu}_{S_B}(2) \cong 1.57, \ \bar{\mu}_{S_B}(3) \cong 1.00. \tag{22}$$

2.3   Renewal and Availability of a Transport Navigation System

As a critical reliability state has been adopted the state $r = 1$. Next, the number of navigation system's renewals is fixed. It is assumed that, first $N_1 = 5$ system renovations are to the best state $z = 3$. Renewals of the system from the sixth to the eighth ($N_2 = 8$) take place to the state 2, so there are 3 renewals of the system to the state 2.

Table 1. The expected values of the time until *Nth* time the reliability critical state is exceeded by a single master navigation system S_M and a navigation system with a back-up system S_B.

N	$N_1 = 5,$ $N_2 = 8,$ $N_3 = 10$		$N_1 = 4,$ $N_2 = 8,$ $N_3 = 10$		$N_1 = 3,$ $N_2 = 8,$ $N_3 = 10$	
	S_M	S_B	S_M	S_B	S_M	S_B
1	2.22	3.33	2.22	3.33	2.22	3.33
2	4.44	6.67	4.44	6.67	4.44	6.67
3	6.67	10.00	6.67	10.00	6.67	10.00
4	8.89	13.33	8.89	13.33	8.89	13.33
5	11.11	16.67	11.11	16.67	10.44	15.67
6	13.33	20.00	12.67	19.00	12.00	18.00
7	14.89	22.33	14.22	21.33	13.56	20.33
8	16.44	24.67	15.78	23.67	15.11	22.67
9	18.00	27.00	17.33	26.00	16.67	25.00
10	18.51	27.76	17.84	26.76	17.17	25.76
11	19.02	28.52	18.35	27.52	17.68	26.52

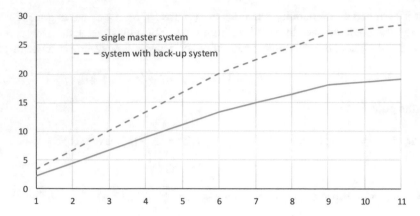

Fig. 2. The graph of the expected values of the time until Nth time the reliability critical state is exceeded by a single master navigation system and a navigation system with a back-up system.

Table 2. The expected values of the number $N(t,1)$ of times that the critical state is exceeded by a single master system S_M and a navigation system with a back-up system S_B, up to time point t.

t [years]	S_M	S_B	t [years]	S_M	S_B	t [years]	S_M	S_B
0	0	0	10	4.50	3.00	20	11.00	6.00
1	0.45	0.30	11	4.95	3.30	21	11.00	6.43
2	0.90	0.60	12	5.40	3.60	22	11.00	6.86
3	1.35	0.90	13	5.85	3.90	23	11.00	7.29
4	1.80	1.20	14	6.43	4.20	24	11.00	7.71
5	2.25	1.50	15	7.07	4.50	25	11.00	8.14
6	2.70	1.80	16	7.71	4.80	26	11.00	8.57
7	3.15	2.10	17	8.36	5.10	27	11.00	9.00
8	3.60	2.40	18	9.00	5.40	28	11.00	10.31
9	4.05	2.70	19	10.97	5.70	29	11.00	11.00

Finally, the last two renewals of the system i.e. the ninth and tenth ($N_3 = 10$), include operations that causes the system to return to the critical state 1. After the system has exceeded the critical state the eleventh time, the system remains in the state of full inability. Then, using the formulas (2) and (6) respectively, the expected values of variables, representing time until successive exceeding the reliability critical state $r = 1$ by the system, can be determined. Their values in both cases, counted in years for a single master navigation system S_M and for a navigation system with a back-up system S_B, are given in Table 1 and presented in Fig. 2.

The results for system S_M and S_B, are also compared in Table 1 for other values of the number of system renewals $N_1 = 4$, $N_2 = 8$, $N_3 = 10$ and $N_1 = 3$, $N_2 = 8$, $N_3 = 10$.

The expected values of the number $N(t,1)$ of times, that a single master navigation system and a navigation system with a back-up system exceed the critical state up to time point t, calculated from (8) and (10), are presented in Table 2.

The expected values of the number $N(t,1)$ of renewals after the exceeding the critical state up to time point t by a single master navigation system and a master navigation system with a back-up system are also compared graphically in Fig. 3.

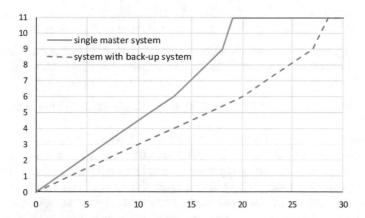

Fig. 3. The graph of the expected values of the number of times a system exceeds the critical state for a single master navigation system and a navigation system with a back-up system.

The number $N(t,1)$ of renewals for the single master system up to time point t, is grater by 50% relative to that for the master navigation system with a back-up system, in case when the system is restored to the best state 3. This difference increases up to 90%, in case when the navigation system without a back-up system can be repaired only to worse reliability state 2 or 1.

3 Conclusion

A multistate approach to the assessment of system reliability allows for a more accurate analysis of the operation of systems and the diagnostics of their work safety. Many technical systems are multistate systems due to the deteriorating technical condition of their components. It is then possible to consider intermediate states of a system and its components, in addition to the state corresponding to the new one and the state of total inability. The number of such states should be determined after consulting with the experts operating the system. Similarly, establishing a critical state of the system, the exceeding of which, does not ensure an adequate level of operational efficiency requires consultation with an expert.

Figure 4 shows the trajectory of transitions between the reliability states, described in Sect. 2.1, for a single master navigation system, and a navigation system with a back-up system, depending on the time expressed in years. Detailed conditions of systems renewals have been adopted as described in Sect. 2.3 ($N_1 = 5$, $N_2 = 8$, $N_3 = 10$). In addition, it is assumed that the system is repaired after exceeding its critical reliability state $r = 1$, and time of renovation is ignored. These results could be

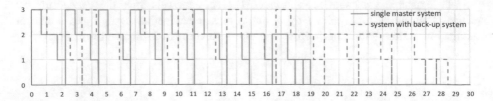

Fig. 4. The trajectory of transitions between the reliability states for a single master navigation system, and a navigation system with a back-up system ($N_1 = 5$, $N_2 = 8$, $N_3 = 10$).

compared with the results under the assumption that the repair occurs after exceeding the state "2".

By analyzing selected scenarios of transitions between system states, for example, time to system failure without repair option can be compared for a single system, a system with reserve, or assuming repair to different reliability states. In practical terms, such an analysis would gain, taking into account the costs of constructing the system, reserves of its components and repair costs. Detailed analysis, based on the history of system failures and renewals, could be used to plan the operation of the system and to avoid exceeding the critical state of the system.

References

1. Xue, J.: On multi-state system analysis. IEEE Trans. Reliab. **34**, 329–337 (1985)
2. Xue, J., Yang, K.: Dynamic reliability analysis of coherent multi-state systems. IEEE Trans. Reliab. **44**, 683–688 (1995)
3. Kołowrocki, K.: Reliability of Large and Complex Systems, 2nd edn. Elsevier, London (2014)
4. Kołowrocki, K., Soszyńska-Budny, J.: Reliability and Safety of Complex Technical Systems and Processes: Modeling – Identification – Prediction – Optimization, 1st edn. Springer-Verlag, London (2011)
5. Guze, S.: Reliability analysis of multi-state ageing series-consecutive m out of n: F systems. In: Proceedings of the European Safety and Reliability Conference ESREL 2009, vol. 3, pp. 1629–1635. CRC Press/Balkema, Prague (2010)
6. Blokus-Roszkowska, A., Dziula, P.: An approach to identification of critical infrastructure systems. In: Simos, T., Tsitouras, C. (eds.) Proceedings of the International Conference of Numerical Analysis and Applied Mathematics ICNAAM 2015, pp. 440006-10–440006-13. AIP Publishing (2016)
7. Badoux, R.A.J.: Availability and maintainability. In: Colombo, A.G., Keller, A.Z. (eds.) Reliability Modeling and Applications, pp. 99–124. Kluwer Academic Publishers Group, Dordrecht (1987)
8. Zio, E.: An Introduction to the Basics of Reliability and Risk Analysis. Series in Quality, Reliability and Engineering Statistics 13. World Scientific Publishing, Singapore (2007)
9. Zio, E., Compare, M.: Evaluating maintenance policies by quantitative modeling and analysis. Reliab. Eng. Syst. Saf. **109**, 53–65 (2013)
10. Pham, H., Suprasad, A., Misra, R.B.: Availability and mean life time prediction of multistage degraded system with partial repairs. Reliab. Eng. Syst. Saf. **56**, 169–173 (1997)

11. Barlow, S.E., Wu, A.S.: Coherent systems with multi-state components. Math. Oper. Res. **3** (11), 275–281 (1978)
12. El-Neweihi, E., Proschan, F., Sethuraman, J.: Multistate coherent systems. J. Appl. Prob. **15** (12), 675–688 (1978)
13. Muhammad, M., Majid, M.A., Mokhtar, A.A.: Reliability evaluation for a multi-state system subject to imperfect repair and maintenance. Int. J. Eng. Technol. **10**(01), 59–63 (2010)
14. Soro, I.W., Nourelfath, M., Ait-Kadi, D.: Performance evaluation of multi-state degraded systems with minimal repairs and imperfect preventive maintenance. Reliab. Eng. Syst. Saf. **95**(2), 65–69 (2010)
15. Nourelfath, M., Chatelet, E., Nahas, N.: Joint redundancy and imperfect preventive maintenance optimization for series–parallel multi-state degraded systems. Reliab. Eng. Syst. Saf. **103**, 51–60 (2012)
16. Kołowrocki, K., Soszyńska, J.: Reliability and availability of complex systems. Qual. Reliab. Eng. Int. **22**(1), 79–99 (2006)
17. Weintrit, A., Dziula, P., Siergiejczyk, M., Rosiński, A.: Reliability and exploitation analysis of navigational system consisting of ECDIS and ECDIS back-up systems. In: Weintrit, A. (ed.) Activities in Navigation. Marine Navigation and Safety of Sea Transportation, pp. 109–115. CRC Press, Leiden (2015)
18. Guze, S., Smolarek, L., Weintrit, A.: The area-dynamic approach to the assessment of the risks of ship collision in the restricted water. Sci. J. Marit. Univ. Szczec. Zesz. Nauk. Akad. Morsk. w Szczec. **45**(117), 88–93 (2016)
19. Dziula, P., Kołowrocki, K., Rosiński, A.: Issues concening identification of critical infrastructure systems within the Baltic Sea area. In: Podofillini, L., Sudret, B., Stojadinovic, B., Zio, E., Kröger, W. (eds.) Safety and Reliability of Complex Engineered Systems ESREL 2015, pp. 119–126. CRC Press/Balkema, Leiden (2015)

Tool for Mutation Testing of Web Services

Ilona Bluemke[✉] and Artur Sawicki

Institute of Computer Science, Warsaw University of Technology,
Warsaw, Poland
I.Bluemke@ii.pw.edu.pl

Abstract. The Web Services technology becomes more and more popular because it allows to easily utilize and integrate existing software applications even working on different platforms, and to create new services. This way of software development causes new issues for Web Service testing to ensure the quality of service that is published. Mutation analysis can be used to measure the adequacy of tests or to reveal errors. In mutation testing the original code is modified using set of mutation operators. For Web Services the mutation operators can modify SOAP messages or WSDL documents. A tool named Exodus, supporting the mutation testing of Web Services is presented. This tool accepts files describing Web Services in WSDL (Web Services Description Language) format and generates altered (mutated) versions of them using some mutation operators. The architecture and the implementation of Exodus tool, which can be integrated with the build process, is shortly presented. Exodus is able to perform fully automated mutation analysis on tests referring to Web service being developed. Exemplary usage of it is also given.

Keywords: Mutation testing · Mutation operators · Web services
WSDL

1 Introduction

Web services are self-descriptive, self-contained applications based on an open standard and currently are widely used in the development of software for the World Wide Web. A Web service is based on Simple Object Access Protocol (SOAP) [1] and XML [2] technology. Web Service relies on a family of protocols to describe, deliver, and interact with each other, such as the Web Service Description Language (WSDL) [3], the Universal Description, Discovery and Integration (UDDI) [4] protocol or the Web Service Inspection Language (WSIL) [5], and the SOAP. Web Service is vendor-neutral, language-agnostic, and platform-independent. Therefore, developers of a Web Service are not able to assume which type of clients will use it, and the developers at the client side, are not aware of which language and platform are used at the server side. A Web Service and its clients can be developed in totally different programming languages and on different platforms.

Web Service can be described by a Web Service Description Language [3] document. WSDL was established as a standard by the W3C [6]. The WSDL document describes Web Service in terms of its interfaces with operations, types of data it takes or returns. Other systems interact with the Web Service in a manner prescribed by its description using SOAP messages.

Testing effort is often a major cost factor during the software development and sometimes consumes more than 50% of the overall development effort [7]. With services the difficulty of testing increases because of the changes that this architectural style induces on both the system and the software business/organization models.

Mutation testing is a software testing technique that was introduced more than forty years ago. The general idea is that the faults are deliberately introduced into the program to create a set of faulty programs called mutants. Each mutant program is obtained by making a small change (e.g. change operator in expression) in the original one. The type of this change is defined by mutation operator (e.g. operator change). To assess the quality of a given set of tests these mutants are executed against the set of input data to see, if the inserted faults can be detected by these tests. An survey on mutation techniques was written by Jia and Harman [8].

Basics information on Web Services testing is presented in the first part of Sect. 2. One of the existing approaches is mutation testing. However mutation testing of Web Services was proposed in 2005 [9], to our best knowledge, there are only two tools supporting it: Solini and Vergilo implemented a tool WSeTT [10], and Bluemke and Grudzinski implemented a simple mutant generator [11]. Both these tools only generate some mutants and can't be integrated with whole design/implementation process.

Our goal was to develop a tool, able to perform fully automated mutation analysis on tests referring to Web service being developed, and which can be integrated with the build process. At the Institute of Computer Science Warsaw University of Technology an application, named Exodus, enabling the mutation testing of Web Services, has been designed and implemented. Exodus can be consolidated with the design/implementation of Web Services through Maven [12], such property was not possible for earlier tools [10, 11]. However mutant generator [11] was implemented at the same institute, we admit that neither its code, nor design were used in Exodus.

The paper is organized as follows. Section 2 identifies key features of Web Services testing, also WSDL is briefly presented. Some mutation operators are also described. Section 3 focuses on the architecture of our tool. Simple example is presented. Finally, Sect. 4 concludes the paper, highlighting some issues and indicates future research directions.

2 Mutation Testing of Web Services

Testing services is widely presented in the literature, primarily in the areas of unit testing of services and orchestrations, integration testing, regression testing, and testing of non-functional properties. Canfora and Di Penta made a survey on testing SOA [13]. They also pointed out several problems which remain open and need additional research work e.g.: improving testability, combining testing and run-time verification, validating fully decentralized systems. Comprehensive and excellent surveys on Web services testing can also be found in [14–17]. Several Web Services testing approaches were developed to address these new challenges, the survey can be found in [14].

Based on the surveyed Web Services testing approaches found in the literature, the existing approaches were divided by Ladan [14] into four classes:

1. WSDL-based test case generation,
2. mutation-based test case generation (e.g. [9, 10]),
3. test modeling (e.g. [18]),
4. XML-based approaches (e.g. [19]).

Hanna and Munro [19] proposed a framework that can be used to test the robustness quality attribute of a Web Service. This framework is based on analyzing the Web Service Description Language (WSDL) document of Web Services to identify which faults could affect the robustness and then test cases were designed to detect those faults.

Bai and Dong [20] also are using the WSDL document to generate tests in a testing framework that includes test case generation, test controller, test agents and test evaluator. An example of a tool (named WSDLTest) for semi-automatic testing of Web Services (but not using mutation approach), based on WSDL documents is described in [21]. In [22] another tool, but of the same name, is described. This tool generates Web service requests from the WSDL schemas and adjusts them in accordance with the pre-condition assertions written by the tester. It dispatches the requests and captures the responses.

Several testing tools, which can be used for testing web services, are available on the market, the most recognizable among them are: HP QuickTest Professional [23], Parasoft SOAtest [24], SOAPSonar [25], SoapUI [26].

In this paper we concentrated on the WSDL-based and mutation - based approach to testing Web Services. Below basics of WSDL are given.

2.1   WSDL

WSDL [3] - Web Services Description Language, is an abstract language specifying location and functionalities offered by a web service. It contains specific information about the web service, for example needed parameters, returned data. XML Schema is used for the presentation of WSDL description containing the messages send and received from the service. To communicate with the web service SOAP messages are exchanged and they are described by WSDL as operations. WSDL describes the format for interfaces, it can also describe the interactivity of a given service. WSDL description only shows the possible but not required interactions. The current version of WSDL, recommended by W3C [6], is 2.0 but the older version - WSDL 1.1 is still quite common. The structures of these versions are shown in Fig. 1.

A WSDL document contains following sections:

- **Definition**: it defines the name of web service, declares multiple name spaces used throughout the remainder of the document, and contains all service elements.
- **Data types**: XML schemas of data types used in messages, when simple types are used the document does not need to have a types section.
- **Message**: an abstract definition of the data, in the form of a message presented either as an entire document or as arguments to be mapped to a method invocation.

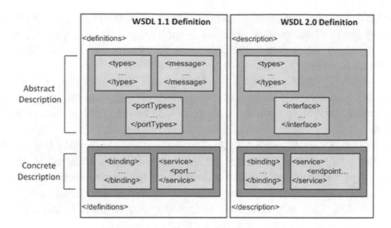

Fig. 1. Structure of WSDL documents (based on [27])

- **Operation:** abstract definition of the operation for a message, such as naming a method, message queue, or business process, that will accept and process the message.
- **Port type**: abstract set of operations mapped to one or more endpoints, defining the collection of operations for a binding; the collection of operations, as it is abstract, can be mapped to multiple transports through various bindings.
- **Binding**: protocol and data formats for the operations and messages defined for a particular port type.
- **Port**: a combination of a binding and a network address, providing the target address of the service communication.
- **Service**: a collection of related endpoints encompassing the service definitions in the file; the services map the binding to the port and include any extensibility definitions.

Additionally, the WSDL specification also defines following sections: documentation (contains comments, documentation and can be included inside any other WSDL element), import (used to import other WSDL documents or XML Schemas).

2.2 Mutation Operators

Nine mutation operators were suggested by Siblini and Mansour in [9]:

- Switch Types Complex Type Element (STCE),
- Switch Types Complex Type Attribute (STCA),
- Occurrence Types Complex Type Element (OTCE),
- Occurrence Types Complex Type Attribute (OTCA),
- Special Types Element Nil (STEN),
- Switch Types Simple Type Element (STSE),
- Switch Types Simple Type Attribute (STSA),
- Switch Messages Part (SMP),
- Switch Port Type Message (SPM).

The first four were implemented (STCE, STCA, OTCE, OTCA) in our Exodus tool. The chosen group of operators is responsible for manipulating complex types part of a WSDL document and they represent possible faults related to Web Service implementation.

The **Switch Types Complex Type Element** (STCE) operator is defined as the operator that will switch elements of the same type (e.g. strings, integers) within a single complex type element that could be defined in the types section of the WSDL document.

The **Switch Types Complex Type Attribute** (STCA) operator is defined as the operator that will switch attributes of the same type (strings, integers) within a single Complex Type element that could be defined in the Types section of the WSDL document. The operation of the Switch Types Complex Type Attribute operator is similar to STCE. The only difference is in the switched parts – in case of STCA, attributes of the same type are switched instead of elements.

The **Occurrence Types Complex Type Element** (OTCE) operator adds or deletes an occurrence of an element in the complexType element in the Types part of a WSDL document. This operator can be applied only if there are several elements in the complex type.

The **Occurrence Types Complex Type Attribute** (OTCA) operator adds or deletes an occurrence of an attribute in the complexType element that could be defined in the Types part of a WSDL document. The operator can be applied only if there is more than one element in the complex type. Similar relation as in STCE and STCA operators occurs in OTCE and OTCA. The difference between these two lies in the part being added or deleted- in case of OTCE it is an element, whereas in OTCA – an attribute.

Other mutations operators were proposed by Solino and Vergilio in [10]. Other approach in mutation testing of Web Services and completely different mutation operators are presented in [29]. The authors propose to mutate SOAP message parameters and create combined mutants, using more than one mutation operator at a time.

3   Exodus Tool

The general idea of mutation testing of Web Services based on WSDL document is presented in Fig. 2. The tester provides a WSDL document with a Web Service to be tested. The mutation tool applies mutant operators selected by the tester and generates mutated WSDL documents. Subset of tests concerning usage of Web Service described by original WSDL is run against each mutant and original. Results of each mutant are compared with those for original WSDL to establish which of them are "killed" and which remain "live". The results are presented to the tester.

The name of our tool - Exodus refers to hero from a series of comics X-Men. Exodus [30] was one of mutants which established brotherhood of mutants. Our application gathers several modules, like in brotherhood. Exodus, supporting mutation testing of Web Services, provides following function:

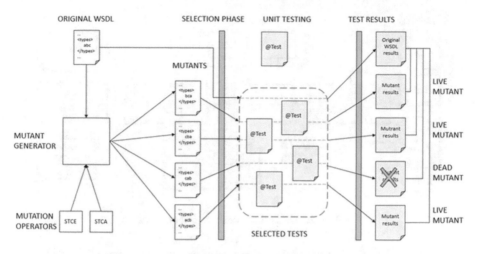

Fig. 2. Mutation testing of web services (based on [28])

- Generates mutants for a given WSDL file,
- Stores these mutants in files,
- Performs mutation analysis for some mutants and tests data,
- Stores the results of analysis
- Presents the results to the tester.

The above listed steps can be executed independently, from the command line, or can be integrated with Maven [12], a tool for building projects.

The architecture of Exodus tool, providing the above listed functions, is presented in Fig. 3 it contains three modules: MUTANT GENERATOR, TEST RUNNER and RESULTS CONVERTER. First of them takes original WSDL document and selected mutation operator as inputs. It outputs generated mutants to a directory (selected by user), where each mutant is stored in a folder, named by operator used to generate it. TEST RUNNER uses this directory structure alongside with selected test classes to run tests and outputs results in XML files. These files can be used for further processing or converted by RESULT CONVERTER to HTML reports which is much more readable for a tester.

4 Example

Simple service converting the currency was prepared to show the usage of Exodus tool. Its interface is given in Fig. 4, while implementation of simple method in Fig. 5.

It was assumed, in implementation, that the default currency type is US dollar. The WSDL document for this service is not presented here as it is too long, but is given in [31]. The mutants were generated (three OTCE and one STCE) and test shown in Fig. 6 executed. In this test, the default type of currency was not included. The results

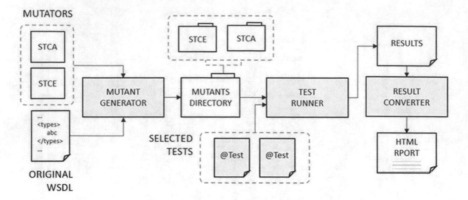

Fig. 3. General structure of Exodus tool

```
1  @WebService
2  public interface CurrencyConverterWebService {
3
4      @WebMethod
5      double convert(
6          @WebParam(name = 'amount') double amount,
7          @WebParam(name = 'from') Currency from,
8          @WebParam(name = 'to') Currency to
9      );
10 }
```

Fig. 4. Interface of the service currency converter

```
1  public class CurrencyConverterWebServiceImpl implements CurrencyConverterWebService {
2
3      // currency conversions loaded from some resource
4      private static final Map<Currency, Map<Currency, Double>> CONVERSIONS = load();
5
6      @Override
7      public double convert(double amount, Currency from, Currency to) {
8          if (from == null) {
9              from = Currency.USD;
10         }
11         if (to == null) {
12             to = Currency.USD;
13         }
14         return amount * CONVERSIONS.get(from).get(to);
15     }
16 }
```

Fig. 5. Implementation of methods of currency converter service

of test are shown in Fig. 7. One OTCE mutant, in which the target currency was omitted, was "live", it means that tests were insufficient, or that this is an equivalent mutant (the results for mutant and original one are the same). After adding the test in Fig. 8 this mutant was also "killed".

```
1    @Test
2    public void shouldConvertPlnToUsd() throws Exception {
3        Client client = clientFactory.client(endpointPublisher.getWsUrl(), WSDL);
4
5        Object[] res = client.invoke('convert', 100.0, Currency.PLN, Currency.USD);
6
7        assertEquals('28.15', String.format('%.2f', (double) res[0]));
8    }
```

Fig. 6. Test of currency converter

```
1    <?xml version="1.0" encoding="UTF-8" standalone="yes"?>
2    <result>
3        <classes>
4            <class>com.gitlab.tempaowca.converter.impl.CurrencyConverterWebServiceTest</class>
5        </classes>
6        <originalWsdl>(...)/cerebro/original.wsdl</originalWsdl>
7        <mutantResults>
8            <mutantResult>
9                <mutant>
10                   <mutator>OTCE</mutator>
11                   <wsdl>(...)/cerebro/OTCE/mutant_3.wsdl</wsdl>
12               </mutant>
13               <outcome>LIVED</outcome>
14               <distinguishingTests/>
15           </mutantResult>
16           <mutantResult>
17               <mutant>
18                   <mutator>STCE</mutator>
19                   <wsdl>(...)/cerebro/STCE/mutant_1.wsdl</wsdl>
20               </mutant>
21               <outcome>KILLED</outcome>
22               <distinguishingTests>
23                   <test>shouldConvertPlnToUsd(...)</test>
24               </distinguishingTests>
25           </mutantResult>
26           <mutantResult>
27               <mutant>
28                   <mutator>OTCE</mutator>
29                   <wsdl>(...)/cerebro/OTCE/mutant_1.wsdl</wsdl>
30               </mutant>
31               <outcome>KILLED</outcome>
32               <distinguishingTests>
33                   <test>shouldConvertPlnToUsd(.)</test>
34               </distinguishingTests>
35           </mutantResult>
36           <mutantResult>
37               <mutant>
38                   <mutator>OTCE</mutator>
39                   <wsdl>(...)/cerebro/OTCE/mutant_2.wsdl</wsdl>
40               </mutant>
41               <outcome>KILLED</outcome>
42               <distinguishingTests>
43                   <test>shouldConvertPlnToUsd(...)</test>
44               </distinguishingTests>
45           </mutantResult>
46       </mutantResults>
47   </result>
```

Fig. 7. Results of testing mutants for currency converter service

```
1    @Test
2    public void shouldConvertUsdToPln() throws Exception {
3        Client client = clientFactory.client(endpointPublisher.getWsUrl(), WSDL);
4
5        Object[] res = client.invoke('convert', 100.0, Currency.USD, Currency.PLN);
6
7        assertEquals('355.18', String.format('%.2f', (double) res[0]));
8    }
```

Fig. 8. Additional test for currency converter

5 Conclusions

Exodus, a tool supporting mutation testing of web services is briefly presented. This tool consists of three modules responsible for appropriate steps of mutation analysis. Each module can be executed from command line or as Maven [12] plug-in. To our best knowledge it is the first tool covering such wide scope of mutation analysis concerning mutation of WSDL documents. Tools for mutations testing of Web Services are rather unique. Solini and Vergilo implemented a tool WSeTT [10] generating mutants for theirs operators. Siblini and Mansour proposed mutation operators [9] and there exist a tool implementing some of theirs operators [11]. These tools are able only to generate mutants and can't be consolidated with the design/implementation of Web Services as Exodus can trough Maven.

Exodus uses a WSDL file describing a Web Service and generates mutants files for STCE, STCA, OTCE, OTCA (described in Sect. 2.2) mutation operators. It produces results as HTML and XML files, which can be used by other tools in the analysis process. It is designed and implemented in such a way, that new functions can be easily added. Many tests, which were prepared during implementation of our tool, can be automatically executed and can facilitate any modification of this tool in future.

References

1. SOAP: Simple Object Access Protocol. http://www.w3.org/TR/soap/. Accessed 17 Mar 2017
2. XML – Extensible Markup Language. http://www.w3.org/XML/. Accessed 17 Mar 2017
3. WSDL. http://www.w3.org/TR/wsdl. Accessed 17 Mar 2017
4. UDDI. http://uddi.xml.org/. Accessed 17 Mar 2017
5. WSIL. www.ibm.com/developerworks/webservices/library/ws-wsilspec.html. Accessed 17 Mar 2017
6. W3C Official Website. http://www.w3.org/. Accessed 17 Mar 2017
7. Elberzhager, F., Rosbach, A., Eschbach, R., Münch, J.: Reducing test effort: a systematic mapping study on existing approaches. Inf. Softw. Technol. **10**(54), 1092–1106 (2012)
8. Jia, Y., Harman, M.: An analysis and survey of the development of mutation testing. IEEE Trans. Soft. Eng. **5**(37), 649–678 (2011)
9. Siblini, R., Mansour, N.: Testing web services. In: 3rd ACS/IEEE International Conference on Computer Systems and Applications (2005)

10. da Silva Solino, A.L., Vergilio, S.R.: Mutation based testing of web services. In: 10th Latin American Test Workshop, pp. 1–6 (2009)
11. Bluemke, I., Grudziński, W.: Mutant generation for WSDL. In: Madeyski, L., et al. (ed.) Software Engineering: Challenges and Solutions, Results of the XVIII KKIO 2016 Software Engineering Conference. Advances in Intelligent Systems and Computing, vol. 504, pp. 103–117. Springer (2017). https://doi.org/10.1007/978-3-319-43606-7_8
12. Maven. https://maven.apache.org/. Accessed May 2017
13. Canfora, G., Di Penta, M.: Service-oriented architectures testing: a survey. In: De Lucia, A., Ferrucci, F. (eds.) Software Engineering, pp. 78–105. Springer (2009)
14. Ladan, M.I.: Web services testing approaches: a survey and a classification. In: Zavoral, F., et al. (eds.) NDT 2010, Part II. CCIS 88, pp. 70–79. Springer, Berlin (2010)
15. Bozkurt, M., Harman, M., Hassoun, Y.: Testing web services: a survey. Technical report, King's College London (2010)
16. Metzger, A.: Analytical quality assurance. In: Papazoglou, M., et al. (eds.) Service Research Challenges and Solutions for the Future Internet, pp. 209–270. Springer (2010)
17. Rusli, H., et al.: A comparative evaluation of state-of-the-art web service composition testing approaches. In: International Workshop on Automation of Software Test, pp. 29–35 (2011)
18. Feudjio, A.G.V., Schieferdecker, I.: Availability Testing for Web Services. Telektronikk (2009). ISSN 0085-7130
19. Hanna, S., Munro, M.: An approach for WSDL-based automated robustness testing of web services. In: Information Systems Development Challenges in Practice, Theory, and Education, vol. 2, pp. 1093–1104. Springer (2008)
20. Bai, X., Dong, W.: WSDL-based automatic test case generation for web services testing. In: IEEE International Workshop on Service-Oriented System Engineering (SOSE 2005), pp. 207–212 (2005)
21. Bluemke, I., Kurek, M., Purwin, M.: Tool for automatic testing of web services. In: Ganzha, M., Maciaszek, L., Paprzycki, M. (eds.) Proceedings of the 2014 Federated Conference on Computer Science and Information Systems. Annals of Computer Science and Information Systems, vol. 3 (2014)
22. Sneed, H.M., Huang, S.: WSDLTest – a tool for testing web services. In: Eighth IEEE International Symposium on Web Site Evolution (WSE 2006), pp. 14–21 (2006)
23. HP QuickTest Professional. http://www8.hp.com/us/en/software-solutions/unified-functional-testing-automation. Accessed May 2017
24. Parasoft. http://www.parasoft.com/soatest. Accessed May 2017
25. SOAPSonar. http://www.crosschecknet.com/products/soapsonar.php. Accessed May 2017
26. SoapUI. http://www.soapui.org/. Accessed May 2017
27. Java Solution Architect: WSDL 2.0 VS WSDL 1.1. http://javasolutionarchitect.blogspot.com/2013/11/wsdl-20-vs-wsdl-11.html. Accessed May 2017
28. MuClipse – The Mutation Testing Process. http://muclipse.sourceforge.net/img/mutation.jpg. Accessed 17 Mar 2017
29. Chen, J., et al.: A web services vulnerability testing approach based on combinatorial mutation and SOAP message mutation. SOCA **8**, 1–13 (2014)
30. Exodus. http://x-men.wikia.com/wiki/Exodus. Accessed 17 Jan 2018
31. Sawicki, A.: Mutation testing of web services. MSc thesis, Institute of Computer Science, Warsaw University of Technology (2017). (in polish)

Associative Probabilistic Neuro-Fuzzy System for Data Classification Under Short Training Set Conditions

Yevgeniy Bodyanskiy[1]([✉]) [iD], Artem Dolotov[1] [iD],
Dmytro Peleshko[2] [iD], Yuriy Rashkevych[3] [iD],
and Olena Vynokurova[1,2] [iD]

[1] Kharkiv National University of Radio Electronics, 14 Nauky av.,
Kharkiv 61166, Ukraine
yevgeniy.bodyanskiy@nure.ua, artem.dolotov@gmail.com,
vynokurova@gmail.com
[2] IT Step University, 83a Zamarstynivs'ka str., Lviv 79019, Ukraine
dpeleshko@gmail.com, vynokurova@gmail.com
[3] Ministry of Education and Science of Ukraine, Kyiv, Ukraine
rashkevyuriy@gmail.com

Abstract. The paper proposes a classifying neuro-fuzzy system intended for operating under short training set and nonconvex classes conditions. The proposed system is constructed of feedforward four-layered architecture that solves task of probabilistic classification and fuzzy associative memory which is based on autoassociative evolving memory on fuzzy basis functions that solves fuzzy classification task. An essential aspect of the considered hybrid system of computational intelligence is its high performance and ease of numerical implementation.

Keywords: Associative probabilistic neuro-fuzzy system · Data classification
Short training data set

1 Introduction

The problem of data clustering based on supervised learning paradigm is among the major ones within the scope of scientific research direction known as data mining, and a great variety of approaches and algorithms to solve it has been suggested [1–3], ranging techniques from purely empirical to strictly mathematical. Here, conventional pattern classification and identification algorithms utilize generally the hypothesis that classes to be recognized are convex and linearly separable [4]. By virtue of the fact that, given real data, the hypothesis fails every so often and classes are not convex typically, conventional approaches become inoperative whereas artificial neural networks come to the forefront for their universal approximation properties [5–9].

However, neural networks are not a cure-all as they call for sufficiently large training sets for their adjustment in case the learning process is based on a particular optimization procedure. In the case that the size of the training set is commensurate

© Springer International Publishing AG, part of Springer Nature 2019
W. Zamojski et al. (Eds.): DepCoS-RELCOMEX 2018, AISC 761, pp. 56–63, 2019.
https://doi.org/10.1007/978-3-319-91446-6_6

with the dimension of feature vector, memory based learning makes it impossible to provide neural network with the required classification quality. Here it appears advisable to utilize memory based learning which implements 'neurons at the data points' principle [10, 11]. Such networks are typified by the probabilistic neural network proposed by Specht [12, 13]. That neural network has proven its efficacy in solving various tasks of classifications diagnostic, pattern recognition etc. [5, 7, 8] due to ease of learning and high performance. While still conventional PNN, same as any other neural network, solves classification task within the scope of crisp approach when each feature vector being analyzed can belong to a single class among possible ones.

However, in numerous tasks concerning chemical and physical experiments, medical and biological information, texts and images etc., feature vector to be classified may belong right to several classes with different membership degree. Usage of hybrid systems of computational intelligence [14–17], among them neuro-fuzzy systems [18, 19] and fuzzy classification and clustering methods [20, 21] in the first place, are seen to have the highest performance in solving the tasks being considered in this paper. It is pertinent to note that both neuro-fuzzy systems learning and fuzzy classification and clustering reside in specific optimization procedures which again call for sufficient size of the training set. It would seem, in this respect, that it may be advantageous to design a hybrid neuro-fuzzy system of computational intelligence that allows of solving problems of probabilistic and fuzzy data classification under short training set and overlapping arbitrary shaped classes conditions.

2 Associative Probabilistic Neuro-Fuzzy System

A probabilistic neural network of Specht and its modifications introduced in [22–24] form the basis for the proposed system. Figure 1 depicts its architecture. Conventional PNN is known to solve Bayes classification problem [4] by restoring data density in each class based on Parzen kernel estimation [25], and, in fact, its architecture is identical to one of radial basis neural network (RBFN) [26] and general regression neural network (GRNN) [27]. Although Bayes classification methods have long been known, neural networks-based implementation of them made it possible to simplify numerical implementation remarkably and improve performance of solving thc problems of pattern recognition, classification, diagnostic etc. It is significant that each input pattern provided for PNN can belong solely to one class with the highest data density about a point of that pattern.

The considered system consists of two parts: feedforward four-layered architecture that solves the task of probabilistic classification, and fuzzy associative memory (FAM) [28–30] which is based on autoassociative evolving memory on fuzzy basis functions [31] that solves fuzzy classification task. A part of the considered system that is responsible for probabilistic classification contains four feedforward layers. The first one is the layer of patterns that is identical to the first hidden layer f PNN and GRNN. The second one is the layer of local adders, the third one contains a single adder for all neurons, and, finally, the output layer is formed by dividers, which number correspond to a number of classes in the training set.

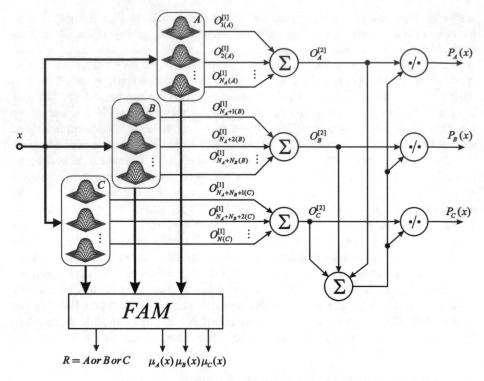

Fig. 1. Associative probabilistic neuro-fuzzy system

Initial information to synthesize a classifying neuro-fuzzy system is a training set of multidimensional input data $x(1), x(2), \ldots, x(N)$ formed of n-dimensional feature vectors with known classification, with that position of a pattern in the initial set, does not matter, i.e. initial information is specified by 'object-property' table form commonly considered in Data Mining.

It is considered that initial information if pre-processed, usually by componentwise mean correction and the following hyperspherical normalization

$$\|x(j)\| = 1, \ \forall j = 1, 2, \ldots, N. \tag{1}$$

In what follows without the loss in generality, we will consider that there are three classes A, B and C in the training set. It is considered also that N_A patterns belong to class A, N_B patterns belong to class B, and N_C patterns belong to class C, i.e.

$$N_A + N_B + N_C = N.$$

Following 'neurons at the data points' concept, number of neurons in the patterns layer is set to N, and each block of the layer is denoted as A, B, and C, contains N_A, N_B, N_C neurons correspondingly. Each input pattern $x(j)$ defines coordinates of centers w_j of kernel activation functions such that

$$w_{ji} = x_i(j); \; j = 1, 2, \ldots, N; \; i = 1, 2, \ldots, n,$$

or in vector form

$$w_j = x(j) = (x_1(j), x_2(j), \ldots, x_n(j))^T.$$

Concurrently with defining centers of activation functions of patterns layer, in fuzzy associative memory, centroids of the formed classes are evaluated

$$w_R = \frac{1}{N} \sum_{l=1}^{N_R} x(l); \; x(l) \in R; \; R = A \text{ or } B \text{ or } C$$

and their radii

$$r_R = \max_l \|x(l) - w_R\|; \; x(l) \in R.$$

At this point learning of neuro-fuzzy system finishes.

When non-classified pattern $x(k)$, $k = N+1$, $N+2, \ldots$, comes to input of the system, neurons of the patterns layer output signal

$$o_j^{[1]}(k) = \exp\left(-\frac{\|x(k) - w_j\|^2}{2\sigma^2}\right), \tag{2}$$

where
$j = 1(A), 2(A), \ldots, N_A(A), N_A + 1(B), \ldots, N_A + N_B(B), N_A + N_B + 1(C), \ldots, N(C)$, σ is a parameter of activation function width.

Notice that due to condition (1), expression (2) may be presented in the simpler form

$$o_j^{[1]}(k) = \exp\left(\frac{w_j^T x(k) - 1}{\sigma^2}\right)$$
$$= \exp\left(\frac{\cos(w_j, x(k)) - 1}{\sigma^2}\right),$$

and since

$$-1 \leq \cos(w_j, x(k)) \leq 1,$$

Values of the output signals of the first hidden layer are defined by inequalities

$$-\frac{2}{\sigma^2} \leq \frac{\cos(w_j, x(k)) - 1}{\sigma^2} \leq 0,$$

$$\exp(-2\sigma^{-2}) \leq o_j^{[1]}(k) \leq 1.$$

The second hidden layer of local adders (one for each class) evaluates sums of output signals of the first hidden layer

$$\begin{cases} 0 < o_A^{[2]}(k) = \sum_{l=1(A)}^{N_A(A)} o_l^{[1]}(k) < N_A, \\ 0 < o_B^{[2]}(k) = \sum_{l=N_A+1(B)}^{N_A+N_B(B)} o_l^{[1]}(k) < N_B, \\ 0 < o_C^{[2]}(k) = \sum_{l=N_A+N_B+1(C)}^{N(C)} o_l^{[1]}(k) < N_C, \end{cases}$$

which then come to final adder of the third hidden layer evaluating scalar value $O_A^{[2]}(k) + O_B^{[2]}(k) + O_C^{[2]}(k)$ that then is transferred to input of dividers of output layer.

It is the output (forth) layer of the system that evaluates pattern $x(k)$ probabilities of membership to a certain class

$$\begin{cases} 0 \leq P_A(x(k)) = \dfrac{O_A^{[2]}(k)}{O_A^{[2]}(k) + O_B^{[2]}(k) + O_C^{[2]}(k)} \leq 1, \\ 0 \leq P_B(x(k)) = \dfrac{O_B^{[2]}(k)}{O_A^{[2]}(k) + O_B^{[2]}(k) + O_C^{[2]}(k)} \leq 1, \\ 0 \leq P_C(x(k)) = \dfrac{O_C^{[2]}(k)}{O_A^{[2]}(k) + O_B^{[2]}(k) + O_C^{[2]}(k)} \leq 1. \end{cases} \tag{3}$$

Concurrently with evaluating probabilities (3), FAM evaluates levels of membership [19, 20] to the same class of the same pattern in fuzzy sense:

$$\begin{cases} 0 \leq \mu_A(x(k)) = \dfrac{\|x(k) - w_A\|^{-2}}{\|x(k) - w_A\|^{-2} + \|x(k) - w_B\|^{-2} + \|x(k) - w_C\|^{-2}} \leq 1, \\\\ 0 \leq \mu_B(x(k)) = \dfrac{\|x(k) - w_B\|^{-2}}{\|x(k) - w_A\|^{-2} + \|x(k) - w_B\|^{-2} + \|x(k) - w_C\|^{-2}} \leq 1, \\\\ 0 \leq \mu_C(x(k)) = \dfrac{\|x(k) - w_C\|^{-2}}{\|x(k) - w_A\|^{-2} + \|x(k) - w_B\|^{-2} + \|x(k) - w_C\|^{-2}} \leq 1. \end{cases} \tag{4}$$

As this takes place, there appears a certain class R associated with $x(k)$ with the highest membership level $\mu_R(x(k))$ and all values of the evaluated membership levels on the FAM output. The considered system has a single free parameter of activation functions width σ. The parameter value is set either based on a rule of thumb over interval 0 to 1 [32], or may be evaluated analytically [21], or adjusts based on recurrent optimization procedure [23], that increases size of training set.

3 Experiments

The effectiveness of proposed associative probabilistic neuro-fuzzy system is examined based on solving classification task using different benchmark and real medical and biomedical data set. In this paper we present the results which are obtained based on data set "Hepatitis" from UCI repository [31].

The data set "Hepatitis" consists of 155 observations and 19 factors, which are taken from the patient card, as follows:

Class	Class: DIE, LIVE	10	LIVER FIRM: no, yes
1	AGE: 10, 20, 30, 40, 50, 60, 70, 80	11	SPLEEN PALPABLE: no, yes
2	SEX: male, female	12	SPIDERS: no, yes
3	STEROID: no, yes	13	ASCITES: no, yes
4	ANTIVIRALS: no, yes	14	VARICES: no, yes
5	FATIGUE: no, yes	15	BILIRUBIN: 0.39, 0.80, 1.20, 2.00, 3.00, 4.00
6	MALAISE: no, yes	16	ALK PHOSPHATE: 33, 80, 120, 160, 200, 250
7	ANOREXIA: no, yes	17	SGOT: 13, 100, 200, 300, 400, 500
8	LIVER BIG: no, yes	18	ALBUMIN: 2.1, 3.0, 3.8, 4.5, 5.0, 6.0
9	PROTIME: 10, 20, 30, 40, 50, 60, 70, 80, 90	19	HISTOLOGY: no, yes

The experiment was carried out 10 times and the results was averaged. The Table 1

Table 1. The comparison results of classification

Neural networks	Accuracy (training set)	Accuracy (testing set)
Associative probabilistic neuro-fuzzy system, 1 learning epoch	98.6%	97%
Cascade neo-fuzzy neural network, 1 learning epoch	98.9%	96%
Cascade correlation neural network, 1 learning epoch	95%	94%
Multilayer perceptron, 25 learning epochs	99.2%	98.5%

shows the comparison results with the existed classification methods.

Thus as it can be seen from experimental results the proposed associative probabilistic neuro-fuzzy system with its learning algorithm provides the best quality of classification among considered on-line mode approaches. Also it can be seen the multilayer perceptron provides better result than proposed approach, but it is trained during 25 epochs.

4 Conclusions

In the paper, a classifying neuro-fuzzy system intended for operating under short training set and nonconvex classes conditions has been designed based on probabilistic neural network and fuzzy associative memory. An essential aspect of the proposed system is its high performance and ease of numerical implementation. Simulation modeling proves efficacy of the proposed approach.

References

1. Han, J., Kamber, M.: Data Mining: Concepts and Techniques. Morgan Kaufman Publishers, Amsterdam (2006)
2. Zaki, M.J., Meira Jr., W.M.: Data Mining and Analysis: Fundamental Concepts and Algorithms. Cambridge University Press, Cambridge (2014)
3. Witten, I.H., Frank, E., Hall, M.A.: Data Mining: Practical Machine Learning Tools and Techniques. Morgan Kaufmann, San Francisco (2011)
4. Fukunaga, K.: Introduction to Statistical Pattern Recognition. Academic Press, New York (1972)
5. Bishop, C.M.: Neural Networks for Pattern Recognition. Clarendon Press, Oxford (1995)
6. Rojas, R.: Neural Networks. A Systematic Introduction. Springer, Berlin (1996)
7. Looney, C.: Pattern Recognition Using Neural Networks: Theory and Algorithms for Engineers and Scientists. Oxford University Press, New York (1997)
8. Callan, R.: The Essence of Neural Networks. Prentice Hall Europe, London (1999)
9. Haykin, S.: Neural Networks. A Comprehensive Foundation. Prentice Hall, Upper Saddle River (1999)
10. Zahirniak, D.R., Chapman, R., Rogers, S.K., Suter, B.W., Kabriski, M., Pyatti, V.: Pattern recognition using radial basis function networks. In: Proceedings of the 6-th Annual Aerospace Applications of AI Conference, Dayton, pp. 249–260 (1990)
11. Nelles, O.: Nonlinear System Identification. Springer, Berlin (2001)
12. Specht, D.F.: Probabilistic neural networks. Neural Netw. **3**(1), 109–118 (1990)
13. Specht, D.F.: Probabilistic neural networks and polynomial ADALINE as complementary techniques for classification. IEEE Trans. Neural Netw. **1**(1), 111–121 (1990)
14. Rutkowski, L.: Computational Intelligence. Methods and Techniques. Springer, Berlin (2008)
15. Kroll, A.: Computational Intelligence. Eine Einfuerung in Probleme, Methoden und technische Anwendungen. Oldenbourg Verlag, Muenchen (2013)
16. Kruse, R., Borgelt, C., Klawonn, F., Moewes, C., Steinbrecher, M., Held, P.: Computational Intelligence. A Methodological Introduction. Springer, London (2013)
17. Du, K.-L., Swamy, M.N.S.: Neural Networks and Statistical Learning. Springer, London (2014)
18. Jang, J.-S.R., Sun, C.-T., Mizutani, E.: Neuro-Fuzzy and Soft Computing – Computational Approach to Learning and Machine Intelligence. Prentice Hall, Upper Saddle River (1997)
19. Bodyanskiy, Y., Vynokurova, O., Setlak, G., Peleshko, D., Mulesa, P.: Adaptive multivariate hybrid neuro-fuzzy system and its on-board fast learning. Neurocomputing **230**, 409–416 (2017)
20. Bezdek, J.C.: Pattern Recognition with Fuzzy Objective Function Algorithms. Kluwer Academic Publishers, Norwell (1981)

21. Hoeppner, F., Klawonn, F., Kruse, R., Runkler, T.: Fuzzy Cluster Analysis Methods for Classification, Data Analysis and Image Recognition. Wiley, Chichester (1999)
22. Bodyanskiy, Y., Shubkina, O.: Semantic annotation of text documents using evolving neural network based on principle "Neurons at Data Points". In: Proceedings of the 4th International Workshop on Inductive Modelling, IWIM 2011, Kyiv, pp. 31–37 (2011)
23. Bodyanskiy, Y., Shubkina, O.: Semantic annotation of text documents using modified probabilistic neural network. In: Proceedings of the 6th IEEE International Conference on Intelligent Data Acquisition and Advanced Computing Systems, 15–17 September 2011, Prague, pp. 328–331 (2011)
24. Bodyanskiy, Y., Pliss, I., Volkova, V.: Modified probabilistic neuro-fuzzy network for text documents processing. Int. J. Compt. **11**(4), 391–396 (2012)
25. Parzen, E.: On the estimation of a probability density function and the model. Ann. Math. Stat. **38**, 1065–1076 (1962)
26. Moody, J., Darken, C.J.: Fast learning in networks of locally-tuned processing units. Neural Comput. **1**, 281–299 (1989)
27. Specht, D.E.: A general regression neural network. IEEE Trans. Neural Netw. **6**(2), 568–576 (1991)
28. Farrell, J.A., Michel, A.N.: A synthesis procedure for Hopfield's continuous-time associative memory. IEEE Trans. Circuits Syst. **37**, 877–884 (1990)
29. Hassoun, M.H., Watta, P.B.: Associative memory networks. In: Handbook on Neural Computation, pp. 1.3:1–1.3:14. IOP Publishing Ltd. and Oxford University Press, Oxford (1997)
30. Cios, K.J., Pedrycz, W.: Neural-fuzzy algorithms. In: Handbook on Neural Computation, pp. 1.3:1–1.3:7. IOP Publishing Ltd. and Oxford University Press, Oxford (1997)
31. Bodyanskiy, Y., Teslenko, N.: Autoassociative memory evolving system based on fuzzy basis functions. Sci. J. Riga Tech. Univ. "Computer Science, Information Technology and Management Sci.", **44**, 9–14 (2010)
32. Murphy, P.M., Aha, D.W.: UCI Repository of Machine Learning Databases. Department of Information and Computer Science, University of California (1994). http://www.ics.uci.edu/mlearn/MLRepository.html

High-Speed Finite State Machine Design
by State Splitting

Damian Borecki[⊠], Valery Salauyou, and Tomasz Grzes

Bialystok University of Technology, Wiejska 45A, 15-351 Bialystok, Poland
pan.d.borecki@o2.pl, valsol@mail.ru, t.grzes@pb.edu.pl

Abstract. A synthesis method of high-speed finite state machines (FSMs) in field programmable gate arrays (FPGAs) based on LUT (Look Up Table) by internal state splitting is offered. Estimations of the number of LUT levels are presented for an implementation of FSM transition functions in the case of sequential and parallel decomposition. Split algorithms of FSM internal states for the synthesis of high-speed FSMs are described. The method can be easily included in designing the flow of digital systems in FPGA. The experimental results showed a high efficiency of the offered method. FSM performance increased by 1.73 times. In conclusion, the experimental results were considered, and prospective directions for designing high-speed FSMs are specified.

Keywords: Synthesis · Finite state machine · High-speed · High performance
State splitting · Field programmable gate array · Look Up Table

1 Introduction

The speed of a digital system and functional blocks depends directly on the speed of their control devices. The mathematical model for the majority of control devices and controllers is a finite state machine (FSM). Because of this, the synthesis methods of high-speed FSMs are necessary for designing high-performance digital systems. In this work we consider the synthesis of high-speed FSMs in field programmable gate arrays (FPGA) based on LUT (Look Up Table).

In [1], a technique for improving the performance of a synchronous circuit configured as an FPGA-based look-up table without changing the initial circuit configuration is presented. Only the register location is altered. In [2], the methods and tools for state encoding and combinational synthesis of sequential circuits based on new criteria of information flow optimization are considered. In [3], the timing optimization technique for a complex FSM that consists of not only random logic but also data operators is proposed. In [4, 5], the styles of FSMs description in VHDL language and known methods of state assignment for the implementation of FSMs are researched. In [6], evolutionary methods are applied to the synthesis of FSMs. In [7], the task of state assignment and optimization of the combinational circuit at implementation of high-speed FSMs is considered. In [8], a novel architecture that is specifically optimized for implementing reconfigurable FSMs, Transition-based Reconfigurable FSM (TR-FSM), is presented. The architecture shows a considerable reduction in area, delay, and power consumption compared to FPGA architectures. In [9], a new model of the

© Springer International Publishing AG, part of Springer Nature 2019
W. Zamojski et al. (Eds.): DepCoS-RELCOMEX 2018, AISC 761, pp. 64–73, 2019.
https://doi.org/10.1007/978-3-319-91446-6_7

automatic machine named the virtual finite state machine (Finite Virtual State Machine - FVSM) is offered. FVSM implemented on new architecture have an advantage on high-speed performance compared with traditional implementation of FSMs on storage RAM. In [10], an implementation of FSMs in FPGA with the use of integral units of storage ROM is considered. Two pieces of FSMs architecture with multiplexers on inputs of ROM blocks which allow reducing the area and increasing high-speed FSM performance are offered. In [11], the reduction task of arguments of transition functions by state splitting is considered; this allows reducing an area and time delay in the implementation of FSMs on FPGA. This paper also uses splitting of FSM states, but the purpose of splitting is an increase of FSMs performance in LUT-based FPGA. The offered synthesis method of high-speed FSMs in FPGA is aimed at practical usage and can be easily included in the general flow of digital system design.

2 Estimations for the Number of LUT Levels for Transition Functions

Let $A = \{a_1, ..., a_M\}$ be the set of internal states, $X = \{x_1, ..., x_L\}$ be the set of input variables, $Y = \{y_1, ..., y_N\}$ the set of output variables, and $D = \{d_1, ..., d_R\}$ the set of transition functions of an FSM.

A one-hot state assignment is traditionally used for the synthesis of high-speed FSMs in FPGAs. Thus, each internal state a_i ($a_i \in A$) corresponds to a separate flip-flop of FSM's memory. A setting of this flip-flop in 1 signifies that the FSM is in the given state. The data input of each flip-flop is controlled by the transition function d_i, $d_i \in D$, i.e. any internal state a_i ($a_i \in A$) of the FSM corresponds with its own transition function $d_i, i = \overline{1, M}$.

Let $X(a_m, a_i)$ be the set of FSM input variables, whose values initiate the transition from state a_m to state a_i ($a_m, a_i \in A$). To implement some transition from state a_m to state a_i, it is necessary to check the value of the flip-flop output for the active state a_m (one bit) and the input variable values of the $X(a_m, a_i)$ set, which initiates the given transition. To implement the transition function d_i, it is necessary to check the values of the flip-flop outputs for all states, such that transitions from which lead to state a_i, i.e. $|B(a_i)|$ values, where $B(a_i)$ is the set of states from which transitions terminate in state a_i, where $|A|$ is the cardinality of set A. Besides, it is necessary to check the values of all input variables, which initiate transitions to state a_i, i.e. $|X(a_i)|$ values, where $X(a_i)$ is the set of input variables, whose values initiate transitions to state a_i, $X(a_i) = \bigcup_{a_m \in B(a_i)} X(a_m, a_i)$.

Let r_i be a rank of the transition function d_i, where

$$r_i = |B(a_i)| + |X(a_i)|. \tag{1}$$

Let n be the number of inputs of LUTs. If the rank r_i for transition function d_i ($i = \overline{1, M}$) exceeds n, there is a necessity to decompose the transition function d_i and its implementation on several LUTs.

Note that by splitting internal states it is impossible to lower the rank of the transition functions below the value

$$r^* = \max(|(Xa_m, a_s)| + 1, m = \overline{1,M}, \; s = \overline{1,M}. \tag{2}$$

In this method, the value r^* is used as an upper boundary of the ranks of the transition functions in splitting the FSM states.

It is well-known that there are two basic approaches to the decomposition of Boolean functions: sequential and parallel. In the case of sequential decomposition, all the LUTs are sequentially connected in a chain.

The n arguments of function d_i arrive on inputs of the first LUT, and the $(n-1)$ arguments arrive on inputs of all remaining LUTs. So the number l_i^s of the LUT's levels (in the case a sequential decomposition of the transition function d_i having the rank r_i) is defined by the expression:

$$l_i^s = int\left(\frac{r_i - n}{n - 1}\right) + 1, \tag{3}$$

where int(A) is the least integer number more or equal to A.

In the case of parallel decomposition, the LUTs incorporate in the form of a hierarchical tree structure. The values of the function arguments arrive on LUTs inputs of the first level, and the values of the intermediate functions arrive on LUTs inputs of all next levels. So the number of LUT's levels (in the case parallel decomposition the transition function d_i having the rank r_i) is defined by the following expression:

$$l_i^p = int(\log_n r_i). \tag{4}$$

It is difficult to predict what type of decomposition (sequential or parallel) is used by a concrete synthesizer. The preliminary research showed that, for example, the Quartus Prime design tool from Intel simultaneously uses both sequential and parallel decomposition. The number l_i levels of LUTs in the implementation on FPGA transition function d_i with the rank r_i can be between values l_i^s and l_i^p, $i = \overline{1,M}$.

Let k be an integer coefficient ($k \in [0, 10]$) that allows adapting the offered algorithm in defining the number of LUT's levels for the specific synthesizer. In this case the number l_i of LUT's levels for the implementation of the transition function d_i having the rank r_i will be defined by following expression:

$$l_i = int\left(\frac{10 - k}{10} l_i^p + \frac{k}{10} l_i^s\right). \tag{5}$$

If $k = 0$, we have $l_i = l_i^p$, i.e. the number l_i of levels corresponds to the fastest parallel decomposition, and if $k = 10$, we have $l_i = l_i^s$, i.e. the number l_i of levels corresponds to the slowest sequential decomposition. The specific value of coefficient k depends on the architecture of the FPGA and the used synthesizer.

The following problem is the answer to the question: when is it necessary to stop splitting the FSM states? In this algorithm, the process of state splitting is finished, when the following condition is met:

$$l_{max} \leq int(l_{mid}), \tag{6}$$

where l_{max} is the number of LUT levels, which is necessary for the implementation of the most "bad" function having the maximum rank; l_{mid} is the arithmetic mean value of the number of LUT levels for all transition functions.

3   Method for High-Speed FSM Synthesis

According to the above discussion, the algorithm of state splitting for high-speed FSM synthesis is described as follows.

   Algorithm 1.

1. The coefficient k ($k \in [0, 10]$) is determined, which reflects the method used by the synthesis tool for the decomposition of Boolean functions.
2. According to (1) ranks r_i ($i = \overline{1, M}$) for all FSM transition functions are defined.
3. On the basis of (3), (4), and (5), for each transition function d_i the number l_i of LUT levels is defined.
4. The values l_{max} and l_{mid} are determined. If condition (6) is met, then go to step 7, otherwise go to step 5.
5. The state a_i, for which $r_i = max$, is selected. If there are several such states, from them the state for which $|A(a_i)| = min$ is selected.
6. The state a_i (which was selected in step 5) is split by means of Algorithm 2 on the minimum number H of states $a_{i\_1}, \ldots, a_{i\_H}$ so that for each state a_{i_h} ($h = \overline{1, M}$) was fulfilled $r_{i\_h} < r^*$, where r^* is defined according to (2); go to step 2.
7. End.

For splitting some a_i state, $i = \overline{1, M}$, which is executed in step 6 of Algorithm 1, Boolean matrix W is constructed as follows. Let $C(a_i)$ be the set of transitions to state a_i. Rows of matrix W correspond to the elements of set $C(a_i)$. Columns of matrix W are divided on two parts according to types of arguments of transition function d_i. The first part of matrix W columns correspond to set $B(a_i)$ of FSM states, the transitions from which terminate in state a_i, and the second part of matrix W columns correspond to set $X(a_i)$ of input variables, whose values initiate the transitions in state a_i. A one is put at the intersection of row t ($i = \overline{t, T}$, $T = |C(a_i)|$) and column j of the first part of matrix W if the transition c_t ($c_t \in C(a_i)$) is executed from state a_j ($a_j \in B(a_i)$). A one is put at the intersection of row t and column j of the second part of matrix W if input variable x_j ($x_j \in X(a_i)$) accepts a significant value (0 or 1) on transition c_t ($c_t \in C(a_i)$). Now the task is reduced to a partition of matrix W on a minimum number H of row minors $W_1, \ldots, W_H$ so that the number of columns, which contain ones in each minor W_h ($h = \overline{1, H}$), do not exceed value r^* defined according to (2). The rows of each minor W_h will define transitions in state a_{i_h} ($h = \overline{1, H}$).

Let w_t be some row of matrix W. For finding the row partition of matrix W on a minimum number H of row minors $W_1, \ldots, W_H$, the following algorithm can be used.

Algorithm 2.

1. Put $h := 0$.
2. Put $h := h + 1$. A formation of minor W_h begins. The row w_t, which has the maximum number of ones, is selected in minor W_h as a reference row. The row w_t is included in minor W_H and the row w_t is eliminated from further reviewing, put $W_h := \{w_t\}, W := W \backslash \{w_t\}$.
3. The rows are added in minor W_h. For this purpose, among rows of matrix W, the row w_t is selected, for which the next inequality is satisfied $|W_h \cup \{w_t\}| \le r^*$, where $|W_h \cup \{w_t\}|$ is the total number of ones in the columns of minor W_h and the row w_t after their joining on OR. If such rows can be selected from several among them, row w_t is selected, which has the maximum number of common ones with minor W_h, i.e. $|W_h \cap \{w_t\}| = max$. The row w_t is included in minor W_h and row w_t is eliminated from further reviewing, put $W_h := W_h \cup \{w_t\}, W := W \backslash \{w_t\}$.
4. Step 3 repeats until at least a single row can be included in minor W_h.
5. If in matrix W all the rows are distributed between the minors, then go to step 5, otherwise go to step 2.
6. End.

We show the operation of the offered synthesis method in the example. It is necessary to synthesize the high-speed FSM whose state diagram is shown in Fig. 1.

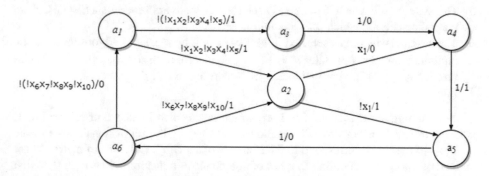

Fig. 1. State diagram of the initial FSM

This FSM represents the machine Moore, which has 6 states $a_1, \ldots, a_6$, 10 input variables $x_1, \ldots, x_{10}$, and one output variable y. The transitions from states a_3, a_4, and a_5 are unconditional, therefore the logical value 1 is written on these transitions as a transition condition. The values of sets $B(a_i)$ and $X(a_i)$, and also ranks r_i of the transition functions for the initial FSM are presented in Table 1, where $\emptyset$ is an empty set. Since for this example we have $max(|X(a_m, a_s)|) = 5$, then (according to (2)) the value $r^* = 6$. Let it is necessary to construct the FSM on FPGA with 6-input LUT, i.e. we have $n = 6$.

Table 1. Values of $B(a_i)$, $X(a_i)$, r_i, l_i^s and l_i^p for the initial FSM

State	$B(a_i)$	$X(a_i)$	r_i	l_i^s	l_i^p
a_1	$\{a_6\}$	$\{x_6, x_7, x_8, x_9, x_{10}\}$	6	1	1
a_2	$\{a_1, a_6\}$	$\{x_1, x_2, x_3, x_4, x_5, x_6, x_7, x_8, x_9, x_{10}\}$	12	3	2
a_3	$\{a_1\}$	$\{x_1, x_2, x_3, x_4, x_5\}$	6	1	1
a_4	$\{a_2, a_3\}$	$\{x_1\}$	3	1	1
a_5	$\{a_2, a_4\}$	$\{x_1\}$	3	1	1
a_6	$\{a_5\}$	$\varnothing$	1	1	1

According to (3) and (4), the values l_i^s and l_i^p are defined for each state (they are presented in the appropriate columns of Table 1). We do not know how the compiler performs a decomposition of Boolean functions, therefore we assume the sequential decomposition (a worst variant) and the value of coefficient k in expression (5) is equal to 10. As a result, the number of LUT levels (which are necessary for the implementation of each transition function) is defined by the value $l_i = l_i^s$. Thus, for our example we have $\text{int}(l_{mid}) = \text{int}(8/6) = 2$.

For this example, we have $l_{max} = l_2^s = 3$, i.e. the condition (9) does not meet for state a_2, since $l_{max} = l_2^s = 3 > \text{int}(l_{mid}) = 2$. For this reason, state a_2 is split by means of Algorithm 2. Matrix W is constructed for splitting state a_2 (Fig. 2).

	a_1	a_6	x_1	x_2	x_3	x_4	x_5	x_6	x_7	x_8	x_9	x_{10}
w_1	1	0	1	1	1	1	1	0	0	0	0	0
w_2	0	1	0	0	0	0	0	1	1	1	1	1

Fig. 2. Matrix W for splitting state a_2

Matrix W has two rows. Row w_1 corresponds to the transition from state a_1 to state a_2, and row w_2 corresponds to the transition from state a_6 to state a_2. The execution of Algorithm 2 leads to a partition of rows of matrix W into two subsets: $W_1 = \{w_1\}$ and $W_2 = \{w_2\}$. So, state a_2 is split into two states a_{2_1} and a_{2_2}, as shown in Fig. 3.

The new values of $B(a_i)$, $X(a_i)$, r_i, l_i^s and l_i^p are presented in Table 2. Now we have $l_{max} = l_{mid} = 1$ and (according to (6)) running of Algorithm 1 is completed.

Thus, for the given FSM by splitting state a_2 we reduced the number of LUT levels from 3 to 1, in the case of sequential decomposition, and from 2 to 1, in the case of parallel decomposition.

4 Experimental Results

The presented method was evaluated against the FSM benchmarks, MCNC [13]. For this purpose, the considered synthesis method was applied to each benchmark of the FSM. Both finite state machines, the initial FSM and synthesized FSM, were described in the Verilog language. Then, standard implementation of FSMs in FPGA by means of CAD Quartus Prime was fulfilled. The number n of the LUT inputs has been set by 4.

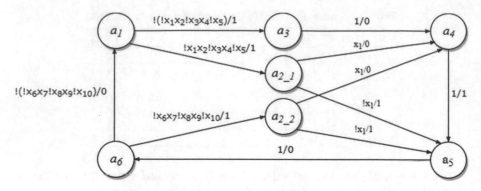

Fig. 3. State diagram of the FSM after splitting state a_2

Table 2. Values of $B(a_i)$, $X(a_i)$, r_i, l_i^s and l_i^p after splitting state a_2

State	$B(a_i)$	$X(a_i)$	r_i	l_i^s	l_i^p
a_1	$\{a_6\}$	$\{x_6, x_7, x_8, x_9, x_{10}\}$	6	1	1
a_{2_1}	$\{a_1\}$	$\{x_1, x_2, x_3, x_4, x_5\}$	6	1	1
a_{2_2}	$\{a_6\}$	$\{x_6, x_7, x_8, x_9, x_{10}\}$	6	1	1
a_3	$\{a_1\}$	$\{x_1, x_2, x_3, x_4, x_5\}$	6	1	1
a_4	$\{a_{2\_1}, a_3\}$	$\{x_1\}$	3	1	1
a_5	$\{a_{2\_2}, a_4\}$	$\{x_1\}$	3	1	1
a_6	$\{a_5\}$	$\varnothing$	1	1	1

The offered method allowed to reduce the rank of transition functions for 21 benchmarks of 48, i.e. in 43.75% of cases.

Table 3 shows results of synthesis of high-speed FSMs by means of presented method, where FSM is the name of benchmark; Ns is the number of the splitted states; M, P, and $lmax$ are the number of FSM's states, FSM's transitions, and the maximum rank of transition functions before application of the synthesis method; $M*$, $P*$, and l_{max}^* are the number of FSM's states, FSM's transitions, and the maximum rank of transition functions after application of the synthesis method; $M*/M$, $P*/P$, and l_{max}/l_{max}^* are relations of the corresponding parameters; mid is the mean parameter value; max is the maximum parameter value.

Table 3 shows that for synthesized benchmarks the average number of splits is equal 3.00, and maximum – 11; the number of FSM states are increased by a factor of 1.68 on the average and maximum by a factor of 3; the number of FSM transitions are increased by a factor of 1.63 on the average and maximum by a factor of 3.36. The use of the synthesis method allows to reduce the rank of transition functions by a factor of 2.40 on the average, and maximum by a factor of 6.33. Thus, the presented method allows considerably to reduce the maximum rank of transition functions by despite increasing the number of FSM states and the transition functions.

The benchmarks were realized on the following families FPGA Arria II, Cyclone V, MAX II, and Stratix V. The results of researches are given in tables, where LE and $LE*$ are the number of the logical elements (area), which necessary for

Table 3. Results of the experimental researches of the synthesis method of high-speed FSMs for benchmarks

FSM	Ns	M	$M*$	$M*/M$	P	$P*$	$P*/P$	l_{max}	$l*_{max}$	$l_{max}/l*_{max}$
BBSSE	11	16	48	3,00	56	188	3,36	7	3	2,33
CSE	1	16	23	1,44	91	154	1,69	8	5	1,60
EX1	1	18	22	1,22	233	379	1,63	8	5	1,60
EX2	1	19	34	1,79	72	72	1,00	6	2	3,00
EX3	1	10	17	1,70	36	36	1,00	3	2	1,50
EX5	1	9	16	1,78	32	32	1,00	3	2	1,50
EX7	1	10	18	1,80	36	36	1,00	4	2	2,00
KEYB	1	19	26	1,37	170	261	1,54	8	4	2,00
PLANET	2	48	51	1,06	115	120	1,04	3	2	1,50
PMA	8	24	39	1,63	73	106	1,45	5	2	2,50
S208	2	18	43	2,39	153	332	2,17	9	5	1,80
S298	1	218	246	1,13	1096	1236	1,13	27	19	1,42
S386	2	13	18	1,38	64	99	1,55	7	3	2,33
S420	2	18	43	2,39	137	296	2,16	9	5	1,80
S820	3	25	36	1,44	232	324	1,40	14	5	2,80
S832	3	25	38	1,52	245	412	1,68	14	5	2,80
S1488	3	48	63	1,31	251	341	1,36	19	3	6,33
S1494	3	48	63	1,31	250	364	1,46	19	3	6,33
SAND	3	32	39	1,22	184	325	1,77	6	5	1,20
SSE	11	16	48	3,00	56	188	3,36	7	3	2,33
STYR	2	30	41	1,37	166	259	1,56	9	5	1,80
mid	3,00			1,68			1,63			2,40
max	11			3			3,36			6,33

realization of the initial FSM and the synthesized FSM; F and $F*$ are the frequency of functioning of the initial FSM and the synthesized FSM; $LE*/LE$ and $F*/F$ are the relations of the corresponding parameters; other parameters have former value.

Table 4 shows that the offered method allows to increase the frequency of FSMs for various FPGA families by a factor of 1.08–1.13 on the average and by a factor of 1.52–1.73 maximum. Using of the presented method also increases the area by a factor of 1.27–1.35 on the average.

Table 5 compares the considered method to the known university programs for FSM state assignment JEDI [14] and NOVA [15], where FJ, FN, and $F*$ are the frequency of the FSM when using the program JEDI, NOVA, and the offered method respectively. Table 5 shows that the offered method allows to increase the maximal frequency of FSMs by a factor of 2.03–2.33 on the average in comparison with the JEDI program and by a factor of 3.68–4.47 on the average in comparison with the NOVA program. The maximum increase in frequency of FSMs makes 3.11 times in comparison with the JEDI program and 8.31 times in comparison with the NOVA program.

Table 4. The implementation results of FSM benchmarks in the Arria II, Cyclone V, MAX II, and Stratix V FPGA families

Family	LE*/LE		F*/F	
	mid	max	mid	max
Arria II	1.31	2.07	1.10	1.54
Cyclone V	1.35	2.14	1.13	1.73
MAX II	1.33	2.03	1.19	1.52
Stratix V	1.27	1.89	1.08	1.64

Table 5. Comparing of the offered method with the JEDI and NOVA programs in case of FSM implementation in the Cyclone V and MAX II FPGA families

FSM	Cyclone V					MAX II				
	FJ	FN	F*	F*/FJ	F*/FN	FJ	FN	F*	F*/FJ	F*/FN
BBSSE	402	119	937	2,33	7,87	311	163	937	3,01	5,75
CSE	300	221	370	1,23	1,67	222	161	370	1,67	2,30
EX1	267	156	352	1,32	2,26	217	129	352	1,62	2,73
EX2	333	188	787	2,36	4,19	302	175	787	2,61	4,50
EX3	280	294	629	2,25	2,14	348	219	629	1,81	2,87
EX5	422	298	587	1,39	1,97	340	206	587	1,73	2,85
EX7	398	308	587	1,47	1,91	320	193	587	1,83	3,04
KEYB	360	264	876	2,43	3,32	282	152	876	3,11	5,76
PLANET	435	215	1030	2,37	4,79	391	124	1030	2,63	8,31
PMA	224	198	575	2,57	2,90	267	116	575	2,15	4,96
S386	408	285	928	2,27	3,26	336	192	928	2,76	4,83
SSE	402	119	937	2,33	7,87	311	163	937	3,01	5,75
mid				2,03	3,68				2,33	4,47
max				2,57	7,87				3,11	8,31

5   Conclusions

Using of the offered method allows considerably to lower the maximum rank of transition functions, however increase a performance of the FSMs is watched not in all cases. It is explained by complexity of the synthesis task of the fast FSMs, for example, in comparison with the task of area reduction. The matter is that performance of the FSMs is influenced by not only results of a logical synthesis, but also results of placement and routing. Besides, performance of FSMs, except transitions functions, is also influenced by complexity of output functions.

The offered method can be also applied to creation of high-speed FSMs in ASIC chips. For this purpose it is enough to define estimates (3) and (4) the number of the circuit levels for specific architecture of ASIC.

Further development of synthesis methods of high-speed FSMs can go on the way of accounting of output function complexity, use of special FSM structural models, architectural FPGA properties, special control of FSM synchronization, etc.

The present study was supported by a grant S/WI/1/2013 from Bialystok University of Technology and founded from the resources for research by Ministry of Science and Higher Education.

References

1. Miyazaki, N., Nakada, H., Tsutsui, A., Yamada, K., Ohta, N.: Performance improvement technique for synchronous circuits realized as LUT-Based FPGA's. IEEE Trans. Very Large Scale Integr. VLSI Syst. **3**(3), 455–459 (1995)
2. Jozwiak, L., Slusarczyk, A., Chojnacki, A.: Fast and compact sequential circuits through the information-driven circuit synthesis. In: Proceedings of the Euromicro Symposium on Digital Systems Design, Warsaw, Poland, 4–6 September 2001, pp. 46–53 (2001)
3. Huang, S.-Y.: On speeding up extended finite state machines using catalyst circuitry. In: Proceedings of the Asia and South Pacific Design Automation Conference (ASAP-DAC), Yokohama, January–February 2001, pp. 583–588 (2001)
4. Kuusilinna, K., Lahtinen, V., Hamalainen, T., Saarinen, J.: Finite state machine encoding for VHDL synthesis. IEE Proc. Comput. and Digit. Tech. **148**(1), 23–30 (2001)
5. Rafla, N.I., Davis, B.: A study of finite state machine coding styles for implementation in FPGAs. In: Proceedings of the 49th IEEE International Midwest Symposium on Circuits and Systems, San Juan, USA, 6–9 August 2006, vol. 1, pp. 337–341 (2006)
6. Nedjah, N., Mourelle, L.: Evolutionary synthesis of synchronous finite state machines. In: Proceedings of the International Conference on Computer Engineering and Systems, Cairo, Egypt, 5–7 November 2006, pp. 19–24 (2006)
7. Czerwiński, R., Kania, D.: Synthesis method of high speed finite state machines. Bull. Pol. Acad. Sci. Tech. Sci. **58**(4), 635–644 (2010)
8. Glaser, J., Damm, M., Haase, J., Grimm, C.: TR-FSM: transition based reconfigurable finite state machine. ACM Trans. Reconfigurable Technol. Syst. (TRETS) **4**(3), 23:1–23:14 (2011)
9. Senhadji-Navarro, R., Garcia-Vargas, I.: Finite virtual state machines. IEICE Trans. Inf. Syst. **E95D**(10), 2544–2547 (2012)
10. Garcia-Vargas, I., Senhadji-Navarro, R.: Finite state machines with input multiplexing: a performance study. IEEE Trans. Comput. Aided Des. Integ. Circuits Syst. **34**(5), 867–871 (2015)
11. Solov'ev, V.V.: Splitting the internal states in order to reduce the number of arguments in functions of finite automata. J. Comput. Syst. Sci. Int. **44**(5), 777–783 (2005)
12. Yang, S.: Logic synthesis and optimization benchmarks user guide. Version 3.0. Microelectronics Center of North Carolina (MCNC), North Carolina, USA (1991)
13. Lin, B., Newton, A.R.: Synthesis of multiple level logic from symbolic high-level description languages. In: Proceedings of the International Conference on VLSI, Munich, pp. 187–196 (1989)
14. Villa, T., Sangiovanni-Vincentelli, A.: Nova: state assignment of finite state machines for optimal two-level logic implementation. IEEE Trans. Comput. Aided Des. Integr. Circuits Syst. **9**(9), 905–924 (1990)

Meta-heuristic Task Scheduling Algorithm for Computing Cluster with 2D Packing Problem Approach

Wojciech Bożejko[1]([✉]), Zenon Chaczko[2], Piotr Nadybski[3], and Mieczysław Wodecki[4]

[1] Department of Automatics, Mechatronics and Control Systems, Faculty of Electronics, Wrocław University of Technology, Janiszewskiego 11-17, 50-372 Wrocław, Poland
wojciech.bozejko@pwr.edu.pl
[2] FEIT Faculty of Engineering and IT, UTS University of Technology Sydney, Sydney, Australia
zenon.chaczko@uts.edu.au
[3] The Faculty of Technical and Economic Science, Witelon State University of Applied Sciences in Legnica, Sejmowa 5A, 59-220 Legnica, Poland
piotr.nadybski@pwsz.legnica.edu.pl
[4] Telecommunications and Teleinformatics Department, Wrocław University of Technology, Janiszewskiego 11-17, 50-372 Wrocław, Poland
mieczyslaw.wodecki@uwr.edu.pl

Abstract. In this paper we present a mathematical model and an algorithm for solving a task scheduling problem in computing cluster. The problem is considered as a 2D packing problem. Each multi-node task is treated as a set of separate subtasks with common constrains. For optimization the tabu search metaheuristic algorithm is applied.

1 Problem Description

Technologies based on distributed systems, especially built with the use of cloud computing model consequently become more and more popular every year. Despite the fact that the computing power is nowadays relatively cheap and much more accessible than a few years ago, there are still some applications where performance of the average processor for desktop or even platform server is insufficient. As an example we can mention scientific researches, advanced computer modeling and simulations or weather forecasting. While the process of designing and constructing advanced processing unit is an expensive task itself, one of the easiest way to gain more computing power is to use several or more widely available on the market components and multiplying their computing power. This idea is commonly used and as a result we expect to obtain the trend of designing multi-core processors and multiprocessors systems.

A computer cluster is a set of connected computers (nodes) that work together and generally can be viewed by its users as a single system. What is important,

© Springer International Publishing AG, part of Springer Nature 2019
W. Zamojski et al. (Eds.): DepCoS-RELCOMEX 2018, AISC 761, pp. 74–82, 2019.
https://doi.org/10.1007/978-3-319-91446-6_8

in computing clusters each node can perform the same tasks and they are controlled and scheduled by dedicated software (e.g. OpenPBS, Slurm). Since maintaining of a cluster is expensive, the efficient usage of available resources is very important. In this paper we consider the new approach for the task scheduling algorithm in High-Performance Cluster (HPC) environment. Promising results gained during computing experiments presented in [5] were the direct reason to continue research in this field. The original idea was improved and it is presented in this paper.

The first models of multiprocessor computations refer to the chemical industry [1], projects scheduling [19] and tasks scheduling in computer systems [2]. In the work [15] it was proved that for two machines the problem with the criterion C_{max} is NP-hard. Approximation algorithms for problems with different number of machines and additional limitations are considered in the works [3,14]. An extensive review of the literature is presented in the monograph [9].

2   Model

Assumption. The system described above can be presented more generally as follows:

There is a set of n tasks $J = \{1, 2, \ldots, n\}$ and computing nodes $M = \{1, 2, \ldots, m\}$. Each computing node is a server with identical hardware configuration and can be considered interchangeably. Each task is described by the following list of parameters:

- each task consists $i \in J$ of u_i subtasks, and $1 \leq u_i \leq m$,
- $p_{i,j}$ - time of the resource reservation for subtask j of task i (*walltime*).

Given the above, the following conditions can be formulated:

 (i) Each task consists of one or more subtasks.
 (ii) Each subtask must be executed only in one node.
 (iii) Each subtasks of single particular task must be started at the same moment.
 (iv) The machines are homogeneous – it does not matter to the final result which subtask will be run on which machine.
 (v) If the task is finished earlier than expected complete time (walltime), the resources cannot be used for another process.
 (vi) The started task/subtask cannot be interrupted.

Goal: The goal is to schedule all tasks execution on available number of nodes with the shortest possible complete time of all tasks.

Each feasible solution can be presented as a matrix of the size $n \times m \times m$:

$$S = [S_{ij,k}],$$

where $S_{ij,k}$ is a starting point for the subtask j of the task i on the machine k.

In reference to the above:

$$S_{ij,k} = S_{ij,1} = S_{ij,2} =, \ldots, = S_{ij,m} \text{ or } S_{ij,k} = 0, \tag{1}$$

where $i \in J$, $j, k = 1, 2, \ldots, m$.

Let's assume that Φ is as set of all feasible solutions and $S \in \Phi$ is any solution meeting feasibility condition. If a subset of all nodes reserved for subtasks of task i we denote as U_i, then:

$$|U_i| \geq 1, \quad i \in J. \tag{2}$$

If we denote all tasks realized on node k as $Z_k(t)$, and time as t

$$\forall_{k \in \{1,2,\dots,m\}} |Z_k(t)| = 1 \text{ or } |Z_k(t)| = 0. \tag{3}$$

Due to initial assumption for this paper, the minimization of the completion time of all task is the criterion. Therefore, cost function for each matrix S can be written as follows:

$$F_{max}(S) = \max\{S_{ij,k} + p_{i,j} : S = [S_{ij,k}] \in \Phi\}. \tag{4}$$

We are looking for such a solution $S^* \in \Phi$, for which the value of

$$F(S^*) = \min\{F_{max}(S) : S \in \Phi\}. \tag{5}$$

3   Packing Strategy as the Solution for the Problem

Optimization methods are the subject of intensive scientific research [4,6–8,20]. The problem of task scheduling in multi nodes computing cluster can be considered as two-dimensional (2D) packing problem. Packing problems are optimization problems concerned with finding a good arrangement of multiple items in larger containing regions [16]. Packing algorithms are widely used in industry. As an example of their usage we can mention the problem of maximizing the utilization of materials where small improvements of the layout can result in large savings of materials used in production process and finally in production costs reduction. This class of problems requires all given rectangles to be placed orthogonally without overlaps into one rectangular container, called the strip, with a fixed width and variable height so as to minimize the height of the strip [12].

General Idea of the Packing Problem. There is a set of rectangles, where each of them have dimensions $w_i * h_i$ where w_i is a width of the i-th rectangle and h_i is its height. The goal is to pack all rectangles into minimal number of bins with fixed dimensions $W * H$ (classic Two-dimensional bin packing problem – 2BP) or in one "stripe" with fixed width and unlimited height, where goal is to limit the height of the stripe (2-dimensional strip packing problem – 2SP). In the above cited paper, we assume that number of nodes is represented by width of the rectangle while the height corresponds to *walltime* of the task. In the model presented in this article we changed the approach. Now each task is recognized as a set of separate rectangles with $w = 1$ where each of them represents a single subtask of a particular task. The general idea is presented in Fig. 1. We can suspect that this significant change will emphatically affect the results.

There are at least few packing strategies described in the literature. Some of them can be found in [13,18]. Let $\pi = (1, 2, \dots, n)$ be a task permutation to load into a bin. Then we can consider following packing strategies:

– Next-Fit Decreasing Height (NFDH): item (task) j is packed left justified on level s, if it fits. Otherwise, a new level ($s := s + 1$) is created, and j is packed left justified into it.
– First-Fit Decreasing Height (FFDH) strategy: item j is packed left justified on the first level where it fits, if any. If no level can accommodate j, a new level is initialized as in NFDH.
– Best-Fit Decreasing Height (BFDH) strategy: item j is packed left justified on that level, among those where it fits, for which the unused horizontal space is a minimum. If no level can accommodate j, a new level is initialized as in NFDH.

The problem of packing was one of the first problems proved to be NP-complete (Garey and Johnson [10]).

The strategies presented above are deterministic. It means that each time we start the procedure of packing identical elements in identical bin the final result will be exactly the same. The parameter that can affect the final arrangement of the elements is the order in which they are packed.

Algorithm 1. CTS

1: Generate starting permutation and count the value of the cost function, using chosen 2d packing strategy. This is the best already known solution.
2: Change the order of the tasks and count the value of the cost function again.
3: If the new value is better than the best already known, the actual permutation is the best known.
4: Repeat Steps 1–3 desired number of times.

Packing Strategy for CTS. As the model described in this paper is not a classic packing problem none of the strategies listed earlier can be used. Because of that fact the following packing method was proposed:

The presented above packing strategy is being used for cost function calculation for the Tabu Search algorithm [11].

Algorithm 2. Packing strategy

1: Check first row if there is at least as many free cells as nodes needed for the task.
2: For each of the free cells in the row check if there is a free space over it (empty cell or no more already created rows).
3: If the above is true, then reserve cells for the current task and add extra rows if necessary. Else, go to the next row and repeat Steps 1–2.
4: Repeat steps 1 - 3 until there is no more task to schedule.

---

4   Computational Experiments

The algorithm presented above was implemented in C# language and tested on Intel Xeon X5650 (2.67 GHz) processor. All data sets used in the presented below experiments were generated randomly. The method used to generate tasks was similar as described in [17] (Type 1: w_j uniformly distributed random variable from $[\frac{2}{3}W; W]$, h_j uniformly distributed random variable from $[1; \frac{1}{2}H]$). The average values of results are presented in the Table 1 and in Fig. 3. Greedy algorithm used in this experiment was based on First-Fit Decreasing Height (FFDH) strategy. Tabu list in TS-Opt and CTS algorithms was set to 7 while the number of iterations performed each time was 50. The lower bound (LB) was calculated as below:

$$LB = \frac{\sum_{i=1}^{n} p_i * u_i}{m}.$$

The lower bound is calculated as the sum of the areas of particular elements divided by the number of nodes and it is theoretical minimal height. It is a

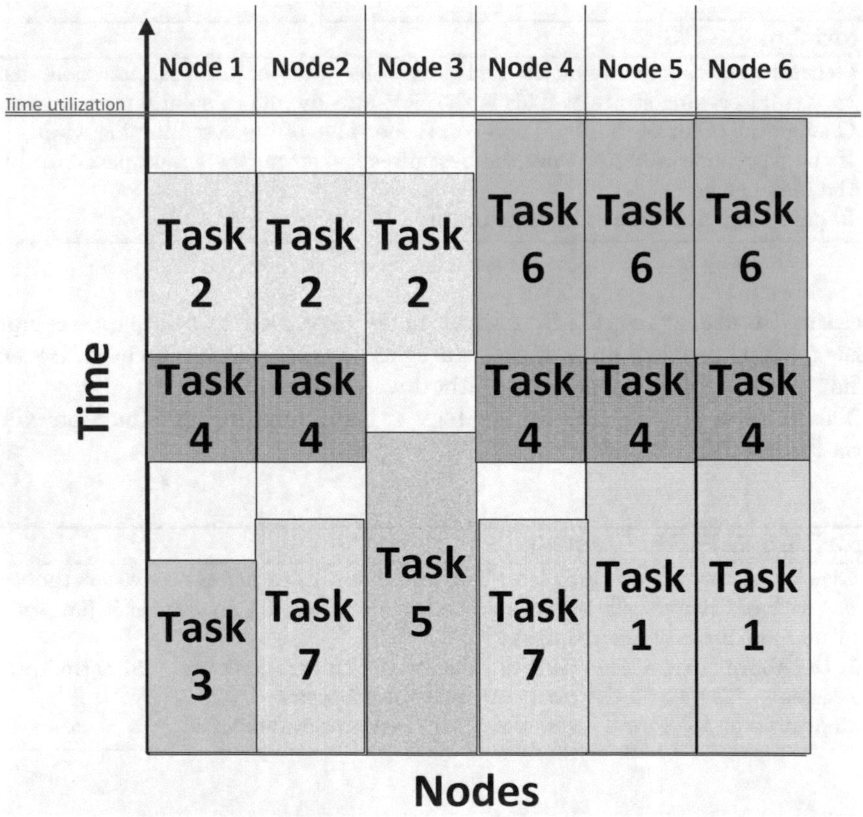

Fig. 1. General idea of task scheduling as a 2SP problem

Table 1. Compartment of efficiency for Greedy and Tabu Search optimized algorithms.

Tasks	Nodes	LB	Greedy	TS-Opt	CTS	Impr.
20	5	59.7	78	65	59	9.23%
30	5	82.8	98	88	73	9.88%
40	5	98.6	125	105	9.52	9.52%
50	5	135	167	152	139	8.55%
50	10	113	155	127	117	7.87%
100	5	242	326	284	256	9.86%
100	10	306	435	332	311	6.33%
500	10	1232	1674	1561	1460	6.47%

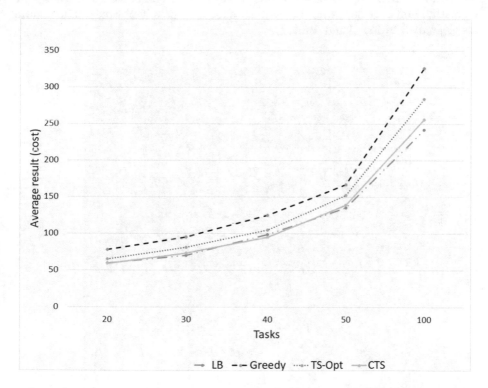

Fig. 2. Comparison of efficiency of Greedy and Tabu Search optimization algorithms (5 nodes).

Table 2. Compartment of computing time in ms. for TS-Opt and CTS algorithms.

Tasks	Nodes	TS-Opt	CTS	Impr.
20	5	1769	1489	15.83%
30	5	7645	5277	30.97%
40	5	18974	9373	50.60%
50	5	58694	24521	58.22%
50	10	143261	38561	73.08%
100	5	847519	145746	82.80%
100	10	6209732	763969	87.70%
500	10	81974765	7365208	91.02%

very optimistic approach, and in many cases impossible to obtain. Due to this fact lower bound cannot be considered as a value of cost function for optimal permutation. In any tested case, the result obtained by the use of the proposed algorithm caused improvement of about 10% when compared to the TS-Opt algorithm and over 30% when set side by side with greedy strategy. The results are presented in Fig. 2 and Table 1.

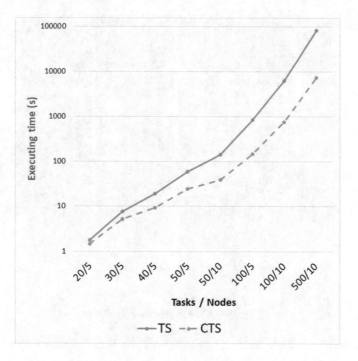

Fig. 3. Comparison of the execution times of the Tabu Search and CTS algorithms.

Tables 1 and 2 description:

- LB - Lower bound,
- Greedy - greedy packing algorithm based on packing strategy,
- TS-Opt - algorithm presented in paper [5],
- Algorithm CTS proposed in this paper,
- Improvement - result of the CTS algorithm compared to TS-Opt.

5   Conclusion

In this paper we proposed an improved algorithm for task scheduling in computing cluster. The presented above computing experiments confirmed that treating task as set of separate subtask with common starting time can improve results of TS-Opt algorithm. What is more, less computational demanding packing strategy resulted in significantly reduced execution time of the algorithm. the presented method can be used for task scheduling in computing cluster environment where there is no extra need to locate task only on neighboring nodes. As there is many potential additional parameters of the model to consider, for example setup times [6], this issue can be continued in the future research.

Acknowledgement. The paper was partially supported by the National Science Centre of Poland, grant OPUS no. DEC 2017/25/B/ST7 /02181.

References

1. Bozoki, G., Richard, J.-P.: A branch-and-bound algorithm for the continous-process job-shop scheduling problem. AIIE Trans. **2**(3), 246–252 (1970)
2. Błażewicz, J., Drabowski, M., Węglarz, J.. Scheduling multiprocessor tasks to minimize schedule length. IEEE Trans. Comput. **35**(5), 389–393 (1986)
3. Błądek, I., Drozdowski, M., Guinand, F., Schepler, X.: On contiguous and non-contiguous parallel task scheduling. J. Sched. **18**(5), 487–495 (2015)
4. Bożejko, W., Uchroński, M., Wodecki, M.: Block approach to the cyclic flow shop scheduling. Comput. Ind. Eng. **81**, 158–166 (2015)
5. Bożejko, W., Nadybski, P., Wodecki, M.: Scheduling tasks for a computing cluster. In: Knosala, R. (ed.) Innovation in Management and Production Engineering, pp. 524–533. Publishing House of the Polish Society for Production Management (2017). ISBN 978-83-941281-1-1
6. Bożejko, W., Kacprzak, L., Nadybski, P., Wodecki, M.: Multi-machine scheduling with setup times. In: Proceedings of the Computer Information Systems and Industrial Management - 15th International Conference, CISIM 2016, Vilnius, Lithuania, 14–16 September 2016. LNCS, vol. 9842, 300–311. Springer (2016)
7. Bożejko, W., Nadybski, P., Wodecki, M.: Two step algorithm for virtual machine distributed replication with limited bandwidth problem. In: Proceedings of the Computer Information Systems and Industrial Management - 15th International Conference, CISIM 2016, Vilnius, Lithuania, 14–16 September 2016. LNCS, vol. 9842, 312–321. Springer (2016)

8. Cao, D., Kotov, V.M.: A best-fit heuristic algorithm for two dimensional bin packing problem. Electr. Mech. Eng. Inf. Technol. (EMEIT) **7**, 3789–3791 (2011)
9. Drozdowski, M.: Scheduling for Parallel Processing. Springer, London (2009)
10. Garey, M.R., Johnson, D.S.: Computers and Intractability: A Guide to the Theory of NP-Completeness. W. H. Freeman & Co., New York (1979)
11. Glover, F.: Tabu search. Part I. ORSA J. Comput. **1**, 190–206 (1989)
12. Imamichi, T., Kenmochi, M., Nonobe, K., Nagamochi, H., Yagiura, M.: Exact algorithms for the two-dimensional strip packing problem with and without rotations. Eur. J. Oper. Res. **198**(1), 73–83 (2009)
13. Lodi, A., Martello, S., Vigo, D.: Recent advances on two-dimensional bin packing problems. Discret. Appl. Math. **123**, 379–396 (2002)
14. Lin, J.F., Chen, S.J.: Scheduling algorithm for nonpreemptive multiprocessor tasks. Comput. Math. Appl. **28**(4), 85–92 (1994)
15. Lloyd, E.: Concurrent task systems. Oper. Res. **29**(1), 189–201 (1981)
16. Lodi, A., Martello, S., Monaci, M.: Two-dimensional packing problems: a survey, Dipartimento di Elettronica, Informatica e Sistemistica, University of Bologna, Viale Risorgimento 2, 40136 Bologna, Italy, March 2001
17. Lodi, A., Martello, S., Vigo, D.: Recent advances on two-dimensional bin packing problems. Discret. Appl. Math. **123**(1–3), 379–396 (2002)
18. Pietrobuoni, E.: Two-dimensional bin packing problem with Guillotine restrictions, Ph.D., Universit'a di Bologna (2015)
19. Vizing, V.: About schedules observing deadlines. Kibernetika **1**, 128–135 (1981)
20. Wodecki, M.: A branch-and-bound parallel algorithm for single-machine total weighted tardiness problem. J. Adv. Manuf. Technol. **37**(9–10), 996–1004 (2007)

Local Optima Networks in Solving Algorithm Selection Problem for TSP

Wojciech Bożejko[1,3]([✉]), Andrzej Gnatowski[1,3], Teodor Niżyński[1,3],
Michael Affenzeller[2,3], and Andreas Beham[2,3]

[1] Department of Automatics, Mechatronics and Control Systems,
Wrocław University of Science and Technology,
11–17 Janiszewskiego St., 50-372 Wrocław, Poland
{wojciech.bozejko,andrzej.gnatowski,teodor.nizynski}@pwr.edu.pl
[2] Heuristic and Evolutionary Algorithms Laboratory,
University of Applied Sciences Upper Austria,
Softwarepark 11, 4232 Hagenberg, Austria
{michael.affenzeller,andreas.beham}@fh-hagenberg.at
[3] Institute for Formal Models and Verification, Johannes Kepler University Linz,
Altenberger Straße 69, Linz, Austria

Abstract. In the era of commonly available problem-solving tools for, it is especially important to choose the best available method. We use local optima network analysis and machine learning to select appropriate algorithms on the instance-to-instance basis. The preliminary results show that such method can be successfully applied for sufficiently distinct instances and algorithms.

Keywords: Algorithm selection problem · Fitness landscape analysis
Local optima networks · Travelling salesman problem

1 Introduction

NP-complete problems are among the most researched problems in operational research. Since optimal solutions are usually unobtainable in acceptable time for the sufficiently large real-world instances, heuristic algorithms are commonly used. There is a wide variety of metaheuristic algorithms, successfully employed to solve many different optimization problems such as: tabu search (TS) [2–4], genetic algorithm (GA) [8] or simulated annealing (SA) [5]. However, there is no agreement on which algorithm is the best for each task. For example, for the flexible job shop scheduling problem, review [6] describes at least 14 categories of optimization algorithms, such as: evolutionary algorithms, hybrids, tabu searches, particle swarm optimizations, neighbor searches, etc. Moreover, within each category one can find multiple different variants of the base algorithm (e.g., for GA and the travelling salesman problem (TSP), refer to survey [21]).

The mentioned challenge is known in literature as the algorithm selection problem (ASP) [15]. In order to make an informed choice, one must obtain as

© Springer International Publishing AG, part of Springer Nature 2019
W. Zamojski et al. (Eds.): DepCoS-RELCOMEX 2018, AISC 761, pp. 83–93, 2019.
https://doi.org/10.1007/978-3-319-91446-6_9

much information about the task being solved, as practically possible. To provide the data describing the instances, fitness landscape (FL) analysis (FLA) is usually employed. The gathered data is usually utilized to solve ASP, using machine learning techniques or even simple methods based on correlation. A comprehensive survey over ASP can be found in [10]. In [14], ASP was researched on the example of the quadratic assignment problem. Two different metaheuristic algorithms were considered: robust taboo search and variable neighborhood search. Several well-known classification methods were used to determine the best algorithm for each tested instance, among others: one variable rule learner (OneR), sequential minimal optimization (SMO) of a support vector machine, gaussian processes (GProc) and linear regression (LR). The features used were chosen from the set of 34 different FL measures. Results were promising with up to 80% of correctly classified instances. Unfortunately, the time required for measuring the features was longer then the combined time of the optimization algorithm runs. Under such conditions, direct empirical evaluation of the algorithms would be also possible, negatively affected the practicality of the proposed solution. The ASP for QAP was also tackled by Smith-Miles [17], where a meta-learning framework based on simple feed forward neural networks [16] and self-organizing feature maps were used to analyze and predict with up to 94% accuracy the performance of three metaheuristic algorithms. In [9], six metaheuristics were tested for the protein structure prediction problem. Correlation was used to identify the relationship between several FL measures and performance of the algorithms.

While the mentioned results suggest that FL analysis is a potent tool for solving ASP, due to an enormous amount of the raw data, it is hard to process and share even a sampled FL between researchers. The concept of local optima networks [13] is a promising solution to this problem, at the same time providing an interesting look at the properties of the instances. LON is a network consisting of locally optimal solutions (nodes) and probabilities of navigating a search process between them (edges). The LON model provides a way to compress the information about the search space and was successfully applied to gathering the data over various optimization problems, such as: TSP [11,12], QAP [8,13,19], or NK [13,20]. However, LON analysis has not yet become as recognized method of obtaining the instance features for ASP specifically, as fitness landscape analysis.

In this paper we investigate a viability of local optima networks analysis in the context of the algorithm selection problem on the example of the travelling salesman problem. The solving algorithm were taken from the Google Optimization Tools (OR-Tools); while test instances were both generated randomly and chosen from TSPLib.

2   Basic Concepts

2.1   Travelling Salesman Problem

In this paper, we use the symmetric TSP as a benchmark problem. Since TSP is a widely known in optimization community, it will be only briefly described. For more details, refer to [1]. In TSP, there is a set of cities $\mathcal{M} = \{1, 2, \ldots, n\}$, that must be visited by a *travelling salesman*. A solution is represented by a permutation of cities from the set, constituting a tour. The tour must be closed and pass through each city exactly once. In the researched variant of TSP, the distance between each pair of cities is equal in both directions. The problem is to find an order of visiting the cities, minimizing the length of the tour.

2.2 Fitness Landscape

A fitness landscape, as described in [18], is a triple $\{S, V, f\}$, where:

- S is a search space, consisting of all solutions. Since usually $|S|$ grows exponentially with the problem size, for larger instances it is impossible to analyze all the solutions. Therefore, in practical applications, various sampling methods are utilized.
- V is a neighborhood function, $V : S \rightarrow \mathcal{P}(S)$. For any given solution $s \in S$, function V assigns a set $V(s)$, consisting of the neighbors of s.
- f is a fitness function, $f : S \rightarrow \mathbb{R}$. For any given solution $s \in S$, function f assigns a real number evaluating a "quality" of the solution. In this paper, we assume that the values of f are to be minimized.

2.3 Local Optima Network

LON is a network of nodes symbolizing problem solutions, connected by edges with weights reflecting the probabilities of traversing between them, using given search operator.

Nodes. The nodes are local optima, i.e. in their neighborhoods there are no solutions with a lower value of the fitness function. Formally, a solution $s \in S$ is a local optimum if and only if $\forall a \in V(s)\ \big(f(a) \geq f(s)\big)$. We use the classic 2-change neighborhood, also utilized in the tested metaheuristic algorithms. As it is impossible to list all local optima of a reasonably big instance, we used a sampling method, described in Sect. 3.1. The set of LON nodes is denoted by N_{LON}.

Edges. There are at least two edge models for LON: basin-transition and escape edges [13]. We selected escape edges, as it is easier to estimate their weights. A directed edge (s, t) between local optima s and t exists only if t can be obtained by applying a kick-operator on s, followed by a hill stepping algorithm (here—first-improvement 2-opt). The weight of an edge is a number, reflecting the probability of transition from s to t; and is estimated during sampling process. The set of LON edges is denoted by E_{LON}.

2.4 Algorithm Selection Problem

The algorithm selection problem is a problem of selecting, from a given set, the best algorithm on an instance-by-instance basis. In this paper, as a method of comparing the algorithms, we measure the best result obtained after a fixed calculation time.

Algorithm 1. Method of sampling LON nodes, based on [8]

Data : I_{nmax}, the desired number of nodes;
I_{natt}, the number of node generation attempts;
a TSP instance
Result: N_{LON}, the set of LON nodes

```
1  N_LON ← {};
2  for i = 1, 2, ..., I_nmax do
3      for i = 1, 2, ..., I_natt do
4          s ← generateRandomSolution();
5          s ← 2-opt(s);
6          if s is a local optimum then
7              if s ∉ N_LON then
8                  N_LON ← N_LON ∪ {s};
9                  break
```

---

3 LON Extraction and Analysis

3.1 Sampling Method

Due to a very large search space, the LON nodes and edges are obtained by a sampling method, similar to the one described in [8]. The method was slightly modified to fit a smaller computational budget. Instead of searching for elite local optima (ones that cannot be improved by a two following swap moves), we confine to the solutions that cannot be improved by only one move. Moreover, in our method, the desired number of LON nodes is a parameter, while in [8], the number of LON node generation attempts.

The Algorithm 1 describes the method of obtaining LON nodes. First, a random solution $s \in S$ is generated. Then, the solution is optimized by a greedy descend algorithm—2-opt (a classic heuristic for TSP [7]). If the solution cannot be further improved by the 2-opt, and is unique; it becomes a node. Otherwise, another random solution is generated. Parameters I_{nmax} and I_{natt} determine the desired number of nodes in LON and the number of attempts to generate each node.

Algorithm 2 summarizes the sampling process of LON edges. The method was described in [8], however we also measure the number of kick moves that led to the solution not present in LON. This parameter indicates how well solution space was sampled by the Algorithm 1. For each node $s \in N_{LON}$ in LON, a kick-move is applied to the related solution. The kick is defined as $k = 2$ random 2-change moves performed one by one. The obtained tour s' is optimized by a first improvement descending algorithm (2-opt modified so that in each iteration the first-improving move is chosen, instead of the best-improving one; this strategy performed better in [8]). If the solution can be found in N_{LON}, the edge (s, s') is added to the set of edges. Otherwise, the move is reported to lead to an unknown local optimum (with the larger computational budget, these local optima could be added to N_{LON}). The process is repeated I_{eatt}-times for each node. The weight of an edge (s, s') is equal to the number of additions of (s, s') to E_{LON} during sampling process. Therefore, the bigger the weight of an edge s, s' is, the more probable the transition between solution s and s' is.

Algorithm 2. Method of sampling LON edges, based on [8]

Data : N_{LON}, the set of LON nodes;
I_{eatt}, the number of random kicks applied to each node;
a TSP instance

Result: E_{LON}, the set of LON edges;
the weights of LON edges

1 $E_{LON} \leftarrow \{\}$;
2 set the weight of each possible edge to 0;
3 **foreach** $s \in N_{LON}$ **do**
4   **for** $i = 1, 2, \ldots, I_{eatt}$ **do**
5          $s' \leftarrow$ applyRandomKick(s);
6          $s' \leftarrow$ firstImprovement_2-opt(s');
7   **if** $s' \in N_{LON}$ **then**
8              $E_{LON} \leftarrow E_{LON} \cup \{(s, s')\}$;
9   increase the weight of $\{(s, s')\}$ by 1;
10   **else**
11   increase the number of kick moves from s leading to solution not present in N_{LON} by 1;

3.2 Measures

We measured various LON parameters, including:

edgeToNode — the edge to node ratio, $edgeToNode = \frac{|E_{LON}|}{|N_{LON}|}$,
escRate — the average number of kick moves required to leave a local optimum,
numSubSinks — the number of subsinks. Solution $s \in N_{LON}$ is a subsink if and only if it has no outgoing edges to the solutions with the lower value of the fitness function, $\forall (s, i) \in E_{LON} \left(f(i) \geq f(s) \right)$,
distLO — the average distance form each node to the node s^* with the lowest value of the fitness function. The distance between any two nodes is defined as reciprocal of the corresponding edge weight. The nodes not connected to s^* are omitted,
conRel — the number of nodes connected to s^* to the number of nodes not connected to s^* ratio,
assortativity — the measure of preference for LON nodes to be connected with similar nodes (nodes with similar number of in-going or out-going nodes, values of fitness function),
clustering — global clustering coefficient, calculated with the graph-tools package for Python.

4   Experiments

4.1   Empirical Setting

Test Instances. The most popular test instances for TSP are gathered in TSPLib[1]. We chose 19 relatively small ones: `eil51`, `berlin52`, `st70`, `pr76`, `eil76`, `rat99`, `rd100`, `kroA100`, `kroB100`, `kroC100`, `kroD100`, `kroE100`, `eil101`, `lin105`, `pr107`, `pr124`, `ch130`, `pr136`, `pr144`, with the sizes varying from 51 to 176 cities. To further diversify the dataset, we generated random instances consisting of 30, 50 and 100 uniformly distributed cities (rnd30, rnd50, rnd100); 30 for each instance size.

Algorithms. We chose a commonly available set of metaheuristic algorithms, Google Optimization Tools (OR-Tools)[2]. OR-Tools allowed the use of various algorithms without implementation concerns. Moreover, the toolbox provides automatic mode, which is supposed to choose the appropriate algorithm for a given task. The initial idea was to compare the selection mechanism, to the one proposed in this paper. For each problem instance, we used the following search options, with default settings: automatic (Auto), greedy descent (GD), guided local search (GLS, most efficient for solving vehicle routing problems according to OR-Tools documentation), simulated annealing (SA), tabu search (TS), objective tabu search (OTS).

To evaluate the algorithms, we launched each one for 1 s for each instance (the short calculation time is due to the small size of the problems). We calculated the relative quality of the obtained solutions from equation

$$\Delta(s) = \frac{f(s) - f(s^*)}{f(s^*)} \cdot 100\%, \quad \text{assuming that } \forall s \in S \ \left(f(s) > 0 \right),$$

where s^* is the best known solution for the instance and f is the fitness function (see Table 1).

The results eliminated the possibility of comparing our method of solving ASP to the solution from OR-Tools. Automatic mode yield nearly the same results as GD, one of the worst performing algorithms (no differences between Auto and GD in Table 1). Moreover, GLS algorithm provided the best results for the vast majority of instances. Thus, the task of choosing the best algorithm for TSP from those available in OR-Tools is trivial.

This fact, however, does not eliminate the possibility of using LON analysis to select the appropriate algorithm from a smaller algorithm portfolio. For the ASP to be meaningful, the portfolio should contain complementary solving methods. Let us consider a pair of algorithms A and B. The number of instances in which algorithm A outperformed B is denoted by $w(A, B)$, while the number of draws, by $d(A, B)$. We defined that A and B are complementary, when $w(B, A) + w(A, B) \geq 1.5 \max\{w(B, A), w(A, B)\}$ and $d(A, B) \leq w(A, B) + w(B, A)$. The results of the described pairwise comparison is presented in Table 2.

---

[1] https://www.iwr.uni-heidelberg.de/groups/comopt/software/TSPLIB95/.
[2] https://developers.google.com/optimization/.

Table 1. Average relative quality of the solutions obtained by the algorithms.

Instances	Average Δ [%]					
	Auto	GD	GLS	SA	TS	OTS
rnd30	2.525	2.525	0.000	2.513	0.152	0.206
rnd30, 50, 100	2.304	2.304	0.093	2.408	0.532	0.381
TSPLib	3.527	3.527	0.172	3.458	2.822	1.063

Table 2. Pairwise comparison of the algorithms performance.

Dataset	Result	Algorithm pairings														
		1v2	1v3	1v4	1v5	1v6	2v3	2v4	2v5	2v6	3v4	3v5	3v6	4v5	4v6	5v6
TSPLib	1st won	0	0	**4**	**4**	3	0	**4**	**4**	3	**11**	**10**	**8**	2	1	1
	Draw	19	5	**7**	6	5	5	**7**	6	5	2	2	2	15	10	11
	2nd won	0	14	**8**	9	11	14	**8**	9	11	6	**7**	9	2	8	7
rnd	1st won	0	0	**34**	19	15	0	**34**	19	15	69	48	**36**	1	1	8
	Draw	90	25	**28**	27	15	25	**28**	27	15	14	32	**34**	59	30	43
	2nd won	0	65	**28**	44	60	65	**28**	44	60	7	10	**20**	30	59	39

1-Auto, 2-GD, 3-GLS, 4-SA, 5-TS, 6-OTS. Results suggesting that the algorithms are complementary
are stated in bold. 1v2 column contains the comparison of the best results of algorithms 1 and 2, 1v3
algorithm 1 and 3, etc.

4.2   Local Optima Networks

Local optima networks were generated with the method described in Sect. 3.1.
For the random instances with $n = 30$ and $n = 50$ cities, we sampled LONs
with $I_{nmax} = 1000$ nodes and $I_{eatt} = 10000$ edge-creating attempts. The values
were tuned so that the sets of the local minima of the smallest random instances
are thoroughly sampled. For rnd100 and for TSPLib instances, the sampling
parameters were set for $I_{nmax} = 10000$ and $I_{catt} = 10000$.

One can define many different measures, capturing specific aspects of LON.
However, some measures are correlated, causing the need for a feature selection.
To eliminate redundant data, we calculated Spearman correlation coefficients
for each pair of LON measures (listed in Sect. 3.2) and each test instance. The
correlation matrix is shown in Fig. 1.

The values of assortativity measures correlates with each other in a significant way, therefore only one is used for further analysis. There is also a clear
correlation between the size of the instance n and most of the other measures.
This is due to the lesser variation in the value of the measures among instances
of the same size. Such phenomenon is probably caused by the bias induced by
the LON sampling method. For example, an average *edgeToNode* ratio for random instances with $n = 30$ and 1000-node LON is 282, while for $n = 50$–30.
For $n = 100$ and 1000 nodes, almost no edges (that are not self-loops) are
present. Increasing the size of LON to 10000 sampled nodes yields in average
edgeToNode $= 2.6$; while the intuition suggests, that with the increase of n,
edgeToNode should also increase. The problem was also signalized e.g. in [12],

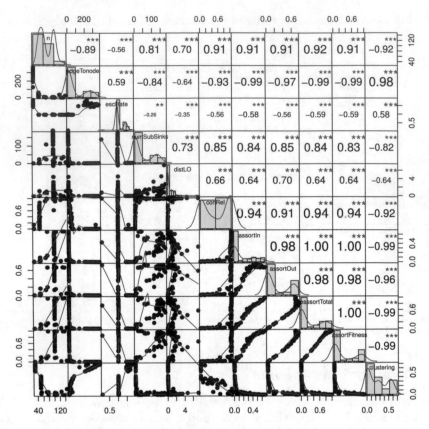

Fig. 1. Correlation matrix for all LON measures. Lower triangle: relationship scatter-plots. Diagonal: histograms. Upper triangle: Spearman correlation coefficient. P-values are indicated by asterisks: *** for $p < 0.001$, ** for $p < 0.01$ and * for $p < 0.1$.

where the influence of sampling effort on the values of measures is shown. Unfortunately, there is no commonly accepted sampling method so far, that we know of, to avoid such an effect.

4.3   Algorithm Selection Problem for TSP

As the ASP algorithm portfolio, we took pairs of complementary algorithms from Table 2. For each pair of algorithms A-B, the problem was to predict if algorithm A outperforms B, B outperforms A, or if they will provide the same results (3-class classification); using features from Sect. 3.2. We chose classifiers from WEKA[3] for this task. The software allows an easy use of multiple machine learning classifiers, from the simple ones, like NaiveBayes or DecisionStump, to more the sophisticated MultilayerPerceptron or RandomForest. As a comparison baseline, zeroR classifier was used—a simple method which always predicts the

---

[3] https://www.cs.waikato.ac.nz/~ml/weka/.

majority category class (as a side effect, it also indicates how unbalanced the dataset is). The percentage of correctly predicted instances was measured.

The results of the experiments are presented in Table 3. For the TSPLib instances, and the algorithm pair GLS-TS, the differences between the prediction performances of zeroR and other classifiers were lower than 5%. However, for the GD (or Auto) and SA algorithms, classifier scored 57.9% correct predictions, while zeroR scored only 42.1%. For the GLS-OTS pair, zeroR was outperformed by 15.8%, 57.9% to 42.1%. Also worth mentioning are the predictions for the SA-OTS and TS-OTS pairs: 73.7% and 68.4% respectively, while zeroR scored 52.6% and 57.9%. Unfortunately, these pairs of the algorithms were not complementary. The classifier was only able to determine whether the algorithms would give the same result or whether the better-performing one would win. Randomly generated instances proved to be much harder to be classified. Only for one pair of the algorithms (not complementary ones), a classifier outperformed zeroR by at least 5%. Probably the differences between instances were too subtle to be captured by a simple LON sampling method used in this paper.

Table 3. Classifiers performance comparison

Dataset	Classifier	Percentages of correctly classified instances for algorithms pairs									
		2v3	2v4	2v5	2v6	3v4	3v5	3v6	4v5	4v6	5v6
TSPLib	best other	73.68	**57.89**	**52.63**	57.89	**68.42**	52.63	**57.89**	78.95	**73.68**	**68.42**
	zeroR	73.68	42.11	47.37	57.89	57.89	52.63	42.11	78.95	52.63	57.89
rnd	best other	61.67	48.33	40.00	**60.00**	73.33	68.33	51.67	86.67	51.67	61.67
	zeroR	63.33	46.67	40.00	55.00	73.33	68.33	51.67	86.67	60.00	58.33

2-GD, 3-GLS, 4-SA, 5-TS, 6-OTS. 2v3 column contains the comparison of the zeroR performance and the best other classifier for the pair of algorithms 2 and 3, 2v4 for then algorithms 2 and 4, etc. The predictions better by at least 5% than zeroR are stated in bold. Results for the complementary algorithm pairs are underlined.

5   Conclusions

We presented a method of solving the algorithm selection problem for the travelling salesman problem. As our main contribution, we investigated using local optima network analysis to provide feature extraction of the problem instances. The initial results suggests, that such method can be successfully applied, provided well diversified instances and complementary solving algorithms. Unfortunately, the computation time required for sampling LON is still longer than that of the solving the instance itself. An interesting observation was the poor quality of the automatic algorithm selection mechanism provided by the researched OR-Library, leaving much room for improvement.

In further studies, we plan to investigate the relationship between the LON sampling method and the values of LON measures. Due to the intense computational demand, this part of LON analysis, in our opinion, requires more attention from the research community.

References

1. Applegate, D.L., Bixby, R.E., Chvatal, V., Cook, W.J.: The Traveling Salesman Problem: A Computational Study. Princeton Series in Applied Mathematics. Princeton University Press, Princeton (2006)
2. Bożejko, W., Gnatowski, A., Idzikowski, R., Wodecki, M.: Cyclic flow shop scheduling problem with two-machine cells. Arch. Control Sci. **27**(2), 151–167 (2017)
3. Bożejko, W., Gnatowski, A., Niżyński, T., Wodecki, M.: Tabu search algorithm with neural tabu mechanism for the cyclic job shop problem. In: Rutkowski, L., Korytkowski, M., Scherer, R., Tadeusiewicz, R., Zadeh, L.A., Zurada, J.M. (eds.) Artificial Intelligence and Soft Computing, pp. 409–418. Springer, Cham (2016)
4. Bożejko, W., Gnatowski, A., Pempera, J., Wodecki, M.: Parallel tabu search for the cyclic job shop scheduling problem. Comput. Ind. Eng. **113**, 512–524 (2017)
5. Bożejko, W., Pempera, J., Wodecki, M.: Parallel simulated annealing algorithm for cyclic flexible job shop scheduling problem. In: Rutkowski, L., Korytkowski, M., Scherer, R., Tadeusiewicz, R., Zadeh, L.A., Zurada, J.M. (eds.) Artificial Intelligence and Soft Computing, pp. 603–612. Springer, Cham (2015)
6. Chaudhry, I.A., Khan, A.A.: A research survey: review of flexible job shop scheduling techniques. Int. Trans. Oper. Res. **23**(3), 551–591 (2016)
7. Croes, G.A.: A method for solving traveling-salesman problems. Oper. Res. **6**(6), 791–812 (1958)
8. Iclanzan, D., Daolio, F., Tomassini, M.: Data-driven local optima network characterization of QAPLIB instances. In: Proceedings of the 2014 Conference on Genetic and Evolutionary Computation, GECCO 2014, pp. 453–460. ACM Press, New York (2014)
9. Jana, N.D., Sil, J., Das, S.: Selection of appropriate metaheuristic algorithms for protein structure prediction in AB off-lattice model: a perspective from fitness landscape analysis. Inf. Sci. **391–392**, 28–64 (2017)
10. Kotthoff, L.: Algorithm selection for combinatorial search problems: a survey. In: Bessiere, C., De Raedt, L., Kotthoff, L., Nijssen, S., O'Sullivan, B., Pedreschi, D. (eds.) Data Mining and Constraint Programming: Foundations of a Cross-Disciplinary Approach, pp. 149–190. Springer, Cham (2016)
11. Ochoa, G., Veerapen, N.: Additional dimensions to the study of funnels in combinatorial landscapes. In: Proceedings of the 2016 on Genetic and Evolutionary Computation Conference, GECCO 2016, pp. 373–380. ACM Press, New York (2016)
12. Ochoa, G., Veerapen, N.: Mapping the global structure of TSP fitness landscapes. J. Heuristics 1–30 (2017)
13. Ochoa, G., Verel, S., Daolio, F., Tomassini, M.: Local optima networks: a new model of combinatorial fitness landscapes. In: Richter, H., Engelbrecht, A. (eds.) Recent Advances in the Theory and Application of Fitness Landscapes, pp. 233–262. Springer, Heidelberg (2014). Chapter 9
14. Pitzer, E., Beham, A., Affenzeller, M.: Automatic algorithm selection for the quadratic assignment problem using fitness landscape analysis. In: Proceedings of the 13th European Conference on Evolutionary Computation in Combinatorial Optimization, EvoCOP 2013, pp. 109–120. Springer, Heidelberg (2013)
15. Rice, J.R.: The algorithm selection problem. Adv. Comput. **15**, 65–118 (1976)
16. Smith-Miles, K.A.: Neural networks for prediction and classification. In: Wang, J. (ed.) Encyclopaedia of Data Warehousing and Mining, pp. 865–869. Information Science Publishing (2006)

17. Smith-Miles, K.A.: Towards insightful algorithm selection for optimisation using meta-learning concepts. In: IEEE International Joint Conference on Neural Networks (IEEE World Congress on Computational Intelligence), pp. 4118–4124. IEEE, June 2008
18. Stadler, P.F.: Fitness landscapes. In: Lässig, M., Valleriani, A. (eds.) Biological Evolution and Statistical Physics, pp. 183–204. Springer, Heidelberg (2002)
19. Thomson, S.L., Ochoa, G., Daolio, F., Veerapen, N.: The effect of landscape funnels in QAPLIB instances. In: Proceedings of the Genetic and Evolutionary Computation Conference Companion, GECCO 2017, pp. 1495–1500. ACM Press, New York (2017)
20. Tomassini, M., Verel, S., Ochoa, G.: Complex-network analysis of combinatorial spaces: the NK landscape case. Phys. Rev. E **78**(6), 066114 (2008)
21. Vaishnav, P., Choudhary, N., Jain, K.: Traveling salesman problem using genetic algorithm: a survey. Int. J. Sci. Res. Comput. Sci. Eng. Inf. Technol. **3**(3), 105–108 (2017)

Robustness of the Uncertain Single Machine Total Weighted Tardiness Problem with Elimination Criteria Applied

Wojciech Bożejko[1]([✉]), Paweł Rajba[2], and Mieczysław Wodecki[3]

[1] Department of Automatics, Mechatronics and Control Systems,
Faculty of Electronics, Wrocław University of Technology,
Janiszewskiego 11-17, 50-372 Wrocław, Poland
wojciech.bozejko@pwr.wroc.pl
[2] Institute of Computer Science, University of Wrocław,
Joliot-Curie 15, 50-383 Wrocław, Poland
pawel@cs.uni.wroc.pl
[3] Telecommunications and Teleinformatics Department,
Faculty of Electronics, Wrocław University of Science and Technology,
Wybrzeże Wyspiańskiego 27, 50-370 Wrocław, Poland
mieczyslaw.wodecki@pwr.wroc.pl

Abstract. In the paper we investigate an uncertain single machine total weighted tardiness problem where uncertainty of times of jobs execution and jobs deadlines are modeled by random variables with a normal distribution. We check the influence of applying elimination criteria to the robustness of the determined solutions. Results are very promising and encourage further research.

Keywords: Elimination criteria
Single machine total weighted tardiness · Uncertain parameters
Tabu search · Normal distribution

1 Introduction

Based on the practice of management, it is widely believed that in a large part of cases we have to deal with uncertain data. Sometimes parameter values change during the realization, which usually leads to a significant deterioration of efficiency due to optimality loss. Uncertainty may present a random nature and it is then assumed that numerical data values are realizations of random variables with known or unknown parameters. This leads to interesting and difficult probabilistic optimization models often based on significant generalizations of classical problems known in the literature.

The single machine scheduling problems with different cost objective functions, despite the simplicity of formulating, belong mostly to the class of the

© Springer International Publishing AG, part of Springer Nature 2019
W. Zamojski et al. (Eds.): DepCoS-RELCOMEX 2018, AISC 761, pp. 94–103, 2019.
https://doi.org/10.1007/978-3-319-91446-6_10

most difficult (NP-hard) problems of combinatorial optimization. Optimal algorithms (usually based on dynamic programming methods or branch and bound constraints [15]) or approximate algorithms can be used to solve them. Due to the exponentially increasing computation time, optimal algorithms applications are limited. In practice, metaheuristic algorithms based primarily on methods of local optimization are used (tabu search, simulated annealing, etc., see [2–7,9]).

In this paper we consider a single machine scheduling problem with costs sum minimization of delayed tasks. Uncertain processing times and due dates are represented by random variables with a normal distribution. We present features that allow us to eliminate certain solutions that greatly accelerate the tabu search algorithm. The main goal is to design algorithms that determine robust solutions resistant to disturbances of parameters appearing in the actual realization. This allows to create schedules in which potential disruptions occurring during production are taken into account. Uncertainty in optimization problems has not been investigated very widely so far, but the area is getting more and more attention [1,8,12,13].

2   Deterministic Scheduling Problem

Let $\mathcal{J} = \{1, 2, \ldots, n\}$ be a set of jobs to execute on a single machine. At any given moment a machine can execute exactly one job and all jobs must be executed without preemption. For each task $i \in \mathcal{J}$ let p_i be a *processing time*, d_i be a *due date* and w_i be a weighted cost function for tardy jobs.

Every sequence of jobs execution can be presented as a permutation $\pi = (\pi(1), \pi(2), \ldots, \pi(n))$ of items from the set $\mathcal{J}$. Let Φ be a set of all such permutations. For each $\pi \in \Phi$ let denote $C_{\pi(i)} = \sum_{j=1}^{i} p_{\pi(j)}$ as a completion time of a job $\pi(i)$. Then

$$T_{\pi(i)} = \max\{0, C_{\pi(i)} - d_{\pi(i)}\} \tag{1}$$

is the *tardiness* of a job $\pi(i)$ and

$$w_{\pi(i)} \cdot T_{\pi(i)} \tag{2}$$

the *penalty* for a delay (or cost of the job execution). On the Fig. 2 an example graph of the cost function is presented. We can observe that as long as completion time is less than the deadline (d_i), there is no penalty related to the task execution. However, if the deadline is exceeded, the cost is increasing proportionally to the length of the delay (T_i) (Fig. 1).

For a sequence $\pi \in \Phi$

$$W(\pi) = \sum_{i=1}^{n} w_{\pi(i)} \cdot T_{\pi(i)}, \tag{3}$$

is a cost for a delayed execution of jobs from the set $\mathcal{J}$ (i.e. *cost of permutation* π).

In the *Total Weighted Tardiness problem* (*TWT* in short) the goal is to find a sequence which is minimizing the cost, i.e. a permutation $\pi^* \in \Phi$ where

$$W(\pi^*) = \min\{W(\pi) \colon \pi \in \Phi\}. \tag{4}$$

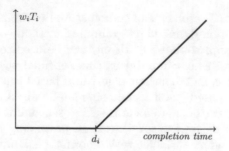

Fig. 1. Cost function representing tardiness of the jobs execution

TWT problem is usually denoted as $1||\sum w_i T_i$ and belongs to the NP-hard problems class (Lenstra et al. [10]). A more comprehensive review of methods and algorithms solving the problem can be found in [2].

3   Probabilistic Jobs Times

Scheduling problems with random parameters are modeled mainly by applying random variables with normal, exponential and uniform distributions (see Pinedo [12], Van den Akker i Hoogeveen [14], Jang [8], Shaked [1,13]).

In order to simplify further considerations we assume w.l.o.g. that at any moment the considered solution is the natural permutation, i.e. $\pi = (1, 2, \ldots, n)$. Moreover, if X is a random variable, then F_X denotes their cumulative distribution function.

In this section we consider a TWT problem with uncertain parameters. We investigate two variants: (a) uncertain processing times and (b) uncertain due dates.

3.1   Random Processing Times

Random processing times are represented by random variables with normal distribution $\tilde{p}_i \sim N(p_i, p_i)$, $i \in \mathcal{J}$. Other parameters, i.e. due dates d_i and cost function weights w_i are deterministic. Completion times $\tilde{C}_i$ are random variables:

$$\tilde{C}_i \sim N\left(p_1 + p_2 \ldots + p_i, a \cdot \sqrt{p_1^2 + \ldots + p_i^2}\right),\tag{5}$$

and tardiness

$$\tilde{T}_i = \begin{cases} \tilde{C}_i - d_i, & \text{if } \tilde{C}_i > d_i, \\ 0, & \text{if } \tilde{C}_i \leq d_i. \end{cases}\tag{6}$$

For each permutation $\pi \in \Phi$ the cost in the deterministic model is defined as $W(\pi) = \sum_{i=1}^n w_{\pi(i)} \cdot T_{\pi(i)}$. A corresponding cost in the random model (defined as (3)) is the following random variable:

$$\tilde{W}(\pi) = \sum_{i=1}^n w_i \tilde{T}_i.\tag{7}$$

In order to compare the costs of permutations from the set Φ we apply the following comparison function:

$$\mathcal{W}(\pi) = E(\tilde{\mathcal{W}}(\pi)) \tag{8}$$

where $E(\tilde{\mathcal{W}}(\pi))$ is the expected value of the random variable $\tilde{\mathcal{W}}(\pi)$. Below one can find the main theorem required to calculate the comparison function $\mathcal{W}(\pi)$. Proofs of the theorems and supporting lemmas can be found in [1]. Let $\mu = p_1 + p_2 + \ldots + p_n$ and $\sigma = \sqrt{p_1^2 + p_2^2 + \ldots + p_n^2}$.

Theorem 1 ([1]). *If the task completion times are independent random variables normally distributed $\tilde{p}_i \sim N(p_k, a \cdot p_i)$ $(i = 1, 2, \ldots, n)$, then the expected value of tardiness (6) of task $i \in \mathcal{J}$ is*

$$E(\tilde{T}_i) = (1 - F_{\tilde{C}_i}(d_i)) \left(\frac{\sigma}{\sqrt{2\pi}} e^{\frac{-(d_i - \mu)^2}{2\sigma^2}} + (\mu - d_i) \left(1 - F_{N(0,1)}(\frac{d_i - \mu}{\sigma}) \right) \right).$$

3.2   Random Due Dates

Random due dates are represented by random variables with a normal distribution $\tilde{d}_i \sim N(d_i, d_i))$, $i \in \mathcal{J}$. Other parameters, i.e. processing times p_i and cost function weights w_i are deterministic. Tardiness is a random variable

$$\tilde{T}_i = \begin{cases} C_i - \tilde{d}_i & \text{if } C_i > \tilde{d}_i, \\ 0, & \text{if } \tilde{C}_i \le d_i. \end{cases} \tag{9}$$

In this variant of the problem we apply the comparison function (8) defined in the previous section. However, expected values and standard deviations formulas need to be adjusted to follow changed definition of $\tilde{T}_i$ random variable. In order to do it we use the following theorem derived in [1]:

Theorem 2 ([1]). *If the expected due dates are independent random variables normally distributed $\tilde{d}_i \sim N(d_i, c \cdot d_i)$, then the expected value of tardiness (9) of the task $i \in \mathcal{J}$ is*

$$E(\tilde{T}_i) = F_{N(0,1)} \left(\frac{C_i - \mu}{\sigma} \right) \left(C_i F_{N(0,1)} \left(\frac{C_i - \mu}{\sigma} \right) \right.$$
$$\left. + \frac{\sigma}{\sqrt{2\pi}} e^{-\frac{(C_i - \mu)^2}{2\sigma^2}} - \mu F_{N(0,1)} \left(\frac{C_i - \mu}{\sigma} \right) \right).$$

The *TWT* problem in both variants (i.e. with random processing times and random due dates) is to find a permutation for which the comparison function (8) is minimal in the set Φ. In short we denote the probabilistic version of the problem (with random processing times) as *TWTP*. Likewise the deterministic version, the problem belongs to the class of NP-hard problems.

3.3   The Method of Execution Acceleration

Finding a solution in the $TWTP$ is based on a tabu search algorithm which is described in details in [2]. In short, the algorithm starts with an initial permutation, then it generates a neighborhood and searches for the best element which is a starting point for the next iteration. The neighborhood is of the size $O(n)$ elements and this size is the main factor of the overall execution time. Having that, we propose *a heuristic partial order relationship* on the set $\mathcal{J}$ in order to reduce the size of the neighborhood and by that speed up the execution time. We introduce the following proposition in different variants corresponding to different problem variants considered in this paper.

Proposition 1. *If processing times $\tilde{p}_i$ in TWTP are random variables and for different tasks $r, j \in \mathcal{J}$ we have*

$$E(p_r) \leqslant E(p_j) \wedge w_r \geqslant w_j \wedge d_r \leqslant d_j, \tag{10}$$

then when generating neighborhood we skip permutations where the task r precedes the task j.

Proposition 2. *If due dates $\tilde{d}_i$ in TWTP are random variables and for different tasks $r, j \in \mathcal{J}$ we have*

$$p_r \leqslant p_j \wedge w_r \geqslant w_j \wedge E(d_r) \leqslant E(d_j), \tag{11}$$

then when generating neighborhood we skip permutations where the task r precedes the task j.

Conditions (10) and (11) are verified before the algorithm execution. After the verification of all pairs a transitive relation closure is completed.

In the further part results from experimental experiments are presented where we refer both to the execution time and the robustness of solutions.

4   Computational Experiments

In this section we describe a method of generating random test data as well as results of comparing solutions calculated by an algorithm without any acceleration method and an algorithm with applied propositions 1 and 2. All tests are executed with a modified version of the tabu search method described in [2]. The algorithm has been configured with the following parameters:

- initial permutation: $\pi = (1, 2, \dots, n)$,
- length of tabu list: n,
- number of algorithm iterations: n,

where n is the tasks number.

An algorithm without acceleration we denote by $\mathcal{AD}$ and the one with acceleration by $\mathcal{AP}$.

4.1   Test Data Generation

All algorithms were tested on the OR-Library common reference test data available together with the best known solutions [11]. Test data was generated for $n = 40, 50, 100$ (n – jobs' number) in the following manner: processing times p_i and weight costs w_i are drawn randomly from ranges $[1, 100]$ and $[1, 10]$, respectively, and due dates d_i are drawn randomly from a range $[P(1 - TF - RDD/2), P(1 - TF + RDD/2)]$ based on parameters $RDD, TF \in \{0, 2; 0, 4; \ldots; 1, 0\}$ where $P = p_1 + \ldots + p_n$. All values are drawn with the uniform distribution and for every pair of values TF, RDD (there are 25 such pairs) 5 test samples were drawn. In total *the set of deterministic data* Ω for a specific n consists on 125 test samples what give 375 test samples for all n.

In order to verify the robustness of presented methods, two separate sets of disturbed data have been generated from the set Ω: (1) assuming uncertain processing times and applying normal distribution $\tilde{p}_i \sim N(p_i, c \cdot p_i)$, $c \in \{0, 02; 0, 04; 0, 06; 0, 08\}$, (b) assuming uncertain due dates and applying normal distribution $\tilde{d}_i \sim N(d_i, c \cdot d_i)$, $c \in \{0, 02; 0, 04; 0, 06; 0, 08\}$. A set of disturbed data we denote as $\tilde{\Omega}$.

Let $\delta = ((p_1, w_1, d_1), \ldots, (p_n, w_n, d_n))$ be an instance for a scheduling problem and let $\mathfrak{D}(\delta)$ be a set of disturbed data generated from δ by modifying processing times or due dates. By a disturbance we refer to replacement processing times or due dates $(p_i, i = 1, \ldots, n)$ (or $(d_i, i = 1, \ldots, n)$) for a randomly drawn values. Disturbed data $\gamma \in \mathfrak{D}(\delta)$ is of the form $\gamma = ((p'_1, w_1, d_1), \ldots, (p'_n, w_n, d_n))$ (or $\gamma = ((p_1, w_1, d'_1), \ldots, (p_n, w_n, d'_n))$) where processing times p'_i ($i = 1, \ldots, n$) (or due dates d'_i ($i = 1, \ldots, n$)) are a realization of the random variable $\tilde{p}_i$ (or $\tilde{d}_i$).

The basic robustness coefficient is defined as a relative distance between investigated and reference solutions. More precisely, having value W as a solution derived by the investigated algorithm and a reference value W^*, relative error $\delta = \frac{W - W^*}{W^*} 100\%$ and it expresses in how many percents the investigated solution W is worse than the reference solution W^*.

Since we want to make an evaluation based on a set of disturbed data, we introduce the following extension to the basic robustness coefficient. For s instances of disturbed data, let $W_1, \ldots, W_s$ be a set of values obtained by the investigated algorithm and let $W_1^*, \ldots W_s^*$ be a set of reference values. We calculate average values from sets $W_1, \ldots, W_n$ and $W_1^*, \ldots W_n^*$, then paste those values into the basic formula obtaining a value which we refer to further as *the generalized relative error*:

$$\Delta = \frac{\frac{W_1 + \ldots + W_n}{n} - \frac{W_1^* + \ldots + W_n^*}{n}}{\frac{W_1^* + \ldots + W_n^*}{n}} = \frac{(W_1 + \ldots + W_n) - (W_1^* + \ldots + W_n^*)}{W_1^* + \ldots + W_n^*}$$

Let introduce the following definitions: ψ as test data instance, $\mathfrak{D}(\psi)$ a set of disturbed data obtained from ψ by modifying processing times (or due dates) according to the assumed distribution. Moreover, we have:

- A_{ref} – algorithm to find a reference value of the problem, preferably an optimal solution,

– A – investigated algorithm, i.e. $\mathcal{AD}$ or $\mathcal{AP}$,
– $\pi_{M,x}$ – a solution obtained by algorithm $M \in \{A, A_{ref}\}$ for an instance x,
– $W(\pi_{M,x}, y)$ – a cost of instance y obtained by applying the solution $\pi_{M,x}$.

Then

$$\Delta(A, \psi, \mathfrak{D}(\psi)) = \frac{\sum_{\varphi \in \mathfrak{D}(\delta)} W(\pi_{A,\psi}, \varphi) - \sum_{\varphi \in \mathfrak{D}(\delta)} F(\pi_{A_{ref},\varphi}, \varphi)}{\sum_{\varphi \in \mathfrak{D}(\delta)} W(\pi_{A_{ref},\varphi}, \varphi)},$$

is the *robustness* of the solution $\pi_{A,\psi}$ (obtained by the algorithm A for instance ψ) on the set of disturbed data $\mathfrak{D}(\psi)$.

Let Ω be a set of samples of the investigated problem. The formula

$$\mathbb{S}(A, \Omega) = \frac{1}{\Omega} \sum_{\psi \in \Omega} \Delta(A, \psi, \mathfrak{D}(\psi)) \tag{12}$$

is the *robustness coefficient* of the algorithm A on the set Ω. The less the value is, the more robust the solution is, i.e. random disturbances of data in real environment have no significant impact on the cost of the solution.

All experiments have been conducted on a personal computer with a processor Intel Xeon 4.0 GHz and all algorithms have been implemented in Java. Experiments have been conducted to verify how applying elimination criteria impacts both the robustness of solution and the time needed to obtain the solution. We expected to have lower time of calculations while keeping the similar level of robustness of the obtained solutions.

4.2 Robustness

In the Table 1 we present the relative distances from solutions established by both algorithms to the best known solution.

Table 1. Relative distances between robustness coefficient of algorithms $\mathcal{AD}$, $\mathcal{AP}$ ($\mathbb{S}(A, \Omega)$) and reference values

N	Random p_i		Random d_i	
	$\mathcal{AD}$	$\mathcal{AP}$	$\mathcal{AD}$	$\mathcal{AP}$
40	0,71	0,71	1,9	1,9
50	0,68	0,68	2,1	2,1
100	0,54	0,54	2,1	2,1
Average	0,64	0,64	2,0	2,0

Due to the fact that there are actually no differences between values for $\mathcal{AD}$ and $\mathcal{AP}$ for both variants of the problem, we can immediately draw a conclusion that applying elimination criteria defined in Propositions 1 and 2 has no impact on the solutions robustness.

4.3   Acceleration of Calculations

Now let's consider how applying the elimination criteria accelerates the calculation time and, very related, the neighborhood size reviewed in the calculations. The results are presented in Tables 2 and 3.

Table 2. Time and neighborhood size of algorithms $\mathcal{AD}$ and $\mathcal{AP}$ for random p_i

N	Time			Neighborhood		
	$\mathcal{AD}$	$\mathcal{AP}$	Diff	$\mathcal{AD}$	$\mathcal{AP}$	Diff
40	7757,5	5936,8	31,3	3900000	2944618,5	32,5
50	17153,3	13243,8	29,4	7656250	5766283,3	32,8
100	205104,5	148153,5	38,4	61875000	44432184,3	39,3
Average	76671,8	55778	33	24477083,3	17714362	34,9

Table 3. Time and neighborhood size of algorithms $\mathcal{AD}$ and $\mathcal{AP}$ for random d_i

N	Time			Neighborhood		
	$\mathcal{AD}$	$\mathcal{AP}$	Diff	$\mathcal{AD}$	$\mathcal{AP}$	Diff
40	5188,3	4066,8	27,6	3900000	2991787	30,4
50	12391,5	9693,3	27,9	7656250	5854248,8	30,8
100	185614,5	134235	38,3	61875000	44253317,3	39,8
Average	67731,4	49331,7	31,3	24477083,3	17699784,4	33,7

According to Tables 2 and 3 we can observe that applying the presented method reduces by almost 35% the permutations needed to be reviewed in the execution process as well as it reduces the execution time by about 32%. On the Figs. 2, 3, 4 and 5 a split on different disturbance factor is visualized.

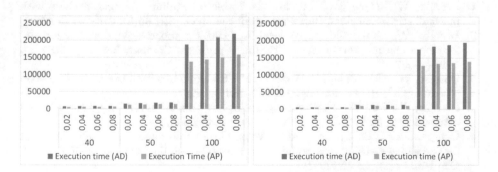

Fig. 2. Execution time of algorithms $\mathcal{AD}$ and $\mathcal{AP}$ for random p_i

Fig. 3. Execution time of algorithms $\mathcal{AD}$ and $\mathcal{AP}$ for random d_i

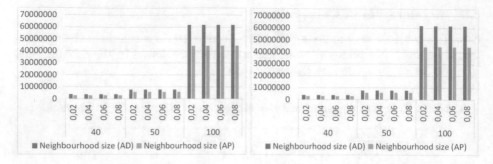

Fig. 4. Neighborhood size during the execution of algorithms $\mathcal{AD}$ and $\mathcal{AP}$ for random p_i

Fig. 5. Neighborhood size during the execution of algorithms $\mathcal{AD}$ and $\mathcal{AP}$ for random d_i

5   Summary

In the paper we have investigated a probabilistic single machine total weighted tardiness problem where uncertainty of jobs execution times and jobs deadlines are modeled by random variables with a normal distribution. We have presented a modified tabu search and applied a method of elimination criteria to reduce the size of neighborhood reviewed during the algorithms' execution and, by that, the execution time. Conducted computational experiments confirm the expectations and provide details to assess the effectiveness. At the same time we observed that robustness has not been altered by the proposed acceleration method.

Acknowledgement. The paper was partially supported by the National Science Centre of Poland, grant OPUS no. DEC 2017/25/B/ST7/02181, and Wrocław University of Science and Technology and RUDN University, agreement no. O/01800/524/2017, project no. 45WB/0001/2017.

References

1. Bożejko, W., Rajba, P., Wodecki, M.: Stable scheduling of single machine with probabilistic parameters. Bull. Pol. Acad. Sci. Tech. Sci. **65**(2), 219–231 (2017)
2. Bożejko, W., Grabowski, J., Wodecki, M.: Block approach-tabu search algorithm for single machine total weighted tardiness problem. Comput. Ind. Eng. **50**(1/2), 1–14 (2006)
3. Bożejko, W., Uchroński, M., Wodecki, M.: Block approach to the cyclic flow shop scheduling. Comput. Ind. Eng. **81**, 158–166 (2015)
4. Bożejko, W., Wodecki, M.: Parallel genetic algorithm for minimizing total weighted completion time. LNCS, vol. 3070. pp. 400–405 (2004)
5. Congram, R.K., Potts, C.N., Van de Velde, S.L.: An iterated dynasearch algorithm for the single-machine total weighted tardiness scheduling problem. INFORMS J. Comput. **14**(1), 52–67 (2002)
6. Crauwels, H.A.J., Potts, C.N., Van Wassenhove, L.N.: Local search heuristics for the single machine total weighted tardiness scheduling problem. INFORMS J. Comput. **10**(3), 341–350 (1998)

7. Grosso, A., Della, C.F., Tadei, R.: An enhanced dynasearch neighborhood for single-machine total weighted tardiness scheduling problem. Oper. Res. Lett. **32**(1), 68–72 (2004)
8. Jang, W., Klein, C.M.: Minimizing the expected number of tardy jobs when processing times are normally distributed. Oper. Res. Lett. **30**, 100–106 (2002)
9. Kirlik, G., Oguz, C.: A variable neighbourhood search for minimizing total weighted tardiness with sequence dependent setup times on a single machine. Comput. Oper. Res. **39**(7), 1506–1520 (2012)
10. Lenstra, J.K., Rinnoy Kan, A.G.H., Brucker, P.: Complexity of machine scheduling problems. Ann. Discret. Math. **1**, 343–362 (1977)
11. OR-Library. http://www.brunel.ac.uk/~mastjjb/jeb/info.html
12. Pinedo, M.: Stochastic scheduling with release dates and due dates. Oper. Res. **31**(3), 559–572 (2002)
13. Shaked, M., Shanthikumar, J.G. (eds.): Stochastic Order. Academic Press, San Diego (1994)
14. Van den Akker, M., Hoogeveen, H.: Minimizing the number of late jobs in a stochastic setting using a chance constraint. J. Sched. **11**, 59–69 (2008)
15. Wodecki, M.: A branch-and-bound parallel algorithm for single-machine total weighted tardiness problem. J. Adv. Manuf. Technol. **37**(9–10), 996–1004 (2007)

An Empirical Evaluation of Risk
of Underpricing During Initial Public Offering

Alexandr Y. Bystryakov[✉], Tatiana K. Blokhina,
Elena V. Savenkova, Oksana A. Karpenko,
and Nikolay S. Kondratenko

Economics Department, Peoples' Friendship University of Russia,
RUDN University, ul. Miklukho-Maklaya 6, 117198 Moscow, Russia
{bystryakov_aya, blokhina_tk, savenkova_ev,
karpenko_oa}@rudn.university, kondrat33@bk.ru

Abstract. The companies are increasingly choosing the initial public offering (IPO) as a way out of investment. In recent years, five Russian venture companies have conducted successful IPO. However, most of the companies which went through the initial public offering suffered from undervaluation of shares and had a risk of lower returns and underpricing on the first day of IPO. The article confirms the hypothesis of high efficiency of IPO venture capital companies. These results confirm the prevailing opinion that IPOs of venture companies are also less risky than those of non–venture capital companies.

Keywords: Initial public offering · Venture capital companies
Non-venture capital companies · Risk of underpricing of shares
Adjusted market return for venture capital companies

1 Introduction

Nowadays there are many different ways to sell the company shares: initial public offering at stock exchange, acquisition, secondary sale, write-off and the return repayment, sale to internal and external management. The most profitable of them is the Initial public offering (IPO) or transactions of merger and acquisition (M&A). Every company attracts capital for financing of new projects and also for expansion of its activities during its all lifecycle. The main goal of venture investment is an opportunity for investors to sell the shares at the price considerably exceeding the buying price in the future. Thus, venture capitalists are able to pay back investments and to get the desired profit [1]. According to the opinion of managing directors of funds, more than in half of the cases going public brings profit, however the investors originally expected a considerably bigger return, than they managed to receive during an exit [2, 3]. But most of the companies which went through the initial public offering suffered from undervaluation of the shares and underwent the risk of lower returns and underpricing on the first day of IPO. Nevertheless, the IPO of venture capital companies proved to be highly efficient. There is an opinion that the IPO of venture companies is less risky than the IPO of non–venture companies. We will try to check this phenomenon in Russian stock exchange.

© Springer International Publishing AG, part of Springer Nature 2019
W. Zamojski et al. (Eds.): DepCoS-RELCOMEX 2018, AISC 761, pp. 104–112, 2019.
https://doi.org/10.1007/978-3-319-91446-6_11

2 Initial Public Offering (IPO) as a "Gold Standard" of an Exit from Investments

Initial public offering (IPO) is gaining popularity in financing of large innovative companies. Most companies performing initial public offering wish to increase the level of capital and also to create the public market in which founders and shareholders can receive financing for future development. Besides, the companies which have performed IPO can significantly improve the liquidity by promoting reduction of cost of the equity of the issuer. In addition to financial benefits there are also non-financial ones of carrying out the IPO, such as increase in brand recognition, public advertizing and others.

Initial public offering is usually considered as a "gold standard" of an exit from investments. Historically the most successful investments – for example investments of Kleiner Perkins and Sequoia into Google or purchase of Ripple wood of the Long-Term Credit Bank of Japan company (later renamed into Shinsei Bank)–ended with an IPO.

In this research we conditionally divided the companies which effected the IPO with the help of venture-companies and the companies without participation of venture capital (non-venture companies). As a rule, venture capital is invested in a company of high-technology sector at initial stages of development. It represents long-term investments in securities or the entities from high or extra-high risk expecting an extremely big profit [4].

The most considerable event in the sphere of IPO of Russian venture companies is entry of Yandex search engine in NASDAQ in 2011. The value of Yandex IPO was about 1430 million dollars, and capitalization of Yandex constituted 3300 million dollars. On the first day the share price grew by 55%.

In 2013 a large Russian payment service provider Qiwi carried out initial public offering on NASDAQ. In case of the offering price of 17 US dollars for a share and amount of placement of 213 mln. US dollars capitalization of the company constituted more than 880 mln. US dollars (Table 1). Shares of Qiwi were initially acquired by foreign investors, 80% of buyers were American residents.

Table 1. IPO of venture-backed IT companies

Company	Date	Offering price	Closing price	Changes on the first day, %	Volume of IPO, mln. US dollars
Tinkoff Bank	25.10.2013	17,50	18,15	3,71	1087,00
Luxoft	26.06.2013	17,00	20,38	19,88	70,00
QIWI Plc.	03.05.2013	17,00	17,08	0,47	213,00
Yandex N.V.	24.05.2011	25,00	38,84	55,36	1430,00
Mail.ru Group	04.11.2010	27,70	35,33	27,55	912,04

Source: composed by the author according to the data of RTS-MICEX

The same year Tinkoff Credit System bank (Tinkoff Bank) during its IPO at the London exchange attracted 1087 mln. US dollars, and capitalization of the company exceeded 3200 mln. US dollars. The share price during the first day of placement grew by 3,71% (Table 1). Positioning of Tinkoff Bank as an IT company attracted a great number of investors, and demand exceeded the offer.

In 2010 Mail.Ru Group carried out initial public offering on the London Stock Exchange, the amount of the IPO constituted 912,04 mln. US dollars. At the same time early investors sold the most part of shares at 27,7 US dollars, and the company attracted only 84 mln. US dollars. For the first trading day the share price grew by 27,55%. Similar positive results of initial public offering of Russian venture companies allow to make the assumption of the possibility of growth as a result of using initial public offering.

However, more often return during an exit comes from acquisition of one company by another. It is connected with possible underpricing of shares when carrying out IPO. From the scientific point of view, underpricing of shares in case of the IPO is determined as a difference between closing price of the first day and offering price [4]. If this difference is negative, then underpricing phenomenon takes place [5]. The venture-backed companies provided in Table 1 didn't face this phenomenon. The similar situation usually concerns market efficiency of the equity, nevertheless, undervaluation was registered almost in all countries where IPO was performed [6]. Since the 1960s this "underestimated discount" in the United States of America on average constituted about 19%.

According to data of Table 2 underpricing of shares took place in case of IPO of many non-venture companies in the Russian stock market. The average value of underestimation constituted 8,19%. ROS AGRO PLC had the greatest extent of underestimation- 53% and Protek had about 38,05%. The positive surplus took place at TransContainer (4,40%) and Megafon (1,81%). Why are most of non- venture-backed companies exposed to underestimation, while venture companies avoid it and receive a positive surplus of share price? We will answer this question during this research.

Table 2. IPO of non venture-backed companies.

Company	Date	Offering price	Closing price	Changes, in %
OVK NPK	29.04.2015	715	710	−0,70%
AESSEL TPG	12.11.2013	90	72	−20,00%
Moscow Stock Exchange	15.02.2013	129,99	131,5	1,16%
Multisystem	27.12.2012	10,1	10	−0,99%
Megafon	28.11.2012	20	20,36	1,81%
Levengook	27.11.2012	14	14,04	0,29%
MD Medical Group	12.10.2012	12	12,09	0,75%
Utinet.ru	20.07.2011	195	176,11	−9,69%
FosAgro	13.07.2011	14	14	0,00%
Global Ports	29.06.2011	15	12,09	−19,40%
Bank FK Otkrytie	18.04.2011	35	35,08	0,22%

(*continued*)

Table 2. (*continued*)

Company	Date	Offering price	Closing price	Changes, in %
ROS AGRO PLC	08.04.2011	15	7,05	−53,00%
HMS Group	09.02.2011	41,25	40,25	−2,42%
Farmsynthes	24.11.2010	24	–	–
TransContainer	12.11.2010	7,95	8,3	4,40%
RNT	07.07.2010	89	85,19	4,28%
Diod	23.06.2010	33,3	29,8	−10,51%
Protek	27.04.2010	184,8	114,49	−38,05%
Russian aquaculture	12.04.2010	6	5,5	−8,33%
RUSAL Plc	27.01.2010	10	9,5	−5,00%

Source: composed by the author according to the data of RTS-MICEX

3   Literature Overview

Initial public offering has been a subject of leading scientists-financiers research. Within the last four decades they represented various models and theories trying to explain underestimation phenomenon using models of asymmetry of information, hypotheses of dispersion of ownership, institutional explanations of behavioral models. Several hypotheses are based on temporary dynamics or evolution of the markets, others offer new ideas [7].

Research concerning participation in IPO of venture-companies (venture capital companies) is of the greatest interest. IPO investments into the companies founded by venture capitalists are considered by potential investors as investments with lower risk and therefore issuers for attracting investors have an opportunity not to reduce offering share price. The companies founded by venture capitalists, as a rule, show big profitability in the long term [8].

So far the efficiency evaluation and calculation of underestimation of shares in the IPO of venture capital companies has been carried out only on the example of the western companies. However, American scientists Baygrave and Timmins claim that the industry of a venture capital takes various forms around the world, reflecting differences in economic and social structures, legal and financial environment [6]. For example, venture capitalists in Russia are inclined to invest their capital in the companies at late stages whereas the American venture capitalists invest the capital at early stages. Thus, Russian venture capitalists are less inclined to risk than their American colleagues. Earlier works on the results of initial public offerings by foreign authors, for example Ibboston and Jaffe [5] and then Ritter [4] included several investigations using various standard sets of coefficients and periods, testified to that, initial public offerings.

As mentioned by Ritter [4], the profit level on the first day of placement depends on the industry. Later Brave and Gompers [1] claimed that belonging of a firm to a certain industry can also influence the success of IPO.

In order to assess the above requirements, a new methodology was adopted for the study of events, whereby the IPO of companies with venture capital was compared to the IPO of companies without venture capital participation. In other words, this study allows to compare the rate of return of shares of two different groups of companies. The first group consists of companies that used venture capital (venture companies) and the second group consisting of firms that did not attract venture capital. The conducted empirical test aimed to test the phenomenon of participation of venture capitalists in carrying out IPO and assess the impact of venture capital on IPO.

4   Methodology

In this study financial information on conducting initial public offerings by Russian companies was collected. All companies were conditionally divided into two groups: IT companies with venture capital and companies without venture capital. After that a regression analysis was made with identifying possible causes of underpricing during IPO. During the period from 2010 to 2015 22 Russian companies without venture capital and 5 venture-backed IT companies made IPO [9]. The largest number of companies conducted initial public offering in 2010 and 2011, respectively (Table 3).

Table 3.   Venture-backed IT firms and non-venture-backed firms.

	VC-backed	Non-VC-backed
2010	1	7
2011	1	6
2012	0	4
2013	3	3
2014	0	1
2015	0	1

Source: composed by the author
according to the data of RTS-MICEX

So, let's test hypothesis number 1. The share prices of venture companies are higher than those of non-venture companies. This verification will be carried out on the basis of Table 4.

The minimum value of the share price for companies with venture capital is 9 times higher than for companies without venture capital. At the same time, the average value (700) and the median (650) for venture companies are much higher than those of the company without venture capital (560.93 and 388).

T-statistics and P-statistics indicate the statistical significance and dependencies of the criteria. The Mann-Whitney U-criterion was used to estimate the differences between two independent samples. The statistical significance of the differences between the levels of the feature in the samples was proven. According to the data in Table 5, the average value and median of IPO price of venture companies is significantly lower, but standard deviation is much lower. So, hypothesis number 1 received its confirmation on the basis of statistical data and can be adopted as the main one.

Table 4. Descriptive statistics of the share price of venture IT companies and companies not supported by venture capital

Coefficients	Supported by venture capital IT firms	Non-supported by venture capital IT firms	t- statistics	Test by Man-Whitney
Minimum	70,00	8,27	T- statistics – 1.24785. P-statistics – 0.11057. The result is not significant when $p < 0.05$	Z-coefficient – 1.45879. p-coefficient –0.1443. The result is not significant when $p < 0.05$
Maximum	1430,00	2240		
Average value	700,00	560,93		
Median	650,00	388		
Standard deviation	662,50	649,09		

Source: composed by the author according to the data of RTS-MICEX

Table 5. Indicators of profitability of shares of companies with participation of venture capital on the first day of the initial public offering

Company	Offering price	Closing price	Market Index at opening of the first day of IPO	Market Index at closing of the first day of IPO	Corresponding market return IPO, r_{im}	Initial market return, r_i	Adjusted market return ar_i, %
Tinkoff Bank	17,50	18.15	3578,36	3581,3	0,0008	0,0371	3,6278
Luxoft	17,00	20,38	9037,76	9067,27	0,0033	0,1988	19,5558
QIWI Plc	17,00	17,08	1572,71	1576,21	0,0022	0,0047	0,2480
Yandex N.V.	25,00	38,84	1285,37	1275,79	–0,0074	0,5536	56,1053
Mail.ru Group	27,70	35,33	2969,41	3024,72	0,0186	0,2754	25,6825

Source: composed by the author according to the data of RTS-MICEX

Let's formulate hypothesis No.2: the return of IPO of venture companies is higher than for companies without venture capital. In order to examine in more detail the efficiency of IPOs of venture companies and non-venture companies, the indicators of the company's initial return, initial adjusted return and adapted market return should be investigated [10, 11].

These indicators illustrated a change in the share price during the first day of IPO, from the opening to the end of the trading day. In addition, this calculation included the values of the stock indices of those exchanges on which the initial public offering was made at the beginning and end of the trading day [12, 13]. So, we have to test the hypothesis that the indicator of the adapted market return (a_{ri}) for companies with venture capital is much higher.

In the process of analysis, we investigated the following indicators: Initial return of the company –r. This indicator (r) was calculated from the logarithm of profitability obtained by dividing the share price in the last selling period (p_t) and the received dividend (d_t) by the share price of the previous period (p_m) according to the following formula:

$$r_t = \ln((p_t + d_t)/p_m) \tag{1}$$

Initial adapted return.

In order to analyze the effect of undervaluation in IPO we used a number of indicators below: The main one is the initial adapted market return (ar_t), which is defined as difference between return (r_t) and reference return of the relevant reference portfolio (r_m).

$$ar_t = r_t - r_m \tag{2}$$

The initial return of the company (r_i) was calculated as difference between the opening price and the closing price on its first trading day (3).

$$r_i = \text{ closing price} - \text{offering price} \tag{3}$$

The corresponding market return index (r_{im}) was calculated as the ratio of the market index at the end of the trading day (I_c) to the index at the beginning of the trading day (I_o) (4).

$$r_{im} = \frac{Ic}{Io} - 1 \tag{4}$$

This indicator (4) takes into account the change of the stock index during the day and allows to see general situation on the market.

The adjusted market return ar_i is the difference between the initial return and the corresponding market return.

$$ar_i = r_i - r_{im} \tag{5}$$

So, according to the data in Table 5 Adjusted market return of the venture companies which is the main indicator of IPO effectiveness is more than zero. Yandex (56.1053%) received the largest adjusted market return, Qiwi Plus (0.248%) had the smallest, while the average value of the adjusted market return of venture companies was 21.044%.

Table 5 presents data on corresponding market return, initial return and adjusted market return (ar_i).

The average value of the adapted market return for non-venture companies is negative and is about –12.95%, which is extremely low and indicates the company's IPO failure. The most unsuccessful is Rosagroplus' IPO, its adjusted market return is negative and is about –52.4726%, which indicates a loss of the company during IPO. Large Russian issuers, including companies with venture capital, usually prefer foreign trading platforms for IPO, medium-sized companies can rely on Russian stock exchanges, and small businesses can rely primarily on over-the-counter services.

Since 2012, companies with venture capital also have the opportunity to post information on the portal www.IPOboard.ru [15]. Even start-up projects are welcome. The main requirements for companies are: involvement in innovation activities, organizational and legal form of JSC, Web-site, business-plans. It should able to attract capital from 200,000 (300,000) to 4,000,000 (5,000,000) $, and in the future - IPO. In the draft IPO-board, participants are offered different levels of development of companies. For start-up companies, it is recommended to place data about the company and support mentor (mentor, consultant) who does not make financial investments, but evaluates the idea, helps with the development of the business plan and strategy of the company. Mentor helps with holding negotiations and attracting investors. As a rule, a 5% stake in the project is the return of the mentor. The next level is the main one and the work of board conductors is mandatory on it. Companies of this level need large financial investments, it is necessary to attract venture funds. At the end of 2016, 310 investors and 224 innovative companies were registered in the IPOboard system mainly from the sectors of Internet technologies [14, 15].

5 Conclusion

During this period of research IPOs of Russian companies with venture capital turned out to be more successful than IPOs of domestic companies without venture capital, which corresponds to the first and second hypotheses. These results confirm the prevailing opinion that IPOs of venture companies are also less risky than non-venture companies. As noted above, this is due to underestimation of non-venture companies in case the company sells shares for the first time, its true value is unknown to investors, and issuers themselves, as a rule, are not engaged in raising investment attractiveness. However, if the company already has a large investor, including a well-known venture capitalist with good business reputation, issuers do not need to pay so much attention to IPOs. T-statistics and P-statistics indicate the statistical significance and dependencies of the criteria. We proved the statistical significance of the first hypothesis. Hypothesis No.2 was also proved: the return of IPO of companies with venture capital is higher than for companies without venture capital. These results confirm the prevailing opinion that IPOs of venture companies are also less risky than those of non–venture capital companies.

In addition, firms with venture capital can attract more professional underwriters and auditors, as well as large institutional investors. Nowadays, access for innovative companies, including venture capital companies to the financial market is greatly facilitated. The success of initial public offering is determined by the capabilities of each particular trading platform. Here, optimality of placement may be achieved by minimizing costs and maximizing the return on the sale of the company's shares to a wide range of investors.

References

1. Brav, A., Gompers, P.: Myth or reality? the long-run underperformance of initial public offerings: evidence from venture and non-venture capital companies. J. Financ. **52**(5), 1791 (1997)
2. Drake, P.D., Vestsuypens, M.R.: IPO underpricing and insurance against legal liability. Financ. Manag. **22**, 64–73 (1993)
3. Iliev, P., Lowry, M.: Venturing Beyond the IPO: Financing of Newly Public Firms by Pre-IPO Investors (2017). https://ssrn.com/abstract=2766125
4. Ritter, J.R.: The long-run performance of initial public offerings. J. Financ. **46**(1), 3–28 (1991)
5. Ibboston, R.G., Jaffe, J.F.: Hot issue markets. J. Financ. **30**(2), 1027–1042 (1975)
6. Bygrave, W.D., Timmins, J.: Venture Capital at the Crossroads. Havard Business School Press, Boston (1992)
7. Humphery-Jenner, M., Suchard, J.-A.: Foreign VCs and venture success: evidence from China. J. Corp. Financ. **21**(1), 16–35 (2013)
8. Ritter, J., Welch, I.: A Review of IPO Activity, Pricing, and Allocations, NBER Working Papers 8805. National Bureau of Economic Research Inc. (2002)
9. http://www.moex.com/
10. Megginson, W.L., Weiss, K.A.: Venture capitalist certification in initial public offerings. J. Financ. **46**(3), 879–903 (1991)
11. Rauch, C.: Square, Inc. - Business Model, Venture Financing and Valuation (2016). https://ssrn.com/abstract=2716331
12. Mikkelson, W.H., Partch, M.M., et al.: Ownership structure and operating performance of companies that go public. J. Financ. Econ. **44**, 281–307 (1997)
13. Vong, A.P.I., Trigueiros, D.: An empirical extension of Rock's IPO underpricing model to three distinct groups of investors. Appl. Financ. Econ. **19**, 1257–1268 (2009)
14. Gao, S., Hou, T.C.T.: An empirical examination of IPO underpricing between high-technology and non-high-technology firms in Taiwan. J. Emerg. Mark. Financ. (2017)
15. http://www.ipoboard.ru/

Dependability Analysis of Hierarchically Composed System-of-Systems

Dariusz Caban$^{(\boxtimes)}$ and Tomasz Walkowiak

Faculty of Electronics, Wrocław University of Science and Technology,
Wybrzeże Wyspiańskiego 27, 50-320 Wrocław, Poland
{dariusz.caban,tomasz.walkowiak}@pwr.edu.pl

Abstract. A class of Systems-of-Systems (SoS) is considered, where systems are hierarchically composed of subsystems. If a system component fails, the system configuration must change (to a new valid state) to tolerate this fault. The dependability of SoS systems improves significantly due to these fault driven reconfigurations. Two methods are proposed to estimate this improvement. One relies on determining the minimal configurations and applying the k-out-of-n reliability model. The second is based on state-transition analysis, where for each state a valid configuration is searched for. As discussed, this is a more appropriate approach, though more complex computationally. Both approaches require an efficient tool for pre-validating configurations. The domain specific language, proposed in [4], is demonstrated to be useful for this.

Keywords: Systems-of-Systems · Configuration · Dependability analysis

1 Introduction

A class of systems-of-systems (SoS) [1, 7] is considered, where systems are composed of subsystems. The composition is hierarchically oriented; each component system has its parent system (except for the root SoS), which can be a component of another. The structure of the system may change, i.e. component subsystems can be moved to other parents during the system lifetime.

The dominating feature architecture of Systems-of-Systems is their flexibility of reconfiguration. A SoS adapts to the changes (within the system and external to it) by changing its configuration, i.e. its subsystems are relocated within the hierarchy and their parameters modified to ensure best operation. This results in a high resilience to operational faults, which trigger system reconfiguration.

The aim of this paper is to provide some guidelines to predicting the dependability of the considered class of SoS. This is not a trivial task – the traditional approaches to dependability analysis, based on the structural reliability or fault trees, cannot be directly applied. A modified approach, based on k-out-n reliability model, is discussed and its weaknesses are exposed. A much more accurate approach, based on state-transition model with automatically applied configuration constraints is proposed.

Dependability analysis based on configuration constrains requires a formalized method of describing system configuration and the constraints placed on it. The analysis does not assume any particular format of the system configuration, it may be

© Springer International Publishing AG, part of Springer Nature 2019
W. Zamojski et al. (Eds.): DepCoS-RELCOMEX 2018, AISC 761, pp. 113–120, 2019.
https://doi.org/10.1007/978-3-319-91446-6_12

expressed in XML, JSON or any proprietary format, such that constraints can be applied to it. It is assumed that the constraints are defined in the Parameter Description Language proposed by us in [4].

2 Hierarchically Composed Systems-of-Systems

A system-of-systems is composed of a number of subsystems, which combine to realize the required operations. Each subsystem may be composed of further sub-subsystems in the same manner. This leads to a hierarchical composition tree of low-level components that operate without any subsystems. Similar components and subsystems can occur multiple times in the same system. In the paper, we will be using the terms: objects (single instances of components or subsystems) and classes (denoting the type of similar objects). This corresponds with the object-oriented programming paradigm used when developing the systems.

A hierarchically composed SoS (Fig. 1) implies some restrictions placed on the how the subsystems are composed. It is assumed that there is a single top level object (pf predetermined class) that represent the system functionality as a whole. This leads to a simple requirement for system operability – a system is operational if its top level object is operational.

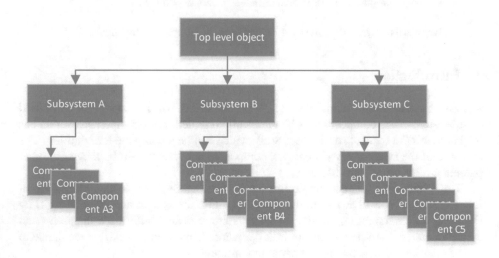

Fig. 1. Hierarchically composed system-of-systems

Hierarchical composition also implies that each subsystem needs objects of specified classes to perform its functions. The actual composition of each subsystem object is determined its configuration. Constraints placed on this configuration determine, among others, if objects of a specific class are required, or optional, or if there must be multiple objects of a class in a subsystem composition.

2.1   Configuration of SoS

The configuration of a system/subsystem/component (in short – object) is described by a set of its parameters, i.e. configurable variables. The names, number and type of parameters is specific to the class of objects. So, each class has a specific set of parameters that may be optional (omitted in a specific object), may be obligatory (at least one parameter must occur in the configuration) or may have multiple instances. The class specifies the names of the parameters, whereas specific objects are characterized by their values (and number of occurrences).

If an object is composed of some other objects, then its configuration is specified not only by the values of its parameters, but also by all the objects it is composed of. Thus, if an object is moved from one subsystem to another, both their configurations change, even if all the parameter values remain the same. Reconfiguration occurs if any parameter values are modified, objects are moved in the composition tree, objects are added, removed from the system, or replaced (e.g. when some objects fail, are repaired or upgraded).

Configurations have restrictions placed on them. These can be very diverse in form. The most common types of them are:

- *Constraints on the number, values and types of parameter* – these restrictions may apply to all the objects of a specified class, or they may depend on the values of other parameters.
- *Constraints on subsystems composition tree* – these restrict the classes of objects that can be the children of a specific parent class. Also, restrictions are placed on the number of child objects of specified classes.
- *Constraints between parameters of objects in the same composition tree branch* – the examples of such constraints are the requirements that some parameters are smaller or larger than some others, possibly in another object. Often the constraint might not be limited to a single object, but to all the children of a parent object: e.g. the limiting value set in the parent object may require that the corresponding parameters in all its children be limited to it.
- *Constraints between parameters of related objects* – the simplest possible constraint of this type is the requirement that all objects of a specific type, having the same value of one parameter, must also have the same value of another parameter. A more complex example of this type of constraints is based on cumulative values of a parameter, e.g. the cumulated sum of the power consumed by objects (one parameter) connected to the same power supply(second parameter) cannot exceed the max power of the supply (third parameter).
- *Constraints on parameter uniqueness* – this type of constraint is very frequent in the systems of systems. A recurrent example of such a constraint is the requirement that the same value (e.g. IP net address) is not used twice anywhere in the system. Constraints between parameters of objects fulfilling some other constraint– a constraint can either be met or be violated, so it can be considered as a logical expression. This can be used as a building block of more complex constraints, having other constraints embedded in it.

The configuration constraints are usually enforced by the components and sub-systems being configured. As proposed in [4] they also need to be expressed in a formal way, e.g. in the PDL language, mainly to enable configuration prevalidation (validation prior to deployment).

This is also the approach proposed for dependability analysis. In this case, each considered system configuration is prevalidated against the constraints expressed in PDL. The configurations that satisfy all the constraints are further called *valid configurations*.

2.2   Faults and Incidents

The considered class of system of systems is hierarchically composed of subsystems. This composition changes during the system lifetime to achieve best performance in the changing requirements of the environment. To this end, there is always a redundancy in the deployed resources, which allows the system to operate correctly in multiple valid configurations. This redundancy may improve the resilience of the system, even if it was not introduced for this purpose.

For a given configuration, a subsystem is operational if all its component systems are operating free of errors. This definition may be applied recursively, i.e. a component subsystem is operational if its components are also operational. This rule also applies to the system as a whole. Of course, this applies to a specific configuration – if the system changes its configuration, it may resume being operational.

There are diverse reasons for a system/component to operate erroneously. These can be categorized as:

- *Hardware faults*: occur relatively rarely due to the advances in technology. However, they require relatively long time to repair (replacement of components usually by an external repair team). Thus, the impact of their occurrence is normally much higher than any other type of fault. The permanent faults manifest themselves very quickly, immediately producing operational errors.
- *Transient hardware faults:* manifest themselves periodically or in specific situations, causing computational errors. These can propagate within the affected component subsystem leading to its failure. In effect, these faults have similar consequences as the software faults.
- *Software faults:* currently the most common reason for erroneous components operation. These faults are not incurred during system operation, they are acquired during software development, and are either resident in the system or injected during system updates. These faults manifest themselves as operational errors only in specific circumstances, usually when the system is reacting to an uncommon event. Human mistakes can have this effect. Intentional attempts at exploiting software faults (in this case called vulnerabilities) constitute the most widespread type of attack on the systems.
- *Malware proliferation:* is another serious threat to the system (undirected attacks that infect the component subsystems). This type of attack can seriously impact system performance – it is required to isolate the affected components from the system until proper recovery is performed.

- *Resource drainage:* can also affect component operability (in case of subsystems interacting with the environment). These attacks, e.g. Denial-Of-Service attacks [8], affect components performance and sometimes lead to their periodical unavailability.

2.3 SoS Reconfiguration

When the system operates in a given valid configuration and an operational error occurs in a subsystem, then all the subsystems using the affected one (including the top level system) also become inoperational. In consequence, the system fails if it is not reconfigured. This type of reconfiguration is called *fault driven*. Besides the fault driven reconfigurations, the system may change for other reasons, usually changes in the requirements for its services, performance optimization, company policy, etc. – in general, *policy driven* reconfiguration.

Fault driven reconfiguration is the basic mechanism responsible for improving dependability of SoS. From the theoretical point of view, the situation is quite clear: when a subsystem/component fails, the set of all valid configurations that do not include this component is determined. If the set is not empty, then the system is reconfigured to one of these valid configurations. Other criteria (e.g. performance) may be applied to choose this valid configuration.

The system fails if there is no valid configuration to switch to. The SoS remains inoperational until a component is recovered (depending on the nature of the fault it may be repaired, replaced, restarted or redeployed). Then, the set of valid configurations is again determined. If it is not empty, the system is reconfigured and it resumes operation.

The techniques employed for changing system configuration are out of scope for these considerations – the SoS may be self-organizing, manually reconfigured or reconfigured according to a fixed strategy [5].

3 Dependability Analysis

Dependability is an integrative concept that encompasses [2]: availability (readiness for correct service), reliability (continuity of correct service), safety (absence of catastrophic consequences), confidentiality (absence of unauthorized disclosure of information), integrity (absence of improper system state alterations), maintainability (ability to undergo repairs and modifications). Of these, operational availability is most adequate for measuring the resilience of the SoS.

The availability function $A(t)$ is defined as the probability that the system is operational at a specific time t. In stationary conditions, most interesting from the practical point of view, the function is time invariant. Thus, the availability is characterized by a constant coefficient, denoted as A. This measure is very interesting, since it has a direct application to system assessment from the business perspective. The asymptotic property of the steady-state availability A:

$$A = \lim_{t \to \infty} \frac{t_{up}}{t} \tag{1}$$

gives a prediction of the total system uptime t_{up}:

$$t_{up} \cong A \cdot t. \tag{2}$$

The availability of the SoS depends on the statistics of error occurrence in the component subsystems and on the redundancies in the components. Since the number of system configurations in a typical SoS is very large, the analysis is nontrivial. Two approaches are discussed, the first is relatively simple, but it may significantly over-estimate the availability. The second attempts to improve the estimate, but requires more complex analysis.

3.1 Approach Using k-out-of-n Reliability Model

The SoS is composed of a number of subsystems of different classes. Let's denote the number of components of class K as N_K. Errors may occur in each of the components, thereafter the component must be restored to its operational state. Errors in a component subsystem may have source in its operation or in the components it is composed of. The availability of a component is the probability that it is inoperational at a time instance due to faults connected with this subsystem and not its components. This availability is denoted as a_K and is assumed time invariant.

Now, if a valid configuration requiring minimal resources is known, the system-of-systems availability can be estimated. Assuming the minimal number of class K components is n_K (i.e. n_K is the total number of components of class K in the minimal configuration), the system availability is assessed as:

$$A = \prod_K P_{N_K}^{n_K}, \tag{3}$$

where $P_{N_K}^{n_K}$ is the probability that the number of operational components of class K is not less than n_K. This probability is computed using the k-out-of-n reliability model [3], i.e.

$$A = \prod_K \sum_{i=n_K}^{N_K} \binom{N_K}{i} \cdot a_K^i \cdot (1 - a_K)^{N_K - i}. \tag{4}$$

This estimate assumes that there is a single minimal configuration or that all the minimal configurations require the same numbers of components of the different classes. If this is satisfied, the availability is estimated quite accurately. Unfortunately, this is rarely the case. Practical SoS is built of numerous components having a huge number of "minimal" configurations. In this case, n_K is the minimal number of class K components computed over all the minimal configurations. Unfortunately, there is no guarantee that there is any valid configuration that uses only those minimal resources, i.e. the estimate in this case is over-optimistic.

With this approach, there is the obvious problem of determining the minimal configurations. This can be greatly simplified by computing the numbers over the composition tree – if a subsystem K is composed of at least $n_{K'}$ component subsystems of class K', and each of these requires $n_{K''}$ components of class K'', then subsystem

K requires $nK' \cdot nK''$ components of class K'' plus components of this class in other subsystems of class K object. Applying these rules in bottom-up calculations, the minimal numbers of each component class is determined.

3.2 Approach Based on Valid Configurations

In this approach, the operability of component subsystems is treated as a stochastic process, where errors occur randomly and components are restored to the state of operability in random time. In this case, a state-transition model of the system is built, where the state is the vector of the operability states of all its components. Transitions are connected with the occurrence of the various errors and with the time needed to repair the components thereafter.

Each system state is analyzed to determine if there is a valid configuration it can operate in. If such a configuration is found, then it is a state of operability of the system. If there is no valid configuration, then the system cannot maintain its operability by reconfiguration. Then, the state is marked as a fail state.

There are standard techniques for analyzing the state-transition processes. They provide estimates of the probabilities of all the system states. If P_r denotes the probability of state r and R is the set of states of operability, then availability is determined as:

$$A = \sum_{r \in R} P_r . \tag{5}$$

This approach is much more precise than the previously proposed. It takes into consideration all the possible combinations of the fail states and do not rely on the notion of minimal resources. There are two problems: how to apply the state-transition analysis to a system with such a huge state space and how to efficiently determine the set of states of operability.

The problem of the size of the state-transition graph is solved by using the technique of Monte Carlo simulation [6], i.e. the graph transitions are randomly fired (according to their random distribution parameters) and the simulations are repeated in a number of runs to obtain an asymptotic estimate of the state probabilities. The advantage of this method is that it does not require prior determination of the full set of states of operability – verification is done when the simulation reaches a specific state. Then, a valid configuration is searched only for the reached states. Very improbable states are never analyzed.

It should be noted that PDL based configuration pre-validation is used to determine if a valid configuration can be found for a given system state. This pre-validation is first applied to some likely configurations derived from the current one, and if this fails a complete cut and bound search is performed.

4 Conclusions

We demonstrated that SoS reconfiguration capabilities can be used to improve system dependability, if fault driven reconfiguration is implemented. We propose two methods for estimating this improvement. The first is very intuitive (being an enhancement on the prevailing engineering approach of "no single point of failure") and it is simple to use, but it tends to overestimate the achieved dependability. The second requires complex computations involving computer simulation and automated configuration pre-validation. But it overcomes the limitations of the first approach.

Both of the approaches rely on the ability to pre-validate SoS configurations. This is provided by the tools implementing PDL. In large SoS systems, this pre-validation is time-consuming due to the large number of constraints. Actually, a lot of these constraints may be omitted in pre-validation, if this is done only for dependability analysis. Unfortunately, constraints are strongly inter-dependent and this selection is not trivial. This still needs investigation to improve the efficiency of the validation/simulation tool.

References

1. Anderson, P., Herry, H.: A formal semantics for the SmartFrog configuration language. J. Netw. Syst. Manag. **24**(2), 309–345 (2016)
2. Avizienis, A., Laprie, J.C., Randell, B., Landwehr, C.: Basic concepts and taxonomy of dependable and secure computing. IEEE Trans. Dependable Secur. Comput. **1**, 11–33 (2004)
3. Barlow, R.E., Heidtmann, K.D.: Computing k-out-of-n system reliability. IEEE Trans. Reliab. **R-33**, 322–323 (1984)
4. Caban, D., Walkowiak, T.: Specification of constraints in a system-of-systems configuration. In: Zamojski, W., et al. (eds.) Advances in Dependability Engineering of Complex Systems. Advances in Intelligent Systems and Computing, vol. 582, pp. 89–96. Springer, Cham (2018)
5. Delaet, T., Anderson, P., Joosen, W.: Managing real-world system configurations with constraints. In: Proceedings of the Seventh International Conference on Networking, pp. 594–601 (2008)
6. Fishman, G.: Monte Carlo: Concepts, Algorithms, and Applications. Springer, New York (1996)
7. Jaradat, R.M., Keating, C.B., Bradley, J.M.: A histogram analysis for system of systems. Int. J. Syst. Syst. Eng. **5**(3), 193–227 (2014)
8. Mirkovic, J., Dietrich, S., Dittrich, D., Reiher, P.: Internet Denial of Service: Attack and Defense Mechanisms. Prentice Hall, Englewood Cliffs (2004)

Estimation of Travel Time in the City Using Neural Networks Trained with Simulated Urban Traffic Data

Piotr Ciskowski[1(✉)], Grzegorz Drzewiński[2], Marek Bazan[1],
and Tomasz Janiczek[1]

[1] Department of Computer Engineering, Faculty of Electronics,
Wrocław University of Science and Technology,
ul. Janiszewskiego 11/17, 50-372 Wrocław, Poland
`piotr.ciskowski@pwr.edu.pl`
[2] CyberTech Students Research Group, Faculty of Electronics,
Wrocław University of Science and Technology, ul. Janiszewskiego 11/17,
50-372 Wrocław, Poland

Abstract. The problem of travel time estimation by neural nets based on traffic data is considered in the paper. After a successful preliminary research on using neural networks to predict travel time based on real data is recalled, the next step of our research is presented, which utilizes urban traffic simulations in SUMO simulator as training data generator for training neural networks.

Keywords: Intelligent Transportation System · Neural nets
Travel time estimation · Traffic simulation

1 Introduction

Intelligent Transportation Systems (ITS) have become a standard in modern cities. They involve many types of technology to collect and analyze enormous amounts of data in order to provide solutions and functionalities useful in transport management. Traffic light control, entrance regulation, vehicle recognition may be given as examples. Their main aim is to improve the quality of public communication, and to support safe and secure traffic in the city.

Travel time is one the most valuable information for road users. Therefore reliable methods of its estimation and prediction for urban areas are required. The difference between estimation and prediction is that the former is based on completed trips on prescribed routes, while for the latter we calculate the travel time from some set of traffic quantities measured on certain points of the city, along the analyzed route. The prediction process includes the time variable as an input, along with time dependent quantities from the past and for current interval. The output is the travel time forecast for some interval in the future. The discussion on differences between these two approaches for travel time may be found in [3]. Travel time prediction using neural networks is a subject of a classical papers [8–10], while travel time estimation using neural networks is undertaken in [1, 2, 11–14].

© Springer International Publishing AG, part of Springer Nature 2019
W. Zamojski et al. (Eds.): DepCoS-RELCOMEX 2018, AISC 761, pp. 121–134, 2019.
https://doi.org/10.1007/978-3-319-91446-6_13

Road traffic simulation tools were used for travel time prediction in [6] for arteries and in [4] for the whole city. The comparison between time series models with simulation approaches may be found in [5]. To our knowledge the usage of road traffic simulation and neural networks for travel time estimation was presented only in [15], where travel time prediction is considered in an interurban scenario with the use of multilayer perceptron neural networks trained with data acquired from macrosimulation of the highway between two urban agglomerations. Training set comprises of simulated places of incidents, so that the resulting neural network can predict travel time depending on the place of their occurrence, simulated there by decreased capacity of the road.

In this paper we propose further development of our methodology of using multilayer perceptron neural networks for travel time prediction introduced in [1], where we developed an approach based on real data from the ITS in Wrocław, Poland. The time needed to travel a selected route is estimated by a neural network supplied with traffic intensity data on intersections along that route.

In the current paper the same methodology is applied for the traffic flows generated by microsimulations, which allow us to provide bigger training data sets for neural networks. Our main idea is to eliminate the processes of car plate recognition and car tracking along a specified route, on which the present-day prediction is based. We suggest a method, in which the travel time between two destinations in the city is not measured exactly by cameras, but estimated by neural networks, based on the data about traffic intensity on intersections along the route.

We have already shown in [1] that neural networks are able to provide proper estimation of travel time based on traffic intensity data collected from the ITS databases. That preliminary research has also revealed one important disadvantage of the method. Relatively long intervals of measurements are required in order to collect sufficient amount of data for neural network training. Therefore in this paper we consider city traffic simulation as a source of training data for the neural net.

To our knowledge this is the first paper to use microsumulation and neural networks for travel time estimation in a non-freeway scenario with many intersections, i.e. in the complex urban area. The results presented are preliminary and the method developed in this paper is to be extended in the future research, in which the travel time prediction system will allow to estimate travel time in future timestamps and will enable us to implement a dynamic Dijkstra algorithm for the simulation of the traffic redirection system [16] in an urban scenario. In our work we use the SUMO simulator [19] based on Krauss car following model. Another approach to travel time prediction in complex systems which is based on cellular automata, is presented in [18].

The paper is organized as follows. In Sect. 2 we shall recall the structure of the ITS in Wroclaw, in Sect. 3 we shall recall our methodology of neural estimation of travel time based on traffic intensity data, and present the idea of urban traffic modelling in SUMO simulator, we will show training results based on these simulation, along with the comparison to previous outcomes based on measurements. In Sect. 4 we shall conclude and present possible research directions for the future.

2   Travel Time Estimation by Neural Nets

2.1   Intelligent Transportation System in Wroclaw

The Intelligent Transportation System in Wrocław is an integrated environment consisting of measurement, communication, database and data analysis infrastructure. The system is implemented on Microsoft SQL Server as the data engine and uses various types of sensors to collect data about traffic, from induction loops to digital cameras. 1440 cameras and sensors have been installed on over 220 intersections throughout the city. Several types of cameras are used to measure various aspects of the traffic, of which two types are important in the scope of the presented research: video detection and ARTR.

There are 348 videodetection cameras installed on the majority of road intersections in Wroclaw. Each camera monitors one or two road lanes, each lane assigned to one or two traffic streams (e.g. straight and turn). These cameras detect single vehicles and store the following data in the dedicated data table: camera id, detection and store time, the length, speed and type of the detected vehicle, the lane number. Video detection cameras spot only the arrival of a vehicle. They record general data on each individual vehicle. The vehicles cannot be identified. Based on these data, one can calculate such traffic characteristics as: the intensity and density of traffic flow, or average speed for each traffic stream on every monitored intersection.

The ARTR cameras recognize car plate numbers. There are 51 such cameras, installed along the main traffic routes in the city. When a certain car plate number is recorded at both ends of a route (and usually one or two cameras along the route for verification), the vehicle's travel time is calculated and stored in a dedicated data table. Then it is used to calculate the average travel time along that route. The ARTR cameras are used to calculate average travel time along 5 main routes in the city – in both directions, given 2 variants for each route. Current times are displayed on 13 variable message boards, installed over the roads.

2.2   Current Travel Time Calculation Method, Based on Car Plate Number Recognition

The travel time information system for drivers, already implemented in ITS Wrocław, is based on real travel times recorded by ARTR cameras. Each time a car (identified by its plate number) is recorded at the beginning and at the end of one of the routes defined in the system, the travel time for that route is updated.

The disadvantages of that solution may have already been pointed out in [1]. As the routes defined in the system are relatively long, only a fraction of cars travelling along that routes is captured both at the beginning and at the end, and then considered for travel time calculation. The number of measurements provided by ARTR cameras in travel time data tables is much smaller than the number of records in data tables filled by video detection cameras. During a preliminary research, performed on a snapshot of ITS data tables, covering two weeks in May 2014, only 100 to 500 cars were found traveling entirely one the routes defined in the system. The measurements were asynchronous, that is scattered unevenly over the two week long recording time.

As a consequence, the main disadvantage of the current method is that travel time information stored in the system, although calculated on exact and verified measurements, is quite often based on either a small sample of traffic, or on outdated records.

The number of video detection cameras installed in ITS Wroclaw is much greater than ARTR cameras and they record data continuously all the time. The amount of data that may be used to estimate travel time is therefore much larger than the number of measurements on which it is calculated nowadays. Moreover the information is provided constantly as a continuous and steady stream of data.

2.3   Travel Time Estimation by Neural Nets, Based on Traffic Intensity and Average Speed

The substitution of travel time calculation, implemented nowadays, with travel time estimation, based on the data provided by video detection cameras, was suggested in [1]. The ARTR cameras would still be used as a source of training data – the real travel times, assumed to be correct after verification and filtering.

The idea of using videodetection camera data to train neural nets estimate the travel time, is illustrated in Fig. 1, while a sample route with both types of cameras is presented in Fig. 2.

Fig. 1. The idea of travel time estimation system. A, B, C – ARTR cameras, k1–k4 – video detection cameras

Fig. 2. An example of a travel route (Kochanowskiego - Lotnicza) used for travel time estimation. red pins – ARTR cameras, yellow pins – videodetection cameras, image: Google Earth

We train a separate neural net for each travel route. The data from ARTR cameras (marked as A and C in Fig. 1 and as red pins in Fig. 2) at both ends of the route are used to calculate the real travel times of individual vehicles, and provide the desired output for neural net training. The inputs of the net are supplied with information on traffic density, collected by videodetection cameras on intersections along the analyzed route (marked as k1 to k4 in Fig. 1 and as yellow pins in Fig. 2).

We assume that a separate neural net is needed for each travel route. The data from ARTR cameras (marked as A and C in Fig. 1 and as red pins in Fig. 2) at both ends of the route are used to calculate the real travel times of individual vehicles, and provide the desired output for neural net training. The inputs of the net are supplied with information on traffic density, collected by videodetection cameras on intersections along the analyzed route (marked as k1 to k4 in Fig. 1 and as yellow pins in Fig. 2).

The results of a preliminary research on this estimation method were presented in [1]. We have selected several routes, defined in ITS Wroclaw, and prepared an ASP. NET/Python 3.4 environment to define routes, extract traffic data from ITS database tables, and to preprocess them to serve as a training set for the neural network. We used MATLAB environment to train neural networks. Multi-layer perceptrons with one hidden layer of 10, 5 and 3 neurons were trained using the Levenberg-Marquardt training algorithm. The tanh activation function was used for the hidden layer neurons, while output neurons used linear activation function. The recorded travel times of vehicles were used as the desired outputs. We have examined various combinations of neural nets' inputs. All the nets were supplied with time of the day on the first input. Other inputs included only the traffic intensity (cars/hour) on all intersections along the route, only the average speed (km/h), and the combination of both. The number of nets' inputs was 9 or 17. The data was divided into training, validating and testing set. Training took usually about 10 epochs (using the L-M training algorithm) and was terminated to avoid overfitting. All the networks reached low values of the MSE: between 0.017 and 0.026, while the correlation coefficient R values varied from 0.69 to 0.78. The best results were obtained using small networks with 3 and 5 hidden neurons, supplied with the combination of daytime, traffic intensity and average speed inputs.

3 Using Traffic Simulations to Train the Neural Nets

The results of experiments conducted in [1], while being promising, have shown that a further improvement of training is required. The main limitation of the method is still the number of training examples. The main idea of the suggested solution is to replace travel time calculation with its estimation, what may provide more accurate and up-to-date estimations at the stage of using the calibrated and tested system. However, during the training phase the desired outputs for the neural networks are still extracted from real observations in ITS Wroclaw. During the preliminary research mentioned in previous Chapter, neural networks were trained only on 598 examples collected for the selected route during the two week period. Therefore we suggest to use a traffic simulator to generate an appropriate amount of training data for neural networks.

3.1 The SUMO Traffic Simulator

Urban traffic modeling is based on creating mathematical models of real-world traffic network. The road traffic is often extended by pedestrian movement, bicycles routes, tram lines, trains, ships or even planes. Different methods used for modeling may be divided into micro-, meso- and macroscoping models, may be based on car following models, cellular automation models, agent-based models, and may include specific algorithms, such as queuing, numerical PDE methods, Monte Carlo etc. Both time and space may be modeled in a continuous or discrete manner.

The SUMO (Simulation of Urban MObility) is an open source system for computer simulation of traffic in urban areas. It was released in 2001 and is developed by DLR Institute of Transport Systems (DLR is the German Aerospace Center). It was written in C++ and is distributed under GPL license. It allows modeling complex systems including roads, vehicles, bicycles, public communication and even ships or trains. The platform is highly portable and designed to handle large road networks. It provides microscopic and continuous road traffic simulations. The system delivers tools for modeling the network, to run and visualize real-time simulations. It is equipped with many extensions providing additional functionalities for visualization, automatic trip generation and much more. It provides rich API for Python and C++ to remote control of simulation process.

The SUMO has been already employed in several projects for research and commercial purpose. In 2005, it was used to estimate traffic during Pope's visit in Cologne. Simulations allowed the city government to take actions to reduce queues on principal arterials of the city. It was also used in 2006 to provide forecast about potential bottlenecks and traffic jams while the FIFA World Cup. Several German towns use it to evaluate the performance of traffic lights control systems, what allows them to safely test and implement modern algorithms of optimization transport.

Simulations in SUMO are microscopic, therefore every single vehicle in simulated network is modeled individually and is characterized by a certain position and velocity. In each time step, which is equal to 1 s, speed and position of every object are updated. The simulation is discrete in the domain of time and continuous in the domain of space. The simulator is able to display main traffic characteristics like free and congested flow.

SUMO uses the Gipps car-driven model invented by Kraus and Janz in 1998, which belongs to the Car Following Models family, where in each time step, the speed of a particular car is adapted to the velocity of the leading object. Such a velocity is marked as the safe velocity and is represented by v_{safe} calculated as:

$$v_{safe} = v_l(t) + \frac{g(t) - v_l(t)\tau}{\frac{v'}{b(v)} + \tau} \tag{1}$$

where $v_l(t)$ is speed of the preceding object in time t, $g(t)$ is the gap between vehicles, τ is the driver's reaction factor, and b is the deceleration function.

In order to reflect the physical ability of objects to accelerate, the desired vehicle speed v_{des} is calculated as the minimum function of several values: the car possible maximum velocity, object speed + max. acceleration and safe velocity as:

$$v_{dest}(t) = min[v_{safe}(t), v(t) + a, v_{max}] \qquad (2)$$

The driver inaccuracy is also included in the model. To illustrate human errors and mistakes done while driving, the modeling engine subtracts a random variable from the desired speed:

$$v(t) = max[0, rand[v_{des}(t) - \varepsilon a, v_{des}(t)]] \qquad (3)$$

With assumption that cars cannot drive backwards there is need to add a variant of minimal possible speed as 0 km/h. With this fact taken into account the final velocity equation is given by (3).

The simulation output may be provided in a few different forms. First and most complex is a full log of simulation process, so called the raw output. In this case each simulation step is represented by a set of objects names, positions and speeds. It provides information about all the streets and vehicles. This makes it useful for complex analysis of the simulated situation and may be a solid base for further data processing. Unfortunately, the complexity of data makes it really hard to evaluate. In order to manage the complexity of the generated 'raw' log, SUMO was equipped with additional logging tools and formats, among which the trip info file was most useful in our research. It includes information on each object departed to the simulation. Each record on this list gives information about departure and arrival positions, speed, time and much more including metadata collected from simulations. The simulator offers also the possibility to measure and generate logs about some interesting points of the evaluated simulation. To execute that operator, one needs to insert measuring infrastructure into a simulated network. When simulations finish the system generates the log files with measured data, which may be saved in CSV or XML format.

3.2 Urban Traffic Simulation

The SUMO allows to run simulations of large areas faster than real-time. For this preliminary research we have decided to analyze the traffic on a single main street and the streets directly connected to it. We have chosen one straight street with many intersections also with traffic lights, the Zachodnia Street (see Fig. 3). It is about 1,5 km long and has ten intersections in both directions. The neighboring streets were also modeled and included in the simulation in order to make it realistic.

During several pilot simulations we have observed the way SUMO simulator models the traffic. We have noticed that every time a car starts its designed route it suddenly appears at the starting point. Correspondingly, when it finishes its route it is immediately transferred out of the simulation space. Therefore we have decided that the cars forming background traffic need to have starting and finishing points in some (quite large) distance from the examined street. Otherwise the traffic representation at a measuring point would be wrong. And could lead to unrealistic way of traffic modeling.

Fig. 3. Simulation area - Zachodnia Street in Wrocław

If the start and finish points were placed at one of the analyzed intersections of the main street, then traffic jams would be minimized in an artificial way leading to misrepresentations in the simulation.

Even in the case of such a limited city area, manual designing and tuning of the model would be very long and tedious task. The dedicated editor for creation traffic network is implemented in the SUMO package but it would take a lot of time to create reasonable mapping (presumably with poor accuracy). Fortunately, the SUMO supports the OSM (Open Street Map) map format, which is an open platform for creating and sharing maps. The maps are created by the community and the data about roads, paths, cafes, stations and more are added all over the world, based on satellite imagery, aerial imagery, GPS navigation, and regular maps. OSM maps may be downloaded in the XML format. These are complex files where each object is encoded together with its parameters and position. OSM files may be edited in the special editors like JOSH which is a Free Java implementation. SUMO provides an OSM Web Wizard tool, which is an application designed for generating traffic models directly from OSM. The above mentioned simulation area is presented in Fig. 4 as an OSM map in JOSH editor.

The SUMO network format is a XML based file. It could be edited in any text editor but the dedicated tool allows to work comfortably in a graphical environment. The prepared map model may be viewed and edited in the SUMO NetEdit application.

The most fundamental structures of the road network are nodes and edges. The nodes are labeled with *id* and represented as points in two-dimensional space with *x* and *y* coordinates. The edges are described by three parameters: *id*, start node and end node. Each edge may also contain an internal object called *lane*, which determines how many route lines are included in a particular edge. It is very important while creating multi traffic lanes street and junctions.

Fig. 4. Simulation area as an OSM map

The graphical form of the simulated area in NetEdit editor is presented in Fig. 5. The map imported from the OSM format needed some manual corrections, as well as defining traffic light cycles. In order to collect input data for the neural net, i.e. traffic intensity measurements on each lane of consecutive intersections, we have defined the induction loop detectors at each lane, marked in yellow in Fig. 6. The detectors count vehicles passing over the loop and save the information to the output file over requested period. The most significant value stored in the file is a *flow* which determines the number of cars passing in each hour. In our simulations we have used the detector period of 100, so the number of cars was aggregated every 100 s. In real Wroclaw ITS

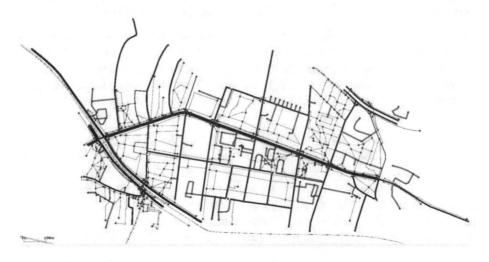

Fig. 5. Simulation area in NetEdit editor. Each red dot represents a node.

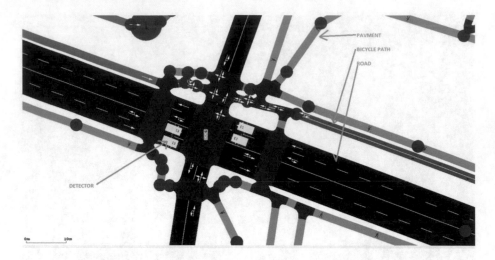

Fig. 6. A model of an intersection with infrastructure elements

databases the information on each passing car is collected and aggregation may be performed with any accuracy, the most widely used being 15 min.

3.3   Training Data Generation

As mentioned earlier in Sect. 2, the idea of this research is to use neural nets to estimate the travel time based on traffic intensity (density, flow) data measured on the consecutive road intersections along the route. The purpose of the traffic simulation using SUMO, is to collect large amounts of data, so that the neural network may learn the dependence of the travel time on various combinations of traffic intensity.

We have assumed that each simulation should take at least one-hour period. We have generated simulation schemas, which included two types of cars:

- background cars, travelling around the simulation area, and generating various road traffic conditions, that is various values of the flow (measured by induction loops installed on the intersections), to be used as neural networks inputs,
- test cars, travelling along the analyzed route, triggered every 100 s, while their arrival time being used to calculate their travel time, that is the desired value of neural networks' outputs.

Simulations may be triggered by SUMO GUI where the whole process is visualized and there is possibility to monitor a flow of cars, while it is also possible to run the using python scripts as a background process. The SUMO GUI is a useful tool for defining and checking simulation scenarios and for initial research of the simulation results. It allows us to find a specified car or an infrastructure element. It also provides a possibility to track a selected car. It proved helpful while designing the simulation model and checking the correctness of simulation parameters. Another useful part of

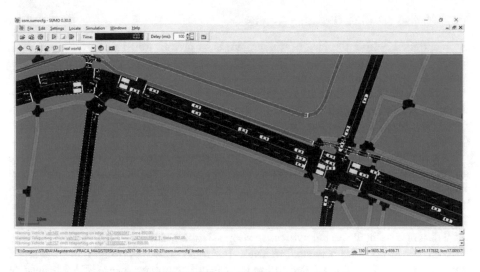

Fig. 7. SUMO GUI view with running simulation

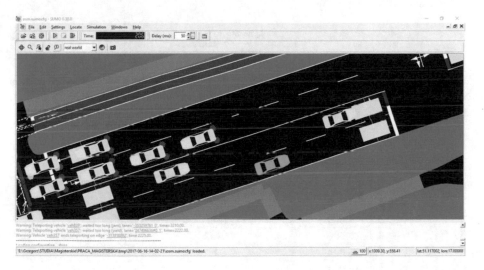

Fig. 8. SUMO GUI - Green car is a test car, yellow cars provide random traffic

the simulator was the time scaler, which allows to control time flow during the simulation and pause it on demand. It allows even to perform the simulation step by step, what may be useful in a detailed examination of the evaluated model. Two examples of SUMO GUI environment during simulation are presented in Figs. 7 and 8.

3.4   Neural Network Training Results

As mentioned before, the idea of this project is to predict travel time of a route from traffic density at intersections along that path. In case of the street presented above, the neural net was supplied with data from 44 detectors.

After several experiments, we have decided to use neural nets with 10 hidden neurons. We have used three training methods: the Levenberg-Marquardt (LM) algorithm, Bayesian Regularization Backpropagation (B-R) and Scaled Conjugate Gradient (SCG).

As the idea of the project is to use large simulated training datasets, we have generated three datasets with different number of training vectors. At first we performed 100 simulations, each one about 1 h long, which resulted in 3600 training examples. Then we used more than 12 thousand training vectors coming from 334 simulations. The largest training set included 126 thousand training records from 3500 simulations.

Training results for neural networks using different training dataset and trained with different algorithms, are presented in Table 1. As the networks' output was the travel time in seconds, we have decided to use the Mean Percentage Error (MPE) instead of Mean Square Error, as the measure of performance. The percentage error for each training example is given by

$$PE = \frac{|error|}{target} \cdot 100\%, \tag{4}$$

while the mean percentage error is averaged over all testing cases. The error was calculated on a testing set, not used during training and validation, made of 15% of all available data.

Table 1. Neural network training results

	Number of training vectors	Training method	MPE
1.	3 600	L-M	15.2%
2.	3 600	B-R	13.8%
3.	3 600	SCG	17.0%
4.	12 024	L-M	14.79%
5.	12 024	B-R	14.81%
6.	12 024	SCG	14.81%
7.	126 000	L-M	17.06%
8.	126 000	B-R	17.50%
9.	126 000	SCG	18.14%

4 Conclusions

The paper presents the next step of our research on neural estimation of travel time in the city based on traffic intensity information along the route. The results presented in Sect. 3 are promising. Traffic simulation is a powerful tool for generating training datasets for neural nets. Further investigation is needed regarding the process of data generation, e.g. the analysis of traffic intensity distribution (input data for the nets) from the design of experiments perspective. Simulation of larger areas is also planned, adaptation of traffic light control programs, as well as the verification of simulation results with real travel times extracted from OTS databases.

Acknowledgements. This work was partially supported from grant no 0401/0114/16 at Wrocław University of Science and Technology.

References

1. Ciskowski, P., Janik, A., Bazan, M., Halawa, K., Janiczek, T., Rusiecki, A.: Estimation of travel time in the city based on intelligent transportation system traffic data with the use of neural networks. In: Zamojski, W., Mazurkiewicz, J., Sugier, J., Walkowiak, T., Kacprzyk, J. (eds.) Proceedings of the Eleventh International Conference on Dependability and Complex Systems DepCoS-RELCOMEX, 27 June–1 July 2016, Brunów, Poland, Advances in Intelligent Systems and Computing, vol. 470, pp. 85–95. Springer, Cham (2016). ISSN:2194-5357
2. Halawa, K., Bazan, M., Ciskowski, P., Janiczek, T., Kozaczewski, P., Rusiecki, A.: Road traffic predictions across mayor city intersections using multilayer perceptrons and data from multiple intersections located in various places. IET Intell. Transp. Syst. **10**, 469–475 (2016)
3. Carrascal, M.: A review of travel time estimation and forecasting for Advanced Traveller Information Systems, Master thesis. Komputazio Zientzlak eta Adimen Artifiziala Saila, Departamento de Ciencias de la Computacion Inteligenzia Artificial, Uni. del Pai Vasco (2012)
4. Liu, Y., Lin, P., Lai, X., Chang, G., Marquess, A.: Developments and applications of a simulation-based online travel time prediction system for Ocean City, Maryland. Transp. Res. Board **1959**, 92–104 (2006). https://doi.org/10.3141/1959-11
5. Hu, T.-Y., Ho, W.-M.: Travel time prediction for urban networks: the comparisons of simulation-based and time-series models. In: 17th ITS World Congress on Transportation Research Board, Busan, South Korea, 25 October 2010–29 October 2010
6. Hu, T.Y., Ho, W.M.: Simulation-based travel time prediction model for traffic corridors, simulation-based travel time prediction model for traffic corridors. In: 2009 12th International IEEE Conference on Intelligent Transportation Systems, ITSC 2009 (2009)
7. Jiang, Z., Zhang, C., Xia, Y.: Travel time prediction model for urban road network based on multi-source data. In: 9th International Conference on Traffic and Transportation Studies, Shaoxing, Zhejiang Province, China (2014)
8. Kisgyörgy, L., Rilett, L.R.: Travel time prediction by advanced neural networks. Period. Politech. Ser. Civ. Eng. **46**(1), 15–32 (2002)
9. Gurmu, Z.K., Fan, W.D.: Artificial neural network travel time prediction model for buses using only GPS data. J. Public Transp. **17**(2), 3 (2014)

10. Innamaa, S.: Short-term prediction of travel time using neural networks on an interurban highway. Transportation **32**(6), 649–669 (2005)
11. Jindal, I., Qin, T., Chen, X., Nokleby, M.S., Ye, J.: A Unified Neural Network Approach for Estimating Travel Time and Distance for a Taxi Trip. arXiv (2017, published)
12. Zheng, F., van Zuylen, H.: Urban link travel time estimation based on sparse probe data. Transp. Res. Part C Emerg. Technol. **31**, 145–157 (2013)
13. Zhan, X., Hasan, S., Ukkusuri, S.V., Kamga, C.: Urban link travel time estimation using large-scale taxi data with partial information. Transp. Res. Part C: Emerg. Technol. **33**, 37–49 (2013)
14. Zhang, W.: Freeway Travel Time Estimation Based on Spot Speed Measurements, Ph.D. Dissertation. Virginia Polytechnic Institute and State University, Virginia (2006)
15. Mark, C.D., Sadek, A.W., Rizzo, D.: Predicting experienced travel time with neural networks: a PARAMICS simulation study. In: Proceedings of the 7th International IEEE Conference on Intelligent Transportation Systems (2004)
16. Bazan, M., Janiczek, T., Halawa, K., Dudek, R., Rudawski, Ł.: Design and development of a road traffic redirection system. Arch. Transp. Syst. Telemat. **10**(1), 3–8 (2017)
17. Bazan, M., Ciskowski, P., Dudek, R., Halawa, K., Janiczek, T., Kozaczewski, P., Rusiecki, A.: Multithreaded enhancements of the Dijkstra algorithm for route optimization in urban networks. Arch. Transp. Syst. Telemat. 9(2) (2016)
18. Małecki, K.: Graph cellular automata with relation-based neighbourhoods of cells for complex systems modelling: a case of traffic simulation. Symmetry **9**(12), 322 (2017)
19. Krajzewicz, D., Erdmann, J., Behrisch, M., Bieker, L.: Recent development and applications of SUMO - simulation of urban mobility. Int. J. Adv. Syst. Meas. **5**(3\4), 128–138 (2012)

Fairness in Temporal Verification
of Distributed Systems

Wiktor B. Daszczuk$^{(\boxtimes)}$ (iD)

Institute of Computer Science, Warsaw University of Technology,
Nowowiejska Str. 15/19, 00-665 Warsaw, Poland
wbd@ii.pw.edu.pl

Abstract. The verification of deadlock freeness and distributed termination in distributed systems by Dedan tool is described. In Dedan, the IMDS formalism for specification of distributed systems is used. A system is described in terms of servers' states, agents' messages, and actions. Universal temporal formulas for checking deadlock and termination features are elaborated. It makes possible to verify distributed systems without a knowledge of temporal logic by a user. For verification, external model checkers: Spin, NuSMV and Uppaal are used. The experience with these verifiers show problems with strong fairness (compassion), required for model checking of distributed systems. The problems outcome from busy form of waiting in some examples. The problem is solved by own temporal formulas evaluation algorithm, using breadth-first search and reverse reachability. This algorithm does not require to specify compassion requirements for individual events, as it supports strong fairness for all cases. Thus it is appropriate for verification of distributed systems.

Keywords: Distributed system specification · Model checking
Deadlock and termination detection · Strong fairness · Compassion

1 Introduction

Nowadays, many computer solutions are based on a distributed system, especially in Internet of Things (IoT) paradigm, where autonomous controllers negotiate their coordinated behavior. Dependability of such systems is highly increased if the coordination protocols are formally verified, for example using model checking technique.

In Institute of Computer Science, Warsaw University of Technology, a verification system Dedan using IMDS formalism (Integrated Model of Distributed Systems [1, 2]) is under development. The IMDS formalism is based on a specification of a system as a set of servers' states and agents' messages. The formalism is founded on a basic observation of distributed systems operation: a *server* gets a *message*, accepts it, executes a procedure serving the message, changes its *state* and issues a *next message* which continues a distributed computation. We treat the messages as carriers of agents: distributed computations, travelling between servers. A system configuration is defined as a set of *current states* of all servers and *pending messages* of all agents.

In the IMDS formalism, a modeled system may be decomposed into a set of resident *server processes* or a set of travelling *agent processes*. Therefore, a *communication*

© Springer International Publishing AG, part of Springer Nature 2019
W. Zamojski et al. (Eds.): DepCoS-RELCOMEX 2018, AISC 761, pp. 135–150, 2019.
https://doi.org/10.1007/978-3-319-91446-6_14

duality is exploited, since server processes communicate by messages, while agent processes communicate by means of servers' states. Several other verification features are included in Dedan, like observation of a global transition graph or simulation over this graph, conversion of a model to a Petri net, etc. IMDS is used primarily for describing a behavior of distributed systems, especially for finding deadlocks and checking of distributed termination. In Dedan verification environment [3], the model checking technique [4] is used for this purpose. The Dedan program is built in such a way that the specification of temporal formulas and temporal verification are hidden to a user for the purpose of automatic verification. The reason is that model checking techniques are seldom known by the engineers. Temporal verification under Dedan was initially planned to be performed in commonly used model checking programs: *Spin* [5] – for verification in LTL (linear time logic), *NuSMV* [6] – for verification in CTL (computation tree logic, but LTL is also implemented), *Uppaal* [7] – for verification in TCTL (Timed CTL, after equipping actions and message transfer with time constraints).

The choice of the verifiers was based on the two aspects: they are popular and free (at least for academic usage). For each of the three environments, a conversion of IMDS specification to the input form of a verifier was developed: *Promela* for Spin, *SMV* for NuSMV and *Timed Automata XML* for Uppaal. After verification, generated counterexamples are converted to a graphical form of sequence diagram of states and messages. For automatic deadlock and termination checking, the universal temporal formulas for communication deadlocks between servers, resource deadlocks between agents and agents' termination were defined [2]. Universal formulas do not depend on a structure of a verified system, the verification is performed automatically by the Dedan program. Unfortunately, in some examples the outlined verifiers reported false deadlocks (in terms on IMDS deadlock definition). The identified reason lays in troubles with fairness conditions that should be imposed on processes in distributed environment. Fairness fails to be held in all three mentioned verification environments in given cases.

To solve this problem, an own algorithm of temporal formulas evaluation is developed. This algorithm is not universal – it verifies only the formulas for communication deadlock detection, resource deadlock detection and distributed termination checking. The algorithm is an improvement of one described in [8]. In this paper an overview of IMDS is given in Sect. 2 (formal definition is in [1, 2]). The benchmark system for temporal verification is discussed in Sect. 3. Fairness constraints required for verification of distributed systems are described in Sect. 4. A conversion of the benchmark to Promela and the result of verification under Spin are shown in Sect. 5. The same for NuSMV is covered in Sect. 6 and for Uppaal in Sect. 7. The overview of an own evaluation algorithm is given in Sect. 8. Conclusions and further work are covered in Sect. 9.

2 Integrated Model of Distributed Systems (IMDS)

IMDS is defined formally in [1]. Here we use simplified version of IMDS, without dynamic process creation [2], to ensure the finiteness of a verified system. A system is compound on a finite set of servers' states P and a finite set of messages M. The *actions* are the relation $\Lambda \subseteq (M \times P) \times (M \times P)$, which denotes that a message meets a server's state and an action replaces the message and the state with a next message and a next state.

A system *configuration t* is a set of current servers' states and pending messages. An action transforms a configuration, containing its input message and input state, to a next configuration containing output state and output message. A system starts with its *initial configuration t_0*. A *Labeled Transition System* (LTS) is built over configurations as LTS nodes, with actions as LTS transitions. An initial configuration t_0 is a root of LTS. Interleaving semantics is assumed, i.e., one action is executed at a time. The LTS contains all possible behaviors of a distributed system and it is a Kripke structure for temporal verification.

A sequence of messages forms a distributed computation called *agent*. An agent travels between servers invoking their actions. Agent-terminating actions are defined in $(M \times P) \times (P)$ (a next message is missing in an action, only a next state is present).

For recognition of the processes in the system, the states are defined as pairs *(server,value)* and the messages are defined as triples *(agent,server,service)*, where a server is a target of the message and a service identifies an procedure offered by the server. In an action, a server component of an input message and of an input state must be equal. The pair *matches* if an action is defined for it. A *server process* groups all actions of a server while an *agent process* groups all actions of an agent. Thus, the system may be decomposed either to server processes (*server view*) or to agent processes (*agent view*). This is a *communication duality* in IMDS.

Note that every server performs an action autonomously (only the server's state and the messages pending at this server are considered). Also, the communication is asynchronous: a server process sends a message to some other server process or an agent process sets the server state regardless of the current situation in other servers. As a result, we may call the process autonomous and asynchronous.

The deadlocks and termination in a system are defined as follows:

- in the sever view – a communication deadlock of a server process – when there are messages pending at the server, but no matching pair of any message with the server's current state will occur: the state will not ever change;
- in the agent view – a resource deadlock of an agent process – when an agent's message is pending at a server but it will never match any current or future state of this server;
- in the agent view – distributed termination is the inevitability of reaching a terminating action by an agent.

The universal temporal formulas that locate the mentioned features are formally defined in [2]. They are not related to the features of a verified system, thus they need not be specified by a user. The temporal logic and temporal formulas evaluation are hidden in the verification program Dedan.

3 The Example

A basic *bounded buffer* example consists of a K-element buffer and two sets of users: *producers* and *consumers*. A proper specification does not lead to a deadlock. Among several versions of bounded buffer, in one of them users are modified to play roles of both producers and consumers (see the source code below). The action of producing/consuming is chosen in nondeterministic way. In such a system, a deadlock may occur if all users decide to read from an empty buffer, and there is no user to put an element into the buffer. Another deadlock occurs if all users decide to write to a full buffer, so it is no way to make a room in the buffer.

To avoid such situations, a construct of *butlers* from Dijkstra's solution of dining philosophers solution is used. Yet, there is a difference: in Dijkstra's solution the butlers inactively prevent all the philosophers from taking their left forks and from taking their right forks (the last philosopher simply waits). In our solution, the butlers cannot hold the users because a rejected user should switch to the opposite decision: to produce instead of to consume or reverse. Therefore, a reply *may_do* or *cannot_do* is issued to a user. Such a system should not fall into a deadlock if it is fair: after many attempts to choose putting and being rejected, a user must at last choose getting (and vice-versa). Unfortunately, all three external verifiers fail to give proper result – in the server view they report a deadlock in the sense of IMDS deadlock definition. This means that the verifiers report some states in the system in which messages are pending at servers but they will never cause any action to be executed. In the next sections we will show this malfunction in all the three verifiers.

The system in IMDS notation is listed below. The buffer capacity K is assumed 2, and the number of users N is 2. The servers of producers-consumers are $S[1]$ and $S[2]$, and the agents are $A[1]$ and $A[2]$, the butlers are *get_b* and *put_b*, the buffer is *buf*.

```
 1. system bounded_buffer ;
 2. #DEFINE N 2
 3. #DEFINE K 2
 4.
 5. server: buf (agents A[N]; servers S[N]),
 6. services {put, get},
 7. states {elem[0..K]} ,
 8. actions {
 9. <i=1..N><j=0..K-1>
10.    {A[i].buf.put, buf.elem[j]} -> {A[i].S[i].ok_put, buf.elem[j+1]},
11. <i=1..N><j=1..K>
12.    {A[i].buf.get, buf.elem[j]} -> {A[i].S[i].ok_get, buf.elem[j-1]},
13. };
14.
15. server: S (agents A; servers buf, put_b:butler, get_b:butler),
16. services {doSth, may_do, cannot_do, ok_put, ok_get, go_neutral},
17. states {neutral, prod, cons},
18. actions {
19.              {A.S.doSth, S.neutral} -> {A.put_b.wants_do, S.prod},
20.              {A.S.doSth, S.neutral} -> {A.get_b.wants_do, S.cons},
21.              {A.S.cannot_do, S.prod} -> {A.S.doSth, S.neutral},
22.              {A.S.cannot_do, S.cons} -> {A.S.doSth, S.neutral},
23.              {A.S.may_do, S.prod} -> {A.buf.get, S.prod},
24.              {A.S.may_do, S.cons} -> {A buf.put, S.cons},
25.              {A.S.ok_put, S.prod} -> {A.put_b.done, S.prod},
26.              {A.S.ok_get, S.cons} -> {A.get_b.done, S.cons},
27.              {A.S.go_neutral, S.prod} -> {A.S.doSth, S.neutral},
28.              {A.S.go_neutral, S.cons} -> {A.S.doSth, S.neutral},
29. };
30.
31. server: butler (agents A[N]; servers S[N]) ,
32. services {wants_do, done},
33. states {want[0..N-1]},
34. actions {
35. <i=1..N><j=0..N-2>
36.              {A[i].butler.wants_do, butler.want[j]} ->
37.                 {A[i].S[i].may_do, butler.want[j+1]},
38. <i=1..N>
39.              {A[i].butler.wants_do, butler.want[N-1]} ->
40.                 {A[i].S[i].cannot_do, butler.want[N-1]},
41. <i=1..N><j=1..N-1>
42.              {A[i].butler.done, butler.want[j]} ->
43.                 {A[i].S[i].go_neutral, butler.want[j-1]},
44. };
45.
46. servers buf, S[N], get_b: butler, put_b: butler ;
47. agents A[N] ;
48.
49. init -> {
50. <j=1..N>    S[j](A[j], buf, put_b, get_b).neutral,
51.             buf(A[1..N], S[1..N]).elem[0],
52.             get_b(A[1..N], S[1..N]).want[0],
53.             put_b(A[1..N], S[1..N]).want[0],
54. <j=1..N>    A[j].S[j].doSth,
55. }.
```

The notation of IMDS code above is intuitional (the example is in the server view, the agent view of the example is included in [9]). Server types (lines 5,15,31) are defined with states (l.7,17,33), services (l.6,16,32) and actions (l.8-12,18-28,34-43). Formal parameters (l.5,15,31) specify agents and other servers used. Then, server and agent variables are declared (l.46,47). At the end, servers and agents are initialized (l.49-54).

The system for K = 3 is illustrated informally as the automata types (*buf, S, butler*) in Fig. 1, because the specification in IMDS may seem exotic to an unskilled reader. The names of the automata implementing server types are inside rounded boxes. Initial states of the automata are appointed by bold, not anchored arrows. Some transitions are with repeaters of the form $<i = 1..N>$. Multiple transitions (but not those caused by repeaters) have double arrows. The transitions implement the actions of IMDS in such a way that input and output states of a transition are the states of the automaton, input message is an input symbol enabling a transition while output message is an output symbol of the transition (automata are Mealy-style). The output messages are directed to given servers, but they may not be accepted immediately – the messages may be pending in IMDS. Note the nondeterministic choice in transitions outgoing from the *neutral* state, it is implemented in lines 19 and 20 of the source code.

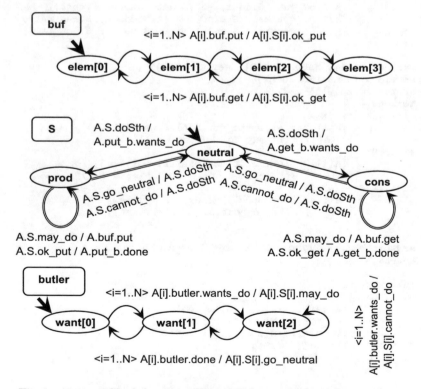

Fig. 1. Automata-like informal model of IMDS *bounded buffer* system, K = 3

In the Dedan program, temporal verification for deadlock detection and termination checking was planned using external model checkers: Spin for LTL, NuSMV for CTL and Uppaal for TCTL. Dedan converts IMDS specification to input formats of the three programs: Promela, SMV and Timed Automata XML. A counterexample generated in a case of a deadlock (or a witness of distributed termination) is presented by Dedan program in readable form of sequence diagram-like graph (Fig. 2).

4   Nondeterminism and Fairness

Three kinds of IMDS nondeterminism take place in the described model:

- Nondeterminism *between servers* models distribution. Servers make their decisions independently, therefore if two or more servers have their actions enabled (their current states and pending messages *match*); any one of them may execute its action first. For example, in initial configuration {(*buf,elem*[0]), (*S*[1],*neutral*), (*S*[2], *neutral*), (*get_b,want*[0]), (*put_b,want*[0]), (*A*[1],*S*[1],*doSth*), (*A*[2],*S*[2],*doSth*)} actions are enabled in both servers *S*[1], *S*[2]. Any of the two severs may be chosen to execute an action.
- Nondeterminism *in server* means that the current state of this server may match more than one pending message (of different agents). The server should chose the action to execute in nondeterministic way. In a *butler* server: it may have more than one request (*wants_do* message) and it should grant the agents the access to the buffer fairly (lines 35-37).
- Nondeterminism *in agent* is a situation in which two or more actions are possible in the current state of the server appointed by the current message of the agent. This models a nondeterministic choice in the agent's calculation. An example is a decision of every agent *A*[1] and *A*[2]: to *prod* or to *cons* (lines 19,20).

In all three cases of nondeterminism, a choice should be done fairly, which means that if the same nondeterministic choice is to be decided infinite number of times, every option of the choice should be also selected infinite number of times. A verifier should be fair in any one of the three types of nondeterministic choice: between servers, in server and in agent.

A system may be automatically converted to the agent view by the Dedan program, but this is beyond the present paper. The universal temporal formulas used for deadlock detection and termination checking are based on observation of pending messages and executed actions.

For the purpose of the verification, atomic Boolean formulas (labeling the nodes of the LTS) are defined: D_s, E_s, D_a, E_a, F_a. D_s – true in all configurations where at least one message is pending at the server s; E_s – true if an action in the server s is executed; D_a – true in all configurations where a message of the agent a is pending; E_a – true in all configurations where an action is executed, having a message of the agent a on input; F_a – true if a terminating action is executed in the context of the agent a.

The temporal formulas for deadlock and termination detection are, for LTL and CTL:

- communication deadlock in a server s: LTL – $\Diamond\Box\,(D_s \wedge \neg E_s)$, CTL – **EF AG** $(D_s \wedge \neg E_s)$; there are messages pending at the server s, but the current state does not match any of the pending messages, and no matching message may occur at the server in the future;
- resource deadlock in an agent a: LTL – $\Diamond\Box\,(D_a \wedge \neg E_a)$, CTL – **EF AG**$(D_a \wedge \neg E_a)$; the agent's message is pending at a server and neither the current state of this server matches the message nor such a state may occur in the future;
- termination of an agent a: LTL – $\Diamond\,(F_a)$, CTL – **AF**(F_a); there is no message pending with the agent a identifier.

Fairness in nondeterministic choices reflects the system distribution: every server acts autonomously, and anything happening in other servers should not influence the possibility of a server to make a progress (to execute an enabled action). The same concerns independence of agents which model distributed computations: an agent may influence other agents only by setting given values of servers' states. It cannot enable or inhibit actions of other agent in any other manner. Both these features reflect asynchrony of distributed system: an action in the server is enabled or denied on a basis on local features (D_s), which are its state and messages waiting locally, without any knowledge of the global state or synchronization with other servers. Similarly, each agent makes progress based on its current message and the current state of the server to which this message is addressed (D_a), regardless of the features such as the global state or current messages of other agents.

In model checking, several notions of fairness are introduced, primarily to model independence of parallel processes and fair scheduling in operating systems. The basic notions are justice (weak fairness) and compassion (strong fairness) [10]. A just scheduler executes a process infinitely often while a compassionate scheduler ensures that a process that is enabled infinitely often is executed infinitely often. Justice and compassion may be attributes of the whole verified system. Alternatively, to avoid a computation in which a given transition is forever ignored, individual transitions can be marked as just or compassionate [11]. Also, compassion applied to a given pair of events $q1$, $q2$ may be formulated as temporal formulas, which say that if a preceding event p occurs for multiple times, then both events $q1$ and $q2$ must occur in the future: LTL – $\Box\,(p \Rightarrow \Diamond q1)$; $\Box\,(p \Rightarrow \Diamond q2)$; CTL – **AG**$(p \Rightarrow$ **AF**$q1)$; **AG**$(p \Rightarrow$ **AF**$q2)$.

For temporal verification of the example system, strong fairness (compassion) is needed. If an agent periodically tries to perform *get*, and it is consequently rejected by the server *get_b*, is should at last switch to *put* (as both *get* and *put* are possible from the perspective of the *neutral* state of a user's home server).

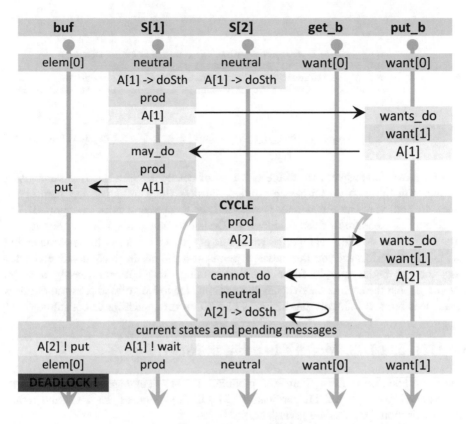

Fig. 2. Sequence diagram of the false deadlock identified by Spin in bounded buffer model

5 LTL Model Checking in Spin

For verification in linear temporal logic (LTL), a commonly used model checker Spin [5] was used. The LTL deadlock detection formulas are presented in Sect. 4.

The model is converted from IMDS notation to Promela [5], the input language of Spin. Every server is implemented as Promela process, and communication is performed by means of asynchronous channels. Pending messages are implemented as variables, for example in the *buf* server: *buf_A_1_service* and *buf_A_2_service*. The execution of an action is implemented as a sequence of statements guarded by a proper value of a pending message and a proper state (line 2). For catching the execution of an action, the variable *buf_action* is set to *true*, then to *false* (l.3,4). A next message is issued to the *A_1_S_1* channel (l.4). After the definition of *buf*, the LTL formula detecting a deadlock in the server is given (l.9).

```
1. do
2.              :: ( buf A_1_service==put) && ( buf_state==elem_0 ) ->
3.              buf action=1; buf_state=elem_1 ;
4.              buf A_1_service=none; buf_action=0; A_1_S_1 ! ok put
5.              //analogous statement implements every action of server buf
6.              ...
7. od
8. ...
9. ltl buf_dd {!( < >[](( buf_c_A_1 || buf_c_A_2 ) && buf_action==0))}
```

Spin gives the answer *true* to the question of deadlock. Spin has two versions of dealing with fairness: no fairness and weak fairness (they influence the verification time). In both versions a deadlock is reported, as compassion is not supported. The graphical representation of the deadlock in Dedan sequence diagram is presented in Fig. 2. The false deadlock is reported when one of the users sent *put* (the putting action has not been performed yet: the message pends at *buf*) and the second one endlessly asks the put butler (*put_b*) for a grant to putting. The butler endlessly answers *cannot_do*, but the user never switches to receiving due to lack of compassion. Fairness in agent is broken: infinitely producing is chosen, never switching to consuming.

6   CTL and LTL Model Checking in NuSMV

The same bounded buffer system was modeled in SMV language, the input form of NuSMV verifier [6], for CTL verification. The CTL version of deadlock and termination detection formulas are given in Sect. 4.

In SMV every server is modeled as a module (equivalent to a process in Spin). Actions are based on changing the values of the module's variables. The variables used for a server implementation are, on the example of the *buf* server: the variable *state*: {*elem_0,elem_1,elem_2*} (line 7) stores a state of the server, messages are pending in variables *A_1_service* and *A_2 service* (lines 3,4), the variable *phase* (line 8) defines a phase of an action execution (receiving a message – *rec*, choosing of an action – *A_1_send*, *A_2_send*, sending message and changing the server's state – *change*), and for every agent a variable for storing a next message (*A_1_to_S_1_mes*, *A_2_to_S_2_mes* – lines 5 and 6):

```
 1. MODULE buf (inistate, A_1_mes, A_2_mes)
 2. VAR
 3.             A_1_service: {none, put, get};
 4.             A_2_service: {none, put, get};
 5.             A_1_to_S_1_mes: {none, doSth, ok_put, ok_get};
 6.             A_2_to_S_2_mes: {none, doSth, ok_put, ok_get};
 7.             state: {elem_0 , elem_1 , elem_2};
 8.             phase: {rec, A_1_send, A_2_send, change};
 9. ASSIGN
10.             init (A_1_service):=none;
11.             next (A_1_service):= case
12.               phase=rec & A_1_mes=put: none, put;
13.               phase=rec & A_1_mes=get: none, get;
14.               phase=change & A_1_to_S_1_mes!=none: none;
15.               TRUE: A_1_service ;
16.             esac ;
17. ...
```

Communication channels are not included in SMV language and should be modeled as additional modules which store pending messages until they are accepted.

The NuSMV CTL formula for deadlock detection in the server *buf* follows:

```
NAME buf_dd:=AG((buf.A_1_service!=none | buf.A_2_service!=none) ->
              AF buf.phase!=rec )
```

Fairness between servers is natural in NuSMV, but compassion is not provided generally, it must be modeled as COMPASSION statements. These statements tell which options should be selected at least once if enabled infinitely frequently. The example shows nondeterminism in sever: when a message *put* from *A_1* is pending in *rec* phase infinitely often, then it must eventually cause a message to be sent in the context of *A_1* (*phase* is changed to *A_1_send*), similarly for *A_2*:

```
COMPASSION (phase=rec & A_1_mes=put & state=elem_0, phase = A_1_send)
COMPASSION (phase=rec & A_2_mes=put & state=elem_0, phase = A_2_send)
```

This looks appealing, but after invocation of NuSMV a message says that using COMPASSION statements do not guarantee to be held in CTL verification (indeed, the result is similar to that of Spin). Following the advice above, we tried to verify deadlock freeness in LTL, which is also possible in NuSMV. The example LTL formula (agent *A_1* deadlock checking) is:

```
LTLSPEC NAME A_1_dd :=
        G ((buf.A_1_service!=none | proc_1.A_service!=none) ->
        F (buf.A_1_service=none & proc_1.A_service=none))
```

The result is even worse, because using LTL formulas the verification seemed to last forever and we interrupted it. It comes from compassion requirement, as there are 15 COMPASSION statements which guarantee fair nondeterminism in the example.

7 CTL Model Checking in Uppaal

The last external model checker used is Uppaal [7]. The verifier is addressed to check CTL formulas with real-time constraints (TCTL). We do not use time constraints in the example, therefore the model and the verification concern "ordinary" CTL (without time constraints). The fragment of the code for the *buf* server in the input form of Uppaal (XML Timed Automata) is presented below. The model has the form of the automaton, with Uppaal locations being the *buf* states, and Uppaal transitions being the actions. The channels between servers are implemented as simple variables of corresponding types, holding the pending messages. States of servers are implemented as locations. Every action has the form a transition, defined between locations, with a guard being a message pending at a channel (l.4). An execution of a transition causes a "consumption" of an input message (l.6) and issuing an output message (l.7).

```
1. <transition >
2.         <source ref="id0" />
3.         <target ref="id1" />
4.         <label kind="guard">A_1_buf == buf_put</label>
5.         <label kind="assignment">
6.             A_1_buf = none ,
7.             buf_A_1_S_1 = S_ok_put
8.         </label>
9. </transition >
```

The CTL temporal formula in Uppaal version is (the operator $p \longrightarrow q$ stands for $\mathbf{AG}(p \Rightarrow \mathbf{AF}q)$): A_1_buf != none --> A_1_buf == none.

The documentation and literature on Uppaal does not refer to fairness in verification. Indeed, Uppaal reports a deadlock in verification and gives the counterexample similar to Spin (Fig. 2). However, the simulation of the model under Uppaal shows normal operation, in which the variable *A_1_buf* changes to *put* or *get* and back to *none*. During simulation, compassion is ensured manually by the designer.

8 Fair Verification Algorithm

The experience with commonly used verifiers: Spin, NuSMV and Uppaal shows that the verification with non-busy form of waiting gives proper results. Yet, the fairness problems with busy form of waiting should be defeated. There are algorithms for LTL and CTL model checking with strong fairness [10]. Model checking with strong fairness requires explicit specification of the events, which should be treated as compassionate.

In the paper, we present an own algorithm for CTL model checking, which supports compassion without specification of compassionate events. Only a limited set of temporal formulas is needed in Dedan for deadlock and termination detection, they are listed in Sect. 4. The algorithm is based on "Checking By Spheres" algorithm (CBS [8]), with some improvements. The basic CBS algorithm is a breadth-first search, constructing consecutive spheres of nodes of the reachability space, distant from the initial node for 1, 2, ..., etc. transitions. The algorithm finishes if a next sphere is empty (all nodes are visited) or if a node satisfying a given condition is found.

In [8] a temporal verifier of the whole CTL is covered, with some extensions for configuration quantification and component-aware operators. For verification in Dedan, the verifier was reduced to the two formulas necessary for IMDS deadlock and termination detection: **EF AG** φ (see Sect. 4 – resource deadlock and communication deadlock) and **AF** φ (see Sect. 4 – agent termination). First of all, it is checked if the configuration space is purely cyclic, or alternatively if nodes outside the cyclic part exist. In the purely cyclic space, the fairness condition causes all execution paths to visit any configuration in the space infinitely often. In the other case, fairness causes any path to settle down in a strongly connected subgraph without escape from it (we call it ending strongly connected subgraph – or ending subgraph in short). The verification of **EF AG** φ, illustrated in Fig. 3, is based on reverse reachability. The formula φ is $(D_s \wedge \neg E_s)$ – communication deadlock or $(D_a \wedge \neg E_a)$ – resource deadlock. The larger cloud is a cyclic part, the smaller clouds are ending subgraphs. The chessboard-filled circle is the initial configuration. First, all nodes fulfilling φ are identified (circles with dashed filling in Fig. 3a). Then, past of these nodes is found (dotted filling of the ending subgraphs and the cyclic part in Fig. 3b). If any ending subgraph is left (subgraph with grilled filling in Fig. 3b), **EF AG** φ is *true* – a deadlock occurs.

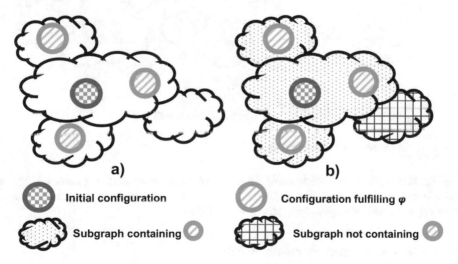

Fig. 3. Illustration of the verification of formula **EF AG** φ: (a) finding nodes fulfilling φ, (b) calculating the past of the nodes fulfilling φ

The formula **AF** φ (where $\varphi = (F_a)$ is a termination condition) is verified in three cases (Fig. 4). Again – the larger cloud is a cyclic part, the smaller ones are ending subgraphs. Initial configuration is the circle with chessboard filling.

1. The graph is cyclic or not (therefore the edge of ending subgraph in Fig. 4a is dotted), but there is a border surrounding the initial configuration, every node at the

border satisfying φ (Fig. 4a). Fairness forces any path to cross the border – the formula is *true*.

2. The graph is purely cyclic, and there is at least one configuration satisfying φ (Fig. 4b). The fairness condition forces any path to reach this configuration – the formula is *true*.

3. The graph contains ending subgraphs – fairness causes any path to fall into one of these subgraphs, and then to visit any configuration in this subgraph infinitely often. If there exists a configuration fulfilling φ in every ending subgraph (Fig. 4c) – the formula is *true*.

The result of verification of bounded buffer example, using Dedan built-in verifier, reports no deadlock, both in server processes verification and in agent processes.

Note that the proposed verification method does not affect a system specification, the difference lies in a temporal formula evaluation algorithm. The same system may be verified using internal Dedan verifier which supports compassion, or using one of external verifiers, without compassion.

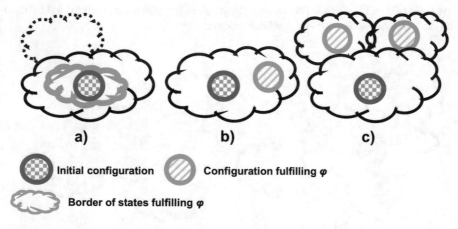

Fig. 4. Illustration of the verification of **AF** φ formula: case (a) the border of states fulfilling φ surrounding the initial state, case (b) cyclic space with a state fulfilling φ, case (c) ending strongly connected subgraphs, every with a state fulfilling φ

9   Conclusions and Further Work

The Dedan program supports an engineer in verification of distributed systems for deadlocks freeness, without any knowledge on temporal logics and model checking. If a communication deadlock occurs, a sequence diagram of messages is generated, leading from the initial configuration to the deadlock. In a case of resource deadlock, a sequence diagram of changes of servers' states and messages inside agents is generated. For a partial deadlock, the servers/agents taking part in the deadlock are shown.

Dedan was planned to specify distributed systems, leaving the verification to the external model checkers: Spin, NuSMV and Uppaal. This worked well in a case of processes waiting inactively for enabling conditions. Yet, in a case of active waiting, all three model checkers found false deadlocks. This was because of failures in fairness. The own compassion-aware verifier was built into Dedan, improved to cover a limited set of formulas presented in this paper. The algorithm does not require to specify the points that should be treated as compassionate: it supports strong fairness for every shape of a verified system. The single specification is used for both internal compassion-aware verification and external verification without compassion.

The Dedan program works well and is used in Operating Systems Laboratory in ICS, WUT. Students verify their synchronization projects against deadlocks and repair wrong solutions, checking the correctness with Dedan. The program is available at [3]. Several examples of IMDS system specification (including "two semaphores" and "bounded buffer") are at [9]. The next steps of the Dedan development are:

- Extension of modeling in IMDS to Timed Automata [12], in which time constraints will be added to actions and message passing. This will allow to check for deadlocks in real time-dependent systems. Uppaal program will be used for TCTL verification, and then the Dedan built-in verifier will be modified to cover time-dependent deadlocks.
- Partial order reductions and symbolic verification will be applied in built-in Dedan verifier for large models verification.

New features like assertion checking and invariant discovery will be added.

References

1. Chrobot, S., Daszczuk, W.B.: Communication dualism in distributed systems with petri net interpretation. Theor. Appl. Inform. **18**, 261–278 (2006). arXiv: 1710.07907
2. Daszczuk, W.B.: Communication and resource deadlock analysis using IMDS formalism and model checking. Comput. J. **60**, 729–750 (2017). https://doi.org/10.1093/comjnl/bxw099
3. Dedan. http://staff.ii.pw.edu.pl/dedan/files/DedAn.zip
4. Baier, C., Katoen, J.-P.: Principles of Model Checking. MIT Press, Cambridge (2008). ISBN 9780262026499
5. Ben-Ari, M.: Principles of the Spin Model Checker. Springer, London (2008). https://doi.org/10.1007/978-1-84628-770-1
6. Cimatti, A., Clarke, E., Giunchiglia, E., Giunchiglia, F., Pistore, M., Roveri, M., Sebastiani, R., Tacchella, A.: NuSMV 2: an OpenSource tool for symbolic model checking. In: Brinksma, E., Larsen, K.G. (eds.) CAV 2002. Computer Aided Verification. LNCS, vol. 2404, Copenhagen, Denmark, 27–31 July 2002, pp. 359–364. Springer, Heidelberg (2002). https://doi.org/10.1007/3-540-45657-0_29
7. Behrmann, G., David, A., Larsen, K.G., Pettersson, P., Yi, W.: Developing UPPAAL over 15 years. Softw. Pract. Exp. **41**, 133–142 (2011). https://doi.org/10.1002/spe.1006
8. Daszczuk, W.B.: Evaluation of temporal formulas based on "checking by spheres." In: Proceedings Euromicro Symposium on Digital Systems Design, Warsaw, Poland, 4–6 September 2001, pp. 158–164. IEEE (2001). https://doi.org/10.1109/dsd.2001.952267

9. Dedan Examples. http://staff.ii.pw.edu.pl/dedan/files/examples.zip
10. Pnueli, A., Sa'ar, Y.: All you need is compassion. In: 9th International Conference on Verification, Model Checking, and Abstract Interpretation, VMCAI 2008, San Francisco, CA, 7–9 January 2008, pp. 233–247. Springer, Heidelberg (2008). https://doi.org/10.1007/978-3-540-78163-9_21
11. Gómez, R., Bowman, H.: Discrete timed automata. Technical report 3-05-2005, Canterbury, UK (2005). https://kar.kent.ac.uk/14362/1/TR305.pdf
12. Bérard, B.: An introduction to timed automata. In: Control of Discrete-Event Systems, pp. 169–187. Springer, London (2013). https://doi.org/10.1007/978-1-4471-4276-8_9

Dual-Processor Tasks Scheduling Using Modified Muntz-Coffman Algorithm

Dariusz Dorota[(⊠)]

Cracow University of Science and Technology, Cracow, Poland
ddorota@pk.edu.pl

Abstract. This paper presents a new method of scheduling with using a modification of Muntz-Coffman algorithm. This novel method takes into account two-processor tasks. As a specification of system graph is used, which consider one-processor tasks too, and attribute of the divisibility/indivisibility of tasks. Scheduling of these tasks is preparing on NoC (Network on Chip) architecture which consists of two and three processors. In this paper algorithm of scheduling tasks using new approach to prioritize and prepare ranking of tasks on chosen architecture are presented.

Keywords: Dual-processor task · Scheduling task · Divisibility of the task

1 Introduction

In recent years, embedded systems equipped with multiple processors have become more and more popular [1, 2]. Examples are mobile phones, tablets or many IoT devices. Multiprocessor systems allow you to increase the speed of embedded systems, while applying other constraints such as the demand for power of the system or the used area [3]. One of the most important issues when creating multi-processor embedded systems is scheduling tasks. The scheduling process consists of such allocation of tasks to resources, that the overall system will be executed as soon as possible, while maintaining the other predefined restrictions [4]. The scheduling methods are used in many areas, due to the high complexity of this process, computer methods are used to implement the scheduling process.

In computer science, the most obvious application of scheduling are operating systems, where scheduling comes down to performing tasks in the right order using the available resources (processors). Over the years, during the development of computer science and all kinds of computer systems, the scheduling methods have evolved and are constantly evolving in order to perform actions on the delivered equipment (processors) in the most effective way possible. The problem of task scheduling will be presented below, starting with the problem formulation and ending with the ways of representing tasks.

© Springer International Publishing AG, part of Springer Nature 2019
W. Zamojski et al. (Eds.): DepCoS-RELCOMEX 2018, AISC 761, pp. 151–159, 2019.
https://doi.org/10.1007/978-3-319-91446-6_15

2   The Issue of Scheduling

Scheduling is a process that you can meet every day and in almost all areas. The importance of scheduling gains in IT where its most obvious use is the area of operating systems, especially real-time operating systems [4–6]. In the field of operating systems, scheduling comes down to performing tasks in the right order using the available resources (processors), while maintaining the other predefined restrictions [7]. Multiprocessor systems allow increasing the speed of embedded systems, while applying other constraints such as the demand for system power or used area [1, 3, 6, 7]. Over the years, during the development of computer science and all kinds of computer systems, the scheduling methods have evolved and are constantly evolving in order to perform actions on the delivered equipment in the most effective way possible.

Among the algorithms of scheduling, deterministic and stochastic algorithms can be distinguished [4, 5]. Both in the first and in the second case, it is possible to distinguish the scheduling on the single machine and on the parallel machines [4, 5]. The scheduling of tasks for single-machine systems consists of allocating tasks on the processor so that the time requirements imposed on individual tasks are met. The minimization of resources is performed in the process of co-design, where scheduling is one of the elements of creating the entire system. In this work, one of the possible elements of the co-design process is considered without considering the cost of minimization. Typically, the scheduling process considers minimizing many parameters at the same time, which is called multi-criterial optimization. In this work, priority is first of all one parameter, time of system performance, which is one of the most important from of the perspective of the implementation of the entire system. A single task was defined as T_i. The task set can be denotated as $T = \{T_1, T_2, \ldots, T_N\}$. A single processor is denotated as P_j, while a set of processors denotated as $P = \{P_1, P_2, \ldots, P_M\}$. Embedded processors may use universal processors or hardware modules that can generally be described as a computational element (PE).

The presented approach assumes that there is a database containing computing elements for implementation in SoC (System on Chip) systems: (a) tasks execution times, (b) modules surface. In the proposed method two basic types of PE can be distinguished: (a) Universal programmable processor (PP), (b) Specialized hardware modules (HC). Each of universal programmable processors can perform all the tasks that are compatible with them. PPi is determined by memory space occupied by the required task, while SPP_i determines the area of PPi processor. In contrast, PPi, HC (hardware module) can perform only one task, but there is a of hardware implementations of the same task.

Additional designations used in the work:

S_j – the earliest possible time to start task Tj
C_j – the earliest possible end time of the task Tj
t_i – time of task "i" completion
d_i – required end time for task "i".

The earliest possible end time of task j can be specified by the formula:

$$C_j = S_j + t_j$$

In order to be able to rank tasks in a single and multiprocessor system, the following condition must be met:

$$C_j \leq d_j$$

2.1 Task Representation in Embedded Systems

This work presents an algorithm for dependent tasks. There are several ways to represent the second type of tasks, including the task graph, SystemC, etc. [8, 9]. In this work, an acyclic task graph was used for the system specification. The input tasks are specified by the graphs proposed by the author. Acyclic directed graph G = {V, E} can be used to describe the proposed embedded system. Each node in the graph represents a task, while the edge describes the relationship between related tasks (time of transmission between tasks). Each of the edges in the graph is marked by a label d, where each of its indices defines the tasks that connect. In this paper author use Gantt chart to represent scheduled tasks in chosen architecture. The graphs of the tasks used for the system specification are presented below on Figs. 1 and 2.

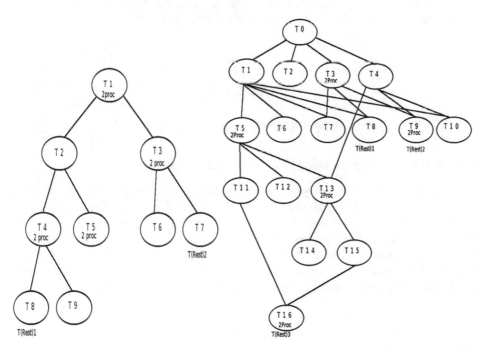

Fig. 1. Exemplary graph 1.   **Fig. 2.** Exemplary graph 2.

3   Dual-Processor Tasks

Currently presented tasks in the scheduling process are performed on a single machine and on parallel machines. A novelty in the approach presented in the article is the use of two-processor tasks, i.e. the need to perform a specific task/tasks on two processors at the same time.

The system's reliability is the motivation to introduce and use such tasks. Such reliability can be achieved by using redundancy for critical tasks and thus testing the correctness of the implementation of the selected task on the second processor. In the proposed approach, only selected tasks are two-processor tasks in order not to significantly increase the cost of the entire system. Reliability plays a particularly important role in systems requiring a high degree of operational correctness, especially in real-time systems, used in industries, where both temporal determinism and correct operation of the whole system play a significant role.

3.1   Presentation of Two-Processor Tasks in Graph

As already mentioned in the work, the task graph has been adopted as input data for the system under consideration. The only change in relation to the one-processor tasks is to add annotations in the task graph indicating that the task must be completed as a two-processor task. This was accomplished by adding the description "2 Proc" in the representation of a given task, which means the necessity of its simultaneous implementation on two machines. Assuming that T_i is a task with a number "i" in a task graph, then the task $T_{i/2}$ means that the task has a number and is two-processor. Using the graphic symbol, such a task is presented as follows (Fig. 3):

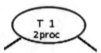

Fig. 3. Graphical representation of two-processor task

3.2   Scheduling Process Using Dual-Processor Tasks

The process of scheduling two-processor tasks follows the principles introduced in the Muntz-Coffman algorithm with the changes presented in this article. Task priorities are determined on the basis of the adopted algorithm with the reservation that two-processor tasks have the appropriate multiplier. The use of this approach allows for the promotion of dual-processor tasks from other tasks, of course, the order conditions and time constraints must be met here. As an example, you can specify task prioritization according to the modified Muntz-Coffman algorithm (Table 1):

Table 1. Exemplary prioritization for tasks shown in Fig. 1.

Number	Time	Level of task/time to ending							
T1	1	9	0	0	0	0	0	0	0
T2	1	7	7/1	0	0	0	0	0	0
T3	2	6	6/2	6/1	0	0	0	0	0
T4	2	6	6/2	6/2	6/1	0	0	0	0
T5	2	2	2/1	2/1	2/1	2/1	0	0	0
T6	2	2	2/2	2/2	2/1	2/1	0	0	0
T7	1	1	1/1	1/1	1/1	1/1	1/1	1/1	1/1
T8	2	2	2/2	2/2	2/2	2/2	2/2	2/1	0
T9	1	1	1/1	1/1	1/1	1/1	1/1	1/1	1/1

4 Muntz-Coffman Algorithm and Its Modifications

The best example of the need to develop the scheduling methods is the constantly growing market of embedded systems or IoT, where miniaturization and system performance become significant.

Such systems require the use of multiprocessor architecture in order to speed up system operation in a significant way using parallel operations. An important issue in systems of this type is scheduling tasks. Scheduling tasks for multiprocessor architectures has been considered in many works [10, 11]. An important issue is to consider the attribute of divisibility/indivisibility of tasks in the scheduling process [12]. An important aspect is the consideration of two-processor tasks in the scheduling process [13–15]. According to the author's knowledge, the first applications of dual-processor tasks were presented in [14]. The scheduling of two- and several-processor tasks has been presented in [14, 15]. In the system presented in [14], independent tasks are considered. This approach allows for any scheduling of tasks so as not to exceed the predetermined time for the entire system. The issue of divisibility of tasks dealt with in earlier author's works is extremely interesting. Dividend in scheduling tasks for two processors was presented in [17]. The algorithm proposed by Muntz-Coffman is the optimal algorithm for scheduling tasks divided on two machines [16]. It is an extension of previous works by its authors. The proposed algorithm performs task scheduling according to the following procedure:

1. Setting levels of tasks not performed.
2. Assignment of tasks with the highest levels to processors, assuming that if the number of tasks is greater than the processors - assign each task (B = m/a) of computing power, where is the number of tasks with the highest level. Performing task simulation until the task with the lowest level would be per formed earlier than the task with a higher level. Then go back to step 1. Repeat the procedure until all tasks have been completed.
3. Apply the Mc Noughton algorithm to obtain optimal ranking.

The following work presents an algorithm that performs scheduling of two-processor tasks based on the modified Muntz-Coffman algorithm. According to previous studies presented by the author, regarding the attribute of divisibility, also in this case the use of the divisibility attribute resulted in beneficial effects. According to the current state of knowledge, apart from [14, 15], the Muntz-Coffman algorithm and its modifications have not been considered for the task scheduling process, especially for dual-processor tasks. Based on the effects of the work presented above, as well as the author's previous experience with the issue of scheduling, two-processor tasks with the attribute of divisibility were proposed here. The procedure of the modified Muntz-Coffman algorithm for two-processor tasks is given below:

1. Load the system specification in the form of a task graph
2. Set levels for all non-completed tasks:
    1. The task level p_x is the sum of the task execution times t_i on the longest path for the selected task $P_x = A_x = \max\left(\sum_{i=x}^{n} t_i\right)$, where $A_x = \max(A_{j=1}^k)$, is the time of the longest path for task x, for the initial task x is selected $\max(A_{j=1}, \ldots, A_k)$
    2. If the selected T_i task at time t_i is a two-processor task:
        1. $p_x = t_x * 2$
3. Select the task with the highest level
    1. If there are two (or more) tasks at the same level:
        1. The priority is a dual-processor task, (the exception are dependencies which condition the execution of subsequent tasks, i.e. enabling the implementation of successors, e.g. the need to perform a task/tasks not to exceed the time limits T (**Rest X**, where x is the number of the next time limit)
        2. if there are two (or more) two-processor tasks at the same level, the first task should be higher in the hierarchy (or with the lower number of the task)
        3. Delete the selected task from the set $(P_a, P_b, \ldots, P_z)$
4. Simulate the task execution on the selected processor/processors per unit of time
5. If after the simulation the processor is available in the selected time unit:
    1. Select the next task (go to step 1)
6. If there are still unassigned tasks in the graph, go back to step 1
7. Exit the algorithm

5   Result and Analysis

This chapter presents the results obtained during experiments that reach task scheduling with the attention of two-processor tasks. Experiments were carried out on a selected number of graphs for systems composed of two and three processors. Exemplary ordering of tasks in the target system are shown in Figs. 4 and 5. The target system is implemented on a predefined multiprocessor architecture based on the NoC network. The target architecture is generated in the first step of creating the system, assuming that along with the specification given in the form of a graph of tasks, the number of computational elements on which the system is to be implemented is determined. The table below (Table 2) presents the calculated priorities of tasks in individual steps that directly affect the way of ordering tasks in the target system, and thus have an impact on the duration of the entire system (Fig. 6).

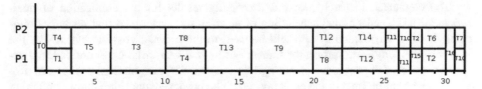

Fig. 4. Scheduling task in 2-processor architecture from specification in Fig. 1

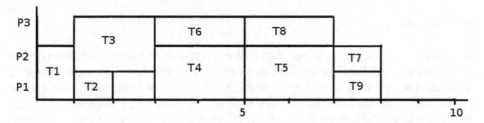

Fig. 5. Scheduling Task in 3-processor architecture from specification in Fig. 1

Table 2. Exemplary task scheduling times for selected 2 and 3 processor architectures

Graph number	Count of tasks	Two-proc tasks	Count of proc.	Time of system executions
G1	10	4	3	10
G2	17	6	2	32
G3	13	5	2	21
G4	13	5	2	24
G5	9	3	2	9
G6	9	4	2	17
G7	9	4	3	13
G8	7	3	2	8
G9	7	3	3	6

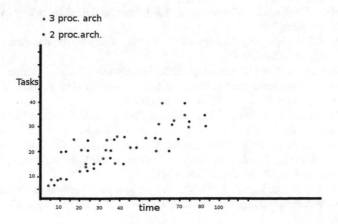

Fig. 6. Comparition of tasks execution for 2 and 3-processors architecture

The conducted research provides promising results for the application of dual-processor tasks and optimal scheduling of tasks in the system, so that resources allocated to the implemented system will be used to the greatest possible extent. For the sake of simplicity, inter processor transmissions between tasks were omitted. Subsequent research will also focus on taking into account the transmission, as they can also affect the execution time of the entire system. The proposed algorithm allows obtaining satisfactory results for both two-processor and three-processor architectures.

6 Summary

The presented work will consider the problem of scheduling tasks in multiprocessor embedded systems based on the NoC network. The system specification is presented using the task graph, generated by author. Like the previous ones, this author's work also considers the attribute of divisibility of tasks. The scheduling process presented in the article is a modification of the approach used in the Muntz-Coffman algorithm. A novelty in relation to scheduling works is the consideration of two-processor tasks here. This is to ensure the reliability of the system. The obtained results of the experiments confirm that the introduction of dual-processor tasks only affects the execution time of the entire system. However, in areas that require reliability and correct operation of the system, the benefits are disproportionate to the longer time of the task. Certainly, more research will be carried out in this interesting direction. Future work will concern the introduction of three-processor tasks as well as consideration of inter processor transmissions in the NoC network for which the ordering is performed.

References

1. Kopetz, H.: Real-time Systems: Design Principles for Distributed Embedded Applications. Springer, US (2011)
2. Lee, E.A., Seshia, S.A.: Introduction to Embedded Systems: A Cyber-Physical Systems Approach. MIT Press, Cambridge (2011)
3. Rajesh, K.G.: Co-synthesis of hardware and software for digital embedded systems, Ph.D. thesis, 10 December 1993
4. Smutnicki, C.: Algorytmy szeregowania zadań. Oficyna Wydawnicza Politechniki Wrocławskiej, Wrocław (2012)
5. Pinedo, M.L.: Scheduling Theory, Algorithms and Systems. Springer, Heidelberg (2008)
6. Tynski, A.: Zagadnienie szeregowania zadań z uwzględnieniem transportu. Modele, własności i algortmy, Ph.D. thesis, Wrocław (2008)
7. Ost, L., Mandelli, M., Almeida, G.M., Moller, L., Indrusiak, L.S., Sassatelli, G., Moraes, F.: Power-aware dynamic mapping heuristics for NoC-based MPSoCs using a unified model-based approach. ACM Trans. Embed. Comput. Syst. (TECS) 12(3), 75 (2013)
8. Khan, G.N., Iniewski, K. (eds.): Embedded and Networking Systems: Design, Software, and Implementation. CRC Press, Boca Raton (2013)
9. Eles, P., Peng, Z., Kuchcinski, K., Doboli, A.: System level hardware/software partitioning based on simulated annealing and tabu search. Des. Autom. Embed. Syst. 2(1), 5–32 (1997)

10. Błądek, I., Drozdowski, M., Fuinand, F., Schepler, X.: On contiguous and non-contiguous parallel task scheduling. J. Sched. **18**, 487–495 (2015)
11. Popieralski, W.: Algorytmy stadne w optymalizacji problem przepływowego szeregowania zadań, Ph.D. Thesis (2013)
12. Dorota, D.: Scheduling tasks in embedded systems based on NoC architecture using simulated annealing. In: Dependability and Complex Systems DepCoS-RELCOMEX, Brunów, Poland (2017)
13. Błażewicz, J., Drozdowski, M., Węglarz, J.: Szeregowanie zadań wieloprocesorowych w systemie jenorodnych duoprocesorów. Zeszyty Naukowe Politechniki Śląskiej (1988)
14. Błażewicz, J., Drabowski, M., Węglarz, J.: Scheduling multiprocessor tasks to minimize schedule length. IEEE Trans. Comput. **35**, 389–393 (1986)
15. Błażewicz, J., Drozdowski, M., Guinand, F., Trystam, D.: Scheduling a divisible task in a two-dimensional toroidal mesh. Discret. Appl. Math. **94**, 35–50 (1999)
16. Muntz, R.R., Coffman, E.G.: Optimal preemptive scheduling on two-processor systems. Trans. Comput. **18**, 1014–1020 (1969)
17. Błażewicz, J., Cellary, W., Słowiński, R., Węglarz, J.: Badania operacyjne dla informatyków. Warszawa, WNT (1983)

Comparison of Results Computations of Meta-Heuristic CAD Algorithms for Complex Systems with Higher Degree of Dependability

Mieczyslaw Drabowski[✉]

Cracow University of Technology, Warszawska 24, 31-155 Kraków, Poland
drabowski@pk.edu.pl

Abstract. The paper presents comparisons of results obtained by implemented meta-heuristic algorithms, which were discussed and presented within series proceedings on 9th, 10th, 11th and 12th International Conferences on Dependability and Complex Systems. These Conferences were held at Brunowe Palace in 2014–2017.

Keywords: Complex system · Scheduling · Partition · Allocation
Dependable · Optimization · Algorithms: genetic · Simulated annealing
Neuronal net · Ant Colony Optimization · Tabu search · CAD tools

1 Introduction

Since 2014, every year, I have had the honor to present the problems of high-level design of complex systems with a high degree of dependability. All these researches are presented in conference proceedings of the International Conferences on Dependability and Complex Systems DepCoS-RELCOMEX organized by Wrocław University of Science and Technology. Proceedings were published by Springer's "Advances in Intelligent Systems and Computing" series.

So, then: In 2014, a new system model was presented, which was based on a modified computer system model for tasks scheduling. In this model, above all, the following were added: searches, in addition to the best scheduling of tasks, also the best resource configurations taking into account, along with the time optimization criteria, also the cost and power consumption criterions. In this model were added also, so-called multiprocessor tasks, mainly two-processor, as a tasks of self-testing the system. These tasks enable checking the proper operation of the hardware and software, as well as damage analysis and the diagnosis of faults.

The objective of computer-aided design of complex computer systems (i.e. that contain sets of processors and additional resources, that compute great number of programs) is to find the optimal solution, in accordance with the requirements and constraints imposed by of the stated specification of the system. The following optimality criteria are usually taken into consideration: the speed of action, the cost of the implementation, the power of consumption and the degree of dependability.

W. Zamojski et al. (Eds.): DepCoS-RELCOMEX 2018, AISC 761, pp. 160–172, 2019.
https://doi.org/10.1007/978-3-319-91446-6_16

The identification and partitioning of resources between various implementation techniques is the basic matter of automatic aided design. Such partitioning is significant, because every complex system must be realized as result of hardware implementation for its certain tasks and of software implementation for other. Additionally scheduling problems are one of the most significant issues occurring in design of operating procedures responsible for controlling the allocations of tasks and resources in complex systems.

The new model and new construction methods software and hardware components are developed jointly coherently connected to each other, what the final solution reduced costs and increases the speed of action, differently than before. This model and methods for systems with a high degree of dependability were presented in [1].

The resources distribution is to specify, what hardware and software are in system and to allocate theirs to specific tasks, before designing execution details. Another important issue that occurs in designing complex systems is assuring their fault-free operation. Such designing concentrates on developing dependable and fault-tolerant architectures and constructing dedicated operating procedures for them. In this system an appropriate strategy of self-testing during regular exploitation must be provided. The general model and new concept of parallel to tasks scheduling and resources partition for complex systems with higher degree of dependability was presented in detail in [1]. We proposed in these papers the following schematic diagram of a coherent process of synthesis for systems of faults tolerant, Fig. 1, which in turn increases the degree of dependability of these systems.

The suggested method of this synthesis consists of the following steps:

1. specification of requirements and constraints for the system,
2. specification of all tasks,
3. assuming the initial values of resource set,
4. defining testing tasks and the structure of system, self testing strategy selection,
5. scheduling of tasks,
6. the evaluation the operating speed and system cost, multi-criteria optimization,
7. the change should be followed by a modification of the resource set, a new system partitioning into hardware and software parts and an update of structure and test tasks (go to step 5).

Consequently, we get a self-testing, self-diagnostic and potentially also the fault tolerant system. This model defined various problems of resource allocation, scheduling tasks also allocation of tasks and resources. The individual instances of these problems usually have computational complexity - in their decision-making form - Np-complete [2] and therefore the algorithms solving these problems are based on the heuristic approach [3].

The following meta-heuristic approaches were examined in 2015–2018, which sub-optimally solved the problems of partition resources, scheduling of tasks and allocation of tasks and resources in fault-tolerant structures. They were approaches: a hybrid approach (genetic with simulated annealing) with Boltzmann tournaments was proposed in the publication [4], modifications of neural network Tsang-Wang in [5], the approach based the Ant Colony Optimization in [6]. This years is presented the Tabu Search algorithm.

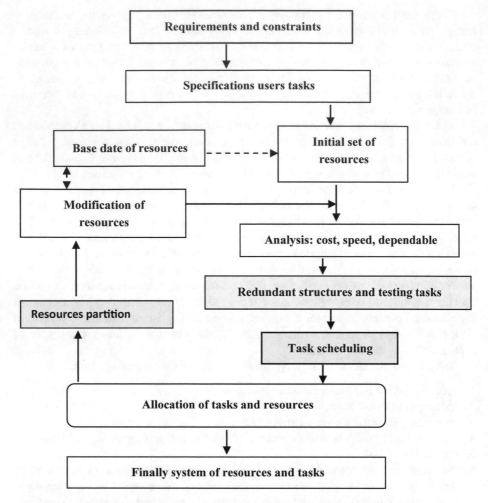

Fig. 1. The process par-synthesis of dependable complex system

The following questions arise: are all meta-heuristics needed for calculations in the design of these systems? Which ones are the best? Which ones give weaker solutions? Attempts to answer such questions are presented in this paper.

2   Comparison Computations

The tasks used during the tests (in which were not passed explicit) are generated as digraphs and they are received as:

- Graphs STG http://www.kasahara.elec.waseda.ac.jp/schedule/
- (rand0008, rand0038, rand0107, rand0174, rand0017, rand0066, rand0106, rand0174, rand0020, rand0095, rand-136) [7]
- Graphs TGFF http://ziyang.eecs.umich.edu/~dickrp/tgff/ [8].

In the examples the following pool of resources [Table 1] available for synthesis is used:

Table 1. Available resources

No.	Type of processor	ID	Speed	Cost	Power consumption
1	General	P1	1	1.00	0.01
2	General	P2	2	1.60	0.02
3	General	P5	5	2.20	0.05
4	General	P10	10	3.70	0.1
5	Dedicated	ASIC1	1	0.50	0.01
6	Dedicated	ASIC2	2	0.75	0.02
7	Dedicated	ASIC3	3	1.00	0.03
8	Dedicated	ASIC4	4	1.25	0.04
9	Dedicated	ASIC5	5	1.50	0.05
10	Dedicated	ASIC10	10	2.75	0.11
11	Memory modules	PAO	1	0.2	0.001

2.1 Dependent Tasks Without Cost of Memory

Minimization of Cost

The best solutions (cheapest structures) are generated by the genetic and ACO algorithms – Fig. 2 Second graph – Fig. 3 – shows that cheaper structures generated by these algorithms can be even quicker.

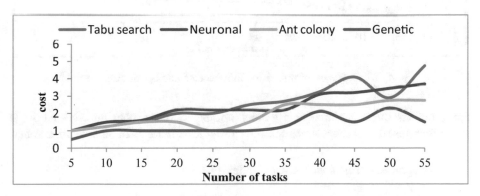

Fig. 2. Minimization of cost – the influence of number of tasks on cost

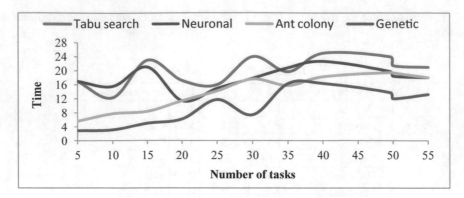

Fig. 3. Minimization of cost – the influence of number of tasks on time

Minimization of Time

The growth of number of tasks causes quite a lot of the differentiation of results for cost. The best solutions generated the genetic algorithm and algorithm of the ants' colony – Fig. 4 Second graph (Fig. 5) presented, that the cheaper structures are additionally much quicker, is them a significantly the shorter time of executing the tasks.

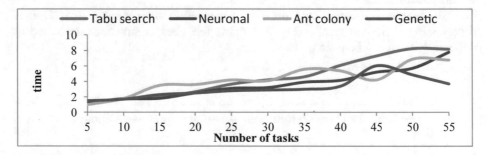

Fig. 4. Minimization of time – the influence of number of tasks on time

For problems under research at a greater number of tasks both genetic and ACO algorithms generate cheaper structures and shorter scheduling than other heuristic ones.

2.2 Independent Tasks with Cost of Memory

Cost Minimization

Charts show that system structures generated by genetic and ant colony algorithms are cheaper and quicker. With an increase in the number of tasks the algorithms use even higher number of resources (Fig. 6) for generating structures what drives the costs up, but simultaneously task performance time diminishes (Figs. 6 and 7).

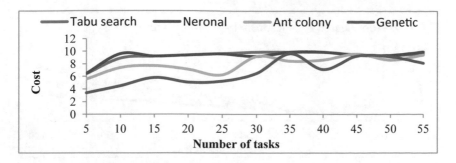

Fig. 5. Minimization of time – the influence of number of tasks on cost

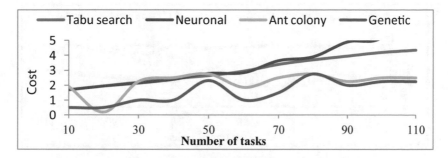

Fig. 6. Minimization of cost – the influence of number of tasks on cost

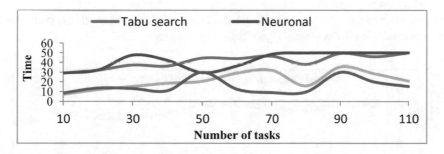

Fig. 7. Minimization of cost – the influence of number of tasks on time

Minimization of Time

ACO and genetic algorithms generate the cheapest solutions, although for speed criterion the advantage, especially of ACO algorithm, is not so explicit – Figs. 8 and 9.

2.3   The Influence of Precedence Constraints on Computation

In this example were compared algorithms: Tabu search, genetic and colony ants. In tests were 45 dependent tasks and the number of edge were changes.

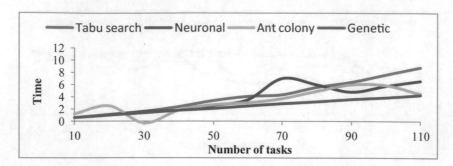

Fig. 8. Minimization of time – the influence of number of tasks on time

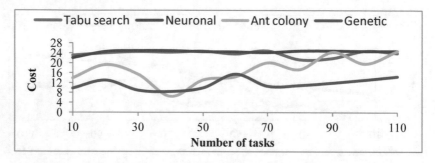

Fig. 9. Minimization of time – the influence of number of tasks on cost

Minimization of Cost

One can observe the advantage of genetic algorithm over other algorithms. Poor results are generated with Tabu search algorithm - Figs. 10 and 11.

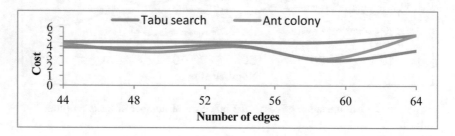

Fig. 10. Minimization of cost – the influence of number of edge on cost

Minimization of Time

Also in this case genetic algorithm obtains better results – Figs. 12 and 13. One should indicate a possible and significant impact of anomalies in task scheduling on the quality of the obtained results of the synthesis [9]. Task scheduling is dominant in a process of coherent synthesis and such anomalies may lead to the fact that the differences of computation results between coherent and non-coherent synthesis may sometimes be

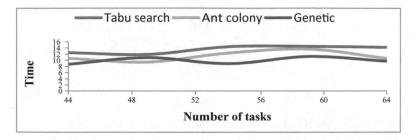

Fig. 11. Minimization of cost – the influence of number of tasks on time

smaller and sometimes may be significant. They appear then certain inequality's in behavior of different algorithms of coherent synthesis maybe, for some starting data. The following examples [10] show a possibility of appearing such anomalies.

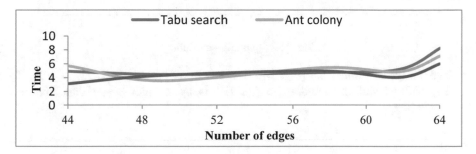

Fig. 12. Minimization of time – the influence of number of edges on time

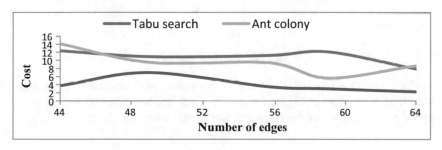

Fig. 13. Minimization of time – the influence of number of edges on cost

Example 1

Take an example of this digraph of tasks [Fig. 14]:

- Optimum scheduling [Fig. 15] for n = 9 tasks of times $t_1 = \{t_1 = 3, t_2 = 2, t_3 = 2, t_4 = 2, t_5 = 4, t_6 = 4, t_7 = 4, t_8 = 4, t_9 = 9\}$ i m = 3 of identical processors at the following priority list $L' = \{T_1, T_2, T_3, T_4, T_5, T_6, T_7, T_8, T_9\}$ is of length equaled 12 units and is as follows:

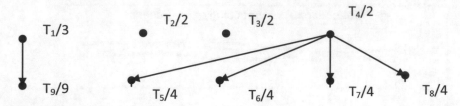

Fig. 14. Digraph of tasks. Task time vector: [3, 2, 2, 2, 4, 4, 4, 4, 9].

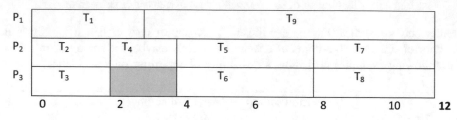

Fig. 15. Optimum scheduling

- The change of priority list [Fig. 16]: new list L = {T_1, T_2, T_4, T_5, T_6, T_3, T_9, T_7, T_8} and scheduling according to this list is by 2 units longer than optimum scheduling:

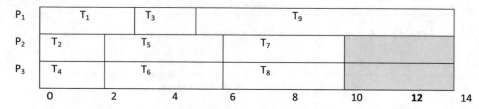

Fig. 16. The first anomaly

- An increase in the number of processors, new number of processors m' = 4 and the length of scheduling is bigger by 3 units than optimum (bigger cost and smaller speed of the system for realization of the same system functions) [Fig. 17]:

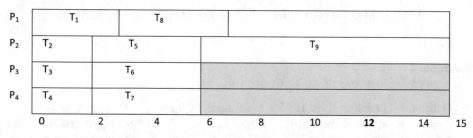

Fig. 17. The second anomaly

- Diminishing of performance time for all the tasks $t_i' = t_i - 1$, and the scheduling is longer than optimum scheduling by 1 unit [Fig. 18]:

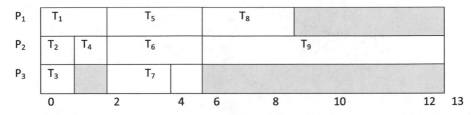

Fig. 18. The third anomaly

- Eliminating of some subsequent constraints – new constraints (smaller of their numbers), and scheduling is longer by 4 units than optimum scheduling [Fig. 19]:

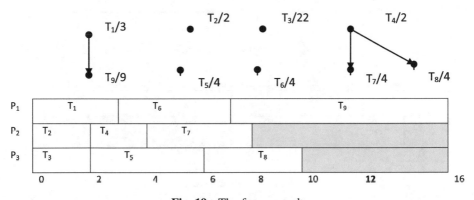

Fig. 19. The four anomaly

Example 2

Take an example of this digraph of tasks – Fig. 20:

- Optimum scheduling for n = 7 tasks of times $t_1 = \{t_1 = 4, t_2 = 2, t_3 = 2, t_4 = 5, t_5 = 5, t_6 = 10, t_7 = 10\}$ i m = 2 of identical processors at the following priority list $L = \{T_1, T_2, T_3, T_4, T_5, T_6, T_7\}$ is of length equaled 12 units ($C_{max} = 19$) [Fig. 21] and is as follows:
- Diminishing of performance time for all the tasks $ti' = t_i - 1$ and the scheduling is longer than optimum scheduling (independently from choice list!) – Fig. 22:

For different problem instances, particular algorithms may achieve different successes; others may achieve worse results at different numbers of tasks. The best option is to obtain results of different algorithms and of different runs.

Results obtained for sets of independent tasks, similarly as in the case of dependent tasks, did not provide a definite answer which of presented methods of optimization

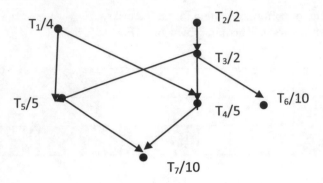

Fig. 20. Digraphs of tasks. Task time vector: [4, 2, 2, 5, 5, 10, 10].

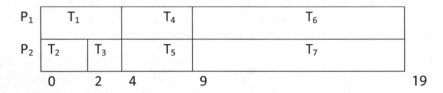

Fig. 21. Optimum scheduling

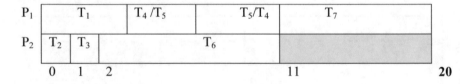

Fig. 22. The anomaly

always gives better results. When analyzing the results of algorithms it is clear that genetic algorithm solutions are the best in terms of time, but worse in terms of cost. The answer to the question, which of the our algorithms gives better solutions, is dependent on the instance of investigated problem.

3   Conclusions

Studies on the synthesis of complex computer systems have practical motivations and are still valid – e.g. [11–13]. This paper is about the problems of design of complex systems with high degree of dependability. Such a design is carried out on a high level of abstraction in which system specification consists of a set of tasks, which should be implemented by a pool of resources and these are listed in the database (container or a catalogues) and are available (exist or them can be fast created). Resources possess certain constraints and characteristics, including speed, power, cost and dependability parameters. The presented solutions are based on meta-heuristic methods: genetic and simulated annealing, neural, Tabu search and ant colony. The algorithms have been implemented and made a number of computational experiments whose selected results are presented in this article.

The implementation of different (four) algorithms made it possible to compare the results obtained by these algorithms and for diversified problem instances. Selected results are presented in this article. For different problem instances, particular algorithms may achieve different successes; others may achieve worse or better results at different numbers of tasks or for other optimization criterions. The recommended option is to obtain of results by different algorithms for different their actions and the compare these solutions. We can then set the ranking of solutions and choose the best. We receive then the best possible design of a complex system with an increased level of dependability.

References

1. Drabowski, M., Wantuch, E., Deterministic schedule of task in multiprocessor computer systems with higher degree of dependability. In: Zamojski, W., Mazurkiewicz, J., Sugier, J., Walkowiak, T., Kacprzyk, J. (eds.) Proceedings of the Ninth International Conference on DepCos-RELCOMEX, pp. 165–175. Springer (2014)
2. Garey, M., Johnson, D.: Computers and Intractability: A Guide to the Theory of NP-Completeness. Freeman, San Francisco (1979)
3. Dorigo, M., Di Caro, G., Gambardella, L.M.: An algorithms for discrete optimization. Artif. Life 5(2), 137–172 (1999)
4. Drabowski, M.: Boltzmann tournaments in evolutionary algorithm for CAD of complex systems with higher degree of dependability. In: Zamojski, W., Mazurkiewicz, J., Sugier, J., Walkowiak, T., Kacprzyk, J. (eds.) Proceedings of the Tenth International Conference on DepCos-RELCOMEX, pp. 141–152. Springer (2015)
5. Drabowski, M.: Modification of neural network Tsang-Wang in algorithm for CAD of systems with higher degree of dependability. In: Zamojski, W., Mazurkiewicz, J., Sugier, J., Walkowiak, T., Kacprzyk, J.: 11th International Conference on DepCos-RELCOMEX, pp. 121–133. Springer (2016)
6. Drabowski, M.: Adaptation of Ant Colony Algorithm for CAD of complex systems with higher degree of dependability. In: Zamojski, W., Mazurkiewicz, J., Sugier, J., Walkowiak, T., Kacprzyk, J. (eds.) 12th International Conference on DepCos-RELCOMEX, pp. 141–150. Springer (2017)
7. http://www.kasahara.elec.waseda.ac.jp/schedule/
8. http://ziyang.eecs.umich.edu/~dickrp/tgff/
9. Błażewicz, J., Drabowski, M., Węglarz, J.: Scheduling multiprocessor tasks to minimize schedule length. IEEE Trans. Comput. 35(5), 389–393 (1986)
10. Błażewicz, J., Ecker, K., Pesch, E., Schmidt, G., Węglarz, J.: Handbook on Scheduling. Springer, Berlin (2007)

11. Kapadia, N., Pasricha, S.: A system-level cosynthesis framework for power delivery and on-chip data networks. IEEE Trans. Very Large Scale Integr. Syst. **24**(1) (2016)
12. Ou, J., Prasanna, V.K.: Energy Efficient Hardware-Software Co-Synthesis Using Reconfigurable Hardware (2017). ISBN 9781138112803
13. Bezati, E., Casale Brunet, S., Mattavell, M., Janneck, J,W.: High-level synthesis of dynamic dataflow programs on heterogeneous MPSoC platforms. In: International Conference on Embedded Computer Systems: Architectures, Modeling and Simulation (SAMOS) (2016)

Short-Term and Long-Term Memory in Tabu Search Algorithm for CAD of Complex Systems with Higher Degree of Dependability

Mieczyslaw Drabowski[✉]

Cracow University of Technology, Warszawska 24, 31-155, Kraków, Poland
drabowski@pk.edu.pl

Abstract. The paper introduces a proposal of a new algorithm for Computer Aided Design (CAD) of complex system with higher degree of dependability. This algorithm optimizes tasks scheduling, resources partition and the allocation of tasks and resources. Complexity of computation for decision versions of these optimization problems is NP-complete, so solved efficiently e.g. by meta-heuristic algorithms. Presented in this paper algorithm is based on Tabu Search method. It may have a practical application in developing tools for rapid prototyping of such systems. The Tabu Search method uses memory structures. This structures describe the found solutions or provide sets of rules for consideration only new solutions.

Keywords: Complex system · Scheduling · Partition · Allocation
Dependable · Optimization · Movement · Neighborhood · Short-term memory
Long-term memory · Aspiration · CAD tools

1 Introduction

This article is a continuation of the research considered by the author and published in the conference proceedings of previous editions of this conference.

The starting point for these researches was the model of a complex systems with higher degree of dependability. It was presented in [1], like problems generated by this model and its computational complexity.

Due to computational complexity to NP-complete problems, they were solved by meta heuristic approaches, which the author studied and presented at the further conferences: i.e. a hybrid approach (genetic with simulated annealing) with Boltzmann tournaments was proposed in the publication [2], modifications of neural network Tsang-Wang in [3], the approach based the Ant Colony Optimization in [4]. The current paper presents the Tabu Search algorithm.

All of these approaches give different results and there is no best approach. The quality of the results in different approaches depends on the specific input and the instance of the problem.

In these approaches, we applied the following method for the synthesis of complex systems with the increase dependability [1]. The proposed method of synthesis consists of the following steps:

© Springer International Publishing AG, part of Springer Nature 2019
W. Zamojski et al. (Eds.): DepCoS-RELCOMEX 2018, AISC 761, pp. 173–183, 2019.
https://doi.org/10.1007/978-3-319-91446-6_17

A. specification of requirements and constraints for the system,
B. specification of all tasks,
C. assuming the initial values of resource set,
D. defining testing tasks and the structure of system, self-testing strategy selection,
E. scheduling of tasks,
F. the evaluation the operating speed and system cost, multi-criteria optimization,
G. the change should be followed by a modification of the resource set, a new system partitioning into hardware and software parts and an update of structure and test tasks (go to step E).

The objective of these papers is to present of meta-heuristic approaches to the solution problems of dependable complex systems design, i.e. a simultaneously solution to tasks scheduling and resource assignment problems. The set of tasks consists with tasks of self testing and tasks users. A collection of resources creates the redundant structure with high dependability. The development of these algorithms allowed us to compare the results achieved by these various meta-heuristics. We plan to present them in the next, prepared article.

2   Adaptation of Tabu Search Algorithm to Solve the Problems of Synthesis

2.1   Example Actions of Algorithm for Scheduling Problem

The general idea of Tabu Search method is presented in [5, 6] and it is as follows:

```
Algorithm TS:                                              {
generate a single initial solution
set tabu list length, t
until termination criteria is met                      {
   for neighborhood i = 1 to N              {
      make a move to a neighboring solution
      evaluate
      record in ranked order, j = 1 to N
      ++ i                                               }
   for j = 1 to N                        {
      if solution j is not tabu, break to NEXT
      or  if solution j is better than   best so far, break to NEXT
      ++ j                              }
   NEXT: add solution j to tabu list
   if tabu list length > t, remove oldest solution from   it
   if solution j is better than  best so far
   best  so far= solution j                            }
   return  best so far                                 }
```

This method uses computational memory in controlling current process of searching for a solution and additionally defines such components as: movement, neighborhood, aspiration. Let's consider a general optimization problem which is to extreme (e.g. of minimization) the purpose function (criteria function): *ext {f(z): z∈Z}*, where *Z* is the set of all acceptable solutions. Movement *v* is a function transferring the elements of set $Z(v)$ in Z, where $Z(v) \subset Z$ is a set of the acceptable solutions in relation to which it is possible to use movement, but the movement does not cause the new solution to get out of the set of acceptable solutions. Usually it comes to equality $Z(v) = Z$, which means that movement is a function transferring one solution into another e.g. in permutation, it transfers an element from position i to position j. It is remarkable that the obtained, as a result of movement, solution is not necessarily better than the solution in relation to which the movement was used. The functional neighborhood $O(z)$ of a solution is a set of all the movements which, used in relation to a solution z, will not cause taking newly obtained solutions out of the set of acceptable solutions i.e. $O(z) = \{v \in V: \in Z (v)\}$, where V is a set of all the possible movements.

The neighborhood $S(z)$ of a solution is a set of solutions into which it is possible to transfer solution z with help of movements which belong to functional neighborhood $O(z)$ of solution z, i.e. $S(z) = \{v(z): v \in O(z)\}$.

Short-term memory (tabu lists) contains attributes of forbidden movements and/or solutions. This memory is realized on principle of FIFO queue of a definite length. During subsequent algorithm iterations, the attributes of the performed movement and/or the obtained solution are introduced on the list end; if the number of elements on the list reaches the established limit, the top element introduced earlier is removed from the list. The main goal of using short-term memory is to prevent algorithm from falling into cycles and searching the same solutions again and again.

Long-term memory contains the algorithm states coming from the definite moments of algorithm work. The algorithm state consists of base zb solutions, short-term memory and the information about the movements performed by the algorithm from the solution. This memory is realized in shape of a stack of the definite length. When the number of states exceeds the established value, the last state located on it is removed. The state is saved by memory if a better solution than the one on top of the stack has been found (as the first located in the stack state is the algorithm state with an initiating solution). When the number of returns to the states exceeds an established value, the state is removed from the stack. Aspiration function is a function which allows removing the state from the movement or a forbidden solution in justified cases. For instance, the value of criteria function of the best solution found so far can be an aspiration function. In this case, a forbidden solution zz, fulfilling the inequality $f(zz) < f(z)$ becomes non-forbidden. The other criteria of removing tabu state from a solution or movement are: the movement or solution with tabu state is the best in the neighborhood, the solution differs significantly from the current one (the variety increases), the solution is similar to the current one (intensity of searching increases), a powerful impact on the change of possible movement structure. Algorithm tabu search starts computation from a certain starting base solution $zb \in Z$ [7]. It performs searching the base neighborhood $S(z)$ with help of movements $v \in O(z)$, in order to find the best solution z^* in this neighborhood. Solution z^* becomes base solution zb for the next iteration of algorithm. Such procedure is repeated until it reaches the assumed condition to finish algorithm performance. To avoid falling into cycles (i.e. avoid searching the same sequences of solutions again and again), after movement performance, its attributes are recorded and/or generated solutions are introduced onto tabu lists. While searching neighborhood $S(z)$, we do not consider solution z^* to be the base one for the next iteration if on tabu lists there are attributes of this solution or the movement leading to it. If algorithm performs a certain number of iterations without improvement of criteria function value from the point of recreating the state, its state is recreated from long-term memory and computation is continued. The information about previous movements from this state is used, so that they should not be repeated.

Example:

Below, we present a simple example of the use of Tabu Search method as a solution to the classic problem of optimizing preemptable and independent task scheduling, to minimize the task schedule length (C_{max}) on identical processors. It is assumed that two identical processors and five independent tasks are available [8]. The following times (in conventional units) of individual tasks are accepted: $t_1= 1, t_2= 1, t_3= 1, t_4= 2, t_5= 1$. The movement is described by 6 parameters: $\lambda^1, \lambda^2, \lambda^3, \lambda^4, \lambda^5, \lambda^6$, where λ^1 – stands for the number of processor, on which the chosen task to transfer is being carried out, λ^2 – the number of position in scheduling, on which the chosen task to transfer is located, λ^3 – the number of task to transfer during movement, λ^4 – the number of processor, onto which the tasks will be transferred, λ^5 – the number of position in scheduling, onto which the task will be transferred, λ^6 – the number of task, which is located on λ^5 position and λ^4 processor. On the tabu list there are movement attributes: name tag of task to transfer in the movement (λ^3) and the position, onto which this task will be transferred (λ^5). The example assumes that the length of tabu list equals five and the following initial solution is accepted. The main purpose of Tabu Search method is to use experience from previous searching for an optimal solution in currently conducted searching. Four general requirements for computation memory should be taken into consideration during implementation: (1) Quality – capability of differentiating between solutions and movements. Prognosticating that their acceptance will lead to the improvement not only on the basis of criteria function value of a specific solution but also on the previously saved run of algorithm. (2) Induction – considered as the impact of choices taken during the searching process on the quality and on the structure as well; to save information about the impact of the taken choices while considering an exact solution is an important aspect of learning and being able to make decisions in the future. (3) Frequency –understood as saving numbers of returns to a certain state of algorithm and effects which they bring on the improvement for the goal of a purpose function. (4) Innovation – understood as saving and fixing the state, "forbidden" for the last performed movements and visited solutions; the movements have been executed recently and repeating their performance does not lead to a considerable change. The realization of algorithm was shown in Fig. 1.

An important element which has an impact on the effectiveness of an algorithm is to come close to the balanced state between an intensive approach to a small set of selected, best prognosticated solutions or movements and the variety of search, continuous visiting new sets of solutions where an optimal solution may be found. During algorithm work one can see the following behaviors: (1) Falling into oscillations – going among the experiences of extreme criteria function values. (2) Walking step by step and tunneling – passing slowly between solutions which are close to each other and avoiding movements which lead to worse solutions. (3) Returns to the best solutions – repeated returns to the saved in long-term memory best solutions found so far. (4) Restart – another beginning of algorithm performance with a new starting solution.

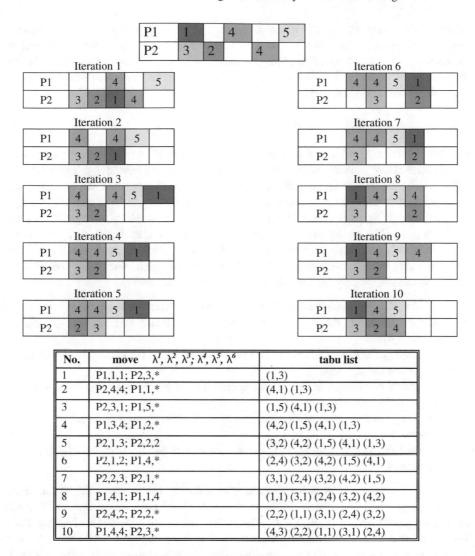

Fig. 1. The example of algorithm realization

2.2 Partition of Resources

For resources partition problem the initial solution is generated randomly or by means of other heuristic algorithms. It is the first correct solution and it is not optimal. Further optimization is conducted by Tabu Search algorithm. For the solution of this problem, it is assumed that a movement is the swapping of one of resources for the other with simultaneous transfer of all performed tasks, which use this resource. Movement is realized as follows: (1) From resources so far allocated to tasks, the resource to be swapped is drawn (let it be PO resource). (2) From the set of available resources, a resource is drawn (let it be PN resource). (3) Tasks allocated to PO resource will be transferred and allocated to PN resource. In case when empty processor (0) is drawn from the resources set, all tasks, which are being carried out on the processor chosen at the first step, will be removed to other processors. Movement parameters are stored in the short-term memory, i.e. on the tabu list. For this method the optimization criteria are the cost and speed of system.

2.3   Task Scheduling

For task scheduling problem, the initial solution is generated randomly or by means of other heuristic algorithms and further optimization is conducted by Tabu Search algorithm. The obtained initial solutions meet all the requirements which are specified for input data. In the algorithm which solves optimization problems of task scheduling, the movement is defined as task transfer (or an exact part – quantum – of task) carried out from one of the processors onto another processor and onto a different position in the schedule. In case when we want to transfer a selected task (or its part) onto the position where there is another task, the tasks will be swapped. Such a movement was described unambiguously by the following six attributes:

1. Task identifier to transfer tasks during movement – variable *task_from.*
2. Processor identifier on which a selected task is being carried out – variable *processor_from.*
3. The number of task position from pt. 1 in the schedule on the processor from pt. 2 – variable *quantum_from*.
4. The number of position in the schedule onto which we want to transfer a selected task – variable *quantum_to*.
5. Processor identifier onto which we want to transfer a selected task – variable *processor_to*.
6. Task identifier which is located in position from pt. 4 and the processor from pt. 5 – variable *task_to.*

For non-preemptive task scheduling problems during movement – the whole tasks are transferred and for *preemptable* tasks – their selected parts are transferred. *Quantum_to* and *quantum_from* variables define the starting position of task execution in the schedule. It is necessary to specify the above mentioned attributes for the movement performance in an exact schedule, but Tabu Search method alone does not determine which factors should be taken into consideration in defining movements, what features they should have and what criteria for comparing them are. In task scheduling problems for various types of input data, movement definition and individual parts of it have significantly different meaning for determining whether a certain movement is identical with another one. For example, in scheduling unit-time and nonpreemptable tasks on parallel identical processors, the position of an exact task is insignificant; however, when tasks are preemptable, the position, onto which a certain task will be transferred, contains important information, determining subsequent movements and the quality of the obtained solution. In the developed approach, the function which compares movements is flexible, its operation is dependent on parameters defined by input data inserting. In tasks scheduling problems the neighborhood is a set with a great number of elements. Moreover, the ways of generating them differ from one another depending on the type of a problem. Obtaining complete content of the environment in the middle of every iteration causes significant extending of algorithm work; therefore, determining a new movement is partially random. At the beginning, the task to be moved is drawn; next, processors are chosen on which a random task is carried out. Among many processors only one is drawn. In case of preemptable tasks, only a part of task to be moved is drawn. The target place of task moving is determined by drawing a random processor and its position in the schedule. Verifying whether a certain movement belongs to the environment occurs during its generating, but complete surrounding content is never known. The consequences of this solution of generating movement are double. In states, when actual solution is distant from an optimal one and many possible movements exist, algorithm quickly approaches an optimal solution. However, if an optimal solution is close and only few possible movements exist, many wrong moves are generated and algorithm signifi-cantly slows down. Short-term memory consists of all newly made movements. For the sake of constant number of elements on the list, each new element added causes automatic removal of the element from the end of the list. The number of elements on the list is the algorithm parameter. A list of generated solutions was used as long- term memory. Information obtained on its basis: total number of iterations executed from the beginning of algorithm action, number of iterations obtained without better solution, the number of recent returns to initial solution, present value of parameter which determines diversity in environment searching. To accelerate the work of algorithm, the initial solution and the best solution so far, are retained during algorithm action.

2.4   Algorithm of Synthesis

In Tabu Search algorithm for optimization of resource partition and task scheduling, like in both previous algorithms, the initial solution is generated randomly or by means of other heuristic algorithms. Initial solutions, obtained in this way are not optimal, but they meet all the requirements specified in input date; then, further optimization is performed by Tabu Search algorithm. In the algorithm which is a coherent approach to task scheduling optimization and resource partition problems, the movement is defined as a transfer of specified part of task (or the whole task), which is being carried out on one of the processors, to another processor and to another position in the schedule. In comparison with task scheduling algorithm, the difference is that movements without specifying tasks and positions are possible; like in tabu search algorithm for resource partition, the exchange of processors for other ones in the resource set is possible. Tasks, which are carried out on a swapped processor, are transferred to a new processor taken from the resource set or to other active processors. Such a situation is likely to happen when tasks are dependent and a new processor is slower than the previous one or in case of problems with additional resources. In case of coherent approach to task scheduling and resource partition, functional surrounding as well as neighborhood are double sets, separate for task scheduling and resource partition. New movement cre-ation is partially random. The mechanisms are parallel to those in algorithms for resource partition and for task scheduling problems. First, movements for task scheduling are generated, after they have been made and after some possible returns to the best solution so far found, movement for resource partition problem is generated.

This kind of solution permits a coherent resource partition and various task schedules verification for these task resources partitions.

Short-term memory consists of all recently made movements. Apart from modification characterized in previous chapters, here new modifications are introduced. Tabu list associated with resource partition possesses mechanisms which allow combining each element from the list of movements for resource partition with the list of movements for task scheduling. Subsequent searching the list is limited to finding the movements for dividing and to checking whether the movement for scheduling exists on the joined list. It is a simple mechanism which allows to retain movements efficiently and effectively in memory and to search them. A list of generated solutions was used as a long-term memory. Pieces of information obtained on the basis of values of these variables are as follows: the total number of iterations performed from the beginning of algorithm action, the number of iterations without acquiring a better solution, the number of recent returns to the initial solution, the present value of parameter which describes diversity in neighborhood searching. The criteria, which are being considered, constitute concurrently the total cost and scheduling length C_{max}.

Selected Algorithm Parameters
Length of tabu list (short-term memory) – in shape of FIFO list – has an impact on numbers of iterations and calculation time. Too short list of tabu results in frequent returns to the same wrong solutions,whereas too long one leads to the constraints of solution space and long time of iterations' performance connected with longer time of list search and more frequent movement generating. Generally, the length of tabu list should be increased together with the increase of the problem size. Length of long-term memory – in shape of LIFO list which contains selected states of algorithm – increases the efficiency of algorithm. The appointed values have found application in subsequent computation experiments.

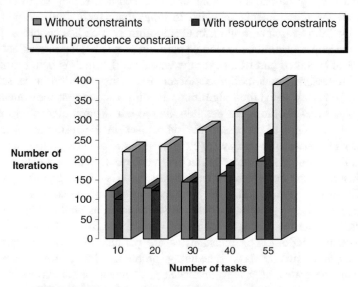

Fig. 2. Impact of problem type on the number of iterations

The increase in computational complexity of the problem brings about an increase in the number of iterations (Fig. 2) which are necessary to achieve the satisfactory result. Algorithm takes the most effective characteristics for the problems with limited resources. It is not necessary to execute additional iterations to check the accordance of the received result with the requirements of the resource constraints. Resource constraints cause the increase in the scheduling length C_{max} for this type of a problem; thus, less iteration are necessary to achieve a satisfactory result.

3   Selected Computational Experiments with Tabu Search Algorithm in Synthesis

Comparison of Results Received for Partition of Resources and Full Synthesis (with Tasks of Scheduling)
For the generated data computation experiments have been conducted which compared partition of resources and full synthesis. The tasks used during the tests are received as graphs STG [9]. The following results have been obtained:

Minimization of Cost
Figure 3 shows that coherent synthesis achieves better results producing cheaper structures (even above 10%); the income relatively rises for the bigger number of tasks.

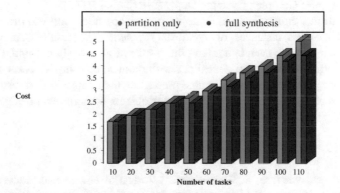

Fig. 3. Comparison between partition of resources and synthesis – minimum of cost

Minimization of Time
Figure 4 shows that in case of the results received for a coherent program, there has been an improvement of the obtained results – for the criterion of time minimization – above 20%. In case of the results obtained for synthesis there has been an improvement, particularly for optimization according to time criterion.

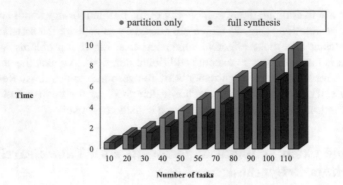

Fig. 4. Comparison between partition of resources and synthesis – minimum of time

4   Conclusions

The computational experiments show, that Tabu Search algorithm can solve resource and task scheduling problems. For a great deal of success, the optimization criterion. The structure of the algorithm (scheduling and allocation of tasks) are the best - are both cheap and fast.

This is the next algorithm used for high-level synthesis of systems with a higher degree of dependability presented as part of the International Conference on Dependability and Complex Systems DepCoS-RELCOMEX.

Computational experiments indicate its proper operation and usefulness for these problems. For some instances of synthesis problems, the results achieved by this algorithm are similar or even better than the results of previously presented algorithms: hybrid - genetic and simulated annealing, ant colonies, or using a neural network.

The author plans to develop an article presenting the comparison of results obtained by this different algorithms in the presented problems of high level design.

References

1. Drabowski, M., Wantuch, E.: Deterministic schedule of task in multiprocessor computer systems with higher degree of dependability. In: Zamojski, W., Mazurkiewicz, J., Sugier, J., Walkowiak, T., Kacprzyk, J. (eds.) Proceedings of the Ninth International Conference on Dependability and Complex Systems DepCos-RELCOMEX. Advances in Intelligent Systems and Computing, vol. 286, pp. 165–175. Springer (2014)
2. Drabowski, M.: Boltzmann tournaments in evolutionary algorithm for CAD of complex systems with higher degree of dependability. In: Zamojski, W., Mazurkiewicz, J., Sugier, J., Walkowiak, T., Kacprzyk, J. (eds.) Proceedings of the Tenth International Conference on Dependability and Complex Systems DepCos-RELCOMEX. Advances in Intelligent Systems and Computing, vol. 365, pp. 141–152. Springer (2015)

3. Drabowski, M.: Modification of neural network Tsang-Wang in algorithm for CAD of systems with higher degree of dependability. In: Zamojski, W., Mazurkiewicz, J., Sugier, J., Walkowiak, T., Kacprzyk, J. (eds.) Proceedings of the 11th International Conference on Dependability and Complex Systems DepCos-RELCOMEX. Advances in Intelligent Systems and Computing, vol. 470, pp. 121–133. Springer (2016)

4. Drabowski, M.: Adaptation of Ant Colony Algorithm for CAD of complex systems with higher degree of dependability. In: Zamojski, W., Mazurkiewicz, J., Sugier, J., Walkowiak, T., Kacprzyk, J. (eds.) Proceedings of the 12th International Conference on Dependability and Complex Systems DepCos-RELCOMEX. Advances in Intelligent Systems and Computing. vol. 582, pp. 141–150. Springer (2017)

5. Glover, F., Laguna, M., Taillard, E., de Werra, D.: A user's guide to tabu search. Ann. Oper. Res. **41**, 1–28 (1993). Baltzer, Basel

6. Nowicki, E., Smutnicki, C.: An advanced tabu search algorithm for the job shop problem. J. Sched. **8**, 145–159 (2005)

7. Glover, F., Laguna, M.: Tabu Search. Kluwer Academic Publishers, Norwell (1997)

8. Błażewicz, J., Drabowski, M., Węglarz, J.: Scheduling multiprocessor tasks to minimize schedule length. IEEE Trans. Comput. **35**(5), 389–393 (1986)

9. http://www.kasahara.elec.waseda.ac.jp/schedule/stgarc_e.html

Construction of the Structure Function of Multi-State System Based on Incompletely Specified Data

Andrej Forgac[1(⊠)], Jan Rabcan[1], Elena Zaitseva[1], and Igor Lukyanchuk[2]

[1] University of Zilina, Univerzitna 8215/1, 010 26 Zilina, Slovakia
{andrej.forgac,jan.rabcan,
elena.zaitseva}@fri.uniza.sk
[2] Université de Picardie Jules Verne, 1 Chemin du Thil, 80000 Amiens, France
lukyanc@gmail.com

Abstract. A Multi-State System (MSS) is the type of mathematical model of a system in the reliability analysis in which both the system and its components may experience more than two reliability states. One of the possible approaches of MSS evaluation is based on system representation by a structure function that maps the system components states into the system state/reliability/availability (performance level). But the structure function can be used for completely specified data only. In this paper new method for structure function construction based on incomplete data is proposed. This method is based on the use of decision tree that is inducted according to initial data. A structure function is constructed as decision table that is formed by inducted decision tree.

Keywords: Multi-State System · Structure function
Incompletely specified function · Decision tree

1 Introduction

There two typical mathematical interpretation of investigated system in reliability engineering depending on number of analyzed performance levels [1]. Binary-State System allows analyzing two states in the system functioning only. Another type of is *Multi-State System* (MSS) that permits to consider some states of reliability or availability (performance levels) in the system behavior. One of the possible mathematical description of MSS is structure function that maps the system components states into this system performance level. The advantage of the structure function is the possibility to be formed for the system of any structural complexity [2, 3]. However, this function can be constructed for completely specified data about system behavior. But initial data in reliability analysis is uncertain and incompletely specified as a rule [4, 5]. Therefore, special methods for the structure function based on uncertain and incompletely specified data must be developed.

In this paper, we develop new method for the MSS structure function construction based on uncertain data that is interpreted as *Multiple-Valued Logic* (MVL) function.

© Springer International Publishing AG, part of Springer Nature 2019
W. Zamojski et al. (Eds.): DepCoS-RELCOMEX 2018, AISC 761, pp. 184–194, 2019.
https://doi.org/10.1007/978-3-319-91446-6_18

In this method, the structure function is considered as the classifier that divides all possible components states into some groups and the number of these groups agree with number of system performance levels. This interpretation of the structure function allows us to use typical method for classification of uncertain data from Data Mining [6]. We propose the use *Decision Tree* (DT) [7]. The use of DTs for construction of the structure function assumes induction of a tree based on the data about the real system behavior and these data can be incompletely specified. Structure function values are then defined for all combinations of component states by the DT.

2   Mathematical Representation of Multi-State System

The structure function $\phi(x)$ maps sets of components states into MSS performance levels [1, 3]. The structure function of MSS can be defined as time-depended function and time-independent function for system in stationary state or fixed time. This function for MSS of n components in stationary state or fixed time is defined as [1, 2]:

$$\phi(x_1, \ldots, x_n) = \phi(x) : \{0, \ldots, m_1 - 1\} \times \ldots \times \{0, \ldots, m_n - 1\} \to \{0, \ldots, M - 1\} \quad (1)$$

where $\phi(x)$ is the system state (performance level) from failure ($\phi(x) = 0$) to perfect functioning ($\phi(x) = M - 1$); $x = (x_1, \ldots, x_n)$ is the state vector; x_i is the i-th component state that changes from failure ($x_i = 0$) to perfect functioning ($x_i = m_i - 1$).

The structure function (1) can be interpreted as MVL function if $m_i = m_j = M$ $(i \neq j)$:

$$\phi(x_1, \ldots, x_n) = \phi(x) : \{0, \ldots, m - 1\} \times \ldots \times \{0, \ldots, m - 1\} \to \{0, \ldots, m - 1\} \quad (2)$$

In the paper [3] the methodology to transform of structure function in form (1) into structure function in form (2) is considered. Therefore, any structure function can be interpreted as MVL function. This interpretation allows us to use mathematical methods of MVL in reliability analysis. But the structure function of MSS (and MVL function too) has large dimension. In MVL for the decision of this problem for MVL function the *Multi-Valued Decision Diagram* (MDD) is used [8]. The application of MDD for MSS representation is considered in [9, 10]. Some of methods for qualitative and quantitative analysis of MSS based on system representation by MDD are considered in [9–12]: the methods for fault tree construction [11]; algorithms for the frequency indices calculation [9]; importance analysis of MSS [10, 12].

The MDD of MSS is a directed acyclic graph [12]. For the structure function (2), this graph has m terminal nodes, labeled from 0 to ($m - 1$), representing m corresponding constants from 0 to ($m - 1$). Each non-terminal node is labeled with a structure function variable x_i and has m outgoing edges. The leftmost edge is labeled by "0" and agrees with the component failure. The last m-th outgoing edge is labeled "$m_i - 1$" edge and presents the perfect operation state of the system component. Terminal nodes of the MDD correspond to the performance levels of the MSS. In this case, the non-terminal node outgoing edges are interpreted as component states. The probabilistic interpretation of the MDD and rules for MSS measure calculation are based on the canonical Shannon expansion for MVL-function and considered in detail in paper [8].

In reliability engineering, the methods for MDD construction are proposed for completely specified structure function [8–12]. But initial data in reliability analysis of the real-world system is uncertain [5, 7]: some values of performance levels cannot be obtained. Therefore, we propose new method for the construction of the structure function (2) in form of MDD based on uncertain data.

The structure function (2) in which we do not have all its values defined is called *Incompletely Specified Structure Function* (ISSF) and can be considered as incompletely specified MVL-function [13]. The ISSF of MSS can be considered as classification structure for m sets: known sets of components states (vectors x) are divided into m groups (the number of which corresponds to the number of performance levels). It is typical classification problem in Data Mining [6]. We propose the use of DT for this classification.

3   MDD Construction for Incompletely Specified Structure Function of MDD

The proposed method for MDD construction for ISSF of MSS consists of next steps: (*i*) DT induction based on initial data about MSS reliability behavior and (*ii*) the transformation of the inducted DT into MDD.

3.1   The Correlation of Decision Tree and Structure Function

DT is inducted based on N training samples of objects U = $\{u\}$. Every sample is defined by n input attributes A = $\{A_1, ..., A_n\}$. Each attribute A_i ($1 \leq i \leq n$) measures some feature presented by a group of discrete linguistic terms. We assume that each group is a set of m ($m \geq 2$) values of subsets $\{A_{i,0}, ..., A_{i,m-1}\}$. We assume that each object u in the universe is classified into a set of classes $\{B_0, ..., B_{m-1}\}$. This set describes the class (output) attribute B. The output attribute B has to be determined by values of attributes A_i. DTs can result in a decision rules or a decision table [7]. A decision table indicates all possible values of input attributes and a corresponding value of the output attribute that is calculated according to the DT.

DT consists of 2 types of nodes, namely internal and external. The top-level node is called a root. Each internal node has some successors and it represents a test on individual input attribute and outcome edges of internal nodes mean a possible test result of the attribute test. Internal node also has a subset of samples meeting the tests on attributes from the root to this node. External nodes have not successors and they are called leaves. Each leaf represents individual class defined by output attribute. DT classify samples sequentially from the root to leaf. Classification is based on associated attributes with internal nodes, thus the classifying sample passes gradually from the root to a leaf that determines the value of its output attribute. Important step in the construction of DT is to select the test attribute for its nodes. This attribute should best divide a given set of samples at a given tree level, in terms of the target classification. The goal is to identify those attributes that can divide the records with a maximum reduction of uncertainty. Attributes associated with internal nodes can be determined by different criterions.

Interpretation of DT terminology for the problem of the structure function construction is shown in Table 1. In the case of structure function construction (see Table 1), the input attributes A_i agree with the system components x_i $(i = 1, ..., n)$ and the output attribute B is assigned with the system performance level $\phi(x)$. Each non-leaf node is associated with an attribute $A_i \in A$, or in terms of reliability analysis: each non-leaf node is associated with a component. The non-leaf node of the attribute A_i has m outgoing branches. The s-th outgoing branch $(s = 0, ..., m-1)$ from the non-leaf node A_i agrees with the value s of the i-th component $(x_i = s)$. The path from the top node to the leaf indicates the vector state of the structure function by the values of attributes and the value of the output attribute corresponds to the system performance level. If any attribute is absent in the path, then all possible values of the states are defined for the associated component. Therefore, calculation of the decision table for all possible values of the components states is determined as the truth table of the structure function (2).

Table 1. A correlation of the terminologies of DT and structure function

DT	Structure function
Number of input attribute: n	Number of the system components: n
Attribute A_i $(i = 1, ..., n)$	System component x_i $(i = 1, ..., n)$
Attribute A_i values: $\{A_{i,0}, ..., A_{i,j}, ..., A_{i,mi-1}\}$	The i-th system component state: $\{0, ..., m-1\}$
Output attribute B	System performance level $\phi(x)$
Values of output attribute B: $\{B_0, ..., B_{M-1}\}$	System performance level values: $\{0, ..., m-1\}$

The inducted DT can be transformed in MDD according to typical rules for MDD construction [8] because the DT represents completely specified function. The constructions of MDD based on DT or truth table are investigated problem in MVL (Fig. 1). Therefore, the important step in proposed algorithm is DT induction.

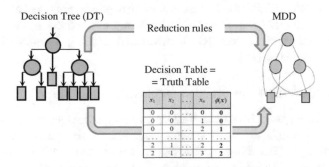

Fig. 1. The MDD construction based on DT

3.2 Decision Tree Induction

The several algorithms for induction of decision trees have been proposed, for instance, ID3, C4.5, CART, CHAID [6, 14, 15]. Algorithm ID3 has been designed by Quinlan in [14]. ID3 selects association attributes to internal nodes by information gain, which is based on entropy. Shannon entropy of set S of examples is defined as:

$$H(S) = -\sum_{j=1}^{C} \frac{k_j}{|S|} * \log_2 \frac{k_j}{|S|}$$

where k_j is number of instances belonging to the j-th class and $|S|$ is the cardinality measurement of set S. Information gain voice that the uncertainty of the set S can be appreciably reduced after splitting by attribute A. Information gain is defined as measurement of entropy before and after splitting of set S by attribute A. It stands for expected entropy reduction. The mathematical form of information gain G(S, A) has following form:

$$G(S,A) = H(S) - H(S|A) = H(S) - \sum_{v \in vals(A)} \frac{|S_v|}{|S|} H(S_v)$$

where S_v is the subset of S where samples have value of attribute A equal to v. Information gain $G(S,A)$ usually prefer attributes with a large number of different values. This problem was improved by Quinlan in [15] in algorithm C4.5 by using gain ratio defined as follow:

$$Gr(S,A) = \frac{G(S,A)}{-\sum_{v \in vals(A)} \frac{|S_v|}{|S|} * \log_2 \frac{|S_v|}{|S|}}$$

If the instances included in an induction process are classified, the classification is very accurate. But if some new instance occurs, the classification of such instance ends with mistake usually. This problem is known as overfitting and the possible way how to reduce its impact is the usage of pruning techniques. Pruning techniques remove subtrees of the tree, which can improve classification accuracy of new instances (unknown during tree induction). In this paper, two pre-pruning techniques are used. Pre-pruning techniques determine leaf nodes during tree induction, therefore the time cost of tree induction is reduced. Implemented pruning techniques use special parameters described in [15].

3.3 Example of MDD Construction for Structure Function

Consider the example for MDD construction based on incompletely specified data. Let us represent a simple laparoscopic surgery procedure for reliability evaluation [16] by MDD. According to [16], this procedure can be interpreted as MSS that consists of four components ($n = 4$): device (a laparoscopic robotic surgery machine [17]), two doctors (anesthesiologist and surgeon) and a nurse. This system as MSS and introduce the numbers of states for every component and number of performance levels of the system. The function of the simple laparoscopic surgery procedure can be interpreted as MVL function of four variables ($n = 4$) with three values ($m = 3$). Let this system has three performance levels: 0 – non-operational (fatal medical error); 1 – partially operational (some imperfection); 2 – fully operational (surgery without any compli- cation). The device (x_1) has three states: 0 – failure; 1 – partially functioning; 2 – functioning. The work of anesthesiologist (x_2), surgeon (x_3) and the nurse (x_4) can be modeled both by 3 levels, i.e.: 0 – (the fatal error); 1 – (sufficient); 2 – (perfect or the work without any complication).

The structure function of the system for analysis this simplified version of a laparoscopic surgery is composed of 81 situations (state vectors). Let us suppose that information about this system is incomplete and represented by 66 states only (Table 2).

Table 2. The structure function of laparoscopic surgery procedure success

x_1x_2	x_3x_4								
	00	01	02	10	11	12	20	21	22
00	0	0	0	0	*	*	0	0	*
01	0	*	0	0	0	0	0	0	0
02	0	0	0	0	0	0	0	0	0
10	0	0	0	0	0	0	0	*	1
11	0	0	0	0	1	*	0	2	2
12	0	0	*	1	1	*	*	2	2
20	0	*	0	0	0	1	0	*	*
21	0	0	1	1	1	2	1	*	2
22	0	0	1	1	*	2	1	*	2

This data can be considered as ISSF. The ISSF can be represented by DT that permits to construct the truth table or MDD of this structure function. Induction of DT is based on data in Table 2. The variables values are interpreted as input attributes and the function values are considered as output attribute. We use ordered type of decision trees [16]. These trees have the same attribute at each level. The DT for the considered example of laparoscopic surgery procedure for reliability evaluation is shown in Fig. 2.

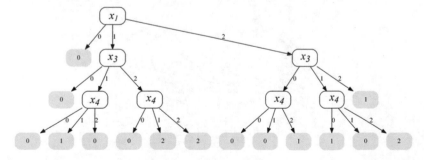

Fig. 2. The DT of the structure function of laparoscopic surgery procedure success

The inducted DT does not include node for component x_2. It is caused by tree pruning according to method of the DT induction. In point of view of the considered problem, it means minimal influence of this component into the system functioning. The truth table of the restored structure function (Table 3) is formed by the DT in Fig. 2 as decision table (the restored values of the function are marked by blue color): the values of output attribute (function value) are calculated for all possible values of inputs attributes (variables). For example, assume that the state vector is $x = (2\ 0\ 0\ 1)$. Analysis based on the DT starts with the attribute A_1 (Fig. 2) that is associated with the first variable x_1 of the structure function. The value of this variable is $x_1 = 2$ for the specified state vector and branch for the attribute value $A_{1,2}$ is considered. This branch is associated with the analysis next attribute A_3 that is agreed with variable x_3. According to the variable value in the state vector ($x_3 = 0$), the output attribute value (the structure function value) is defined as 0 without analysis of other attributes.

Table 3. The restored structure function of laparoscopic surgery procedure success

x_1x_2	x_3x_4								
	00	01	02	10	11	12	20	21	22
00	0	0	0	0	0	0	0	0	0
01	0	0	0	0	0	0	0	0	0
02	0	0	0	0	0	0	0	0	0
10	0	0	0	0	0	0	0	0	1
11	0	0	0	0	1	2	0	2	2
12	0	0	0	1	1	1	1	2	2
20	0	0	0	0	0	1	0	0	1
21	0	0	1	1	1	2	1	2	2
22	0	0	1	1	1	2	1	2	2

The MDD of the structure function for the considered example can be constructed based on the DT through reduction rules for MDD construction (Fig. 3).

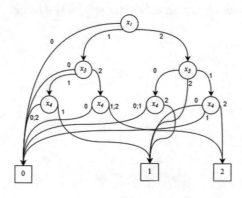

Fig. 3. The MDD of the structure function of laparoscopic surgery procedure success

Therefore, the proposed method allows us to construct MDD for incompletely specified data that is typical in reliability analysis. MDD is used to calculate indices and measures for system reliability. In particular, the probabilities of the performance levels can be computed according to the algorithm in [3, 12]. The scenarios of the system failure are developed based on minimal cuts of the system and the algorithm to define the minimal states through MDD are proposed in [18]. The definition of system critical states for every performance level based on MDD is presented in [12]. For example, in Table 4 the probabilities of the system components states are shown. According to these probabilities, the system performance levels are calculated and presented in Table 4.

Table 4. The system evaluation

States	Components states probabilities				Probabilities of performance levels
	x_1	x_2	x_3	x_4	
0	0.2	0.3	0.1	0.3	0.455
1	0.3	0.4	0.3	0.6	0.404
2	0.5	0.3	0.6	0.1	0.141

4 Algorithm Evaluation

The proposed method is based on the classification procedure. Important characteristic of the classification procedures for incompletely specified data is the accuracy. The DT induction is step that influences the method accuracy. We have implemented a simple case study to verify the proposed method accuracy, that means the accuracy of ISSF restores. To achieve this, the structure functions of MSS are randomly generated and transformed to ISSF by the deleting randomly some values of the functions. The range of deleted values is changed from 5% to 90%. Each transformed structure function can be interpreted as ISSF. We use this data to construct the structure function based on the use of DT induction. DT is induced by the ID3 method presented in [14, 15] and considered in Sect. 3.2 briefly. The structure function construction is implemented according to the concepts introduced in Sect. 3.1. As a result, a single or a small group of state vectors may be misclassified. Therefore, we have to estimate this misclassification using an error rate. The restored structure function and initial complete specified function are compared, and the error rate is calculated as a ratio of erroneous values of the structure function to the dimension of an unspecified part of the function. The experiments have been done for every structure function and fixed values numbers of unspecified function values. The best result has a minimal error rate.

The classification accuracy depends on special parameters α and β, which allows defining the required accuracy and affecting the structure of decision tree as is shown in [7]. A tree branch stops to expand when either the frequency of the branch is below α or when more than β percent of instances left in the branch has the same class label.

These values are thus key parameters needed to decide whether we have already arrived at a leaf node or whether the branch should be expanded further. Decreasing the parameter α and increasing the parameter β allow us to build large FDTs. On one hand, large FDTs describe datasets in more detail. On the other hand, these FDTs are very sensitive to noise in the dataset. The dependency between the classification accuracy and these parameters is shown in Fig. 4. According to this evaluation the best accuracy has FDT with parameters α = 0.05 and β = 0.74 if the considered the structure function with 50 specified values (Fig. 4). The evaluation of the structure function with the best accuracy according to values of parameters α and β (α = 0.05 and β = 0.74) depending the rate of unspecified values is shown in Fig. 5. This analysis discovers that the error rate is dependent on unspecified part of the initial data. This error increases significantly if the unspecified part is less than 10% and most than 85%.

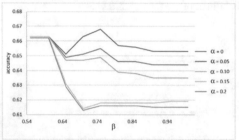
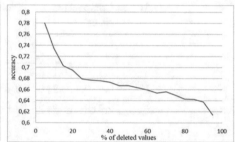

Fig. 4. Accuracy of decision tree according to α and β parameters, when 50% of structure function values has been deleted

Fig. 5. The accuracy of simple laparoscopic surgery procedure depending on number of unspecified values of the structure function

5   Conclusion

In this paper new method for construction of structure function in form of MDD based on incompletely specified data is proposed. The specifics of the reliability analysis (the recovery of uncertain data in accordance with specified data values) was implemented and involve in this method for MSS structure function construction. The principal part of this method is the induction of DT based on uncertain data. The accuracy of the method is defined by the accuracy of DT induction. The evaluation of this accuracy shows that the method has acceptable result if the incompletely specified part of data is more than 10% and less than 85%. The increase of the dimension of specified part causes the decrease in accuracy of the result (Fig. 5).

The structure function of MSS is widely used mathematical description in reliability engineering. There are different methods for calculation of indices and measures for MSS evaluation based on structure function [9, 18, 19]. The structure function representation by MDD is very interesting for reliability analysis because the structure function of MSS has large dimension as a rule. In this paper, we don't consider the problem of MDD optimization and construction of compact diagram [13]. We plan to continue this investigation for the construction of MDD with an optimal number of nodes for the structure function of MSS.

Acknowledgment. This work was partly supported by the grants of VEGA 1/0038/16, VEGA 1/0354/17 and APVV SK-FR-2017-003.

References

1. Murchland, J.D.: Fundamental concepts and relations for reliability analysis of multistate system. In: Barlow, R.E., et al. (eds.) Reliability and Fault Tree Analysis: Theoretical and Applied Aspects of System Reliability, SIAM, Berkeley, pp. 581–618 (1975)
2. Natvig, B.: Multistate Systems Reliability Theory with Applications (2010)
3. Zaitseva, E., Levashenko, V.: Reliability analysis of multi-state system with application of multiple-valued logic. Int. J. Qual. Reliab. Manag. **34**, 862–878 (2017)
4. Lisnianski, A., Frenkel, I., Karagrigoriou, A.: Recent Advances in Multi-State Systems Reliability. Springer, Cham (2018). https://doi.org/10.1007/978-3-319-63423-4
5. Aven, T., Baraldi, P., Flage, R., Zio, E.: Uncertainty in Risk Assessment: The Representation and Treatment of Uncertainties by Probabilistic and Non-Probabilistic Methods (2014)
6. Maimon, O., Rokach, L.: Data Mining and Knowledge Discovery Handbook (2010)
7. Zaitseva, E., Levashenko, V.: Construction of a reliability structure function based on uncertain data. IEEE Trans. Reliab. **65**, 1710–1723 (2016)
8. Miller, D.M., Drechsler, R.: On the construction of multiple-valued decision diagrams. In: Proceedings 32nd IEEE International Symposium on Multiple-Valued Logic, pp. 245–253 (2002)
9. Amari, S.V., Xing, L., Shrestha, A., Akers, J., Trivedi, K.S.: Performability analysis of multistate computing systems using multivalued decision diagrams. IEEE Trans. Comput. **59**, 1419–1433 (2010)
10. Zaitseva, E., Levashenko, V.: Decision diagrams for reliability analysis of multi-state system. In: Proceedings of International Conference on Dependability of Computer Systems, DepCoS - RELCOMEX 2008, pp. 55–62 (2008)
11. Mo, Y.: A multiple-valued decision-diagram-based approach to solve dynamic fault trees. IEEE Trans. Reliab. **63**, 81–93 (2014)
12. Zaitseva, E., Levashenko, V., Kostolny, J.: Multi-state system importance analysis based on direct partial logic derivative. In: Proceedings of 2012 International Conference on Quality, Reliability, Risk, Maintenance, and Safety Engineering, ICQR2MSE 2012, pp. 1514–1519 (2012)
13. Stankovic, R.S., Astola, J.T., Moraga, C.: Representations of Multiple-Valued Logic Functions. Morgan & Claypool Publishers, Princeton (2012)
14. Quinlan, J.R.: Induction of decision trees. Mach. Learn. **1**, 81–106 (1986)
15. Quinlan, J.R.: C4.5: Programs for Machine Learning (Morgan Kaufmann Series in Machine Learning) (1992)
16. Levashenko, V., Zaitseva, E., Kvassay, M., Deserno, T.M.: Reliability estimation of healthcare systems using fuzzy decision trees. In: Proceedings of the 2016 Federated Conference on Computer Science and Information Systems, FedCSIS 2016 (2016)

17. Patel, H.R., Linares, A., Joseph, J.V.: Robotic and laparoscopic surgery: cost and training. Surg. Oncol. **18**(3), 242–246 (2009)
18. Kvassay, M., Zaitseva, E., et al.: Minimal cut vectors and logical differential calculus. In: Proceedings of the International Symposium on Multiple-Valued Logic, pp. 167–172 (2014)
19. Kvassay, M., Zaitseva, E., Kostolny, J., Levashenko, V.: Importance analysis of multi-state systems based on integrated direct partial logic derivatives. In: Proceedings of the International Conference on Information and Digital Technologies (IDT 2015), pp. 183–195 (2015)

Railway Operation Schedule Evaluation with Respect to the System Robustness

Johannes Friedrich[1], Franciszek J. Restel[2(⊠)],
and Łukasz Wolniewicz[2]

[1] Technische Universität Berlin, Salzufer 17-19, 10587 Berlin, Germany
JFriedrich@railways.tu-berlin.de
[2] Wroclaw University of Science and Technology,
27 Wybrzeze Wyspianskiego Str., Wroclaw, Poland
{franciszek.restel,lukasz.wolniewicz}@pwr.edu.pl

Abstract. Due to infrastructure constraints in railway transportation system, each disruption can have a very important influence on the system operation. Therefore, it is important to evaluate transportation processes during their designing phase. The structure of timetable is a key factor, that has effect on the transportation system robustness. A large number of possible cases of train scheduling in the timetable makes it currently impossible to deal with this factor. Therefore, an analysis method to evaluate timetable qualities is needed. The paper is focussed on system robustness due to the timetable structure. A robust timetable leading to a robust transportation system is resistant to strokes from undesirable events. In other words, for a robust timetable there is lack of disruption propagation. The challenge arises how to quantify the robustness of the railway transportation system. Thus, the paper discusses issues related to robustness in context of time reserves. Correct dealing with time reserves is a multidimensional issue. Increasing time reserves increases also the robustness. While, the network capacity decreases. That is why ongoing research results will be useful for timetable designing.

Keywords: Rail transport · Robustness · Timetable

1 Introduction

Due to infrastructure constraints in railway transportation system, each disruption can have a very important influence on the system operation. Therefore, it is important to evaluate transportation processes during their designing phase. The structure of time-table is a key factor, that has effect on reliability, robustness and resilience of a transportation system [14, 23].

Timetable robustness is the ability of a timetable to withstand design errors, parameter variations, and changing operational conditions [19]. Due to [24], robustness in general is the capacity for some system to absorb or resist changes.

In opposite to resilience which is the ability to fast system recovery after collapse, robustness is the ability to keep the system performance after undesirable event occurrence. Thus, an undesirable event affects a train running on a section. The train may be delayed during its remaining running and it may affect other trains. A robust timetable leading to a robust transportation system is resistant to strokes from undesirable events. In other words, for a robust timetable there is lack of disruption propagation.

The challenge arises how to quantify the robustness of the railway transportation system. Robustness is closely related to resource reserves. On one hand, there are physical resources. On the other hand, there are time reserves in the operation schedule. Thus, the paper discusses issues related to robustness in context of time reserves.

2   Reliability, Robustness and Resilience in the Railway Industry – Literature Review

Trains in the system move at intervals. The intervals result from infrastructure boundary conditions. The so-called technical time intervals arise from operating procedures and depend on the local conditions of a station or railway line. The technical time intervals are the minimum spacing between two trains [22].

It is possible to design an operation schedule basing on the minimum value of technical time intervals. In such a case, the infrastructure capacity would be the maximum. While each disruption would shift to all other trains and will be not dumped for the failed train. Influence of operation intensity on disruption propagation was closer described in [13, 27, 29, 30].

According to the world literature, passenger rail traffic is determined by the time-table, the detailed plan of operations [4, 9]. The timetables are usually scheduled taking into account delays. During the planning of traffic on the network, deterministic driving times are assumed.

Disruptions in time are deviations from the planned transportation schedule (timetable). Due to infrastructure constraints in railway transportation system these type of interferences is very important. Deviations can be positive (delay) or negative (untimely) [22]. However, the negative temporary disturbances can be easily ongoing compensated (e.g. through longer stop at a station).

Well drawn up train timetable should guarantee 100% punctuality in normal operating conditions [21]. Travel time should be determined in a manner which takes maximum advantage of the technical parameters of the line, as well as the technical characteristics of the rolling stock used by carriers [2, 12, 26]. In practice, however, situations occur when the operator fails to provide the train service in accordance with the timetable.

If an unwanted event has occurred in the infrastructure subsystem (just before a train entering) this train is primary disrupted. According to the accepted criteria for evaluating disruptions in the time context, delays greater than or equal to 5 min are taken into account [28].

Depending on the location of trains in the timetable, the system will be more or less vulnerable in context of delay propagation. It means, on the next stages of failed train journey and also on other trains as so-called secondary delays. In the worst scenario, one primary disruption of one train may result in disruptions of all other trains. The dispatcher may have also influence on secondary disruption dumping. The recovery of the system, including dispatcher activities, is directly related to the system resilience issue [8]. The delay damping issue is also associated with the so-called robust timetables [6, 9, 18], due time reserves [11] included in planned operation time.

In the paper [5], a dispatching method for recovery of proper operation. Therefore, the discussed issues are related more to resilience than robustness. The most important factors have influence on resilience have been catalogued. The set consists of departure time, overtaking possibility at a station, train order, time reserves and section structure on the line. There are two models discussed, with or without train order changes.

In [7] an indicator for robustness evaluation in critical network points was proposed. It was presented as the inverse value of the shortest time intervals sum. It should allow for finding of weak points in the network.

Application of procedures basing on fully deterministic solutions results in large inaccuracies in technical systems operation [10, 15]. Also in the railway system [17, 20]. However, the most cases of disruptions are random, far from uniform distribution.

Developing methods of timetable designing in relation to reducing occurring delays is an important issue to limit disruptions in the railway system. Advanced models and algorithms are shown in [9, 25].

In [16], a management support model for disturbed traffic was proposed. The model is based on the traffic reorganisation or train cancellation cost in the context of railway staff. In [1] a support model for the decision making process under train traffic disruptions was presented. It provides a thorough approach to the system operation process.

3 Two Trains Meeting Probability - Assumptions

The probability of two trains interfering at any given moment depends on one hand on the infrastructure (given track layout and reserved train paths) and on the other hand on the schedule as well as the punctuality of the trains.

First: the punctuality and the show-up probability of the train must be set in relation for any specific moment. Thus, we need not only the punctuality of trains for specific time corridors (e.g., 95% of the trains arrive within 5 min after the scheduled arrival), but also the distribution of the arrivals (or departures) along a time corridor.

Starting with the arrivals: the probability density curve is shown in Fig. 1, lettering p_{max} the maximum expected probability density of the train arrival (in this case at $t = 0$, meaning the scheduled time of arrival, with t being the difference between the actual arrival and the scheduled arrival). Lettering t_{min} and t_{max} delimit the corridor, in which the train might show up. It can be easily seen that some trains are early and (the most) others are delayed. Time t_t represents the corridor occupation. This influences the number of trains included in the model. For first assumptions, time t_t is therefore limited by t_{min} and t_{max}, respectively. Early arrivals are marked with p_1, late arrivals with p_2. Firstly, it was assumed, the probability density function describing late arrivals can be fit by the exponential distribution.

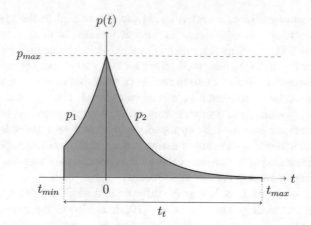

Fig. 1. Probability function of an arriving train

For early arrivals an inverse exponential distribution was assumed. The integral of the combined exponential function between t_{min} and t_{max} is equal to one. However, the functions shown in Figs. 1 and 2 are not proven, but only for problem demonstration basing on [3].

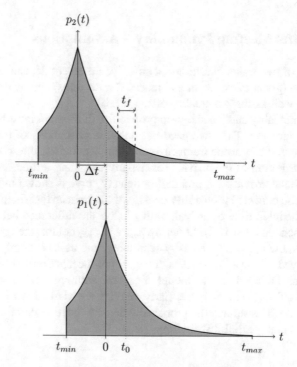

Fig. 2. Interference of two trains

The exact length chosen of t_t (or the limit of t_{max}) settles the question of the train delay probability for higher delays. Highly delayed trains are often not treated in this fashion. Occasionally, they are ended before their scheduled terminus. Thus, t_t is usually set as the train frequency of a single line, so that trains with delays that are so great that they run instead of the next train on the line or even later are excluded. These are usually treated differently and not according to the model's specifics.

Secondly, the schedules of the trains have to be set into relation with each other, so that the difference between the scheduled train arrivals (or departures) defines the interval between the two curves of the two corresponding trains. This is necessary to determine the amount of interference between the two trains: if they interfere even at a single moment, they cannot run at the same time.

Since the interference may last for more than a single moment, an interference corridor (denominated as t_f) needs to be considered. The setting shown in Fig. 2 can be used for any given moment (called t_0). The schedule difference between the two trains is called Δt. In this case, the train of the upper function p_2 interferes during the interval t_f with the train of the lower function p_1 arriving at t_0.

For t_0 the two probability functions need to be changed by the amount of the mutual interference. For example: if both trains use the same tracks, they interfere by the time equal to the blocking time of each of the trains. If one train is always to be preferred, t_f equals the amount of the two blocking times:

- the blocking time of the subordinated train for all arrivals that are now postponed because the preferred train is to be treated without waiting times, and
- the blocking time of the preferred train, because the subordinated train then has to wait (for arrivals after t_0, but still in the interval of t_f).

4   Timetable Description

The basic component of scheduling railway operation processes is the train timetable. The basic method for designing and modelling it is the so-called train diagram. An example was shown in Fig. 3. The train diagram is built in two dimensions. The horizontal axis represents the railway line, that means the driven distance. The vertical axis represents time, increasing from the top to the bottom. Sometimes the axes are swapped. Vertical lines show operating control points or train stops. Horizontal lines show next moments in time.

For the same direction, the space between two signals is called section or block section. At a given moment in time there can be only one train on such a section. For that coordinate system, a train is treated as material point. Train running is shown as angular line.

In relation to real operation, a train enters a section by crossing of the train running line with a vertical line, that represents an operating control point. It leaves the section when it's running line crosses the line that represents the operating control point at the end of the section. Due to the operation processes, previous to the train entering of a section is needed a little time for train route preparation. After the train leaves the block

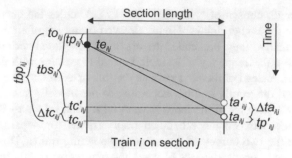

Fig. 3. Train path on a train diagram for one section

section it takes some technical time for cancelling of the train route. Therefore, the occupying time of a section is longer than the physical presence of a train on it.

For exact timetable description, it is necessary to find the characteristic points for the train running on the infrastructure. A section occupation time-window is built on base of the time interval between entering of i-th train on j-th section ($te_{i|j}$) and arrival at the section end ($ta_{i|j}$). The arriving time lettered $ta_{i|j}$ is the planned one, which includes time reserve ($\Delta ta_{i|j}$) added to the arrival ($ta'_{i|j}$, t_{min} from Fig. 1) resulting from the shortest possible running time.

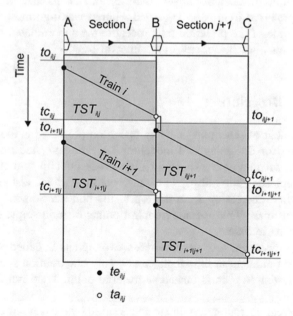

Fig. 4. Train diagram for two trains and two sections with marked TSTs

Between opening of the time-window ($to_{i|j}$) and entering of the i-th train ($te_{i|j}$) a certain time has to pass. This time is needed for train entering preparation ($tp_{i|j}$). On the other hand, also between train arriving ($ta_{i|j}$) and time-window closing ($tc_{i|j}$) a certain time has to pass. This due to need to report that no vehicle has left on the section ($tp'_{i|j}$). The time-window closing lettered $tc'_{i|j}$ represents the fastest possible closing time, due to the shortest train running. Following from this, the planned length of a time-window ($tbp_{i|j}$) is the sum of the shortest section occupation ($tbs_{i|j}$) an additional time reserve ($\Delta tc_{i|j}$). The value of the time reserve is mostly the same like the time reserve for train arriving ($\Delta ta_{i|j}$).

A timetable is designed for a railway line with a few sections, and it locates trains in time and space. Therefore, a train timetable (train diagram) consists of few diagrams that combine a train path and a section. One train placed in time on one section will be called further Train-Section Time-window (TST). Allocation of TSTs in a train diagram are shown in Fig. 4.

The $i|j$-th train-section time-window ($TST_{i|j}$) is described by six parameters.

$$TST_{i|j} = \ <Tr_i, S_j, te_{i|j}, ta_{i|j}, to_{i|j}, tc_{i|j}> \tag{1}$$

The first two ones place the train on a section. It means, the i-th train in the time table (Tr_i) and j-th section (S_j) of the railway line. The next two qualities assign the i-th train entering time on j-th section ($te_{i|j}$) and the arriving time at the end of the section ($ta_{i|j}$). The last two parameters built the time-window. Thus, the opening time ($to_{i|j}$) of j-th section occupation by i-th train, and the closing time ($tc_{i|j}$).

5 Probability of Train Meetings in Case of More Than Two Trains

Each TST may affect other TSTs. All interactions in a time table (Γ) can be seen as set of interactions coming out from all TSTs ($I_{i|j}$).

$$\Gamma = \ <I_{i|j}> \tag{2}$$

Only direct interactions, from a previous TST to others after it will be taken into account. Therefore, a TST can affect only two other TSTs. The next one on the same section ($TST_{i+1|j}$) and the next one on the following section ($TST_{i|j+1}$).

$$I_{i|j} = \ <TST_{i|j} \rightarrow TST_{i+1|j}; TST_{i|j} \rightarrow TST_{i|j+1}> \tag{3}$$

Influence of one TST on another is a function of the time interval between them. For the dependence between TSTs of two different trains one by one on the same section, the time interval has to be calculated by the formula:

$$\Delta t_{i|i+1} = to_{i+1|j} - tc'_{i|j} \tag{4}$$

where $\Delta t_{i|i+1}$ is the time interval between the TSTs for two trains on the same section. It can be assumed, the opening time of next TST ($to_{i+1|j}$) is determined. Thus, the variability is hidden in the closing time of previous TST ($tc_{i|j}$). Therefore, the analysed random variable is the TST closing time lettered as xc. Another example of a probability density function than the shown in Figs. 1 and 2. Due to operation research, it was found that the lognormal distribution is more accurate than the exponential function (Fig. 5). For further consideration of the closing time probability density function, the null point on the time axis was assumed equal to $tc'_{i|j}$. Therefore, negative deviations for TST closing are excluded.

On the time axis with the probability density function for $TST_{i|j}$ closing can be placed the opening time $(to_{i+1|j})$ of the next one $(TST_{i+1|j})$. The possibility exists that previous to the opening $(to_{i+1|j})$ have to be realized some handling actions between train i and train $i + 1$ (for example passenger exchange). The handling actions require time to pass $\Delta th_{i|i+1}$ starting from moment $th_{i|i+1}$. Thus, the probability that $TST_{i|j}$ will affect $TST_{i+1|j}$ is shown in Fig. 5 as marked area under the function curve. This probability is given by the formula:

$$P\left(TST_{i|j} \rightarrow TST_{i+1|j}\right) = 1 - \int_{tc'_{i|j}}^{th_{i|i+1}} f_{i|j}(x_c)dx_c \qquad (5)$$

where $f_{i|j}(x_c)$ is the probability density function for closing time of $TST_{i|j}$ lettered x_c.

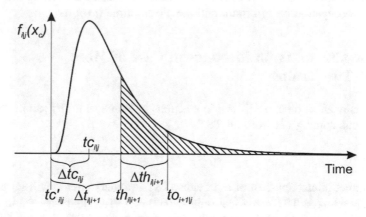

Fig. 5. Example of a probability distribution for deviation from planned TST closing time with marked characteristic times

For the interaction of two TSTs of one train on two sections one by one, the time interval has to be calculated by:

$$\Delta t_{j|j+1} = th_{j|j+1} - ta'_{i|j} \qquad (6)$$

where $\Delta t_{j|j+1}$ is the time interval between the TSTs for the same train on two sections and $\Delta th_{j|j+1}$ is the handling time needed previous to train leaving from a station (for example passenger exchange time). It can be assumed, the train entering time of next TST $(te_{i|j+1})$ is determined. Thus, the variability is hidden in the train arriving time of previous TST ($ta_{i|j}$). Therefore, the analysed random variable is the train arrival time lettered x_a. For further consideration of the arrival time probability density function, the null point on the time axis was assumed equal to $ta'_{i|j}$. Therefore, negative deviations for train arrival are excluded.

On the time axis with the train arrival time probability density function for $TST_{i|j}$ can be placed the train entering time ($te_{i|j+1}$) of the next one ($TST_{i|j+1}$). The possibility exists that previous to the train entering ($te_{i|j+1}$) have to be realized some handling actions between section j and section $j + 1$ (for example passenger exchange). The handling actions require time to pass $\Delta th_{j|j+1}$ starting from moment $th_{j|j+1}$. Thus, the probability that $TST_{i|j}$ will affect $TST_{i|j+1}$ is shown in Fig. 6 as marked area under the function curve. This probability is given by the formula:

$$P\left(TST_{i|j} \rightarrow TST_{i|j+1}\right) = 1 - \int_{ta'_{i|j}}^{th_{j|j+1}} f_{i|j}(x_a)dx_a \tag{7}$$

where $f_{i|j}(x_a)$ is the probability density function for train arrival time of $TST_{i|j}$ lettered x_a.

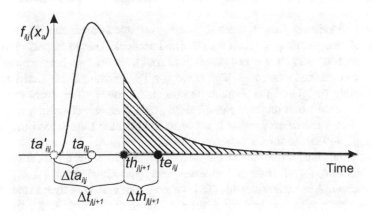

Fig. 6. Example of a probability distribution for deviation from planned train arrival time with marked charactcristic times

It follows, two main variables have to analysed for the train timetable robustness evaluation are the TST train arrival time (x_a) and the TST closing time (x_c). For each formulated interaction, the probability of interaction can be estimated by formulas (5) and (7). All interaction probabilities built a set of probabilities.

$$\Omega = \;<P(TST_{i|j} \rightarrow TST_{i+1|j}); P(TST_{i|j} \rightarrow TST_{i|j+1})> \; = \; <P_k> \tag{8}$$

Having all n interaction probabilities P_k for a timetable, the probability that at least one interaction will occur (P_I) can be calculated as follows:

$$P_I = 1 - \prod_{k=1}^{n}(1 - P_k) \tag{9}$$

On the other hand, the probability that no interaction will occur (P_0) can be also easily calculated.

$$P_I + P_0 = 1 \tag{10}$$

However, one of them or both can be used for characterising of train timetable robustness due to propagation of undesirable event consequences. For the proposed measure, the assumption was done that train sequence is determined and does not change.

6   Train Meeting Probability for a Timetable – Case Study

For a selected railway line in Poland, basing on operational data were identified deviations of train running in relation to defined sections. Due to imperfections of the available operation data, it was not possible to find train arrivals between earliest one ($ta'_{i|j}$) and the planned one ($ta_{i|j}$). The same for TST closing times between earliest possible closing ($tc'_{i|j}$) and the planned one ($tc_{i|j}$). All these values were registered as punctual, that means that they are classified as planned arrival or closing time. Due to that the random variable zero value is equal to the planned train arrival ($ta_{i|j}$) or TST closing ($tc_{i|j}$), and not to the earliest possible times ($ta'_{i|j}$, $tc'_{i|j}$).

It was found, that for all trains the arrival distributions on all sections are quiet similar. Therefore, for all train arrival times was approximated the same probability density function. After analysing the TST closing times, it was found that their distribution is also similar to the arrival time distribution. Therefore, for that random variable was approximated the same probability function. The empirical distribution and the approximated theoretical function are shown in Fig. 7.

In Fig. 7 can be seen probability distributions for train arrival times greater than zero, that means later than the planned time in the timetable. The parameters lognormal distribution were estimated and proven by Chi-squared test at significance level 0.05. Finally for the delay description in the analysed scenario was used the lognormal distribution LN(2.5275, 1.1438).

In Table 1 were placed operation data characteristics for the discussed case. The taken into account TSTs belong to one running direction on a double-track railway line. The probability of punctual train arrival and punctual TST closing was also shown in Table 1.

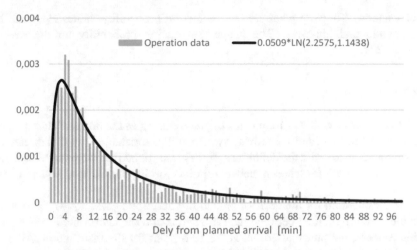

Fig. 7. Empirical distribution for delays and approximated lognormal distribution

Table 1. Operational data characteristics for analysed case

Number of analysed TSTs	Number of identified interactions	$P\left(tc'_{i|j} \leq x_c \leq tc_{i|j}\right),$ $P\left(ta'_{i|j} \leq x_a \leq ta_{i|j}\right)$
188	373	0.9491

To estimate the probabilities of train interactions, formulas (5) and (7) were used with the adaptation resulting from assumptions for Fig. 7. The probability of interaction was calculated for the case as follows:

$$P_k = 0.9491 + 0.509 \cdot \int_0^{t_{ca}} f(x)dx \qquad (11)$$

where $f(x)$ is the probability density function describing delay from schedule, for this operational case $LN(2.5275, 1.1438)$. The same function for both variables, respectively train arrival or TST closing. Upper integral border lettered t_{ca} represents respectively planned opening time of next TST for the following train, or train entering time on the next section. The lower border of the integral is respectively the planned train arrival or TST closing.

After calculation of all interaction probabilities P_k and the probability that at least one interaction will take place (P_I), the probability of no interactions between TSTs was found $P_0 = 0.0318$. After implementation of timetable structure changings, the probability of no interactions has been increased to $P_0' = 0.0373$.

At this moment, no scale for the probabilities was developed. Thus, a few timetables can be compared with respect to robustness, and the best can be chosen for implementation. Knowing what are the interaction probability values P_k, weak links can be found and improved. The highest interaction probability for the case was $P_{kmax} = 0.0294$.

7 Conclusions

After literature review, it was found lack of precise using of the nomenclature related to robustness and resilience of the railway system and its timetable. The first direction for further research is to order the definitions. It was also found that the issue of robustness evaluation is solved for chosen network points or by simulation, which is time-consuming.

A general method for train arrival probability description was shown. The concept limits the train arrival time between the earliest one and the latest one. It was assumed for this concept, the punctual arrival is the zero point for the distribution. After evolution of the concept, more characteristic qualities were introduced, the latest possible arrival was changed into plus infinity, and the zero point of distribution was set in place of the earliest possible arrival.

A unitary part of a timetable was introduced for evaluation process. It was called Train-Section Time-window (TST). Such a TST is described the i-th train in the time table (Tr_i), j-th section (S_j) of the railway line, the i-th train entering time on j-th section $(te_{i|j})$, the arrival time at the end of the section $(ta_{i|j})$, the opening time of the time-window $(to_{i|j})$ and the closing time $(tc_{i|j})$. The identified parameters were the basis for quantitative analysis of the timetable robustness.

Using the mentioned knowledge, the set of interactions between TSTs was defined, and especially all possible interactions starting from one TST. Finally, it was shown how to estimate probabilities of interactions and how to estimate one summarizing probability of interactions or lack of interactions. To show an example, a case study for a chosen railway line was prepared.

The calculated probability of no interactions is very low. This is due to high used capacity of the railway line. Afterwards, the timetable was redesigned. The objective function was to maximize the probability of no interactions. The relative increase of the probability was about seventeen percent. Thus, using the proposed robustness quantification measure, trains can be reorganised to maximise the robustness.

The results are promising. For further studies is planned to prove the method on base of more timetables and more railway lines. Moreover, further research on optimization of timetables due to robustness and resilience is planned.

References

1. Acuna-Agost, R., et al.: SAPI: Statistical Analysis of Propagation of Incidents. A new approach for rescheduling trains after disruptions. Eur. J. Oper. Res. **215**, 227–243 (2011)
2. Bach, L., et al.: Freight railway operator timetabling and engine scheduling. Eur. J. Oper. Res. **241**(2), 309–319 (2015)

3. Beck, C., Briggs, K.: Modelling train delays with q-exponential functions. Phys. A Stat. Mech. Appl. **378**(2), 498–504 (2007). https://doi.org/10.1016/j.physa.2006.11.084

4. Canca, D., et al.: Design and analysis of demand-adapted railway timetables. J. Adv. Transp. **48**(2), 119–137 (2014)

5. Lua, C., Tanga, J., Zhoua, L., Yuea, Y., Huangb, Z.: Improving recovery-to-optimality robustness through efficiency balanced design of timetable structure. Transp. Res. Part C **85**, 184–210 (2017)

6. Dicembre, A., Ricci, S.: Railway traffic on high density urban corridors: capacity, signalling and timetable. J. Rail Transp. Plann. Manag. **1**(2), 59–68 (2011)

7. Andersson, E.V., Peterson, A., Krasemann, J.T.: Quantifying railway timetable robustness in critical points. J. Rail Transp. Plann. Manag. **3**, 95–110 (2013)

8. Enjalbert, S., et al.: Assessment of transportation system resilience. In: Human Modelling in Assisted Transportation. Springer (2011)

9. Hansen, I., Pachl, J.: Railway timetable and traffic: analysis-modelling-simulation. Eurailpress, Zagreb Kliewer. A note on the online nature of the railway delay management problem. Networks 57 (2011)

10. Jodejko-Pietruczuk, A., Werbińska-Wojciechowska, S.: Development and sensitivity analysis of a technical object inspection model based on the delay-time concept use. Eksploatacja i Niezawodność - Maint. Reliab. **19**(3), 403–412 (2017). https://doi.org/10. 17531/ein.2017.3.11

11. Jodejko-Pietruczuk, A., Werbińska-Wojciechowska, S.: Block inspection policy model with imperfect maintenance for single-unit systems. Procedia Eng. **187**, 570–581 (2017). https:// doi.org/10.1016/j.proeng.2017.04.416

12. Ke, B.R., et al.: A new approach for improving the performance of freight train timetabling of a single-track railway system. Transp. Plann. Technol. **38**, 238–264 (2015)

13. Kierzkowski, A., Kisiel, T.: Simulation model of security control system functioning: a case study of the Wroclaw airport terminal. J. Air Transp. Manag. (2016). https://doi.org/10.1016/ j.jairtraman.2016.09.008

14. Kierzkowski, A.: Method for management of an airport security control system. Proc. Inst. Civ. Eng. Transp. (2016). https://doi.org/10.1680/jtran.16.00036

15. Kisiel, T., Zak, L., Valis, D.: Application of regression function - two areas for technical system operation assessment. In: CLC 2013: Carpathian Logistics Congress - Congress Proceedings, pp. 500–505. WOS:000363813400076 (2014)

16. Kroon, L., Huisman, D.: Algorithmic support for railway disruption management. In: Transitions Towards Sustainable Mobility Part 3. Springer-Verlag (2011)

17. Larsen, R., Pranzo, M.: A framework for dynamic re-scheduling problems. Technical report 2012–01, Dip. Ingegneria dell'Informazione. University of Siena (2012)

18. Liebchen, C., et al.: Computing delay resistant railway timetables. Comput. Oper. Res. **37**, 857–868 (2010)

19. Bešinović, N., Goverde, R.M., Quaglietta, E., Roberti, R.: An integrated micro–macro approach to robust railway timetabling. Transp. Res. Part B **87**, 14–32 (2016)

20. Ouelhadj, D., Petrovic, S.: A survey of dynamic scheduling in manufacturing systems. J. Sched. **12**(4), 417 (2009)

21. Pender, B., et al.: Disruption recovery in passenger railways international survey. Transp. Res. Rec. **2353**, 22–32 (2013)

22. Potthoff, G.: Verkehrsströmungslehre (Band 1) – Die Zugfolge auf Strecken und in Bahnhöfen. Transpress (1970)

23. Restel, F.J.: Defining states in reliability and safety modelling. Adv. Intell. Syst. Comput. **365**, 413–423 (2015). https://doi.org/10.1007/978-3-319-19216-1_39

24. Lusby, R.M., Larsen, J., Bull, S.: A survey on robustness in railway planning. Eur. J. Oper. Res. **266**, 1–15 (2018)
25. Schachtebeck, M., Schöbel, A.: To wait or not to wait and who goes first? Delay management with priority decisions. Transp. Sci. **44**, 307–321 (2010)
26. Schöbel, A., Maly, T.: Operational fault states in railways. Eur. Transp. Res. Rev. **4**, 107–113 (2012)
27. Tubis, A., Werbińska-Wojciechowska, S.: Operational risk assessment in road passenger transport companies performing at Polish market. In: European Safety and Reliability ESREL 2017, Portoroz, Slovenia, 18–22 June 2017
28. Vromans, M., et al.: Reliability and heterogeneity of railway services. Eur. J. Oper. Res. **172**, 647–665 (2006)
29. Walkowiak, T., Mazurkiewicz, J.: Soft computing approach to discrete transport system management. In: Lecture Notes in Computer Science. Lecture Notes in Artificial Intelligence, vol. 6114, pp. 675–682 (2010). https://doi.org/10.1007/978-3-642-13232-2_83
30. Walkowiak, T., Mazurkiewicz, J.: Analysis of critical situations in discrete transport systems. In: Proceedings of DepCoS - RELCOMEX 2009, Brunów, Poland, 30 June – 02 July 2009. pp. 364–371. IEEE (2009). https://doi.org/10.1109/depcos-relcomex.2009.39

Networks Key Distribution Protocols Using KPS

Alexander Frolov[1(✉)] and Alexander Vinnikov[2]

[1] National Research University MPEI,
Krasnokazarmennaya, 14, Moscow 111250, Russia
abfrolov@mail.ru
[2] SoftEngineering Ltd., Dokukina, 16, str. 1, Moscow 129226, Russia
al.vin@bk.ru

Abstract. We study cryptographic protocols for the exchange of key information in a computer network consisting of a trusted server T and a set $U = \{1, 2, \ldots, n\}$ of users sharing individual keys of the symmetric cryptosystem $K_{T,i}$, $i = 1, \ldots, n$ with it. We denote by 2^U the set of all possible subsets of the set U, by $\mathbf{P} \subseteq 2^U$ the set of privileged groups P of users, and by $\mathbf{F} \subseteq 2^U$ the set of forbidden user coalitions F. On the server T, by the key pre-distribution scheme (KPS), the preliminary keys (pre-keys) $\mathbf{k}_i$ of the users $i \in P$ are calculated, according to which the users from P can calculate the common working key k_P that is not available to users of any alienated coalition F that does not intersect P. The general structure of protocols on which pre-keys $\mathbf{k}_i$ are computed using upgradable system key information K, delivered to users and implemented for working key computing is substantiated. The protocol uses timestamps providing mutual authentication of users of each privileged group P at a specific moment in time T and security of attacks by members of the forbidden coalition - non-ability to act as a member of a privileged set or impose a compromised key. The protocols inherit the structure and functionality of the underlying protocol with entity authentication using timestamps, extending them to delivering to users the pre-keys and data, allowing to calculate the keys k_P for communications within the privileged groups.

Keywords: Cryptographic protocol · Key predistribution scheme
KPS · Privileged group · Forbidden coalition · Blom scheme
Key distribution pattern · Kerberos

1 Introduction

Key management is one of the main problems of secure communication in computer networks. It includes the problems of generation and distribution of keys, their recording and replacing compromised keys. Currently, the urgent problem of key distribution in sensor networks, combining a plurality of sensors and executive bodies in industrial systems, which security is ensured by cryptographic means [1]. Only these means can provide, for example privacy-preserving smart metering with multiple data consumers [2]. There is wide variety of protocols for distribution of keys generated by the server in its final form using symmetric secret channels. A thorough review and

© Springer International Publishing AG, part of Springer Nature 2019
W. Zamojski et al. (Eds.): DepCoS-RELCOMEX 2018, AISC 761, pp. 209–217, 2019.
https://doi.org/10.1007/978-3-319-91446-6_20

analysis of such protocols is contained in [3, 4]. The most popular appeared Needham-Schroeder protocol, which modification with timestamps underlies the known Kerberos protocol [5]. Currently, protocols in which the process of calculating keys is distributed between the server and the users in order to reduce the amount of secret key information transmitted over channels and saved in sensor nodes is paid attention. This is accomplished by keys pre-distribution schemes (KPS) [6]. The computer network studied in this chapter consists of a trusted server T and a set $U = \{1, 2, ..., n\}$ of users registered on the server, sharing individual keys of the symmetric cryptosystem $K_{T,i}$, $i = 1, ..., n$, with it. Let 2^U be the set of all subsets of the set U. We denote by $P \subseteq 2^U$ the set of privileged groups P of users, and by $F \subseteq 2^U$ the set of forbidden user coalitions F. On the server T, the preliminary keys (pre-keys) k_i of the users $i \in P$ are calculated, according to which the users from P can calculate the common working key k_P that is not available to users of any alienated coalition F that does not intersect P. Pre-keys are computed at the server by a Key Pre-distribution Scheme, denoted in general (P, F)-KPS. Extremely complete overview of studies on the synthesis and analysis unconditionally (i.e., not based on any computational assumption) secure (P, F)-KPS is given in [7]. The most demand are the Blom scheme [8] and the KDP-scheme (Key Distribution pattern) [9]. They or their modifications are implementer in contemporary systems [10, 11]. In a number of works there are described schemes which security depends on the complexity of finding any pre-image of the value of the hash function. [12]. In [13, 14] these approaches are implemented in combination in schemes satisfying mutually complementary correctness conditions. These studies focused on estimates of KPS information rate and KPS security, and it is believed that calculated at server in accordance with (P, F)-KPS pre-keys are delivered to users «off band» - in some secret technique, without using the network. In this chapter, we consider cryptographic protocols by which these keys are delivered to users by appropriate transactions on the same computer network. The variants of the protocols using timestamps are considered. The protocol with timestamps provides for mutual authentication of users of each privileged group P at a specific moment in time T and security of attacks by members of the forbidden coalition - non-ability to act as a member of a privileged set or impose a compromised key. The protocols inherit the structure and functionality of the underlying with entity authentication using timestamps, extending them to delivering to users the preliminary keys and data, allowing to calculate the keys for communications within the privileged groups.

The chapter includes this introduction and five sections. The second section examines the general approach to calculating the KPS-based preliminary keys that are updated in each session. The third section is devoted to the description and analysis of the protocols delivering for users the pre-keys calculated at the server according $(2, 1)$-KPS. The fourth section explores the approaches to building the protocols delivering for users the pre-keys generated on the server according (P, F)-KPS. The fifth section describes a method for implementing the proposed approach to distributed key calculation in a computer network using the Kerberos protocol. In conclusion we formulate general approach to study network key distributing protocols using key pre-distribution schemes discussed in this chapter.

2   Calculating the Upgradeable System of Pre-keys

The pre-keys **k** are computed on the server based on some system secret key information **K** that is updated for each session and fixed **(P, F)**-KPS. In this case, the server uses the public knowledge system **ID** of the pre-keys identifiers id that are characteristic for this **(P, F)**-KPS. It also allows by the secret system key information **K** to calculate the individual secret pre-keys $\mathbf{k} = R_{(P,F)-KPS}(id, \mathbf{K})$ using a public knowledge rule $R_{(P,F)-KPS}$. The system **ID** and this rule constitute a formal description of a particular **(P, F)**-KPS. In $(2, m)$-Blom scheme the secret system key information **K** is the set of $(m + 1)(m + 2)/2$ coefficients of a symmetric polynomial of degree m in each variable

$$f(x,y) = \sum_{i=0}^{m} a_{i,i}x^i y^i + \sum_{j=i+1}^{m} a_{i,j}(x^i y^j + x^j y^i)$$

over finite field F_q, $q > n$; the set **ID** of pre-key identifiers is the set of distinct elements $r \in F_p$; individual pre-key **k** is the set $\mathbf{k} = R_{(2,1)-\text{BlomScheme}}(id, \mathbf{K})$ of coefficients of polynomial $f(x,r)$. Thus, $(2, 1)$-Blom scheme is a system of identifiers and a rule $R_{(2,1)-\text{BlomScheme}}$ for calculating individual pre-keys for a given secret system key information **K**, used as a set of coefficients of a symmetric polynomial. In $(2, 1)$-KDP (n, q)-scheme **K** is the set of binary vectors of length q. The set **ID** includes binary identifiers id of certain subsets $id(\mathbf{K}) \subseteq \mathbf{K}$. These subsets are individual pre-keys, defined by identifiers id: $\mathbf{k} = id(\mathbf{K}) = R_{(2,1)-KDP}(id, \mathbf{K})$. The elements of the set **ID** satisfy the ternary relation: for any i, j, r $id_i \wedge id_j = id_i \wedge id_j \wedge id_r$ implies $r \in \{i, j\}$. Thus, $(2,1)$-KDP (n, q) – scheme is a system **ID** of binary identifiers and a rule $R_{(2,1)-KDP}$ for calculating individual keys for a given secret system key information **K**, as its subsets determined by binary identifiers. We will call the described instruction for calculating the individual keys by the system of identifiers and system key information the *algorithm* **(P, F)**-KPS. It can be used in every session to updatable secret system key information **K** at one and the same system of individual pre-keys identifiers that is responsible for security.

To use individual keys in a computer network, they must be assigned to different users. This is determined by an injective table $I:U \to ID$, $|ID| = n$. Then the individual keys $k_i = R(I(i), \mathbf{K})$ are the pre-keys of users i. In this case, the table **I** and, consequently, the pairs $(i, I(i))$, which are elements of **(P, F)**-KPS, are public knowledge. Each user from the privileged group P by his key k_i and the set of pairs $\{(i, I(i)): i \in P\}$ can calculate the common key k_P of this group according also to the public rule $R' : k_P = R'_{(P,F)-KPS}(k_i, \{(i, I(i)) : i \in P\})$. Thus, to calculate the secret key for privileged group P, it is sufficient to know pre-key k_i of any of its elements and protected from unauthorized modification injection table **I**, as well as a composition of the group or corresponding subtable of **I**. For example, the key $k_{\{i, j\}}$ can be computed by the key k_i obtained according to the algorithm (2.1) - Blom Scheme, and the identifier $I(j) = r_j$ by the rule

$$R'_{(2,1)-\text{BlomScheme}} : k_{\{i,j\}} = R'_{(2,1)-\text{BlomScheme}}(k_i, I(j)) = R'_{(2,1)-\text{BlomScheme}}(k_i, r_j) = f(r_j, r_i).$$

If the pre-key $\mathbf{k}_i$ is obtained by the algorithm $(2, 1)$-KDP (n, q)-scheme, then the common key $k_{\{i,j\}}$ can be calculated from it and id (j) by the rule.

$$R'_{(2,1)-\text{KDP}}$$
$$k_{\{i,j\}} = R'_{(2,1)-\text{KDP}}(k_i, (i, id_i), (j, id_j)) = f(k_i(id_i \wedge id_j)) = \varphi((id_i \wedge id_j)k_i),$$

where $(id_i \wedge id_j)\mathbf{k}_i$ is the subset of the set $\mathbf{k}_i$ corresponding to the binary identifier $(id_i \wedge id_j)$, the function φ maps a particular binary vector to a set of binary vectors, for example, their sum modulo two or the value of the hash function from their concatenation. It easy to see that $(id_i \wedge id_j)\mathbf{k}_i = id_i(\mathbf{K}) \cap id_i(\mathbf{K}) = \mathbf{k}_i \cap \mathbf{k}_j$.

We have shown that in general, the system of pre-keys and then working keys based on the updated system key information can be calculated according to the general scheme taking into account the particularity of the given key pre-distribution scheme.

The protocols considered in the chapter are modifications of the Needham-Schroeder protocol with timestamps [4]. The modification consists in that, in addition to the server, there may be more than two participants and each participant i of the privileged group P is transmitted through a secret channel not a working one, but a pre-key $\mathbf{k}_i$ calculated on the basis of the updated secret system key information $\mathbf{K}$ by the algorithm $(\mathbf{P}, \mathbf{F})$-KPS. In addition, the public knowledge but protected against modification the injection table $\mathbf{I}$ and the composition of the group P or only part $\mathbf{I}_P$ of this table corresponding to this composition are transmitted to group P participants. The receiving of the working keys is confirmed by the challenge-response method by contacts with other participants. As in the basic protocols, the participants i of the network protocols under consideration share the long-term keys $K_{\{T,i\}}$ with the server T.

Describing the protocols, we use notation:

$[<\text{message}>]_k$ is a message protected from modification using the key k,

$\{<\text{message}>\}_k$ is a message encrypted on the key k.

$A \rightarrow B$: $<\text{message}>$ means the transmission of a message from A to B.

M (U) means the substitution of participant U, $U \in \mathbf{U}$ by alienated participant M, $M \in \mathbf{U}$.

U: $<\text{data}>$ means that U has the specified data.

U: $<\text{data1}> \leftarrow <\text{data2}>$ means, that participant has computed $<\text{data1}>$ from $<\text{data2}>$.

t is recipients local time,

$\bar{T}$ is current time, rounded to the accuracy of the regular sessions.

$<\text{data}>^R$ is copy of $<\text{data}>$.

$P(i) \subseteq \mathbf{U}$ is the set of participant identifiers available to participant i.

$K \in_R F_q^\tau$ means the random choice of τ elements of the field F_q.

3   Protocol for Delivering Pre-keys Obtained by (2, 1)-KPS

First consider protocol based on (2, 1)-KPS. It is possible to propose a variant of the protocol in which sessions are performed periodically (see Table 1).

Table 1. Protocol 1

Step	Data and transactions
1.	T: $(2,m)$-KPS, $\mathbf{I}, A, \bar{\mathrm{T}}, \mathbf{K} \in_R F_q^\tau$;
	T: broadcast $B = ([i, \mathbf{I}_i, \bar{\mathrm{T}}, \{k_i\}_{K_{T,i}}]_{K_{T,i}} / i \in \mathrm{U})$;
2.1	for all i: $i, K_{T,i}, (i^R, \mathbf{I}_i, \bar{\mathrm{T}}^R, \{k_i\}_{K_{T,i}}) \leftarrow B$; $k_i \leftarrow \{k_i\}_{K_{T,i}}$,
2.2	if $i = i^R$ and $\bar{\mathrm{T}} = \bar{\mathrm{T}}^R$:
	i: $N_i \in_R F_q$, for $j \in P(i)$:
	$\mathbf{k}_{\{i,j\}} \leftarrow (j, \mathbf{I}_i, \mathbf{k}_i)$;
	$i \rightarrow j$: $(j, i), [i, N_i]_{\mathbf{k}_{\{i,j\}}}$;
2.3	j: $\mathbf{I}_i, \mathbf{k}_{j}, i, \mathbf{k}_{\{i,j\}} \leftarrow (i, \mathbf{I}_i, \mathbf{k}_j)$; i^R, N_i^R;
	if $i = i^R$:
	j:$\mathrm{r} = N_i^R - 1$; $j \rightarrow i$: $[\mathrm{r}]_{\mathbf{k}_{\{i,j\}}}$;
	i: r^R, $\mathrm{r}^R = N_i - 1$?

Consider its functionality.

Step 1.   Server T possessing concrete (2, 1)-KPS, and injection table $\mathbf{I}$ periodically for example every day or every minute generates random secret system key information $\mathbf{K}$, generates and broadcasts $B = ([i, \mathbf{I}_i, \bar{\mathrm{T}}, \{k_i\}_{K_{T,i}}]_{K_{T,i}} / i \in \mathrm{U})$ as the cortege of protected on the keys $K_{\{T,i\}}$ packets for participants i, containing identifier i, the part $\mathbf{I}_i$ of the table $\mathbf{I}$ contained pairs $(j, id(j))$ for participants j from the privileged groups P, $i \in P$, session time $\bar{T}$, and cryptograms $\{k_i\}_{K_{T,i}}$ of their individual pre-keys $\mathbf{k}_i$. Participants through monitoring extract their packages, check its integrity, and, if they have updated time $\bar{\mathrm{T}} = \bar{\mathrm{T}}^R$ update their preliminary and working keys.

Step 2.   Each participant i possesses its identifier i, shared with server T key $K_{T,i}$ performs the following.

2.1. Monitoring broadcast B, it gets its package $[i, \mathbf{I}_i, \bar{\mathrm{T}}, \{k_i\}_{K_{T,i}}]_{K_{T,i}}$ that is protected from modification and extracts a copy i^R of its identifier from it, subtable $\mathbf{I}_i$, cryptogram $\{k_i\}_{K_{T,i}}$ of its pre-key, from which it extracts this pre-key $\mathbf{k}_i$.

2.2. If $i = i^R$ and $\bar{\mathrm{T}} = \bar{\mathrm{T}}^R$, participant i generated the query N_i as a random element of the field F_q. Using triplet $(j, \mathbf{I}_i, \mathbf{k}_i)$, for a given participant j, $j \in P\ (i)$, computes key $k_{\{i,j\}}$ and sends to j a pair of identifiers (i,j) and protected from modification by this key package $[i, N_i]_{k_{\{i,j\}}}$.

Participant j in the step 2.1 received $(j^R, \mathbf{I}_i, T^R, \{k_j\}_{K_{T,j}}) \leftarrow B$; $\mathbf{k}_j$ $\leftarrow \{k_j\}_{K_{T,j}}$.

2.3. Participant j has $\mathbf{I}_i, \mathbf{k}_j, i, k_{\{i,j\}} \leftarrow (i, \mathbf{I}_i, \mathbf{k}_j,); i^R, N_i^R;$
If $i = i^R$ it computes response $\mathrm{r} = N_i^R - 1$ and sends it to participant i transaction $[\mathrm{r}]_{k_{\{i,j\}}}$ protected from modification on the key $k_{\{i,j\}}$.

Finally, participant i verifies response r.

The transaction protection is performed on the key $k_{\{i,j\}}$ computed by the participant i, and the verification occurs on the key $k_{\{i,j\}}$, computed by the participant j. Checking the equality $i = i^R$ for participant j is sufficient to conclude that the specified keys are the same. On the other hand, the transaction $[\mathrm{r}]_{k_{\{i,j\}}}$ is protected on the key computed by participant j, and is checked on the key computed by participant i. Checking the equality $\mathrm{r}^R = N_i{-}1$ for participant i is sufficient to conclude that these keys are the same and that participant j has successfully passed a similar test.

Remark. Steps 2.2' and 2.3 can be executed repeatedly for distinct participants j.

4 About Protocols for Delivering Pre-keys Obtained by (P, F)-KPS

The proposed approach to building network key distribution protocols using (2, 1)-KPS is also applicable in the general case of using **(P, F)**-KPS. Pre-keys are calculated by the system **ID** of pre-keys identifiers characteristic for this (**P, F**)-KPS system and the rule R. They are delivered to users in accordance with the appointment by injective table **I**: U → **ID** and working keys are calculated and verified according to the rule R_1 characteristic for this scheme. For example, for schemes (k,m)-KPS, in contrast to $(2, m)$-KPS, key identifiers derived from subtable $\mathbf{I}_i$ are used to calculate from the pre-key $\mathbf{k}_i$ the working key $k_{\{i, j, ..., s\}}$ of the privileged group $\{i, j, ..., s\}$. To confirm the receipt of working keys, the group members exchange the contents of the group with the corresponding request encrypted on these keys.

5   Using Kerberos for Key Distribution Implementation

In this section, we describe the usage of Kerberos protocol to guarantee safe registration and communication of users with the service T in an unsafe environment, where
the transmitted packets can be intercepted and modified. To register in a network, users
communicate to trusted Key Distribution Center (consisting of Ticket Granting Service
and Authentication Service). The KDC maintains a database of credentials of all users
and services (called principals). For each principal the KDC keeps a symmetric key
known only to this user and the KDC itself. This cryptographic key is used for communication between the user and the KDC. In most practical implementations of the
Kerberos protocol, this encryption keys are based on the hashed users' passwords. In
the process of initial registration, the user sends the encrypted message, containing the
time marker and the principal's name to the KDC. The KDC decrypts and checks the
message, and sends a new key and the Ticket-Granting Ticket (TGT) to the user. This
symmetric key will be used for all further interactions between the client and the KDC.
TGT contains encrypted copy of that key, so that KDC would know, that client,
sending TGT while using the key, is correctly identified.

Next, the client communicates with the KDC service again to request access to the
service T using its TGT and time marker, which are encrypted using the received key.
The KDC creates a pair of tickets, one for the client, one for the server T. Each ticket
contains the name of the principal requesting access, the name of the request recipient
T, the time marker, and the lifetime of the ticket. Both tickets will also contain a new
symmetric key $K_{T,i}$, $i \in \mathbf{U}$, which will be known to both the client and the server
T. This key will provide the possibility of a secure interaction between them. The KDC
encrypts the server ticket using the server's long-term key and puts the server ticket
inside the client ticket, which also contains the $K_{T,i}$. After receiving the ticket, the client
encrypts the time marker using the $K_{T,i}$, sends it with the server ticket to the server
T. Now both the client and the server have $K_{T,i}$. Therefore, the server can be sure that
the client is correctly identified, because $K_{T,i}$ was used to encrypt the time marker. If
server needs to reply to the client, the server will use $K_{T,i}$ and the client will know that
the server is correctly identified.

The interaction of the server T with the client is carried out through the User
Datagram Protocol (UDP) sockets and encrypted with the key $K_{T,i}$, obtained by Generic
Security Services Application Program Interface (GSSAPI) [15]. UDP mechanism
doesn't require connection negotiations and allows for multicast and broadcast operations in which the same message is transmitted to several recipients in a single
transmission. The server receives a connection from a user and keeps it alive after the
correct identification. When every user from the privileged group P gets identified,
server starts delivering pre-keys obtained by (2, 1)-KPS using periodically updated
secret system key information $\mathbf{K}$ and session time $\overline{T}$. The packets $[i, \mathbf{I}_i, \overline{T}, \{k_i\}_{K_{T,i}}]_{K_{T,i}}$
are serialized to text, encrypted with $K_{T,i}$ and delivered to each user i accordingly. The
user verifies the decrypted message integrity by deserializing it and updates his preliminary and working keys.

6 Conclusion Remarks

This chapter proposes the general structure of the network key distribution protocols using the key pre-distribution scheme. The security of a key distribution scheme is provided theoretically by the cryptographic method of constructing it as a system ID of identifiers id of individual keys and the rule R calculating the individual key $\mathbf{k} = \mathrm{R}(id, \mathbf{K})$ using the updatable key information $\mathbf{K}$. The assignment of individual keys to users $i \in \mathbf{U}$ is represented by an table $\mathbf{U} \rightarrow \mathbf{ID}$. The delivery of individual keys that are updated in each session in combination with this table or its fragments to users is provided by network variants of a known authentication protocol with timestamps. The final working keys are calculated using individual pre-keys and injections according to rule specific for given type of key pre-distribution scheme. The distribution of long-term keys $K_{T,i}$, in the network can be realized using the Kerberos AS and TGS service functions, and the server T functions can be implemented by the Kerberos S application service.

Acknowledgement. The work was supported financially by the Russian Foundation for Basic Research, the project 17-01-00485a.

References

1. Li, X., Li, D., Wan, J., Vasilakos, A.V., Lai, C.F., Wang, S.: A review of industrial wireless networks in the context of industry 4.0. Wirel. Netw. **23**, 23–41 (2017)
2. Rottondi, C., Verticale, G., Capone, A.: Privacy-preserving smart metering with multiple data consumers. Comput. Netw. **57**, 1699–1713 (2013)
3. Stinson, D.R.: Cryptography: Theory and Practice, 3rd edn. CRC Press, Boca Raton (2006)
4. Wenbo, M.: Modern Cryptography: Theory and Practice. Hewlett-Packard Company/Prentice Hall, Inc., Upper Saddle River (2003)
5. MIT Kerberos Documentation - Complete reference - API and datatypes. https://web.mit.edu/kerberos/krb5-1.14/doc/appdev/refs/index.html
6. Akhbarifar, S., Rahmani, A.M.: A survey on key pre-distribution schemes for security in wireless sensor networks. Int. J. Comput. Netw. Commun. Secur. **2**(12), 423–442 (2014)
7. Stinson, D.R.: On some methods for unconditionally secure key distribution and broadcast encryption. In: Designs, Codes and Cryptography. Kluwer Academic Publishers, Norwell (1997)
8. Blom, R.: Nonpublic key distribution. In: Advances in Cryptology - Proceedings of EUROCRYPT 1982, pp. 231–236. Plenum, New York (1983)
9. Dyer, M., Fenner, T., Frieze, A., Thomason, A.: On key storage in secure networks. J. Cryptol. **8**, 189–200 (1995)
10. Shruthi, P., Nirmala, M.B., Manjunath, A.S.: Secured modified Bloom's based q-composite key distribution for wireless sensor networks. Int. J. Adv. Comput. Theory Eng. (IJACTE) **2**(5), 2319–2526 (2013)
11. Eschenauer, L., Gligor, V.D.: A key-management scheme for distributed sensor networks. In: CCS 2002 Proceedings of the 9th ACM Conference on Computer and Communications Security, pp. 41–47. ACM Press, New York (2002)
12. Ramkumar, M.: Symmetric Cryptography Protocols. Springer, Heidelberg (2014)

13. Frolov, A.B., Shchurov, I.I.: Non-centralized key pre-distribution in computer networks. In: IEEE Proceedings of International Conference on Dependability of Computer Systems DepCos-RELCOMEX 2008, Szklarska Poreba, Poland, pp. 179–188. Computer Society Conference Publishing Services, Los Alamitos, California, Washington, Tokyo (2008)
14. Frolov, A., Zatey, A.: Probabilistic synthesis of KDP satisfying mutually complementary correctness conditions. In: 2014 Proceedings of International Conference on Advances in Computing, Communication and Information Technology, Birmingham, UK, 16–17 November, pp. 75–79 (2014)
15. Generic Security Services Application Program Interface – Python GSSAPI wrapper. https://pypi.python.org/pypi/gssapi

Neurons' Transfer Function Modeling with the Use of Fractional Derivative

Zbigniew Gomolka(✉)

Faculty of Mathematics and Natural Sciences, Department of Computer Engineering, University of Rzeszow, ul. Pigonia 1, 35-959 Rzeszow, Poland
zgomolka@ur.edu.pl

Abstract. The paper presents new approach the idea of artificial neural network (ANN) modelling and design with the use of calculus of finite differences proposed by Dudek-Dyduch [5, 8] and then developed jointly with Tadeusiewicz [5] and others. Previously such neural nets were applied mainly to the extraction of different features i.e. edges, ridges, maxima, extrema and many others that can be defined with the use of a classic derivative of any order and their linear combinations. The author extend this method of ANN modelling by the use of fractional derivative theory for transfer function modelling. Different types of fractional derivatives and their numerical accuracy have been presented. Finally it was shown that the discrete approximation of fractional derivative of some base functions allows for modelling the transfer function of a single neuron for various characteristics. In such an approach, the smooth control of a derivative order allows the neuron dynamics to be modeled without direct modification of the source code in the IT model. The novel approach universalizes the model of the artificial neurons.

Keywords: Neuron · Transfer function · Fractional derivative

1 Introduction

The Back Propagation algorithm (BP) is the most popular algorithm for minimizing the error function that is used in supervised methods of learning algorithms. Its characteristic element is the neuron transfer function, whose algebraic form implies the IT implementation of a learning algorithm. The derivative of the neuron transfer function is a key limitation here that forces a change of source code in each algorithm when different types of functions like *sigmoid, tanh,* or *Gauss* are implemented [8, 9]. Early models of the first perceptrons described in the work of Minsky and Papert [16, 17] and recently in Convolutional Neural Networks and Cellular Neural Networks hold a common assumption that the transfer function of a single neuron is a constant along the learning process. The model of recurrent networks and deep networks proposed recently by the Schmidhuber [25] research team also assumes pre-defined transition function. This is a strong limitation of the algorithm's flexibility, which results from the structure of the module calculating the gradient of the network error. The derivative of the fractional order, paradoxically discussed firstly in 1695 by Hospital and Leibnitz [22, 23, 28] in their correspondence has become, in current research in the field of

© Springer International Publishing AG, part of Springer Nature 2019
W. Zamojski et al. (Eds.): DepCoS-RELCOMEX 2018, AISC 761, pp. 218–227, 2019.
https://doi.org/10.1007/978-3-319-91446-6_21

complex systems, a new direction of exploration that provides hope of overcoming this limitation. There is also some doubt about the traditional model of feed forward ANN types which have to assume rigid dynamics of a single neuron along the whole process of training. Based on the knowledge and experience of the author in the field of neural networks modeling, a mechanism for the smooth selection of the transition function was proposed that would enable the classical BP algorithm to work under fluent changes of the transfer function controlled by order of fractional derivative of the presumed base function. This theme is the guiding principle of this work, which is a report of the initial stage of research carried out at the Center for Innovation and Transfer of Natural Sciences and Engineering Knowledge at the University of Rzeszow. To provide the reader with a better understanding we recall here briefly the mathematical model which we assumed as the starting point for further consideration in this work. The mechanism of integer discrete derivative measurement is widely used in digital signal processing and neural network modelling. The zero crossings detection in the task of edges searching in the image, the detection of extrema in the task of axial lines searching are just two examples of its applications. The backward difference of discrete function $f(k)$ defined for a real variable k defined for interval $[k_0, k]$, where $0 \leq k_0 \leq k$, of the integer order n is described by the formula:

$$_{k_0} \Delta_k^{(n)} f(k) = \sum_{i=k_0}^{k} a_{i-k_0}^{(n)} f(k + k_0 - i), \; n \geq 0 \tag{1}$$

where adjacent coefficients $a_i^{(n)}$ are defined as follow:

$$a_i^{(n)} = \begin{cases} 1 & for & i = 0 \\ (-1)^i \frac{n(n-1)(n-2)...(n-i+1)}{l!} & for & i = 1, 2, 3, ..., N \end{cases} \tag{2}$$

In this way, the mechanism of weight distribution might be obtained for the modelling of a neural network of homogeneous type. Such networks have been successfully used by Dudek-Dyduch [5] and Gomolka [9] for interferometry images analysis and the efficiency of this type of modelling has been presented in [8]. This technique of discrete measurement we can apply simply to many classes of signals including those performed by ANNs [3, 4, 11, 12, 14, 19, 26, 27].

2   Assumptions About Fractional Derivatives Models

The Grünwald-Letnikov fractional derivative (GL) is the extended definition firstly proposed independently by Anton Grünwald and Aleksey Letnikov [1, 2, 6, 7] in 1867–1868. Based on the definition of the integer derivative and the fractional derivative [10, 13, 15, 20, 21, 24], the GL derivative is described by the formula:

$$_{x_o} D_x^v f(x) = \lim_{h \to 0} \frac{1}{h^v} \sum_{j=0}^{[(x-x_0/h)]} (-1)^j \binom{v}{j} f(x - jh) \tag{3}$$

where $\begin{pmatrix} v \\ j \end{pmatrix}$ denotes Newton binomial, v – order of fractional derivative of $f(x)$, x_0 – the interval range, h - step of discretization. For the given discrete function $f(x)$ of a real variable x defined on the interval $\langle x_0, x \rangle$ where $0 \leq x_0 \leq x$, we assume the backward difference of fractional order v $_{x_0}\Delta_x^{(v)}$ where $v \in R^+$:

$$_{x_0}\Delta_x^{(v)} f(x) = \sum_{i=0}^{\lfloor (x-x_0)\rfloor /h} a_i^{(v)} f(x - ih) \tag{4}$$

or in the equivalent form as:

$$_{x_0}\Delta_x^{(v)} f(x) = \begin{bmatrix} a_0^{(v)} & a_1^{(v)} & \cdots & a_{\lfloor (x-x_0)/h \rfloor}^{(v)} \end{bmatrix} \begin{bmatrix} f(x) \\ f(x-h) \\ \vdots \\ f(x_0) \end{bmatrix} \tag{5}$$

where consecutive coefficients $a_i^{(v)}$ are defined as follow:

$$a_i^{(v)} = \begin{cases} 1 & for & i = 0 \\ (-1)^i \frac{v(v-1)(v-2)...(v-i+1)}{i!} & for & i = 1, 2, 3, \ldots, N \end{cases} \tag{6}$$

N represents the number of discrete measurements. In practice it is often assumed that $x_0 = 0$, so we have:

$$_0\Delta_x^{(v)} f(x) = \sum_{i=0}^{\lfloor \frac{x-x_0}{h} \rfloor} a_i^{(v)} f(x - ih) \tag{7}$$

Consequently the progressive difference might be defined in the following form:

$$_x\Delta_\infty^{(v)} f(x) \sum_{i=0}^{\infty} a_i^{(v)} f(x + ih) \tag{8}$$

or in the vector notation form:

$$_x\Delta_\infty^{(v)} f(x) = \begin{bmatrix} a_0^{(v)} & a_1^{(v)} & \cdots & a_\infty^{(v)} \end{bmatrix} \begin{bmatrix} f(x) \\ f(x+h) \\ \vdots \\ f(\infty) \end{bmatrix} \tag{9}$$

Let's assume for some N that $f(x) = 0$ for $x \leq x\_N \leq \infty$, where $x_N = x + Nh$. Then we receive:

$$_x\Delta_{x_N}^{(v)} f(x) = \sum_{i=0}^{\infty} a_i^{(v)} f(x + ih) = \sum_{i=0}^{\lfloor x_N - x/h \rfloor} a_i^{(v)} f(x + ih) \tag{10}$$

or in the form of a product of vectors:

$$_x\Delta_{x_N}^{(v)} f(x) = \left[a_0^{(v)} a_1^{(v)} \quad \cdots \quad a_{(x_N-x)/h}^{(v)} \right] \begin{bmatrix} f(x) \\ f(x+h) \\ \vdots \\ f(x_N) \end{bmatrix}$$ (11)

Sample graphs of $a_i^{(v)}$ coefficients were shown in Fig. 1.

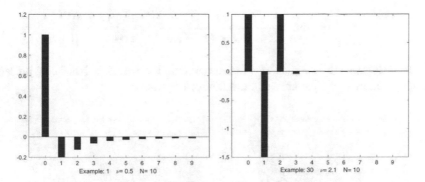

Fig. 1. The exemplary series of $a_i^{(v)}$ coefficients

To determine the variability of the values of coefficients $a_i^{(v)}$ for the value of $N > 10$ an asymptote $1/i$ and $-1/i$ markers were introduced, which allows the stability of the solution for determining the value of subsequent coefficients to be assessed. As can be seen, this sequence is more strongly convergent to 0, which means that successive coefficients meet the condition $\left| a_{i-1}^{(v)} \right| > \left| a_i^{(v)} \right|$ for $i > n$, where n denotes the considered integer part of derivative $n = v$ (Fig. 2).

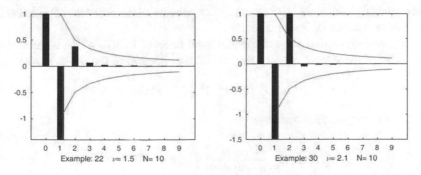

Fig. 2. Asymptotic convergence of coefficients $a_i^{(v)}$

For the purposes of the simulation stage of the work, the model of Riemann-Liouville fractional derivative (FDRL) [15, 18] was adopted for further consideration. The definition of the Reimann-Liouville integral (RLI) [23, 29] will be assumed:

$$_{x_0}I_x^{-v}f(x) = \frac{1}{\Gamma(-v)}\int_{x_0}^{x}(x-\tau)^{-v-1}f(\tau)d\tau \tag{12}$$

where $\Gamma(-v)$ denotes the Gamma function. It is necessary to recall some specific features regarding the accuracy of determining the Gamma function $\Gamma(z)$. The value of this function can be determined using the relationship below:

$$\Gamma(z) = \lim_{i\to+\infty}\frac{i!i^z}{z(z+1)(z+2)\ldots(z+i)} = \frac{1}{z}\prod_{i=1}^{\infty}\frac{\left(1+\frac{1}{i}\right)^z}{\left(1+\frac{z}{i}\right)} \tag{13}$$

At the Fig. 3 the accuracy of Gamma function calculation has been presented assuming variance of i parameter in the range: $i = 1{:}500$.

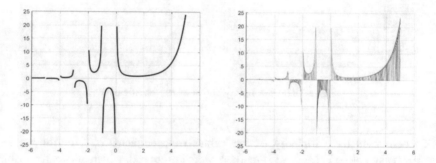

Fig. 3. The basic form of the Gamma function according to i value

The accuracy of the numerical determination of this function for different values i is presented in Fig. 4. The large number of series expressions in relationship (13) significantly increases the accuracy of calculations, at the same time increasing the computational complexity of the algorithm. Computational inaccuracies that are entered directly influence the accuracy of determining non-zero coefficients of derivative or integral of fractional order.

Next, we define a derivative of the fractional order of FDRL, which is calculated as a derivative of the integer order of the non-integer integral of the $(n - v)$ order [1, 9]:

$$_{x_0}D_x^{v}f(x) = D^n{}_{x_0}I_x^{n-v}f(x) \tag{14}$$

where $n - 1 < v \leq n$. Then we have:

$$_{x_0}D_x^{v}f(x) = \frac{1}{\Gamma(n-v)}\frac{d^n}{dx^n}\left[\int_{x_0}^{x}(x-\tau)^{n-v-1}f(\tau)d\tau\right] \tag{15}$$

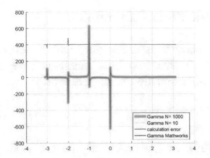
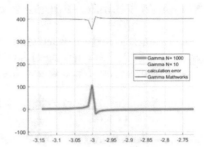

Fig. 4. Examples of determining the Gamma function values for different values of coefficients N

For simplicity, it can be assumed that $n = \lceil v \rceil$, i.e. the nearest integer greater than v. Eventually the Caputo derivative of the fractional order $v > 0$ of the real function $f(x)$ is defined [13]:

$$x_0 D_x^v f(x) = \frac{1}{\Gamma(n-v)} \int_{x_0}^{x} (x-\tau)^{n-v-1} f^{(n)}(\tau) d\tau \qquad (16)$$

where: x_0, x – integration limits, fulfilling the condition $-\infty < x_0 < x < \infty$ and $n - 1 < v < n, \in \mathbb{Z}$. Generalizing the above definitions, we introduce the term diferintegral which is the result of the following notation convention:

$$x_0 D_x^v f(x) = x_0 I_x^{-v} f(x) \qquad (17)$$

where: $x_0 D_x^v$ - derivative of order $v > 0$, $x_0 I_x^{-v}$ – integral of order $-v \le 0$, x_0, x – integration limits or derivation respectively. It is usually assumed $x_0 = 0$. Meerschaert and Tadjeran show [9] that the presented standard GL formula is often unstable for time dependent problems. Hence, they proposed the following shifted GL formulas for the left (18) and right (19) Riemann-Liouville derivatives (RLD) in order to overcome the instability [2, 10, 18]:

$$x_0 D_x^v f(x) = \sum_{i=0}^{m+1} a_i^{(v)} f(x_{m-i+1}) \qquad (18)$$

$$x D_{x_N}^v f(x) = \sum_{i=0}^{M-m+1} a_i^{(v)} f(x_{m+i-1}) \qquad (19)$$

where: M and m denotes range of the interval and amount of shifts respectively. Adopted models of fractional derivatives were used in the ensuing stages of the work to determine the discrete approximation of the derivative of the single neuron transition function.

3 Neuron's Transfer Function with Fluent Dynamics

Considering sigmoid and gauss as two exemplary transfer functions which most often appear in the ANN learning algorithms [8, 9, 22], two base functions have been assumed, a sigmoid-like function and Gauss-like function respectively (20):

$$y_S^v = \log(1 + e^x), \; y_G^{v+1} = \frac{1}{e^{x^2}} \tag{20}$$

Assuming the relationship (3), (5) and (20), we have (similarly for Gauss too):

$$_{x_0}\Delta_x^{(v)} f_S(x) = \begin{bmatrix} a_0^{(v)} a_1^{(v)} & \cdots & a_\infty^{(v)} \end{bmatrix} \begin{bmatrix} \log(1 + e^x) \\ \log(1 + e^{x-1}) \\ \vdots \\ \log(1 + e^{x-x_0}) \end{bmatrix} \tag{21}$$

By adopting such a model of the neuron transition function it is possible to control its shape by manipulating the order of the derivative. In addition, for a given shape of such a function it is possible to calculate its derivative to obtain information about the gradient of error function. This is a key advantage of the proposed method, which allows us to use those functions for which it is not possible to determine the algebraic derivative of the fractional order.

Figure 5 shows an exemplary course of a sigmoidal function for the v coefficient located close to the node of the integer value. Since the log base function is differentiable, it is possible to obtain a series of transfer functions for the particular neuron by changing the value of the v coefficient only. The bottom part of this figure shows also the accompanying error due to the accuracy of discrete derivative approximation.

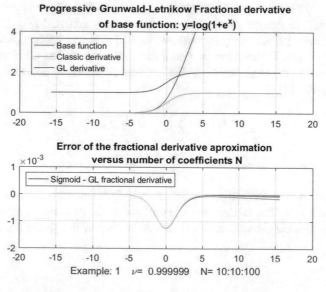

Fig. 5. The accuracy of the fractional derivative approximation

4   Results

A series of simulation experiments have been carried out for various vector lengths of coefficients of GL non-integral order derivative. The upper row of the Fig. 6 shows exemplary characteristics assuming the length of fractional coefficients vector $\vec{a}^{(v)}$, $n = 250$ and input base function in the form of $f_S^{vs(0)} = log(1 + e^x)$. Similar experiments were conducted with the use of Riemann-Liouville derivative-integral obtaining analogous fluency of hypothetical neuron transfer function shaping.

The accuracy of the discrete approximation of a derivative is a key factor deciding about the possibility of using such an approach to obtain various transition functions for a single neuron or a network of such neurons. In the experiments conducted, it was experimentally confirmed that the fractional-order derivative mechanism allows the smooth selection of the form of the transition function by the fractional value of the order of the derivative used. At the current stage of research it has been shown that it is possible to use the proposed method to obtain a new algorithm for BP ANN models. Although several methods of calculating a discrete approximation of a fractional derivative have been presented in the paper, as a part of continuing the research, several tasks may be indicated. Some comparisons considering the accuracy of the numerical calculations have to be made according to the type of progressive, backward or symmetric measurement technique strategies, especially when the local dynamics of

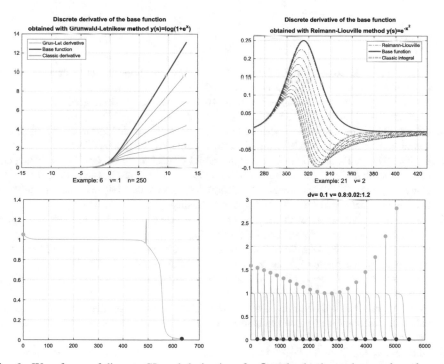

Fig. 6. Waveforms of discrete GL and derivatives for fluently shaping order v and used vector of fractional coefficients n.

the functions become substantially high. We also predict that the appropriate choice of the type of the base transfer function will avoid the instability of those calculations which are close to asymptotes of Gamma function (FDRL and Caputo case and Gamma argument is close to 0^+). The bottom row of Fig. 6 presents the exemplary training trial within the classic BP network (left side) while at the right side the consecutive net trials with the same set of weights upon the fractional shape of the neurons transfer functions is presented. The green circles denote training starts while the red ones denote successful training ends. As can be seen, the net converges to the final SSE goal for the majority of values of v. The essential achievement of the proposed method is the fact that by using model of fractional derivative it is possible to carry out an effective learning process with gradient methods for the transition function with varying dynamics. This property of the proposed approach allows the use of various transition functions without the need to modify the IT model of the neural network. This approach is a completely new original study, which is a novel approach to constructing learning algorithms. Tests with the artificially obtained sigmoid like neuron for the fixed order of v under the ANN while solving the XOR problem have been successful and these are very promising results for further investigations. Although this approach has many advantages nevertheless some of its limitations have to by underlined. Based on the observations of several experiments it has been noted that the speed of the learning with a fractional derivative might be substantially decreased by the number of fractional coefficients. The accuracy of the net efficiency strongly depends on the computational accuracy of derivative approximation.

References

1. Cao, L.D., Rodriguez-Lopez, R.: From fractional order equations to integer order equations. Fract. Calc. Appl. Anal. **20**(6), 1405–1423 (2017)
2. Ding, H., Li, C.: High-order algorithms for Riesz derivative and their applications. Numer. Methods Partial Differ. Equ. **33**(5), 1754–1794 (2017)
3. Drałus, G.: Global models of dynamic complex systems – modelling using the multilayer neural networks. Ann. UMCS Sectio AI Inform. **7**(1), 61–71 (2007)
4. Dudek-Dyduch, E.: Algebraic logical meta-model of decision processes - new metaheuristics. In: Artificial Intelligence and Soft Computing, ICAISC, pp. 541–554 (2015)
5. Dudek-Dyduch, E., Tadeusiewicz, R.: Neural networks indicating maxima and ridges in two-dimensional signal. In: EAANN, FAIS - Helsinki, pp. 485–488 (1995)
6. Garrappa, R.: Numerical evaluation of two and three parameter Mittag-Leffler functions. SIAM J. Numer. Anal. **53**(3), 1350–1369 (2015)
7. Ghosh, U., Sarkar, S., Das, S.: Solution of system of linear fractional differential equations with modified derivative of Jumarie type. Am. J. Math. Anal. **3**(3), 72–84 (2015)
8. Gomolka, Z.: Neural networks in the analysis of fringe images. Ph.D. thesis (2000). (in Polish)
9. Gomolka, Z., Dudek-Dyduch, E., Kondratenko, Y.P.: From homogeneous network to neural nets with fractional derivative mechanism. In: ICAISC 2017. Lecture Notes in Computer Science, vol. 10245 (2017)
10. Ding, H., Li, C., Chen, Y.: High-order algorithms for Riesz derivative and their applications. J. Comput. Phys. **293**, 218–237 (2015)

11. Kondratenko, Y.P., Sidenko, I.V.: Design and reconfiguration of intelligent knowledge-based system for fuzzy multi-criterion decision making in transport logistics. J. Comput. Optim. Econ. Financ. **6**, 229–242 (2014)
12. Kwater, T., Bartman, J.: Application of artificial neural networks in non-invasive identification of electric energy receivers. In: PAEE, Koscielisko, pp. 1–6 (2017)
13. Luo, W.-H., Li, C., Huang, T.-Z., Gu, X.-M., Wu, G.-C.: A high-order accurate numerical scheme for the caputo derivative with applications to fractional diffusion problems. Numer. Funct. Anal. Optim. **39**(5), 600–622 (2018)
14. Mazurkiewicz, J.: Dependability metrics for network systems—analytical and experimental analysis. In: Dependability Engineering and Complex Systems. Advances in Intelligent Systems and Computing, vol. 470 (2016)
15. Ortigueira, M.D., Machado, J.T.: What is a fractional derivative? J. Comput. Phys. **293**, 4–13 (2015)
16. McClelland, J.L.: Explorations in Parallel Distributed Processing: A Handbook of Models, Programs, and Exercises. MIT Press, Cambridge (2015)
17. Minsky, M., Papert, S.: Perceptrons: An Introduction to Computational Geometry, Expanded Edition Paperback. The MIT Press, Cambridge (1987). ISBN 0-262-63022-2
18. Moret, I.: Shift-and-invert Krylov methods for time-fractional wave equations. Numer. Funct. Anal. Optim. **36**(1), 86–103 (2015)
19. Olchawa, A., Walkowiak, T.: Automatic reconfiguration in the response to network incidents by neural networks. In: Al-Dahoud, A. (ed.) ICIT 2009, Amman, Jordan, 3–5 June 2009, pp. 83–89. Al-Zaytoonah University of Jordan, Amman (2009)
20. Ortigueira, M.D.: Riesz potential operators and inverses via fractional centered derivatives. Int. J. Math. Math. Sci. **2006**, 1–12 (2006). Article ID 48391
21. Oustaloup, A., Levron, F., Mathieu, B., Nanot, F.M.: Frequency-band complex noninteger differentiator: characterization and synthesis. IEEE Trans. Circ. Syst. I: Fundam. Theory Appl. I. **47**, 25–39 (2000)
22. Petraš, I.: Fractional Derivatives, Fractional Integrals, and Fractional Differential Equations in Matlab. In: Computer and Information Science – Engineering Education and Research Using MATLAB, pp. 239–264 (2011). ISBN 978-953-307-656-0
23. Podlubny, I.: Matrix approach to discrete fractional calculus. Fract. Calc. Appl. Anal. **3**, 359–386 (2000)
24. Sheng, H., Li, Y., Chen, Y.Q.: Application of numerical inverse Laplace transform algorithms in fractional calculus. J. Franklin Inst. **348**, 315–330 (2011)
25. Schmidhuber, J.: Who Invented Backpropagation? (2014) (updated 2015). http://people.idsia.ch/~juergen/who-invented-backpropagation.html
26. Twaróg, B., Gomółka, Z., Żesławska, E.: Time analysis of data exchange in distributed control systems based on wireless network model. In: Lecture Notes in Electrical Engineering, vol. 452, pp. 333–342 (2018)
27. Twarog, B., Pekala, R., Bartman, J., Gomolka, Z.: The changes of air gap in inductive engines as vibration indicator aided by mathematical model and artificial neural network, DCDS - Series A Issue Supplement, September 2007
28. Vinagre, B.M., Podlubny, I., Hernández, A., Feliu, V.: Some approximations of fractional order operators used in control theory and applications. Fract. Calc. Appl. Anal. **3**, 231–248 (2000)
29. Xue, D.: Computational aspect of fractional-order control problems. In: Tutorial Workshop on Fractional Order Dynamic Systems and Controls, Proceedings of the WCICA 2010 (2010)

The Implementation of an Intelligent Algorithm Hybrid Biometric Identification for the Exemplary Hardware Platforms

Zbigniew Gomolka[1](✉) ⓘ, Boguslaw Twarog[1] ⓘ,
and Ewa Zeslawska[2] ⓘ

[1] Department of Computer Engineering, Faculty of Mathematics and Natural
Sciences, University of Rzeszow, ul. Pigonia 1, 35-959 Rzeszow, Poland
{zgomolka, btwarog}@ur.edu.pl
[2] Department of Applied Information, Faculty of Applied Informatics,
University of Information Technology and Management in Rzeszow,
ul. Sucharskiego 2, 35-225 Rzeszow, Poland
ezeslawska@wsiz.rzeszow.pl

Abstract. The area of security of biometric access control systems is a rapidly
growing field of scientific studies, diversely applicable in banking, electronic
payment, etc. The paper presents the implementation of an intelligent algorithm
hybrid biometric identification with the use of VistaFA2, IriTech and Futronic
scanners. The system uses the biometric reading of human iris, face and fin-
gerprints. An intelligent module determines a similarity measure of a processed
hybrid feature vector to the consecutive records in the access database. This
approach helps to increase the reliability of identification systems and reduces
the risk of counterfeits and intrusion into restricted access resources.

Keywords: Biometric identification · Hybrid systems · Biometric scanners

1 Introduction

Providing fast and secure access to information, places, and equipment connected to the
authorisation of access, is one of the most important security problems. The case of
controlling restricted access to expensive and specialized industrial machinery that
requires special power and privileges from users provides a good example. The effects
of unauthorized access to resources may have tremendous financial and humanitarian
consequences. Thus, biometrics is seen as the first line of defence against one of the
most significant threats to the modern world - terrorism. Nowadays, the majority of
biometric products deployed use information from single biometric devices in the task
of verification or identification. For huge biometric systems there is a demand to ensure
high system accuracy and safety under larger population coverage and demographic
diversity, varied deployment environments, and more demanding performance
requirements. At present, one-type modality biometric systems are finding it difficult to
meet these requirements, and the solution discussed in this paper is to integrate additional sources of information to strengthen the decision process. The system which

© Springer International Publishing AG, part of Springer Nature 2019
W. Zamojski et al. (Eds.): DepCoS-RELCOMEX 2018, AISC 761, pp. 228–237, 2019.
https://doi.org/10.1007/978-3-319-91446-6_22

combines information from multiple biometric channels, algorithms, sensors, and other components constitutes an accurate solution to that problem. Besides improving the accuracy, the idea of fusing multiple sources of biometric data has several advantages, such as increasing population coverage, deterring spoofing activities and reducing false acceptance and false rejection decisions. The continuous growth of the research and commercialization of scientific solutions in this area confirms that this trend will be continued. Therefore, here we propose an intelligent hybrid biometric identification system which uses a multimodal biometric authentication approach fusing fingerprint, facial geometry and iris. The practical implementation of such systems should take into account both software and hardware platform requirements and limitations formatted by standards for reading, processing and writing biometric information in accordance with applicable ISO norms [1–3, 14, 16]. The measurements and applied algorithms must take into account a number of optical noises that should be effectively eliminated, ensuring the high reliability of results. This approach helps to increase the reliability of identification systems and reduces the risk of counterfeits and intrusion into restricted access resources [5, 6, 18]. The research and experimental part of this work have been conducted with instruments and software at the Laboratory of Cognitive Science Research at the Centre for Innovation and Transfer of Natural Sciences and Engineering Knowledge, University of Rzeszow.

2 Software and Hardware Assumptions

To ensure compatibility of the developed application with hardware platforms and operational systems we used standards for biometric data acquisition and exchange. The standards both available on the market and developed are described in [13, 14, 16]. MegaMatcher SDK v. 5.0 development environment provided by Neurotechnology was used for the IT part of the system. As a development platform, MS Visual Studio and C # were chosen. The protocols used for reading and writing biometric information provided by MegaMatcher were used to construct the main application module. This product involves scalable technology that ensures high reliability and speed of biometrical identification even when using large databases. High productivity and efficiency are supported by specialized algorithms that encompass fingerprint, face, iris and voice recognition engines. This allows integrators to use various algorithms in order to ensure better identification results, or any of these engines separately. The fault-tolerant scalable software allows the performance of fast parallel matching, processes a large number of identification requests and handles databases with data of big size. Optionally, MegaMatcher environment also supports server software for local biometrical systems and cluster software for large-scale biometrical products development. The system is compatible with the standard of WSQ by American Advancing Biometric Federal Bureau of Investigation [1]. We use the following software tools: FingerMatcher, FaceMatcher and IrisMatcher. They are designed to take advantage of information flow in different types of computer architectures and biometric scanner types [13]. We have used the Windows 7 operating system. According to above assumptions we choose the scanners listed below (see Fig. 1.):

- **Vista-FA2-01, Vista Imaging Inc VistaFA Bio Camera** – a device for multimodal scanning with the function of image acquisition at 640 × 480 pixel resolution, it incorporates a 3MP full 24bit color face camera to facilitate facial image captures with a flash illumination feature. The integrated microphone & speaker provides 16-khz 16-bit audio I/O, ISO/IEC 19794-6 compliant iris images, ISO/IEC 19794-5 compliant face images (see Fig. 1(a)).
- **IriShield USB MK 2120U** – is a ready-to-use single iris capture camera, suitable for using with smartphones, tablets or other handheld devices. Each eye is illuminated by infrared LED, thus irises can be captured in various indoor and outdoor environments. The captured iris images are compliant with the ISO/IEC 19794-6 standard (see Fig. 1(b)).
- **FUTRONIC Fingerprint Scanner** – a fingerprint scanning window size 16 × 24 mm acquires images at resolution 480 × 320 pixel, 500 DPI with color depth 8 bit 256 grayscale, with a Live Finger Detection (LFD) feature (see Fig. 1(c)).

(a)   (b)   (c)

Fig. 1. Devices used in the experimental stage of investigation

Exemplary biometric data obtained with the above devices are presented in Fig. 2. Different circumstances of measurement cause proper or improper registration of data. These examples show the influence of different factors such as: the scene illumination, finger placement and pressure or influence of eye content levels on the possible noisy

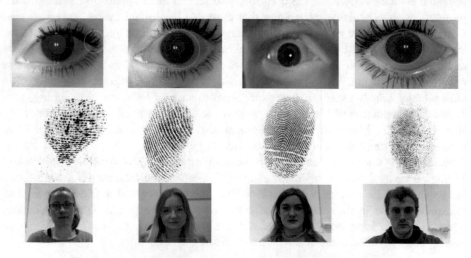

Fig. 2. Proper and improper data acquisition of exemplary subjects

performance of the scanners. The appropriate measurement conditions play a crucial role in data acquisition [11, 15, 17, 19].

3 Biometric Intelligent Hybrid System Proposal and Implementation

Although there are several scenarios of biometric data fusion, most considerations about multimodal biometric systems integrate data at the different stages. It may consider comparison of score levels or weighted decision voting strategy. Both circumventions offer a strong compromise between the ease in combining the data and better information content, because it is a relatively straightforward way to combine the scores generated by different matchers. Therefore, we proposed the flow of the processed signals in the biometric intelligent hybrid system as shown in Fig. 3 below:

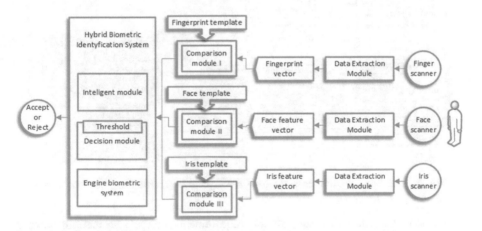

Fig. 3. The flow of the signal and its fusion in the proposed system

The fusion effort in this system is commensurate to three separate verification processes joined together at the intelligent output module which is responsible for generating the ultimate accept or reject decision. In a typical classification approach, one can construct a feature vector with individual comparison scores, and it is then classified into accept or reject classes. A classification approach might use a decision tree, a Support Vector Machine (SVM) or Linear Discriminate Analysis (LDA) algorithm to classify the vector as imposter or genuine. Assuming the hybrid approach, one com-bines individual comparison scores to generate a single scalar score to render the final decision. The intelligent hybrid approach for consolidating comparison scores has given a superior performance record versus the other approaches [4–7, 20].

The organizational schema (see Fig. 4) shows the design of the class structure used in the facial processing channel in the Hybrid Biometric Identification System. Analogously, the information is processed in the other two biometric channels.

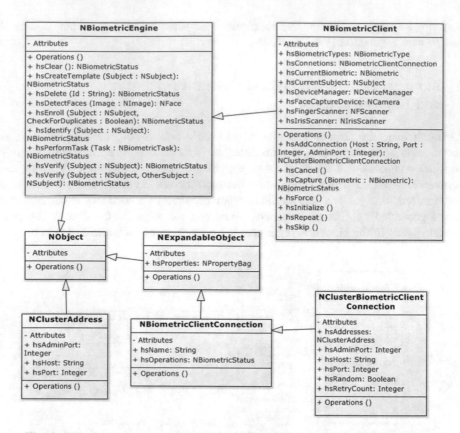

Fig. 4. The class structure project within the Hybrid Biometric Identification System

Such a structure takes into account the requirements of the information exchange protocol described in [1, 8, 9, 10, 12, 13].

An example of invoking new Devices API for an exemplary capturing of the fingerprint image acquired from a FUTRONIC scanner is presented below. First, the code block initializes the licensing module which opens the signal transporting channel. Then, after allocation in memory, hsDeviceManager retrieves the biometric data from the input.

```
try{
  if (!NLicense.ObtainComponents("/local", 5000,
"Devices.FingerScanners")){…}
  NDeviceManager hsDeviceManager =
new NDeviceManager(NDeviceType.FingerScanner, true,
false, null);
  foreach(NFScanner hsScanner in hsDeviceManager.Devices){
    NImage hsImage = hsScanner.Capture(-1);
    hsimage.Save(filename);}}
catch (Exception ex){…}
finally{
  NLicense.ReleaseComponents(hsComponents);}
```

A typical identification system for client workstation data flow diagram of the application (Fig. 5):

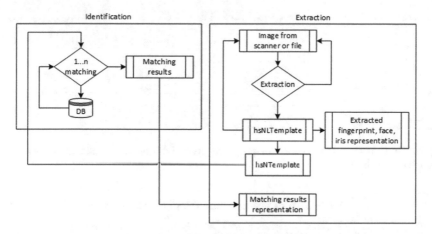

Fig. 5. Decision-making sub-module in the intelligent biometric identification system

Figure 6 shows the Graphic User Interface of the proposed Hybrid Biometric Identification System. Three sections provide the user with online biometrics previews and biometric identification system controls. To simplify the appearance of the interface, we omitted several components that incorporate the technical details of the measurement (irises and finger). Matching threshold is linked to the false acceptance rate (FAR, different subjects erroneously accepted as the same) of the matching algorithm. The higher the threshold is, the lower the FAR and the higher FRR are (false rejection rate, same subjects erroneously accepted as different) and vice versa. The matching threshold for NMatcher has been determined with the following specification (Table 1):

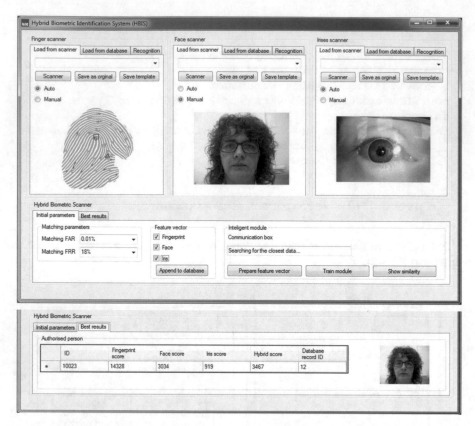

Fig. 6. Graphic User Interface of the proposed Hybrid Biometric Identification System with exemplary match results

Table 1. FAR and matching threshold dependencies

FAR (false acceptance rate)	Matching threshold (score)
100%	0
10%	12
1%	24
0.1%	36
0.01%	48
0.001%	60
0.0001%	72
0.00001%	84
0.000001%	96

or using this formula: $hsTreshold = -12 \log_{10}(FAR)$; where FAR is NOT percentage value (e.g. 0.1% FAR is 0.001). The higher score value at the particular channels presented in the GUI suggests the higher probability that feature collections are obtained from the same person. The proposed intelligent module combines the environment functionality i.e. templates can be complex, can have information of multiple modalities and multiple records. The matching is performed at the lowest level (on records) and resulting scores fused to return a single score. In the proposed software model the confidence score is provided but it might be conveniently changed to the simple answer mode: accepted, rejected.

```
for (int i = 1; i < hsTemplates.Length; i++)
{
  int hsScore = hsMatcher.IdentifyNext(hsTemplates[i]);
  if (hsScore > hsThreshold)
  {
    hsScoreList.rows[0].cells[1].value = hsScore;
    hsScoreList.rows[0].cells[5].value =
hsDatabase.ID[i].value;
  }
}
hsMatcher.IdentifyEnd();
```

As the final result of the complex inference process, the smart module indicates a record in the database of authorized persons. The part of the code presented above performs a simple search for the maximum similarity score value, at the adjacent channels.

4 Conclusion

The proposed intelligent hybrid system for biometric identification is an initial project aimed at increasing the reliability and security of classic unimodal identification systems. The MegaMatcher SDK 5.0 development environment is a convenient programming tool for designing client applications that can use single or multiple authentication methods. The proposed system, in its original form, allows the acquisition of iris, face and dactyloscopic images. The key element of the system is the intelligent module block which implements the fusion of component similarities computed in three different information channels. Taking into account the observations made on the effectiveness of the integration of the various feature vectors, it was discovered that the multimodal system should meet the following reliability criteria:

- universality: means that anyone should have that characteristic;
- uniqueness: the existence of two people with identical characteristics has a very small probability;
- stable characteristics: a characteristic does not change over time;
- quantification: a characteristic can be measured in a quantitative form.

Acknowledgment. The research and experiments were conducted in the Laboratory of Cognitive Science Research, Computer Graphics and Digital Image Processing Laboratory and Real Time Diagnostic Systems Laboratory at the Centre for Innovation and Transfer of Natural Sciences and Engineering Knowledge, University of Rzeszow as a result of EU project "Academic Centre of Innovation and Technical-Natural Knowledge Transfer" based on "Regional Operational Program for Subcarpathian Voivodship for years 2007–2013" Project No. UDA-RPPK.01.03.00-18-001/10-00.

References

1. Advancing Biometric Federal Bureau of Investigation FBI Biometric Specifications. https://www.fbibiospecs.cjis.gov. Accessed 20 Apr 2017
2. Jain, A.K., Klare, B.: Unsang park: face recognition: some challenges in forensics. To appear in the 9th IEEE International Conference on Automatic Face and Gesture Recognition, Santa Barbara, CA (2011). https://doi.org/10.1109/fg.2011.5771338
3. Burge, M.J., Bowyer, K.: Handbook of Iris Recognition. Springer, New York (2013)
4. Dudek-Dyduch, E.: Algebraic logical meta-model of decision processes - new metaheuristics. In: Artificial Intelligence and Soft Computing. Lecture Notes in Computer Science, ICAISC 2015, Zakopane, Poland, vol. 9119, pp. 541–554 (2015)
5. Hentati, R., Hentati, M., Abid, M.: Development a new algorithm for iris biometric recognition. Int. J. Comput. Commun. Eng. **1**(3), 283–286 (2012)
6. Mahesh Naidu, K., Govindarajulu, P.: Biometrics hybrid system based verification. (IJCSIT) Int. J. Comput. Sci. Inf. Technol. **7**(5), 2341–2346 (2016)
7. Kosiuczenko, P.: Specification of invariability in OCL. Softw. Syst. Model. **12**(2), 415–434 (2013). https://doi.org/10.1007/s10270-011-0215-y
8. Kosiuczenko, P.: On the validation of invariants at runtime. Fundam. Inform. **125**(2), 183–222 (2013)
9. Kulkarni, K., Shet, R., Iyer, N.: Hybrid primary and secondary biometric fusion. International Journal of Computer Applications (0975 – 8887), National Conference on Electronics and Computer Engineering (2016)
10. Madeyski, L., Kawalerowicz, M.: Software engineering needs agile experimentation: a new practice and supporting tool. In: Software Engineering: Challenges and Solutions, pp. 149–162 (2016). https://doi.org/10.1007/978-3-319-43606-7_4
11. Ochocki, M., Kołodziej, M., Sawicki, D.: Identity verification algorithm based on image of the iris, Institute of Theory of Electrical Engineering, Measurement and Information Systems, Warsaw University of Technology (2015). (in Polish)
12. Sadowska, M., Huzar, Z.: Semantic validation of UML class diagrams with the use of domain ontologies expressed in OWL 2. In: Software Engineering: Challenges and Solutions, pp. 47–59 (2016). https://doi.org/10.1007/978-3-319-43606-7_4
13. Neurotechnology. http://www.neurotechnology.com/. Accessed 20 Apr 2017
14. NIST national Institute of Standards and Technology. NIST Biometric Image Software (NBIS) (2017). https://www.nist.gov/services-resources/software/nist-biometric-image-software-nbis. Accessed 20 Apr 2017
15. Suganya, S., Menaka, D.: Performance evaluation of face recognition algorithms. Int. J. Recent Innov. Trends Comput. Commun. **2**(1), 135–140 (2014). ISSN: 2321-8169
16. Orandi, S., Libert, J., Garris, M., Byers, F.: JPEG 2000 CODEC Certification Guidance for 1000 ppi Fingerprint Friction Ridge Imagery (2016). http://dx.doi.org/10.6028/NIST.SP.500–300. Accessed 20 Apr 2017

17. Verma, P., Dubey, M., Verma, P., Basu, S.: Daugman's algorithm method for iris recognition – a biometric approach. IJETAE **2**(6), 177–185 (2012)
18. Vishi, K., Yayilgan, S.Y.: Multimodal biometric authentication using fingerprint and iris recognition in identity management. In: 2013 Ninth International Conference on Intelligent Information Hiding and Multimedia Signal Processing (IIH-MSP 2013), pp. 334–341 (2013)
19. Li, X., Yin, Y., Ning, Y., Yang, G., Pan, L.: A hybrid biometric identification framework for high security applications. Front. Comput. Sci. **9**(3), 392–401 (2015). https://doi.org/10.1007/s11704-014-4070-1
20. Zarzycki, H., Czerniak, J.M., Lakomski, D., Kardasz, P.: Performance comparison of CRM Systems dedicated to reporting failures to IT department. Software Engineering: Challenges and Solutions, pp. 133–146 (2016). https://doi.org/10.1007/978-3-319-43606-7_4

Usability, Security and Safety Interaction: Profile and Metrics Based Analysis

Oleksandr Gordieiev[1]([⊠]), Vyacheslav Kharchenko[2],
and Kostiantyn Leontiiev[3]

[1] Banking University, 1 Andriivska Street, Kyiv, Ukraine
alex.gordeyev@gmail.com
[2] National Aerospace University «KhAI», 17 Chkalova Street, Kharkiv, Ukraine
V.Kharchenko@csn.khai.edu
[3] RPC Radiy, Kropyvnytskyi, Ukraine
ksleontiev@gmail.com

Abstract. Attributes of information systems quality described in standard ISO/IEC25010 (2010) are analysed. Some of them are contradictory, dependent and competing. Two of the most competing pairs characteristics are (1) usability and security (U&Sec), (2) usability and safety (U&Saf). The article considers two main aspects of U&Sec interaction called «usable security» and «secure usability». Collaboration and competition of pair of characteristics are discussed as well. Case study is represented by U&Sec interaction for university web-site.

Keywords: Usability and security interaction · Usable security
Secure usability · Usability and safety interaction · ISO/IEC25010
ISO/IEC25023

1 Introduction

1.1 Motivation

Information systems are characterized by a set of characteristics/attributes that are defined by international standards. One from such standards is new and powerful standard ISO/IEC 25010 «System and software quality model» [1], which includes two different models: «Product quality model» and «Model quality in use». The models complete each other. First model defines the following 10 characteristics of information systems: functional suitability, performance efficiency, compatibility, usability, reliability, security, maintainability, portability. Second model consists of following characteristics: effectiveness, efficiency, satisfaction, freedom from risk, and context coverage. Such nomenclature was formed as a result of their evolution during about 60 years [2].

Certain characteristics (subcharacteristics) of information systems interact at each other. I.e. there are situations when strengthening (weakening) of one of the characteristics requires or generates strengthening (weakening) of another or even a group of information systems. In the article, we will consider a couple of the most important,

W. Zamojski et al. (Eds.): DepCoS-RELCOMEX 2018, AISC 761, pp. 238–247, 2019.
https://doi.org/10.1007/978-3-319-91446-6_23

mutually influence and competitive pairs characteristics such as usability and security (U&Sec), and usability and safety (U&Saf).

1.2   State of Art

First of all, let us describe U&Sec and U&Saf pairs attributes [1, 2]. Usability is a degree to which a product or system can be used by specified users to achieve specified goals with effectiveness, efficiency and satisfaction in a specified context of use. Security is a degree to which a product or system protects information and data so that people or other products or systems have the degree of data access appropriate to their types and levels of authorization. Safety is a degree to which a product or system mitigates the potential risk to economic status, human life, health, or the environment. Note, that instead of safety in [1] use term «freedom from risk» . Relationship between safety and freedom from risk represented in [3]. Information systems must have Usability, Security and Safety characteristics, because they must be comfortable in use and secure for information systems and safe for human life, health, or the environment simultaneously. Depending on field of information systems application, levels of U&Sec (U&Saf) requirements and characteristic values are not the same. In most cases, information systems are more usable, including at the expense of security (safety), or more secure (safety) at the expense their usability.

Problems of U&Sec and U&Saf characteristics interaction are well known, researched and presented in articles and books. Analysis of works in this field allows making some conclusions and dividing related publications on the following groups:

- the majority of works are about concrete problems in U&Sec field and mechanisms for their solutions [4–8]. In particular, in [4] alphanumeric passwords problems are viewed and there are presented ways for their decision;
- following group of works are about general conceptual questions in the U&Sec field [9, 10];
- part of works are about problems and peculiarities of U&Sec interaction on required levels [7], on processed levels [11, 12] and on model levels (including UML models) [13];
- small group includes works about U&S problems for mobile applications [14, 15];
- separate works are about analysis of literature in U&Sec problems field [16];
- some articles are about U&Sec characteristics evolution. Authors of such works represent the evolution and interaction of usability and security characteristics [2, 17];
- articles which describe pairs of U&Sec and U&Saf simultaneously are absent. In general case in existing articles authors considered variants of competition for U&Sec pair [18];
- authors of some articles reviewed of U&Saf pair in combination with another characteristics [19].

1.3   Goal and Structure

Preliminary analysis of works in U&Sec and U&Saf field allows making the following conclusions and determining goal of the paper:

- firstly, characteristics of U&Sec and U&Saf which described in last program engineering standards [1, 20, 21] are one of the results of 40 years evolution [2]. They are represented as a complex of characteristics with a set of depended subcharacteristics;
- secondly, analysis of U&Sec subcharacteristics and metrics did not conduct in existing works [4–17], which describe problems interaction of U&Sec characteristics;
- thirdly, separate subdivision was organized at National Institute of Standards and Technology (NIST) of USA [22], which solves tasks of U&Sec interaction. However, well known works describe, first of all, influence of Usability on Security and did not take into account aspects of influence on the level of their subcharacteristics.

Thus, **goal** of the paper is determination, analysis and assessment of U&Sec interaction on subcharacteristics and metrics levels and also analysis of U&Saf interaction conception.

The paper has the following structure:

- description of "Usable security" and "Secure usability" interaction problem;
- analysis of U&Sec interaction on subcharacteristics level and variants U&Sec subcharacteristics interaction;
- analysis of U&Sec interaction on metrics level;
- analysis of U&Saf interaction conception;
- case study for U&Sec interaction. This section includes analysis and assessment of U&Sec interaction for university web-site, the following section describes case study and last one concludes and presents directions of the future research.

2 Usability and Security

Two possible aspects of research and development (i.e. two sides of the same coin) are existent of usable security and secure usability. Let us consider in more details what are the differences between these two aspects.

2.1 Usable Security

First aspect gives an answer on a question: how to develop functions secure access to resources such as, in order to ensure acceptable/necessary level of usability of user interfaces. In order to link of U&Sec characteristics in the usable security aspect was more understandable, we need to represent example of such interaction. Very often procedure of registration on web-site requires from users to confirm their presence near personal computer. It needs to exclude automatic registration on the Internet. As a rule, web-site offers to users input data for CAPTCHA (Completely Automated Public Turing test to tell Computers and Humans Apart) [23]. In majority of cases, the CAPTCHA is information, which automatically generates on picture of web-page and which necessary should be input to textbox.

Sometimes users have problems with input of information from CAPTCHA (i.e. have problem with Public Turing test), because information which is represented on picture periodically cannot be discernible. (Figure 1). Defect in such identification technique can provoke discomfort for user. To solve such problem user necessary, periodically manually reload the picture of CAPTCHA waiting for recognizable information. User can wait long time of appearance recognizable information. User can also delay or cancel web-site registration procedure. This is an example, when «complex» security kills usability – (cSkU) Information systems developers necessarily take into account such aspect, when they make user interfaces projects. We have to exclude situation, when high level of security «kills» the usability.

Fig. 1. Examples of CAPTCHAs.   **Fig. 2.** More simple Public Turing test.

It should be noted, that subdivision at National Institute of Standards and Technology (NIST) of USA researches such U&Sec problems [22].

2.2   Secure Usability

Second aspect has relationships with development of user interfaces thus, in order to ensure necessary level of information security. Let us describe an example of such interaction between usability and security. Public Turing test can be maximally simple and represents one checkbox element, which necessary will set up in significance «check» (Fig. 2). From usability position such variant of Public Turing test is better than his variant on Fig. 1. But from security position such variant (Fig. 2) is worse, because as against previous variant (Fig. 1) such variant is more simply pass (by software bots) during automatic registration without user. In other words, in such a context there is another competition. This is situation, when «simple» usability «kills» security – sUkSec).

2.3   Assessment Criteria and Metrics

Thus, U&Sec characteristics really have interconnection in the form of two aspects and formally differences can be described through «castle» of objective function and limitations.

- in first case, it is necessary to ensure the required level of usability (U_{req}), at that maximize of security (Sec_{max}), i.e. Sec → max, $U \geq U_{req}$;
- in second case, it is necessary to ensure the required level of security (Sec_{req}), at that maximize of usability (U_{max}), i.e. U → max, $Sec \geq Sec_{req}$.

- We pay attention, that U&Sec characteristics and their subcharacteristics described in article as their interpretation in group of standards ISO 25000.
- Examined positions can be represented out in detail as:
- security – is combination of following subcharacteristics [1]: confidentiality (Cong), integrity (Integr), non-repudiation (N-rep), accountability (Acc) and authenticity (Aut):

$$Sec = \{conf, \ Integer, \ N-rep, \ Acc, \ Aut\}; \tag{1}$$

- usability is combination of following subcharacteristics [1]: appropriateness recognizability (AppRec), learnability (Learn), operability (Oper), user error protection (UEP), user interface aesthetics (UIA), accessibility (Acs):

$$U = \{AppRec, \ Learn, \ Oper, \ UEP, \ UIA, \ Acs\}. \tag{2}$$

Let us consider interaction between U&Sec subcharacteristics. We have received a set of variants of U&Sec subcharacteristics interaction because of U&Sec subcharacteristics analysis. Set of variants of U&Sec subcharacteristics represents in [18]. All comments for variants of interaction of U&Sec subcharacteristics have included in [18] too. If to compare data from tables [18] we can conclude that sets of variants of interaction of U&Sec subcharacteristics and their metrics are not identical, but very similar. Some interactions were changed in the subcharacteristics context. Such result is obvious, because U&Sec metrics interact with subcharacteristics.

3   Usability and Safety

3.1   Interconnection and Competition of Characteristics

Usability and safety characteristics have connection, which basis on their mutual competition. Let us review structure of safety characteristic which represent in standard [1] for more precision determination. Such characteristic is not included in model «Product quality model» (Fig. 3). It is represented in model «Quality in use» and has name «Freedom from risk» . It is degree to which a product or system mitigates the potential risk to economic status, human life, health, or the environment (NOTE: risk is a function of the probability of occurrence of a given threat and the potential adverse consequences of that threat's occurrence. Structural elements of characteristic «freedom from risk» represent in [1] following view:

- Economic risk mitigation - degree to which a product or system mitigates the potential risk to financial status, efficient operation, commercial property, reputation or other resources in the intended contexts of use;
- Health and safety risk mitigation - degree to which a product or system mitigates the potential risk for people in the intended contexts of use;
- Environmental risk mitigation - degree to which a product or a system mitigates the potential risk to property or the environment in the intended contexts of use.

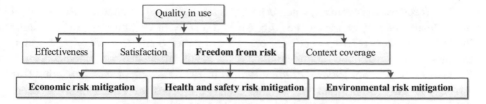

Fig. 3. Structure of freedom from risk characteristic.

Let us represent some well-known examples of interconnection of usability and safety characteristics, i.e. such examples, in which problems in user interfaces usability give problems in safety and sometimes make tragedies. This are the following examples [24]:

- First example with radiation therapy machine Therac-25. Problems with program code and user interfaces created tragedy – at least six accidents between 1985 and 1987, in which patients were given massive overdoses of radiation;
- Second example connection with problem in infusion pumps. According to FDA [25] reports between 2004 – 2009 years as results problems with program code and user interfaces was fixed following incidents: 56000 problems in work of infusion pumps which include 710 deaths and 87 recalls;
- Third example. According to FDA [25] data «An internal FDA study some years ago showed that 44% of medical device recalls were the result of design problems and more than 1/3 involved the device-user interface (device, it's labelling or instructions for use)» .

Similarly, it is with variants of influence competing pairs Usability and Security (secure usability and usable security). Let us determine such variants for pairs of Usability and Safety characteristics. Thus, existence of several variants influence of Usability and Safety:

1. Usable safety. Such variant exists, when user interfaces have to maximal correspond to safety characteristic (Saf_{max}) for required/acceptable usability (U_{req}), i.e. Saf max, $U \geq U_{req}$;
2. Safe usability. This is the opposite case, when user interfaces have to maximal correspond to safety characteristic (U_{max}) for required/acceptable usability (Saf_{req}), i.e. $U \rightarrow$ max, Saf $\geq Saf_{req}$.

3.2   Types of Characteristics Interconnection

It is worth noting, that existence of variant direct (Fig. 4) and indirect (Fig. 5) interconnection (competition) of characteristics. Direct interconnection is competition of some characteristic without intermediary (i.e. characteristic). Example direct variant of interconnection is competition usability and safety characteristics. For indirect interconnection characteristics have influence at each other through characteristic – intermediary.

Let us review the example, when usability influence to safety characteristic through intermediary security characteristic (Fig. 6). Bad quality of usability as result of faults in during creation of user interfaces can fetch to bad quality of security. Further, through subcharacteristic confidentiality can create influence on subcharacteristic «economic risk mitigation» of «freedom from risk» («safety») characteristic. This article has only description of variants interconnections of characteristics. More detailed analysis and their representation can be researched later.

4 Case Study

We will represent simple example of U&Sec interaction. First of all, worth noting, that metrics U&Sec equal to subsubcharacteristics (i.e. U&Sec subcharacteristics of second level). In this case, with usage of calculated significances, from U&Sec metrics, in author`s opinion, it is possible to do quantitative analysis of U&Sec interaction.

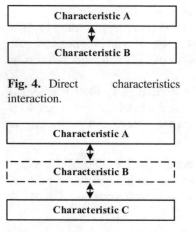

Fig. 4. Direct characteristics interaction.

Fig. 5. Indirect characteristics interaction.

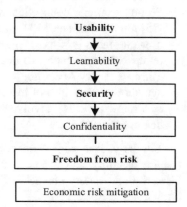

Fig. 6. Example of indirect characteristics interaction

We will do such analysis for separate subcharacteristics of U&Sec characteristics. For example, we will consider interaction of Operability and Confidentiality subcharacteristics on basis of such interaction with metrics. For that see [18], which includes the description of metrics and required primitives for calculation. Object of our research will be web-site of Banking University (http://ubs.edu.ua/en/), which is on the stage of the development. Calculated metrics of significances for web-site before making changes in this web-site (i.e. before testing) are represented in Table 1.

Table 1. Metrics significances.

Subcharacteristics/metrics		1	2
Operability	Operational consistency	0,3	0,1
	Message clarity	0,8	1
	Customizing possibility	0,6	0,8
Confidentiality	Access controllability	0,6	0,8

1. Metrics significances before make changes (i.e. before testing);
2. Metrics significances after make changes.

For calculation of single significance for Operability subcharacteristic we are using additive convolution, in which weighting coefficients for significances of metrics can be assessed by the following way:

$$Ch_k = Subch_i * WC_i + Subch_{i+1} * WC_{i+1} + \ldots + Subch_n * WC_n, \qquad (3)$$

Ch_k – characteristic, WC_i – weighting coefficient, $SubCh_i$ – subcharacteristic.
In result of calculation, we give following significances:

– before making changes
  $Operability_{before} = 0,3 * 0,33 + 0,8 * 0,33 + 0,6 * 0,33 = 0,099 + 0,264 + 0,198 = 0,561;$
– after making changes
  $Operability_{after} = 0,1 * 0,33 + 1 * 0,33 + 0,8 * 0,33 = 0,033 + 0,33 + 0,264 = 0,627.$

Further, we will compare received significances for Operability and Confidentiality subcharacteristics:

– before making changes Operability = 0,561, a Confidentiality = 0,6;
– after making changes Operability = 0,627, a Confidentiality = 0,8.

As a result, we received significances for Operability and Confidentiality subcharacteristics. Such significances increased after making changes in web-site in comparison with before making changes. For Operability the difference equals 0,066 and for Confidentiality – 0,2. Thus, we have confirmation of our supposition about interaction of Operability and Confidentiality characteristics, when increase in the level of one subcharacteristic incurring to increase in the level of other subcharacteristic [18].

5   Conclusions

We have considered two basic aspects of U&Sec interaction: usable security and secure usability. Similar approach can be applied to analyse U&Saf interaction. Differences in such aspects were analysed by using practical examples.

This work includes results of analysis of U&Sec and U&Saf interaction on the level of subcharacteristics.

Results of U&Saf interaction variants analysis allow assessing correlation characteristics and supporting design decisions.

In future authors plan to make complete quantitative analysis of interaction of U&Sec subcharacteristics on the base of calculated metrics values. Such analysis must confirm that variants of interaction of U&Sec subcharacteristics assessment will be correct. Also, we plan to analyze interaction between U&Sec characteristics of information systems and another once, and review in more details direct and indirect characteristics interaction variants.

Practical results of such assessment improve requirements foundation for usability, security, safety and other characteristics and support design decision making for usability and safety critical domains.

References

1. ISO/IEC 25010:Systems and software engineering – Systems and software Quality Requirements and Evaluation (SQuaRE) – System and software quality models, ISO/IEC JTC1/SC7/WG6 (2011)
2. Gordieiev, O., Kharchenko, V., Fominykh, N., Sklyar, V.: Evolution of software quality models in context of the standard ISO 25010. In: Proceedings of the Dependability on Complex Systems DepCoS – RELCOMEX (DepCOS), 30 June–4 July, Brunow, Poland, pp. 223–233 (2014)
3. Lann, D.: What is the Relationship Between Safety and Risk? (2017). http://avatarms.com/safety-risk/
4. Bindu, C.S.: Secure usable authentication using strong pass text passwords. Comput. Netw. Inf. Secur. 3(2015), 57–64 (2015)
5. Alsuhibany, S.A.: A benchmark for designing usable and secure text-based captchas. Int. J. Netw. Secur. Appl. (IJNSA) 8(4), 41–54 (2016)
6. Thorpe, J., van Oorschot, P.C.: Graphical dictionaries and the memorable space of graphical passwords. In: Proceedings of the 13th USENIX Security Symposium, 9–13 August, San Diego, CA, USA, pp. 10–26 (2004)
7. Al-Sarayreh, K.T., Hasan, L.A., Almakadmeh, K.: A trade-off model of software requirements for balancing between security and usability issues. Int. Rev. Comput. Softw. 10(12), 1157–1168 (2016)
8. Evaluating the accessibility, usability and security of Hospitals websites: An exploratory study. In proc. International conference on Cloud System and Big Data Engineering (Confluence-2017), at Noida, Uttar Pradesh, India (2017). https://www.researchgate.net/publication/313841977_Evaluating_the_accessibility_usability_and_security_of_Hospitals_websites_An_exploratory_study
9. Lampson, B.: Privacy and security usable security: how to get it. Commun. ACM 52(11), 25–27 (2009)
10. Payne, B.D., Edwards, W.K.: A brief introduction to usable security. IEEE Internet Comput. 12, 13–21 (2008)
11. Flechais, I., Mascolo, C., Sasse, M.A.: Integrating security and usability into the requirements and design process. Int. J. Electron. Secur. Digit. Forensics 1, 12–26 (2007)
12. Faily, S., Lyle, J., Fléchais, I., Simpson, A.: Usability and security by design: a case study in research and development. In: Proceedings of the NDSS Workshop on Usable Security, San Diego, CA, USA (2015). http://eprints.bournemouth.ac.uk/22053/1/flfs15.pdf

13. DiGioia, P., Douris, P.: Social navigation as a model for usable security. In: Proceedings of Symposium On Usable Privacy and Security (SOUPS), 6–8 July, Pittsburgh, PA, USA, pp. 101–108 (2005)
14. Melicher, W., Kurilova, D., Segreti, S.M., Kalvani, P., Shay, R., Ur, B., Bauer, L., Christin, N., Cranor, L.F., Mazurek, M.L.: Usability and security of text passwords on mobile devices. In: Proceedings of the CHI Conference on Human Factors in Computing Systems (CHI 2016), Santa Clara, California, USA, pp. 527–539 (2016)
15. Boja, C., Doinea, M.: Usability vs. security in mobile applications. In: Proceedings of the IE 2013 International Conference, pp. 138–142 (2013)
16. Nwokedi, U.O., Onyimbo, B.A., Rad, B.B.: Usability and security in user interface design: a systematic literature review. Int. J. Inf. Technol. Comput. Sci. (IJITCS) **8**, 72–80 (2016)
17. Gordieiev, O., Kharchenko, V., Fusani, M.: Evolution of software quality models: usability, security and greenness issues. In: Proceedings of the 19-th International Conference on Computers (part of CSCC 2015), 16–20 July, Zakynthos Island, Greece, pp. 519–523 (2015)
18. Gordieiev, O., Kharchenko, V., Vereshchak, K.: Usable security versus secure usability: an assessment of attributes interaction. In: Proceedings of the 13th International Conference, ICTERI 2017, 15–18 May, Kyiv, Ukraine, pp. 727–740 (2017)
19. Wegge, K.P., Zimmermann, D.: Accessibility, usability, safety, ergonomics: concepts, models, and differences. In: Proceedings of the 4th International Conference on Universal Access in Human-Computer Interaction, UAHCI 2007, 22–27 July, Beijing, China, pp. 294–301 (2007)
20. ISO/IEC 25023: Systems and software engineering – Systems and software Quality Requirements and Evaluation (SQuaRE) – Measurement of system and software product quality, ISO/IEC JTC1/SC7/WG6 (2011)
21. ISO/IEC 25030: Software engineering – Software product Quality Requirements and Evaluation (SQuaRE) – Quality requirements, ISO/IEC (2007)
22. Usability of security team at National institute of standards and technology. http://csrc.nist.gov/security-usability/HTML/about.html
23. Completely Automated Public Turing test to tell Computers and Humans Apart, CAPCHA. http://www.captcha.net/
24. Newman, R.: User Interface Design for Medical Devices - The Relationship Between Usability and Safety. Presentation. 29 April 2016. https://www.slideshare.net/UPABoston/user-interface-design-for-medical-devices-the-relationship-between-usability-and-safety
25. U.S. Department of Health and Human Services. US food & drug administration. https://www.fda.gov/

Business Availability Indicators of Critical Infrastructures Related to the Climate-Weather Change

Sambor Guze[✉]

Gdynia Maritime University, Gdynia, Poland
s.guze@wn.am.gdynia.pl

Abstract. The paper presents the approach to business continuity of critical infrastructures with impact of climate-weather change process. First, the review of known results about business continuity management, reliability, safety and risk analysis is done. In the next step, the climate-weather change process and related reliability indicators are introduced. Moreover, the critical infrastructures' business continuity indicators related to climate-weather change process are defined under assumption about renovation with non-ignored time. Finally, the Port Critical Infrastructure (PCI) is defined. The business availability indicators are applied and calculated for data from experts and under academic assumptions for PCI.

Keywords: Business continuity · Business availability coefficient
Climate impact · Safety · Critical infrastructure

1 Introduction

In these times, the functioning of societies is strongly dependent on various types of technical or economic systems and related services. Their main goal is to ensure continuity for the maintenance of vital societal functions. The system or asset, which is essential to realize this goal is called Critical infrastructure (CI) [7]. The importance of CI is particularly noticeable in the case of negative impact of its damage, destruction or disruption by natural disasters, terrorism, criminal activity or malicious behavior, to the security of the particular country and the well-being of its citizens.

It means, that the complex technical systems are increasingly exposed to threats from disruptive events like unexpected system failures, terrorist attacks, natural disasters, etc. The significance of the latter is increasing due to global climate change [17, 18]. The organizations use risk analysis and management methods to protect them from the potential losses caused by the distinguished above disruptive events. One part of this methods are studies and research on modelling the reliability and safety of complex technical systems [2, 3, 11, 14, 16]. The other way is common application of the safety and risk management methods given exemplary in [1, 22]. Unfortunately, the recovery/ restoration process, is not considered by mentioned methods. In practice, the potential losses that an organization or society might suffer also depend on the recovery process because this can generate a high cost. For these systems, conventional risk analysis and

© Springer International Publishing AG, part of Springer Nature 2019
W. Zamojski et al. (Eds.): DepCoS-RELCOMEX 2018, AISC 761, pp. 248–257, 2019.
https://doi.org/10.1007/978-3-319-91446-6_24

management methods should be extended. Thus, the recovery process should be taken into account as the integrated part of the risk management [8, 14, 20].

The way to integrate the recovery process within the preventive approach to the risk assessment is realized by the Business Continuity Management (BCM) [5, 12, 13]. The comprehensive approach to BCM planning, with focuses on internal and external information security threats are discussed in [5]. Some tools integrate the framework to support the BCM planning, what is shown in [10, 21]. And some results pointed out that BCM not only considers the protection of the system before the crisis, but also the recovery process during and after the crisis [20]. The one of well-known approach is using the system reliability models to BCM planning [8]. In the general approach BCM offers great potential benefits, but the complexity of the problem is such that most currently existing BCM strategies are based on qualitative methods only, what limits practical applications. Unfortunately, few papers concern the quantitative modelling and analysis of business continuity [4, 21].

The article is an attempt to use the Semi-Markov approach [16–19, 21] and availability theory [9, 15, 16] to BCM modeling in the case of critical infrastructures related to climate-weather change process [17, 18].

2   Business Availability Indicators of Critical Infrastructure with Impact of Climate-Weather Change Process

As it was mentioned in Introduction, the very important thing for business continuity of organizations is the recovery/restoration process. In the theory of technical system reliability, there are the research about the renovation with ignored and non-ignored time [14–16]. In this section, we propose to use these results for the reparaible critical infrastructures and their networks. There are considered only the CIs with non-ignored time of renovation.

2.1   Critical Infrastructure with Impact of Climate-Weather Change Process

We consider the critical infrastructure related to the climate-weather change process $C(t)$, $t \in <0, \infty)$, and impacted in a various way at the climate-weather states c_h, $h = 1, 2, \ldots, w$. We assume that the changes of the climate-weather states of the climate-weather change process $C(t)$, $t \in <0, \infty)$, at the critical infrastructure operating area have an influence on the critical infrastructure safety structure and on the safety of the critical infrastructure assets A_i, $i = 1, 2, \ldots, n$, as well. The following climate-weather change process parameters at the critical infrastructure operating area can be identified either statistically using the methods given in [17, 18] or evaluated approximately by experts:

- the number of climate-weather states w;
- the vector of the initial probabilities $q_h(0) = P(C(0) = c_h)$, $h = 1, 2, \ldots, w$, of the climate-weather change process $C(t)$ staying at particular climate-weather states c_h at the moment $t = 0$;

- the matrix of probabilities of transition q_{hl}, h, $l = 1, 2, \ldots, w$, of the climate-weather change process C(t) between the climate-weather states c_h and c_l;
- the matrix of mean values of conditional sojourn times $N_{hl} = E[C_{hl}]$, h, $l = 1, 2, \ldots, w$, of the climate-weather change process C(t) conditional sojourn times C_{hl} at the climate-weather state c_h when the next state is c_l.

The following climate-weather change process characteristics at the critical infrastructure operating area can be either calculated analytically using the above parameters of the climate-weather change process or evaluated approximately by experts [17, 18]:

- the vector of limit values of transient probabilities $q_h(t) = \mathrm{P}(\mathrm{C(t)} = c_h)$, $t \in$ $<0, +\infty)$, $h = 1, 2, \ldots, w$, of the climate-weather change process $C(t)$ at the particular climate-weather states c_h;
- the vector of the mean values of the total sojourn times $\hat{N}_h = E[\hat{C}_h] = q_h\theta$, $h = 1, 2, \ldots, w$, of the total sojourn times $\hat{C}_h$ of the climate-weather change process $C(t)$ at the critical infrastructure operating area at the particular climate-weather states c_h $h = 1, 2, \ldots, w$, during the fixed critical infrastructure operation time θ.

2.2 Critical Infrastructure Reliability Indicators with Impact of the Climate-Weather Change Process

Now, we denote the critical infrastructure conditional lifetime in the reliability state subset $\{u, u+1, \ldots, z\}$, $u = 1, 2, \ldots, z$, while the climate-weather change process C(t), $t \in <0, \infty)$, at the critical infrastructure operating area is at the climate-weather state c_h, $h = 1, 2, \ldots, w$, by $[\overline{T}(u)]^{(h)}$, $u = 1, 2, \ldots, z$, and the conditional reliability function of the critical infrastructure related to the climate-weather change process C(t), $t \in$ $<0, \infty)$, by the vector [17, 18]

$$[\overline{\boldsymbol{R}}(t, \cdot)]^{(h)} = [1, [\overline{\boldsymbol{R}}(t, 1)]^{(h)}, [\overline{\boldsymbol{R}}(t, 2)]^{(h)}, \ldots, [\overline{\boldsymbol{R}}(t, z)]^{(h)}], \tag{1}$$

with the exponential coordinates defined by

$$[\overline{\boldsymbol{R}}(t, u)]^{(h)} = P([\overline{T}(u)]^{(h)} > t | C(t) = c_h) \tag{2}$$

for $t \in <0, \infty)$, $u = 1, 2, \ldots, z$, $h = 1, 2, \ldots, w$.

The reliability function $[\overline{\boldsymbol{R}}(t, u)]^{(h)}$, $t \in <0, \infty)$, $u = 1, 2, \ldots, z$, $h = 1, 2, \ldots, w$, is the conditional probability that the critical infrastructure related to the climate-weather change process C(t), $t \in <0, \infty)$, lifetime $[\overline{T}(u)]^{(h)}$, $u = 1, 2, \ldots, z$, $h = 1, 2, \ldots, w$, in the reliability state subset $\{u, u+1, \ldots, z\}$, $u = 1, 2, \ldots, z$, is greater than t, while the climate-weather change process C(t), $t \in <0, \infty)$, is at the climate-weather state c_h, $h = 1, 2, \ldots, w$.

Next, we denote the critical infrastructure related to the climate-weather change process C(t), $t \in <0, \infty)$, unconditional lifetime in the reliability state subset $\{u, u+1, \ldots, z\}$, by $\overline{T}(u)$, $u = 1, 2, \ldots, z$, and the unconditional reliability function of the critical infrastructure related to the climate-weather change process C(t), $t \in$ $<0, \infty)$, by the vector

$$\overline{R}(t, \cdot) = [1, \overline{R}(t, 1), \ldots, \overline{R}(t, z)], \tag{3}$$

with the coordinates defined by

$$\overline{R}(t, u) = P(\overline{T}(u) > t) \tag{4}$$

for $t \in <0, \infty)$, $u = 1, 2, \ldots, z$.

In the case when the system operation time θ is large enough, the coordinates of the unconditional reliability function of the critical infrastructure related to the climate-weather change process C(t), $t \in <0, \infty)$, defined by (4), are given by

$$\overline{R}(t, u) \cong \sum_{h=1}^{w} q_h [\overline{R}(t, u)]^{(h)} \tag{5}$$

for $t \geq 0$, $u = 1, 2, \ldots, z$, where $[\overline{R}(t, u)]^{(h)}$, $u = 1, 2, \ldots, z$, $h = 1, 2, \ldots, w$, are the coordinates of the critical infrastructure related to the climate-weather change process C (t), $t \in <0, \infty)$, conditional reliability functions defined by (1)–(2) and q_h, $h = 1, 2, \ldots, w$, are the climate-weather change process C(t), $t \in <0, \infty)$, at the critical infrastructure operating area limit transient probabilities at the climate-weather states c_b, $h = 1, 2, \ldots, w$, given by in [17, 18].

Other reliability indicators of the critical infrastructure related to the climate-weather change process C(t), $t \in <0, \infty)$, are:

- the mean value of the critical infrastructure unconditional lifetime $\overline{T}(r)$ up to exceeding critical reliability state r given by

$$\overline{\mu}(r) = \int_0^\infty [\overline{R}(t, r)] dt \cong \sum_{h=1}^{w} q_h [\overline{\mu}(r)]^{(h)}, \tag{6}$$

where $[\overline{\mu}(r)]^{(h)}$ are the mean values of the system conditional lifetimes $[\overline{T}(r)]^{(h)}$ in the reliability state subset $\{r, r+1, \ldots, z\}$ at the climate-weather state c_b, $h = 1, 2, \ldots, w$, given by

$$[\overline{\mu}(r)]^{(h)} = \int_0^\infty [\overline{R}(t, r)]^{(h)} dt, \quad b = 1, 2, \ldots, w, \tag{7}$$

and $[\overline{R}(t, r)]^{(h)}$, $h = 1, 2, \ldots, w$, are defined by (1)–(2) and q_h are given in [19],
- the standard deviation of the critical infrastructure lifetime $\overline{T}(r)$ up to the exceeding the critical reliability state r given by

$$\overline{\sigma}(r) = \sqrt{\overline{n}(r) - [\overline{\mu}(r)]^2},$$ (8)

where

$$\overline{n}(r) = 2 \int_0^\infty t\overline{R}(t, r)dt,$$ (9)

and $\overline{R}(t, r)$ is defined by (3) for $u = r$; and $\overline{\mu}(r)$ is given by (6).

2.3   Critical Infrastructure with Non-ignored Time of Renovation Related to Climate-Weather Change Process

We assume here that the considered critical infrastructures after exceeding the critical reliability state are repaired and that the time of their renovation is not very small in comparison to their lifetimes in the reliability states subsets not worse than the critical one and we may not omit it. Furthermore, we consider the climate-weather change process defined in Subsect. 2.1. Under these assumptions, it is possible to obtain the results formulated for in the following proposition according to results given in [14–16].

Proposition 1. If assets of the multi-state repairable critical infrastructure with non-ignored time of renovation have the exponential reliability functions at the climate-weather states q_h, $h = 1, 2, \ldots, w$, given by (1)–(2), the critical infrastructure reliability critical state is r, $r \in \{1, 2, \ldots, z\}$, and the successive times of the critical infrastructure's renovations are independent and have an identical distribution function with the expected value $\overline{\mu}_o(r)$ and the variance $\overline{\sigma}_o^2(r)$, then:

1. the climate-weather related business availability coefficient of the critical infrastructure at the moment t, $t \geq 0$, for sufficiently large t, is given by

$$\overline{A}_{BC,t}(t, r) \cong \frac{\overline{\mu}(r)}{\overline{\mu}(r) + \overline{\mu}_o(r)}, \quad r \in \{1, 2, \ldots, z\};$$ (10)

2. the climate-weather related business availability coefficient of the critical infrastructure in the time interval $<t, t + \tau)$, $\tau > 0$, for sufficiently large t, is given by

$$\overline{A}_{BC,T}(t, \tau, r) \cong \frac{1}{\overline{\mu}(r) + \overline{\mu}_o(r)} \int_\tau^\infty \overline{R}(t, r)dt, \quad t \geq 0, \ r \in \{1, 2, \ldots, z\};$$ (11)

where $\overline{\mu}(r)$ and $\overline{\sigma}(r)$ are expressed by formulae (6), (8) respectively, and $\overline{R}(t, r)$ is unconditional reliability function of the exponential series critical infrastructure related to the climate-weather change process C(t) in the form:

$$\overline{R}(t,r) \cong \sum_{h=1}^{w} q_h \exp[-\sum_{i=1}^{n} [\lambda_i(r)]^{(h)} t],\tag{12}$$

for $t \geq 0$, $h = 1, 2, \ldots, w$, $r = 1, 2, \ldots, z$.

3   Business Availability Coefficients of the Port Critical Infrastructure

We are taking into account, the Port Critical Infrastructure (PCI) as the important node of country transportation system and trade. We distinguish following three assets in PCI: technical loading/unloading equipment (A_1), hydrotechnical infrastructure (A_2), transport infrastructure (A_3). Further, we assume that Port Critical Infrastructure is a series CI with five safety states ($z = 4$), defined as follows:

- a safety state 4 – PCI operations are fully reliable - z_4,
- a safety state 3 – *PCI* operations are less reliable - z_3,
- a safety state 2 – *PCI* operations are endangered with a minor failure - z_2,
- a safety state 1 – *PCI* operations are endangered with major failure -z_1,
- a safety state 0 – *PCI* is destroyed and dangerous for society and environment - z_0.

Moreover, we assume, that the critical reliability state of the Port Critical Infrastructure is $r = 2$.

Under historical hydro-meteorological data, we distinguish the following w = 4 climate-weather states:

- climate-weather state c_1 – the air temperature from −25 up to −15 or 25 up to 35 and the soil temperature from −30 up to −5 or 20 up to 37;
- climate-weather state c_2 – the air temperature from −25 up to −15 or 25 up to 35 and the soil temperature from −30 up to −5 or 20 up to 37 and strong wind;
- climate-weather state c_3 – the air temperature from −15 up to 5 or 5 up to 25 and the soil temperature from −5 up to 5 or 5 up to 20 and strong wind;
- climate-weather state c_4 – the air temperature from −15 up to 5 or 5 up to 25 and the soil temperature from −5 up to 5 or 5 up to 20.

Next, in the same way, the values of the climate-weather change process C(t) limit transient probabilities $[q_h]_{1 \times 4}$ at the climate-weather states c_h, $h = 1, 2, 3, 4$ are given based on academic assumptions and expert's opinions:

$$q_1 = 0.011, \ q_2 = 0.002, \ q_3 = 0.019, \ q_4 = 0.968.\tag{13}$$

The CIs' lifetimes in the reliability states are expressed in years and they have the exponential reliability functions (12) with the intensities of departure from the reliability subsets, by the assumption, given by:

- for u = 1

$$[\lambda_i(1)]^{(1)} = 0.00022, \quad [\lambda_i(1)]^{(2)} = 0.00004,$$
$$[\lambda_i(1)]^{(3)} = 0.00038, \quad [\lambda_i(1)]^{(4)} = 0.01936, \quad i = 1,2,3; \tag{14}$$

- for u = 2

$$[\lambda_i(2)]^{(1)} = 0.00055, \quad [\lambda_i(2)]^{(2)} = 0.0001,$$
$$[\lambda_i(2)]^{(3)} = 0.00095, \quad [\lambda_i(2)]^{(4)} = 0.0484, \quad i = 1,2,3; \tag{15}$$

- for u = 3

$$[\lambda_i(3)]^{(1)} = 0.00088, \quad [\lambda_i(3)]^{(2)} = 0.00016,$$
$$[\lambda_i(3)]^{(3)} = 0.00152, \quad [\lambda_i(3)]^{(4)} = 0.07744, \quad i = 1,2,3; \tag{16}$$

- for u = 4

$$[\lambda_i(4)]^{(1)} = 0.0011, \quad [\lambda_i(4)]^{(2)} = 0.0002,$$
$$[\lambda_i(4)]^{(3)} = 0.0019, \quad [\lambda_i(4)]^{(4)} = 0.0968, \quad i = 1,2,3. \tag{17}$$

According to (1)–(2) and (14)–(17), the unconditional reliability function of the Port Critical Infrastructure related to the climate-weather change process C(t), $t \in\ <0,\infty)$, is given by the vector [19]:

$$\overline{\boldsymbol{R}}(t,\cdot) = [1, \overline{\boldsymbol{R}}(t,1), \overline{\boldsymbol{R}}(t,2), \overline{\boldsymbol{R}}(t,3), \overline{\boldsymbol{R}}(t,4)], \tag{18}$$

with coordinates given by

$$\overline{\boldsymbol{R}}(t,1) \cong 0.011\exp[-0.00066t] + 0.002\exp[-0.00012t]$$
$$+ 0.019\exp[-0.00114t] + 0.968\exp[-0.05808t] \tag{19}$$

$$\overline{\boldsymbol{R}}(t,2) \cong 0.011\exp[-0.00165t] + 0.002\exp[-0.0003t]$$
$$+ 0.019\exp[-0.00285t] + 0.968\exp[-0.1452t] \ ' \tag{20}$$

$$\overline{\boldsymbol{R}}(t,3) \cong 0.011\exp[-0.00264t] + 0.002\exp[-0.00048t]$$
$$+ 0.019\exp[-0.00456t] + 0.968\exp[-0.23232t] \ ' \tag{21}$$

$$\overline{\boldsymbol{R}}(t,4) \cong 0.011\exp[-0.0033t] + 0.002\exp[-0.0006t]$$
$$+ 0.019\exp[-0.0057t] + 0.968\exp[-0.2904t] \ ' \tag{22}$$

where $[\overline{\boldsymbol{R}}(t,u)]^{(h)}$ are coordinates of conditional reliability function and q_h have values given in (13). The unconditional reliability function is presented in Fig. 1.

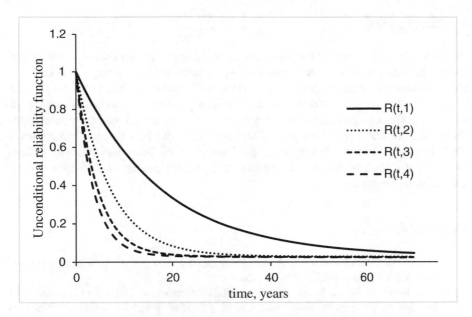

Fig. 1. Port critical infrastructure unconditional reliability function

We assume, that the critical reliability state is $r = 2$, and the Port Critical Infrastructure's renovation time based on experts' opinions has the mean value $\overline{\mu}_o(2) = 0.250 = 3$ months and the standard deviation $\overline{\sigma}_o(2) = 0.125 = 1.5$ months. Thus, the climate-weather related business availability coefficient of the Port Critical Infrastructure at the moment t, $t \geq 0$, for sufficiently large t, according to (10) is given as

$$\overline{A}_{BC,t}(t, 2) \cong \frac{\overline{\mu}(2)}{\overline{\mu}(2) + \overline{\mu}_o(2)} \approx \frac{26.67}{26.67 + 0.25} \approx 0.9907. \tag{23}$$

Moreover, the climate-weather related business availability coefficient of the critical infrastructure in the time interval $<t, t + \tau)$, $\tau > 0$, for sufficiently large t, is given by

$$\overline{A}_{BC,T}(t, \tau, 2) \cong \frac{1}{\overline{\mu}(2) + \overline{\mu}_o(2)} \int\limits_{\tau}^{\infty} \overline{R}(t, 2) dt = 0.0371 \int\limits_{\tau}^{\infty} \overline{R}(t, 2) dt \quad t \geq 0. \tag{24}$$

The obtained result in (23) shows that considered Port Critical Infrastructure has 99% of business availability. It means that port services are available for operators almost consecutive and time of unavailability is very small. Furthermore, it means that PCI is business-efficient and has a short repair/renovation time.

4 Conclusions

The main aim of the paper is the presentation of the approach to business continuity of critical infrastructures with an impact of the climate-weather change process. The climate-weather change process and related reliability indicators have been introduced. Moreover, the critical infrastructures' business continuity indicators related to climate-weather change process have been proposed and defined under the assumption about its renovation with non-ignored time. Finally, the Port Critical Infrastructure (PCI) has been defined. Furthermore, the business availability indicators have been applied and calculated for PCI, according to data from experts and under academic simplified assumptions.

References

1. Aven, T., Cox, L.A.: National and global risk studies: how can the field of risk analysis contribute? Risk Anal. **36**(2), 186–190 (2016)
2. Blokus-Roszkowska, A., Kołowrocki, K.: Reliability analysis of ship-rope transporter with dependent components. In: Nowakowski, et al. (eds.) Safety and Reliability: Methodology and Applications – Proceedings of the European Safety and Reliability Conference, ESREL 2014, pp. 255–263. Taylor & Francis Group, London (2015)
3. Blokus-Roszkowska, A.: Reliability analysis of the bulk cargo loading system including dependent components. In: Simos, T., Tsitouras, C. (eds.) Proceedings of the International Conference of Numerical Analysis and Applied Mathematics 2015 (ICNAAM 2015): 440002-1–440002-4. AIP Publishing, AIP Conf. Proc. 1738 (2016)
4. Boehmer, W., Brandt, C., Groote, J.F.: Evaluation of a business continuity plan using process algebra and modal logic. In: 2009 IEEE Toronto International Conference, Science and Technology for Humanity (TIC-STH), pp. 147–152. IEEE (2009)
5. Brandt, C., Hermann, F., Engel, T.: Modeling and reconfiguration of critical business processes for the purpose of a business continuity management respecting security, risk and compliance requirements at credit Suisse using algebraic graph transformation. In: 13th IEEE Enterprise Distributed Object Computing Conference Workshops, EDOCW 2009, pp. 64–71 (2009)
6. Cerullo, V., Cerullo, M.J.: Business continuity planning: a comprehensive approach. Inf. Syst. Manag. **21**(3), 70–78 (2004)
7. European Union: European Council. Council Directive 2008/114/EC of 8 December 2008 on the identification and designation of European critical infrastructures and the assessment of the need to improve their protection. Brussels (2008)
8. Faertes, D.: Reliability of supply chains and business continuity management. Procedia Comput. Sci. **55**, 1400–1409 (2015)
9. Gamiz, M.L., Roman, Y.: Non-parametric estimation of the availability in a general repairable. Reliab. Eng. Syst. Saf. **93**(8), 1188–1196 (2008)
10. Gibb, F., Buchanan, S.: A framework for business continuity management. Int. J. Inf. Manag. **26**(2), 128–141 (2006)
11. Guze, S.: Reliability analysis of multi-state ageing series-consecutive m out of n: F systems. In: Proceedings of the European Safety and Reliability Conference - ESREL 2009, vol. 3, Prague, pp. 1629–1635 (2010)

12. Herbane, B.: The evolution of business continuity management: a historical review of practices and drivers. Bus. Hist. **52**(6), 978–1002 (2010)
13. International Organization for Standardization (ISO) (2012)
14. Kołowrocki, K.: Reliability of Large and Complex Systems, Elsevier, Amsterdam, Boston, Heidelberg, London, New York, Oxford, Paris, San Diego, San Francisco, Singapore, Sidney, Tokyo (2014)
15. Kolowrocki, K., Soszynska, J.: Reliability, availability and safety of complex technical systems: modelling – identification – prediction – optimization. J. Polish Saf. Reliability Assoc. Summer Saf. Reliability Semin. **4**(1), 133–158 (2010)
16. Kołowrocki, K., Soszyńska-Budny, J.: Reliability and Safety of Complex Technical Systems and Processes: Modeling - Identification - Prediction - Optimization. Springer, Heildeberg (2011)
17. Kołowrocki, K., Soszyńska-Budny, J., Torbicki, M.: Critical infrastructure operating area climate-weather change process including extreme weather hazards. J. Polish Reliab. Reliab. Assoc. **8**(2), 15–24 (2017). Special Issue on EU-CIRCLE Project, Critical Infrastructure Operation and Climate-Whether Change Modelling, Prediction and Data Processing
18. Kołowrocki, K., Soszyńska-Budny, J., Torbicki, M.: Integrated impact model on critical infrastructure reliability related to climate-weather change process including extreme weather hazards. J. Polish Reliab. Reliab. Assoc. **8**(4), 21–32 (2017). Special Issue on EU-CIRCLE Project Critical Infrastructure Impact Models for Operation Threats and Climate Hazards, Part 2. Impact Assessment Models
19. Limnios, N., Oprisan, G.: Semi-Markov Processes and Reliability. Birkhauser, Boston (2005)
20. Snedaker, S.: Business Continuity and Disaster Recovery Planning for IT Professionals. Newnes (2013)
21. Zeng, Z., Zio, E.: An integrated modeling framework for quantitative business continuity assessment. Process Saf. Environ. Prot. **106**, 76–88 (2017)
22. Zio, E.: Critical infrastructures vulnerability and risk analysis. Eur. J. Secur. Res., 1–18 (2016)

Genetic-Algorithm-Driven MIMO Multi-user Detector for Wireless Communications

Mohammed Jasim Khafaji$^{(\boxtimes)}$ and Maciej Krasicki ⓘ

Faculty of Electronics and Telecommunications, Chair of Wireless
Communications, Poznan University of Technology, 60-695 Poznań, Poland
mohammed.j.khafaji@doctorate.put.poznan.pl

Abstract. In the paper, evolutionary optimization strategy, represented by the genetic algorithm (GA) is considered as a multiuser detection (MUD) method for a multiple-input multiple-output (MIMO) wireless system. With the aim to boost lacking GA convergence, Zero-Forcing (ZF) detection is proposed as an initial processing phase. Additionally, a multi-stage GA routine is considered as a method to make the search for data estimates more effective.

Keywords: Multiuser detection · MIMO · Genetic algorithm · Zero forcing

1 Introduction

The development of multiuser detection techniques is one of the most vital activities in communications due to the growth in the end users' demands [1]. Multiuser detection (MUD) is a technique in which a single receiver jointly detects many signals transmitted simultaneously by different users. In other words, it recovers the desired signal(s) from interference and noise [2].

The most promising MUD techniques can exploit high capacity of Multiple-Input Multiple-Output (MIMO) channels. However, high computational complexity of the optimal Maximum Likelihood (ML) receiver is not acceptable from a practical point of view [3, 4]. Apart from the optimal ML algorithm, some linear MIMO MUD solutions, like Zero-Forcing (ZF) or Minimum Mean Square Error (MMSE) have been developed; they can be considered benchmarks for the most recent contributions. In the authors' opinion, one of the most interesting solutions to the MUD problem is the one utilizing methods adopted from natural genetics, exemplified by a genetic algorithm (GA) [5] in the current paper.

The ZF algorithm will play two roles in our research. The first is just the benchmark for the considered GA-based method. But, in the further part, the ZF will be incorporated *into* the GA-based MUD. The reader is kindly asked to keep that ambiguity in mind.

The paper is organized as follows. In Sect. 2, the MIMO MUD problem is characterized. Section 3 is to describe the essentials of GA and how it can be used in the MIMO MUD case. In Sect. 4, several simulation results are presented. Section 5 concludes the work.

© Springer International Publishing AG, part of Springer Nature 2019
W. Zamojski et al. (Eds.): DepCoS-RELCOMEX 2018, AISC 761, pp. 258–269, 2019.
https://doi.org/10.1007/978-3-319-91446-6_25

Regarding notation, superscripts $(\cdot)^{-1}$ and $(\cdot)^H$ indicate the matrix inverse and Hermitian transpose operation, respectively, whereas $\|\cdot\|^2$ is the Frobenius norm.

2   MIMO Multiuser Detection Problem

The system under study is the uplink with four users, one transmit antenna each $(N_T = 4)$. All the users are supposed to transmit their signals synchronously and to be mutually independent. For each user, uncoded bitstream is mapped into QPSK, 16-QAM, or 64-QAM symbols. The signals from all users reach the N_R receive antennas through a multiple-input multiple-output (MIMO) flat-fading Rayleigh channel $(N_{R=4})$, as shown in Fig. 1.

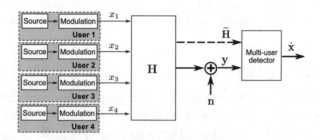

Fig. 1. Multi-user MIMO system model

A vector $\mathbf{y}$ represents the signals received through all the receive antennas at a given time instance:

$$\mathbf{y} = \begin{bmatrix} h_{11} & \cdots & h_{1N_T} \\ \vdots & \ddots & \vdots \\ h_{N_R1} & \cdots & h_{N_RN_T} \end{bmatrix} \begin{bmatrix} x1 \\ \vdots \\ x_{N_T} \end{bmatrix} + \begin{bmatrix} n1 \\ \vdots \\ n_{N_R} \end{bmatrix}$$

where $\mathbf{x}[x_1...x_{N_T}]^T$ is a multisymbol representing QPSK, 16-QAM, or 64-QAM symbols transmitted by N_T users, simultaneously, $\mathbf{H}$ is an $N_R \times N_T$ complex channel gain matrix – it is assumed to be estimated ideally at the receiver side $(\tilde{\mathbf{H}} = \mathbf{H})$. Finally, $\mathbf{n} = [n_1...n_{N_R}]^T$ is a complex Additive White Gaussian Noise (AWGN) sample vector.

The MUD task is to retrieve a multisymbol estimate

$$\dot{\mathbf{x}} = \arg\,\min_{\tilde{\mathbf{x}}}\|\mathbf{y} - \tilde{\mathbf{H}}\,\tilde{\mathbf{x}}\|^2 \tag{2}$$

The search space consists of all possible candidate multisymbols $\tilde{x} = (\tilde{x}_1,...\tilde{x}_{N_T})$: $\tilde{x}_i \in \chi, \forall_i$, where χ is the signal constellation set. The MUD succeeds if $\dot{\mathbf{x}} = \mathbf{x}$.

3 Genetic-Algorithm-Driven MIMO MUD

3.1 Genetic Algorithm Overview and Terminology

Genetic Algorithm (GA) is a search-based iterative optimization method, which utilizes operators acquired from the natural genetics; it emulates the natural biological evolution [6]. GAs operate on a population of possible solutions (individuals) using the principle of survival of the fittest ones. In the most popular GA approach, individuals are called by their chromosomes, *i.e.*, binary vectors, representing unique discrete values of the coordinates through the search space [7].

GA couples individuals and applies some genetic manipulations on their offspring (usually, the best individuals are paired with different individuals many times within one reproduction phase). All the individuals are evaluated regarding their fitness, and only the best ones are kept alive. The result is that the population size is kept constant from one GA iteration to another, and the average population fitness increases, gradually [8].

3.2 Basic GA-Based MUD

In the proposed GA-based MUD, every individual represents a candidate multisymbol $\tilde{x} = (\tilde{x}_1, \ldots \tilde{x}_{N_T})$. Let 2^M be the modulation order. Then, QPSK, 16-QAM, or 64-QAM symbols can be unequivocally addressed by M-bit-long labels $\tilde{\mathbf{b}}_i \overset{\text{def}}{=} \beta(\tilde{x}_i)$ associated with them according to a specific labeling map β (Gray map is used hereinafter). Therefore, the chromosome in the considered optimization problem could be, simply, formed as a concatenation $\tilde{\mathbf{b}} = (\tilde{\mathbf{b}}_1 \ldots \tilde{\mathbf{b}}_{N_T})$ of the labels, assigned to the symbols transmitted concurrently by all users. Note that the chromosome length depends on both the number of users and the modulation order. For 16-QAM case, illustrated in Fig. 2, each label consists of 4 bits, and there are $N_T = 4$ users, so the total chromosome length is 16.

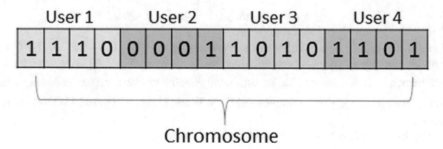

Fig. 2. Chromosome representing a candidate multisymbol for which $\tilde{\mathbf{b}}_1 = (1110), \tilde{\mathbf{b}}_2 = (0001), \tilde{\mathbf{b}}_3 = (1010), \tilde{\mathbf{b}}_4 = (1101)$

It is worth mentioning that the search space of the GA is of a discreet type, since the permitted positions of constellation points are strictly limited. As a consequence, GA does not introduce any discretization loss.

The proposed GA-based MUD algorithm follows the flowchart shown in Fig. 3. The subsequent steps of the algorithm are briefly described below.

Population Initialization: In the basic approach, all initial chromosomes are randomly generated. The performance of the algorithm in finding the optimal solution is highly affected by the population size: a high number gives a better solution but slows the convergence [9]. In the considered scenario, the number of individuals in population (population size) is set to be 50 (it is a moderate value).

Fitness Evaluation: The optimum ML-based decision metric acts as the objective function to evaluate the fitness of each individual.

$$f(\tilde{\mathbf{x}}) = \left\| \mathbf{y} - \tilde{\mathbf{H}}\tilde{\mathbf{x}} \right\|^2 \tag{3}$$

The best chromosome in the generation should have the lowest value of the objective function.

Starting from the second iteration of the GA algorithm, the worst individuals at the current population die in order to keep constant population size through subsequent iterations. The survivors can become the parents for the next step.

Selection: In the current work, the roulette wheel selection rule [10] is applied. In that approach, all individuals have a chance to be picked up, but the probability of being selected is proportional to the fitness of a given individual.

Crossover: This routine means merging two parents to form the offspring. The chromosomes of the children consist of the fractions of both parents' chromosomes. In the current work the single-point crossover method is used. It means that the parents' chromosomes are cut at a random position to create two slices. Respective slices of both parents' chromosomes are exchanged with each other to create 2 children.

Mutation: During the mutation step, some random changes are applied to individual genes of the childrens' chromosomes. It is, simply, to replace 0 by 1 or vice versa on a randomly selected bit position of the chromosome. Not every child undergoes mutation. Instead, it should be a small percent of the newborns [10, 11]. In the current work is assumed to be 0.5%.

Ending Criteria: There are several ending criteria known from the literature. They can base on the number of generations passed from the algorithm start, or consider the so-called stall generations. In the second case, the generation counter resets every time an individual with the best fitness value ever is catched within the current population. In the basic GA-MUD setup, the stopping condition of the GA iterative processing is set in terms of maximum generations' number to 500. The output of the GA is the individual exhibiting the best fitness ever.

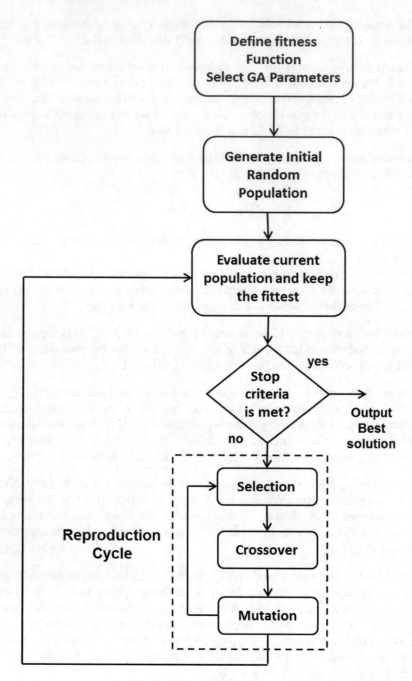

Fig. 3. Proposed GA flowchart

3.3   ZF-GA Multiuser Detection

In this novel approach, Zero-Forcing detector is utilized at the receiver side in order to create a seed individual (chromosome) for the subsequent GA operations. It is expected to enhance the receiver performance by boosting GA convergence. At first, the multisymbol estimate $\dot{\mathbf{x}}_{ZF}$ is determined by ZF detector, and the respective chromosome $\dot{\mathbf{b}}_{ZF}$ determines a seed individual of the initial population, as shown in Fig. 4. All the individuals within the initial population except for the seed, are generated randomly, as for the basic setup, described in Sect. 3.2.

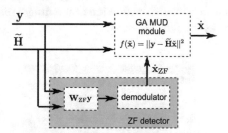

Fig. 4. ZF-GA MUD block diagram

The ZF-based signal estimate is obtained as a linear combination of the signals received by different antennas, transformed by the ZF array weight matrix $\mathbf{W}_{ZF}$ as follows:

$$\dot{\mathbf{x}}_{ZF} = \arg\ \min_{\tilde{\mathbf{x}}} \|\tilde{\mathbf{x}} - \mathbf{W}_{ZF}\mathbf{y}\|^{2}, \tag{4}$$

where $\mathbf{W}_{ZF} = \left(\tilde{\mathbf{H}}^{H}\tilde{\mathbf{H}}\right)^{-1}\tilde{\mathbf{H}}^{H}$, and $\mathbf{W}_{ZF}\tilde{\mathbf{H}}$ gives the identity matrix.

The cost of incorporating ZF into GA-MUD technique is the time and resources consumed to obtain the seed chromosome, which is expected to be small compared with the benefits it brings.

3.4   Multistage GA Approach

In this scenario, we propose a multistage genetic algorithm in order to make the search more effective, thereby diminishing bit error rate (BER). The time consumption is neglected. In the case of a cascade of GA instances, each GA stage works to improve the solution developed by the previous stage, as shown in Fig. 5. The initial population of the first GA stage is random, as for the basic GA-MUD, described in Sect. 3.2. The best individual outputted by the first GA stage, $\dot{\mathbf{x}}^{(1)}$, is the seed for the next stage, and so on. Each time the single seed individual is accompanied by individuals generated randomly. Thanks to the multi-stage approach, there is a possibility to change the search strategy during the algorithm run or to move to another region of the search space, thereby avoiding a stuck at a local minimum.

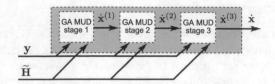

Fig. 5. Three-stage cascade GA MUD block diagram

Another approach is to run 3 parallel GA instances concurrently, with the same (random) intialization population and different settings of GA parameters, as illustrated in Fig. 6. The individual with the minimal value of the fitness function among the three GA instances is treated as the final solution.

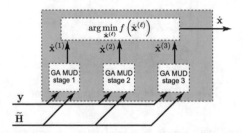

Fig. 6. Block diagram for 3-instance parallel GA scheme

4   Simulation Results

Evaluation of various MUDs is based on their performance, expressed by BER vs. SNR function. Each time, four 1-antenna concurrent transmitters and one 4-antenna receiver are considered ($N_T = N_R = 4$).

4.1   Basic GA-Based MIMO MUD

In Fig. 7, the performance of the GA-based MUD is examined with respect to the modulation order (64-QAM, 16-QAM, and QPSK) and compared with the ZF benchmark. The same GA parameters, specified in Table 1, are applied each time.

The results display poor performance of the 64-QAM system equipped with GA MUD, *i.e.*, error floor at *ca.* $2 \cdot 10^{-2}$ BER level appears, and GA MUD is outperformed by the ZF reference at high SNR. For 16-QAM, error floor occurs at *ca.* $2.5 \cdot 10^{-3}$, but GA is still significantly outperformed by ZF. Finally, in the case of QPSK modulation, the GA performance is comparable with that of ZF. The conclusion is that the basic GA MUD is appropriate for systems transmitting low-order modulations, only.

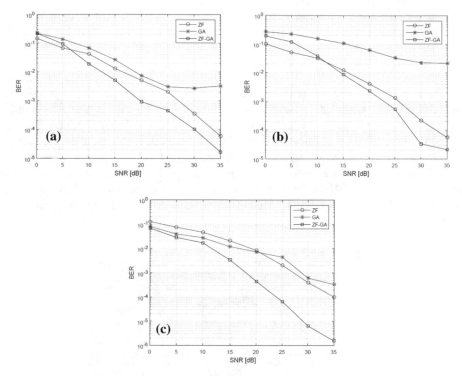

Fig. 7. BER vs SNR for the compared MUDs: basic GA, reference ZF, and the proposed GA-ZF: (a) 64-QAM, (b) 16-QAM, (c) QPSK

Table 1. Basic GA parameters

Parameter	Value
Population size	50
No. of generations	500
Stall generations	500
Mutation probability	0.005
Crossover fraction	0.95
Crossover function	Single-point
Selection function	Roulette wheel

4.2 ZF-GA MIMO MUD

Taking into consideration the conclusion drawn in the previous section, let us use GA to *support* ZF MUD receiver in the way presented in Sect. 3.3, instead of attempting to *replace* it.

As can be seen from Fig. 7, ZF-GA MUD (refer to ZF-GA lines) promises some SNR gain over the regular ZF. The gain is especially high for QPSK modulation (*ca.* 11 dB at 10^{-3} BER level). For 16-QAM as much as 7 dB gain of ZF-GA MUD over regular ZF detector is observed at the BER level of 10^{-3}. For 64-QAM, ZF-GA MUD still exhibits some gain (about 3 dB) over the regular ZF MUD at the BER level of 10^{-3}.

4.3   Multistage GA Approach

As described previously, the performance of GA (used without ZF aid) is poor in the case of high order modulation. Hence, the multistage GA approach is proposed in order to increase the probability of finding the optimal solution. Figure 8 shows the performance of a MUD consisting of 3 GA blocks (used in either parallel- or cascade configuration) for 64-QAM, referred to the one-stage GA. The fitness function and other main settings are the same in all stages (population size is set to 50), as for the reference 1-stage GA MUD. It can be seen that the BER measures are almost identical for all the compared receivers. It means that additional stages do not bring any performance enhancement.

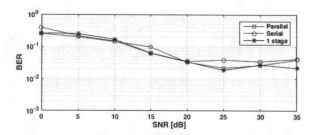

Fig. 8. BER vs. SNR for 3-stage parallel/cascade GA MUD and for the reference 1 stage GA MUD (64-QAM modulation assumed)

Figure 9 presents the results for 3-stage cascade GA MUD with 64-QAM modulation assumed. Two different parameters' settings, detailed in Tables 2 and 3, respectively, are considered for the 3-stage cascade GA MUD. The settings for 1-stage GA MUD were updated according to Table 4 in order to agree with Proposal 2 of 3-stage cascade GA MUD in terms of both the total number of generations.

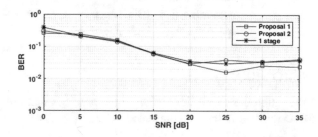

Fig. 9. BER vs. SNR for 3-stage cascade GA MUD, and for 1-stage GA MUD (2 different GA parameters' settings, 64-QAM modulation assumed)

Table 2. 3-stage cascade GA MUD parameters (Proposal 1)

GA Parameters	GA1	GA2	GA3
Population size	50	100	200
No. of generations	150	150	250
Stall generations	150	150	250

Table 3. 3-stage cascade GA MUD parameters (Proposal 2)

GA Parameters	GA1	GA2	GA3
Population size	50	50	50
No. of generations	500	500	500
Stall generations	500	500	500

Table 4. Updated 1-stage GA MUD parameters

Parameter	Value
Population size	150
No. of generations	1500
Stall generations	1500

It can be observed that the 3-stage cascade GA MUD does not improve the system performance significantly, in the case of 64-QAM modulation, regardless of assumed settings. However, it might be symptomatic that enlarging the population size (as in Proposal 1) brings relatively better results than increasing the number of generations considered as Proposal 2.

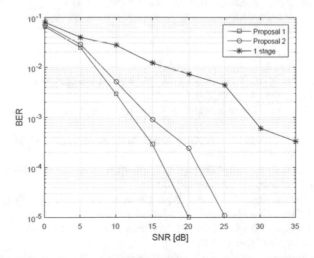

Fig. 10. BER vs. SNR for 3-stage cascade GA MUD, and for 1-stage GA MUD (2 different GA parameters' settings, QPSK modulation assumed)

Figure 10 demonstrates the effect of using 3-stage cascade GA MUD with QPSK modulation. It appears that this time the 3-stage cascade GA MUD brings an excellent (about 15 dB) SNR gain over the one-stage reference. Proposal 1 configuration (with the population size growing from one stage to another) outperforms Proposal 2 (with the extended number of considered generations) by about 2 dB, which shows that it is more effective to assume more (different) individuals existing at any given time than just increasing the number of algorithm iterations. In the second case, it is a higher risk that the fittest individual dominates a small population and, consequently, then no further improvement can be observed.

5   Conclusion

In this paper, the performance of GA-based MIMO MUD has been studied. The GA-based detector exhibits relatively good performance for a low-order modulation case. The genetic algorithm seems to fail in the case of 64-QAM modulation. However, it can cooperate with the ZF combiner, effectively, in such a way that the ZF multi-symbol estimate is "injected" into the initial population in order to boost GA convergence. The proposed combined ZF-GA MIMO MUD exhibits a significant SNR gain over the original ZF reference, especially for QPSK modulation.

In the light of the presented results, the second proposed MIMO MUD design, $i.e.$, the multistage cascade GA, is another attractive solution to the MUD problem, which brings as much as 15 dB gain over the ZF benchmark in the case of QPSK modulation. The advantage of the multi-stage GA MUD over both the ZF-GA MUD and the reference ZF detector is that the first does not require computing matrix inversion.

Acknowledgment. The presented work has been funded by the Polish Ministry of Science and Higher Education within the status activity task "Wireless networks – multiple access, transmission, error protection" in 2018 (Grant No. 08/81/DSPB/8123).

References

1. Tse, D., Viswanath, P.: Fundamentals of Wireless Communication, 1st edn. Cambridge University Press, New York (2005)
2. Barry, J.R., Lee, E.A., Messerschmitt, D.G.: Digital Communication. Springer, Netherlands (2012)
3. Chen, M.: Iterative Detection for Overloaded Multiuser MIMO OFDM Systems. Dissertation, University of York (2013)
4. Berenguer, I.: Advanced Signal Processing Techniques for MIMO Communication Systems. Dissertation, University of Cambridge (2005)
5. Getu, A.: Genetic Algorithm-Based Joint Channel Estimation and Data Detection For Multi-User MIMO. Dissertation, Addis Ababa University (2014)
6. Sivanandam, S.N., Deepa, S.N.: Introduction to Genetic Algorithms, 1st edn. Springer-Verlag, New York (2008)
7. Coley, D.A.: An Introduction to Genetic Algorithms for Scientists and Engineers. World Scientific Publishing Co Inc., Massachusetts (1999)

8. Haupt, R.L., Haupt, S.E.: Practical Genetic Algorithms. John Wiley & Sons, Hoboken (2004)
9. Abramson, M.A.: Genetic Algorithm and Direct Search Toolbox User's Guide. The MathWorks Inc., Massachusetts (2007)
10. Simon, D.: Evolutionary Optimization Algorithms. John Wiley & Sons, Hoboken (2013)
11. Chipperfield, A.J., Fleming, P.J., Pohlheim, H.: A genetic algorithm toolbox for MATLAB. In: Proceedings of the International Conference on Systems Engineering, 6–8 September, pp. 200–207. Coventry, UK (1994)

Monte-Carlo Simulation and Availability Assessment of the Smart Building Automation Systems Considering Component Failures and Attacks on Vulnerabilities

Vyacheslav Kharchenko[1,2] ⓘ, Yuriy Ponochovnyi[3](✉) ⓘ,
Artem Boyarchuk[1] ⓘ, Eugene Brezhnev[1,2] ⓘ,
and Anton Andrashov[2] ⓘ

[1] National Aerospace University KhAI, Kharkiv, Ukraine
{V.Kharchenko,a.boyarchuk}@csn.khai.edu,
e.brezhnev@csis.org.ua
[2] Research and Production Company Radiy, Kropyvnytskyi, Ukraine
a.andrashov@radiy.com
[3] Poltava National Technical University, Poltava, Ukraine
pnchl@rambler.ru

Abstract. The information and control system of smart building is considered as a set of subsystems including a building automation system (BAS) which has three-level structure (automation/control, communication, data base, cloud/management). BAS security and availability during its life cycle are assessed using the Markov models and Monte-Carlo simulation. Markov model is used to calculate BAS availability considering the possibility of recovery and different kinds of the faults. The Monte-Carlo simulation is applied to investigate any flow of intrusions into the BAS by analyzing system availability of a fault/vulnerability to occur during time depending on failure rate of its subsystems. The results of analytical and simulation modeling are compared to assure trustworthiness of availability assessment taking into account attacks on vulnerabilities.

Keywords: Simulation models · Building automation systems
Availability functions · Probability distribution function

1 Introduction

A set of smart buildings subsystems which performs information and control functions is considered as a building automation system (BAS), in other words the automation system of a dwelling, office or public institution [1]. The structure of the BAS of a smart building [2] includes the levels of control functions implemented on the CPU/FPGA, wireless communications and databases located on local servers or cloud data centers (Fig. 1).

Failures of the BAS component can be caused by physical and design defects (reliability factor); attacks on vulnerabilities (information security factor) [3]. Given the

© Springer International Publishing AG, part of Springer Nature 2019
W. Zamojski et al. (Eds.): DepCoS-RELCOMEX 2018, AISC 761, pp. 270–280, 2019.
https://doi.org/10.1007/978-3-319-91446-6_26

criticality of the consequences of failures, it is necessary to justify strategies and parameters for servicing and restoring systems. During the life cycle, software components can be modified (updates, patches). One of the key issues in the development of service strategies is their separate or general implementation, taking into account the factors of reliability and safety [4].

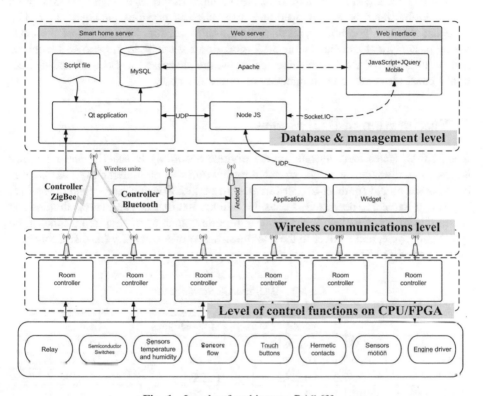

Fig. 1. Levels of architecture BAS [2].

Elimination of design defects and vulnerabilities leads to the changed parameters, in particular, the failure rates. The existing models of systems with variable parameters use full-scale experiment [5], Bayesian methods of investigation [6], Markov and semi-Markov processes [7]. The well-known Markov (semi-Markov) processes are based on the formation of a set of system states, transitions between them, and determination of the transition intensities. This set is more preferable in the case of evaluating the readiness of BAS with regard to various types of failures. In the case when the construction of the Markov model becomes more complicated with the increased number of states, the Monte Carlo method is used [8, 9]. The application of the Monte Carlo method is justified for complex systems with a large number of elements; in which random factors are interrelated in a certain way [10]. In addition, the simulation of random phenomena by the Monte Carlo method is carried out in order to verify the reliability of the results of the analytical Markov model.

The papers [7, 11] consider the Markov BAS models in detail. These models allow you to take into account software failures and attacks on vulnerabilities. However, some limitations of the Markov models do not allow them to be applied to real systems. Therefore, the aim of this article is to develop and study BAS simulation models using the Monte Carlo method.

The paper is structured as a follows. Next Sect. 2 describes a set of basic Markov's models which have been developed and researched in previous publications [7, 11]. BAS simulation algorithm is presented in the third section. The results of the comparing of analytical and simulation BAS modeling are analysed in Sect. 4. Technique and peculiarities of BAS simulation technique are discussed in Sects. 5 and 6. Section 7 concludes and describe direction of future research.

2 Specification of BAS Models

Sets of BAS states and transmission are divided according to security and reliability issues. First, the security part is presented by Nv (number of vulnerability); second one is presented by Nd (number of software/design faults). Specification of the models [7] is given in Table 1. These models consider elimination of faults and vulnerabilities and decreasing of Nv, Nd. Two indicators are important such as the minimum time of the system life cycle, and recover to the maximum value of availability (AMBASconstant) during period (TMBASconstant).

Table 1. BAS models specification

Type of service	Model name	Number of faults	Number of vulnerabilities	Number of services
-	MBAS1	0...Nd	0...Nv	0
Common	MBAS2.1	0...Nd	0...Nv	∞
	MBAS2.2	0...Nd	0...Nv	0...Np
Separate	MBAS3.1	0...Nd	0...Nv	∞
	MBAS3.2	0...Nd	0...Nv	0...Ndp, 0...Nvp

In some cases, the elimination process inside the system will not be able to eliminate the vulnerability or design fault; in this case, the maintenance strategies, which give the support for system to increase the elimination process, have been added. In our case, two types of maintenance strategies are used:

1. The common maintenance, which deals with design fault and vulnerability in same time, and it means that the process of elimination will be sequential between design fault and vulnerability;
2. The separated maintenance, which deals with vulnerability and design fault separately one by one. In next section, we will be describing the characteristics of maintenance strategies for two models: one with common maintenance and another with separated maintenance.

The Markov BAS model includes a marked state and transitions graph. For example, the graph for MBAS1 is shown in Fig. 2. The paper [11] describes the parameters of the given graph in detail, as well as suggests their numerical values.

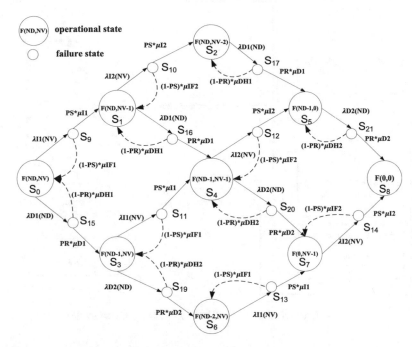

Fig. 2. Marked graph of the MBAS1 analytical model with due regard for software defects and vulnerabilities.

The program represents this graph as a combination of arrays of vertices V and arcs E; G = (V, E).

Graphs of BAS analytical models are formed by special Matlab functions of [7]. For simulation models, we are going to use these functions to preserve the logic of the system functioning.

3   BAS Simulation Algorithm

The key operation of the Monte Carlo method is to draw a random time between neighboring events. This time is distributed according to the exponential law. The Matlab Statistics Toolbox uses the function R = exprnd(mu) [12]. When using, it is necessary to invert the argument of the function as t = exprnd(1/mu).

The flow chart of the BAS simulation algorithm, as exemplified by MBAS1 model conditions, is shown in Fig. 3.

The modeling process is as follows. After running the script, we enter input values of input parameters (intensity of transitions and the dimension of the model). It is also

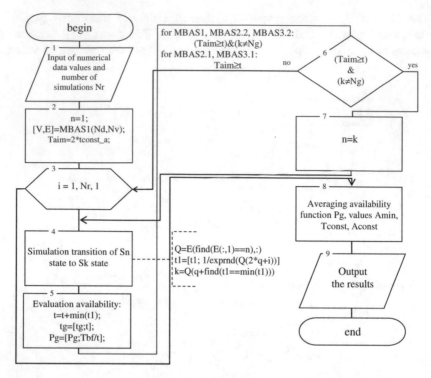

Fig. 3. BAS simulation algorithm with Monte Carlo method.

necessary to enter the parameters `Taim_interval` (the time interval for the system study) and `Nr` (the number of complete drawings of the simulation model).

The drawing using direct graph data is as follows. At time `t = 0`, the system is in the initial state S0, which corresponds to the first element of the array `V`, and the auxiliary variable is assigned value 1. It is possible to make two transitions from the state S0, to the states S9 and S15. Therefore, we enter two elements of the array `E`: `E (1,:)` and `E(19,:)`.

Further, we enter random combinations of time intervals in the vector `t1` defining the transition from the state S0. These combinations are obtained with the help of the function `t1 = exprnd(1/mu)`, where `mu` is the corresponding arc weight of the graph from the array `E`. The transition to the next state is modeled in terms of the variable `k`.

Then, the time `Tbf` is recalculated between failures (since the transition was performed from the operable state S0). In transition from the failure state, we would perform the recalculation of the recovery time `Tbr` and services.

Then assignment `n = k` takes place, and if the system has not reached the absorbing state S8 (for Nd = 2, Nv = 2), or the total time of all drawings has not exceeded `Taim_interval`; we perform the subsequent transition to a new state. If while drawing, the system has entered into an absorbing state, the system availability is equal to 1 for all subsequent time intervals.

4   Compared Results of the BAS Analytical and Simulation Modeling

After finishing the specified number Nr of complete drawings, [tconst] and [Amin] arrays are formed. We process them using the Matlab typical statistical functions (as one-dimensional random variables). However, the availability implementations have two dimensions, namely unique time intervals and their availability values.

We enter the values of the time intervals and the availability functions in the array Ai. At the first stage, we discard the determined values of the elements of the matrix Ai at the points t = 0 and t = Taim_interval (A (t) = 1). To do this, we uniqualize operations Ai = unique(Ai, 'rows').

Further, the elements of the Ai matrix are grouped according to the samples (the size of the grouping sample is determined by the variable n_vibor). For each sample, we have determine the averaged value of the time interval and those of the availability function using the mean (X) function. The results of the processing are shown in Fig. 4 for different dimensions of the n_vibor sample.

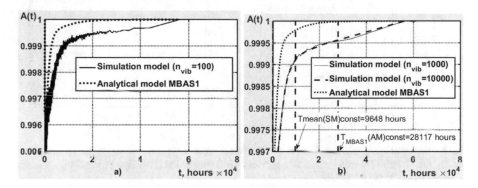

Fig. 4. Compared results of processing the total amount of the Monte Carlo drawings of MBAS1 model with different sample dimensions and the analytical model.

Increased sample dimensions allow us to get a smoother graph. This is convenient for visual comparison of analytical and simulation results. Also in Fig. 4b, the time axis scale shows the location of the values of the T_{MBAS1}(AM)const result indicator of the analytical model (determined with an operational margin of $\xi = 10^{-5}$) and the averaged value tconst (T_{mean}(SM)const in Fig. 4b) of all the complete drawings of the Monte Carlo model. The value of tconst is defined for each simulation. To do this, the time is fixed, after which the value of the availability function will be at least T_{MBAS1}(AM)const $- \xi$.

The value of the TMBAS1const indicator of the analytical model is located approximately in the middle between the values mean (tconst) and max (tconst) of the simulation model (Fig. 4 does not show). Next we interpolate the result of analytical modeling (the availability function) by the set of time intervals of the simulation model.

5   Studying the Results of the BAS Simulation Modeling

The main modeled `tconst` is a random value in the range [367.7 ...... 5.1576e +04]. Its distribution differs from the standard one and is shown in Fig. 4a.

Since the distribution law of the random variable `tconst` is unknown, we use the special function `Mathlab` fitmethis.m [13].

The function fitmethis (tconst) has suggested the gamma distribution:

$$f(t) = \frac{t^{a-1}}{b^a \cdot \Gamma(a)} e^{-\frac{t}{b}}; t \geq 0 \tag{1}$$

The distribution parameters are as follows: a = 2.22245 (with an operational margin of 0.0293806) and b = 4341.19 (with an operational margin of 64.3568). The gamma distribution function is shown in Fig. 5a.

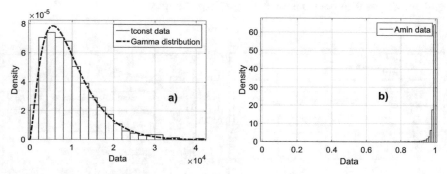

Fig. 5. Probability distribution function of the resulting indicator tconst and its representation by gamma distribution (a) and probability distribution function for drawings of the Amin resulting indicator (b) of MBAS1 model.

It is obvious from Fig. 5a that the difference between the values of T_{MBAS1}const of the analytical model and mean(tconst) and max(tconst) of the simulation model is large. But we know that when determining T_{MBAS1}const, the analytical model uses the operational margin of 10^{-5}. For the time interval mean(tconst) = 9648.1 h, the value of the availability function of the analytical model is Pga = 0.999791. It corresponds to the definition of T_{MBAS1}const with an operational margin of 2.09e−04.

For max(tconst) = 51576 h, the value of the availability function of the analytical model is Pga = 0.999999855652. It corresponds to the definition of T_{MBAS1}const with an operational margin of 1.4435e−07.

The second resultant indicator of the Amin simulation model is a random value in the interval [0.0078 ... 0.9996]. The minimum of the availability function is known to be located at the initial stage of the system operation (near the point t = 0), so it is very difficult to collect reliable statistics of the variation of the random Amin value. The distribution function of the random variable Amin is shown in Fig. 5b.

Amin indicator drawings give a long left tail of values with a low distribution density. They are unlikely to have a useful statistical load. The average value mean (Amin) = 0.9808 relative to the minimum of the availability function of the analytical model $A_{MBAS1}min$ = 0.9964 has an operational margin of 0.0156, or a fractional accuracy of 1.57%.

6 Peculiarities of the BAS Simulation for Graphs Without Absorbing States

Using the proposed algorithm (Fig. 3), we have developed BAS models for graphs without absorbing states:

- taking into account an unlimited number of common services (analog of MBAS2.1 [7]);
- taking into account an unlimited number of separate services (analog of MBAS3.1 [11]).

We are going to consider the key features of such models. To determine the resulting parameter `tconst`, we fix the time of the first "entry" of the system into the state S8. This technique is implemented by using the auxiliary variable `tc`. At the beginning of the tests, `tc = 0`, after transition to the state S8, `tc = 1`. Also, we have changed the condition for the end of the full drawing: if the total time of all drawings exceeds `Taim_interval` (since the system always carries out general service procedures). The results of the simulation modeling of the system with unlimited common services are shown in Fig. 6.

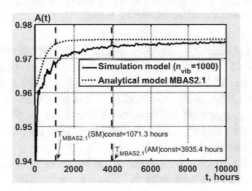

Fig. 6. Compared results of processing the total amount of the Monte Carlo drawings of MBAS2.1 model with the analytical model.

As $t \rightarrow \infty$, the availability curves of the analytical and simulation models approximate. At the end of the research interval (t = 20000 h, Fig. 6 does not show), an absolute accuracy in determining the $A_{MBAS2.1}const$ for the simulation model was 2.0011e−04 (relative accuracy of 0.02%).

The drawings of the resulting indicator `tconst` are within the interval [63.13 … …5773.3 h]. As in the previous case, the value of the indicator $T_{MBAS2.1}$const of the analytical model is located approximately in the middle between the values `mean(tconst)` and `max(tconst)` of the simulation model.

The function `fitmethis(tconst)` has suggested a lognormal distribution:

$$f(t) = \frac{1}{t \cdot \sigma \cdot \sqrt{2\pi}} e^{-\frac{(\ln t - \mu)^2}{2\sigma^2}} \qquad (2)$$

The distribution parameters are mu = 6.75208 (with an error of 0.0208161) and sigma = 0.658262 (with an operational margin of 0.0147302). The function of the lognormal distribution is shown in Fig. 7a.

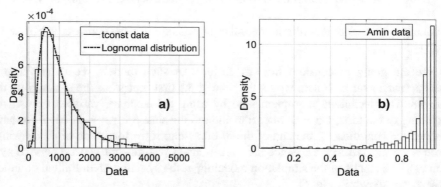

Fig. 7. The probability density distribution function of the resulting indicator tconst and its representation by the lognormal distribution (a) and probability density distribution function of the resulting indicator Amin (b) of the MBAS2.1 model.

For mean(tconst) = 1056.2 h, the value of the availability function of the analytical model is Pga = 0.9751. It corresponds to the definition of $T_{MBAS2.1}$const with an operational margin of 4.7166e−04. For max(tconst) = 5773.3 h, the value of the availability function of the analytical model is Pga = 0.9756. It corresponds to the definition of $T_{MBAS2.1}$const with an operational margin of 5.1651e−07.

The second resulting Amin indicator is within the interval [0.0362 … 0.9794]. The density Amin distribution function is shown in Fig. 7b.

The averaged value mean(Amin) = 0.8761 relative to the minimum of the availability function of the analytical model $A_{MBAS2.1}$min = 0.9619 has an absolute accuracy of 0.0859, or a relative accuracy of 8.6%.

7   Conclusion

The article describes the sequence of the development of the BAS simulation model in detail and verifies the precision (similarity) of the results of the analytical and simulation models. The work evaluates the availability function of BAS in the conditions of changing the system parameters. This function is dynamic in nature. At the first stage, the availability function decreases to its minimum, the value of which is considered as the first result parameter. Further, the availability function seeks its steady state and other estimated parameters are the time of transition of the availability function to the steady state and its value in this state. The values of these resulting parameters are compared for systems without maintenance, with common and separate maintenance for reliability and safety.

The results of simulation modeling for systems with unlimited service have shown satisfactory similarity with the corresponding analytical models. Generally, when using simulation models, the value of the availability function in an established mode can be determined with an operational margin of no more than $2.4 * 10^{-4}$.

In our opinion, we should abandon the simulation models for systems with an exponential distribution of input parameters. The Monte Carlo method does not allow unambiguous determination of the values of the resulting indicators TMBASconst and AMBASmin. However, the application of the Monte Carlo methods is indispensable in the study of models of BAS functioning with non-Markov failure and recovery flows. Such problems go beyond the scope of our paper.

The Matlab programs presented in the work are not intended for integration into the algorithms of the "smart building" system functioning. With their help, at the design stage of the system, reliability and cybersecurity performance as well as selection of the types and frequency of system maintenance during the operational phase are evaluated.

Further research should be directed to the analysis of the influence of input parameters on the resulting indicators of the Markov models of the availability assessment of the BAS structure.

References

1. Brad, B.S., Murar, M.M.: Smart buildings using IoT technologies. Constr. Unique Build. Struct. 5(20), 15–27 (2014)
2. Lobarev, S.: Smart home - the overall architecture of the system. https://habrahabr.ru/post/223163/. Accessed 05 Jan 2018
3. Granzer, W., Kastner, W., Neugschwandtner, G., Praus, F.: Security in networked building automation systems. In: 2006 IEEE International Workshop on Factory Communication Systems, pp. 283–292 (2006)
4. Trivedi, K., Kim, D., Roy, A., Medhi, D.: Dependability and security models. In: 2009 7th International Workshop on Design of Reliable Communication Networks, pp. 11–20 (2009)
5. Cybenko, G., Landwehr, C.: Security analytics and measurements. IEEE Secur. Priv. Mag. 10, 5–8 (2012)
6. Gashi, I., Popov, P., Stankovic, V.: Uncertainty explicit assessment of off-the-shelf software: a Bayesian approach. Inf. Softw. Technol. 51, 497–511 (2009)
7. Kharchenko, V.S., Abdulmunem, A.-S.M.Q., Ponochovnyi, Y.L.: Markov availability model of smart building automation system with separate and common reliability-security related maintenance. Syst. Control Navig. Commun. 4(36), 88–94 (2015)
8. Zio, E.: The Monte Carlo Simulation Method for System Reliability and Risk Analysis. Springer Series in Reliability Engineering (2013)

9. Grosu, R., Smolka, S.: Monte carlo model checking. Tools and algorithms for the construction and analysis of systems, pp. 271–286 (2005)
10. Robert, C., Casella, G.: Monte Carlo Statistical Methods. Springer Texts in Statistics (2004)
11. Kharchenko, V., Ponochovnyi, Y., Abdulmunem, A., Andrashov, A.: Availability models and maintenance strategies for smart building automation systems considering attacks on component vulnerabilities. In: Advances in Dependability Engineering of Complex Systems, pp. 186–195 (2017)
12. Exponential random numbers – exprnd. https://www.mathworks.com/help/stats/exprnd.html. Accessed 05 Jan 2018
13. Francisco de Castro.: FITMETHIS finds best-fitting distribution to data vector. https://www.mathworks.com/matlabcentral/fileexchange/40167-fitmethis. Accessed 05 Jan 2018

A Method for Passenger Level of Service Estimation at the Airport Landside

Artur Kierzkowski[✉], Tomasz Kisiel[iD], and Maria Pawlak

Department of Maintenance and Operation of Logistics Transportation
and Hydraulic Systems, Faculty of Mechanical Engineering, Wroclaw University
of Science and Technology, W. Wyspianskiego 27, 50-370 Wroclaw, Poland
{artur.kierzkowski, tomasz.kisiel}@pwr.edu.pl,
209096@student.pwr.edu.pl

Abstract. This paper presents a conception of the method based on a logistic support model relating to the functioning of a handling agent at the airport. The scope of the model includes passengers at the air terminal in the landside zone. The presented model can be used to perform an assessment of the system in check-in and security control subsystems. The method is designed to predict system-related assessment from the passenger's point of view. Passenger assessment significantly translates into profits made by airports through non-aviation activity. Hence, this is a key assessment factor used to select the appropriate number of resources needed to perform passenger service, depending on the predefined flight schedule. The paper proposes the use of a hybrid of two methods: computer simulation and multiple regression. The research being carried out worldwide is focused on the capacity of systems by analysing subsequent processes of passenger service as independent systems. This paper takes into account the relationships between check-in and security control systems, which affects the total assessment of passenger service in the landside zone. The proposed model may be applied to any airport featuring a centralised structure of the security control system and check-in service performed at basic desks with a process operator.

Keywords: Airport · Check-in · Security control · Level of service

1 Introduction

The sustainable development policy [1] states the necessity to remove bottlenecks in key network infrastructures, including the transport infrastructure. The research completed by [2] unambiguously indicate that an increase in linear and point infrastructure capacity of air transport is required. There are various areas being investigated in scientific literature to find some improvement of the transport infrastructure. The main factors being considered include: reliability of the technical system [3–5], the effectiveness of its operation [6–11] and vulnerability aspects [12]. These issues are often the cause of redundant structures implemented in the system and the dynamic allocation of resources depending on the demand. The situation may take place at any airport where the variable intensity of checking-in passengers corresponds to a various

© Springer International Publishing AG, part of Springer Nature 2019
W. Zamojski et al. (Eds.): DepCoS-RELCOMEX 2018, AISC 761, pp. 281–291, 2019.
https://doi.org/10.1007/978-3-319-91446-6_27

number of open service desks and counters. Dynamic management is aimed at min-imisation of the use of resources, with the appropriate system capacity maintained. System capacity may be directly expressed or referred to the level of passenger service.

Passengers using the service process at the air terminal may assess the process on the basis of numerous factors [13, 14]. It is known that an essential factor that affects the assessment of the level of service is the time spent on queuing and completion of the service process [15]. The airport manager, through the allocation of resources needed to perform tasks, is able to influence the change in the value of these parameters.

One must note that current research concerns individual subsystems. The research is mainly focused on the reduction of a passengers' queuing time [16–18]. The rela-tionships of subsequent subsystems are excluded. For instance, the performance of the check-in process considerably affects the stream of passengers reporting ready for security control. These are key passenger service processes in the landside zone [19]. In this paper, the presented model will take this fact into account because dynamic management of these two subsystems may be used to influence passenger queuing time both during the check-in and during the security control process. In turn, this affects the output assessment of passenger impressions.

Another aspect in favour of system assessment using the level of service indicator is that passenger impressions directly influence the benefits gained from non-aviation activity. The research conducted by [20] unambiguously showed that an increase in passenger assessment by 0.1 on a 1–5 scale contributes to an additional profit of USD $0.80 per passenger.

As presented in this section, the concept of passenger level of service estimation can be used to determine a general assessment of the service system depending on the allocated technical and human resources. The method based on a simulation model, aided by the Monte Carlo simulation, returns an estimated value of the feedback provided by passengers.

2   The Concept of the Level of Service Estimation Method

The core problem is shown in Fig. 1. While feeding input data into the IT system, such as a flight schedule with the number of passengers, it is expected to estimate the assessment of level of service given by passengers with reference to the work schedule of service desks.

The proposed approach requires a method that can be used to predict the value of the output indicator. Due to the complexity of the problem, a computer simulation model is proposed; the model will be used to estimate the LoS indicator by means of a Monte Carlo estimation. The Monte Carlo method is used in stochastic processes, and it is based on n-time performance of a series of independent simulations by choosing the value of random variables X. According to Bernoulli's law of large numbers, it can be assumed that the result can be determined as the expected value with an appropriate number of repetitions [21].

The proposed simulation model will be operated with four modules (Fig. 2). The first module, based on input data, is used to generate passengers at the right instant of time for passenger service modules. Passenger services will be carried out in two

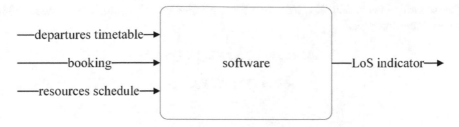

Fig. 1. Concept of the method used to estimate the LoS indicator.

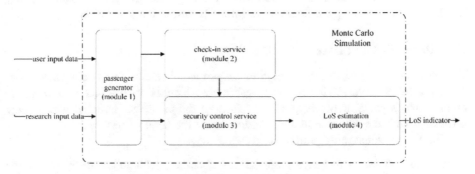

Fig. 2. Structure of the simulation model.

subsystems: check-in (Module 2) and security control (Module 3). Then, the predicted result of the LoS indicator will be calculated in Module 4 based on multiple regression.

2.1 User Input Data

In order to carry out an analysis, the software user is required to enter flight schedule *Fl* given as a set of subsequent flight operations fl_i (1).

$$Fl = \langle fl_{i=1}; \ldots; fl_n \rangle \qquad (1)$$

Each flight-related operation is characterised by a set of attributes (2):

- *dep* – time of aircraft departure;
- *cr* – air carrier index;
- *paxn* – number of departing passengers;
- *ckin* – percentage share of passengers going through check-in.

$$fl_i = \langle dep_i; cr_i; paxn_i; ckin_i \rangle \qquad (2)$$

The user also defines the functioning schedule of the resources dedicated to service individual carriers at check-in desks (3) and stations dedicated to passenger service at the security checkpoint (4). The functioning schedule of stations is entered with five-minute intervals.

$$ckinR_{cr} = \langle X1_{cr}; X2_{cr}; X3_{cr}; \ldots \rangle \tag{3}$$

$$scR = \langle Y1; Y2; Y3; \ldots \rangle \tag{4}$$

where

- $X1_{cr}$; $X2_{cr}$; $X3_{cr}$, ... – number of desks dedicated to passenger service for the carrier cr at subsequent five-minute intervals during the simulation;
- $Y1$; $Y2$; $Y3$; ... – number of desks dedicated to passenger service at the security checkpoint at subsequent five-minute intervals during the simulation.

2.2 Research Input Data

The implementation of the model requires experimental studies on an actual system so that the level of service indicator can be estimated. A study of an actual system can be divided into a stage involving passenger flow at the passenger terminal and a stage associated with the response to the question about the level of service.

The former stage must gather data relating to:

- the moment when passengers appear for check-in, with reference given to the aircraft departure time;
- the moment when passengers (who are not going through the check-in service) appear for security control and with reference given to their aircraft departure time;
- the duration of passenger service at the check-in desk;
- the duration of passenger service in the security control lane;
- the time spent at the passenger terminal between the completed check-in service and the entrance to the security checkpoint queueing system (with a specified moment when check-in service is completed before the scheduled aircraft departure).

Based on the completed study of the actual system, it is necessary to estimate the probability density functions of events that are designated in the following sections as follows:

- $f_{ckin}(x)$ – probability density function of the passenger appearing in the check-in queueing system before the scheduled departure time;
- $f_{sc}(x)$ – probability density function of the passenger appearing in the security checkpoint queueing system before the scheduled departure time (for passengers who did not go through the check-in service);
- $f_{sckin}(x)$ – probability density function of service duration at the check-in desk;
- $f_{ssc}(x)$ – probability density function of service duration in the security control lane;
- $f_{ckintosc}(x)$ – two-dimensional probability density function of the time spent at the terminal between the exit from the check-in system and entrance to the security checkpoint queueing system (with reference of the moment when check-in service is completed to the aircraft departure time).

The other stage of data collection consists in a survey among the passengers as regards the assessment of the level of service. For each passenger, the following data (if exists) must be recorded:

- general assessment of the level of service in the airport's landside zone;
- queueing time before check-in and its assessment;
- check-in service duration its assessment;
- queueing time before security control and its assessment;
- security control duration its assessment.

These will be the basis to determine:

- $LoS_T(x)$ – multiple regression function of the assessment of the level of service in the airport's landside zone;
- $f_{cw}(x)$ – regression function of the assessment of queueing time before check-in;
- $f_{cs}(x)$ – regression function of the assessment of check-in service time;
- $f_{scw}(x)$ – regression function of the assessment of queueing time before security control;
- $f_{scs}(x)$ – regression function of control lane service time.

2.3   Module 1: Passenger Generator

The passenger generator module is necessary to enter passengers appearing at the terminal in the relevant queueing systems. In this case, the stream of appearing passengers is complex. It cannot be described by means of a general probability density function of the time between subsequent appearances. This is a result of a finite stream of appearances with time limitations (time interval when a passenger may appear). In addition, passengers at check-in desks undergo the procedure in a system dedicated to air carriers, which necessitates the division of passengers into various groups in one of service stages.

The passenger generator is used to develop a set of all passengers (5). Next, the variables (6) in Module 1 are assigned to each passenger.

$$PAX = \langle pax_{j=1}; \ldots; pax_k \rangle \tag{5}$$

where

- pax_j – another j-th passenger;
- $k = \sum_{i=1}^{n} paxn_i$ – sum of passengers in all flights.

$$pax_j = \langle pcr_j; pdep_j; pckin_j; trep_j \rangle \tag{6}$$

where

- pcr_j – j-th passenger membership to the carrier index cr;
- $pdep_j$ – time of simulation relating to the departure of the aircraft on which the j-th passenger travels;
- $pckin_j$ – check-in process completion, where 1 – service, 0 – no service;
- $trep_j$ – instant of time when the j-th passenger is generated in the simulation model.

Module 1 of the simulation model is used to generate all passengers according to times $trep_j$. For this purpose, it is necessary to determine the probability density function of passengers appearing for the check-in service and security control service. For passengers whose value $pckin_j = 1$, variable $trep_j$ will be determined based on the formula (7) and, for passengers who are about to undergo security control ($pckin_j = 0$), the formula is (8).

$$trep_j = pdep_j - f_{ckin}(x) \qquad (7)$$

$$trep_j = pdep_j - f_{sc}(x) \qquad (8)$$

where

- $f_{ckin}(x)$ – value of a random variable generated by the pseudorandom number generator for the probability density function of passengers appearing at check-in;
- $f_{sc}(x)$ – value of a random variable generated by the pseudorandom number generator for the probability density function of passengers appearing at security control.

2.4   Module 2: Check-in Service

The proposed model takes into account the structure dedicated to the check-in service. This means that passengers go to dedicated desks, depending on the carrier that is going to provide the transport service. In this case, the structure of the system is parallel. There are Q queues and the number of desks D dedicated to each carrier is variable in time and in accordance with the input data set $ckinR_{cr}$ (3). The structure of the system is shown in Fig. 3.

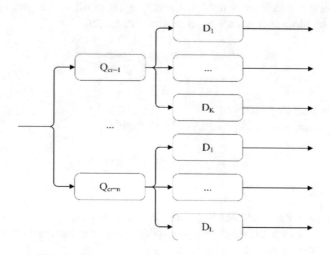

Fig. 3. Structure of the check-in queueing system.

Module 2 receives the generated passengers with the assigned variable $pckin_j = 1$. The principle of the process is in accordance with the FIFO strategy. The passenger is positioned at the end of the Q queue assigned to the carrier cr. The queueing passenger is waiting to be the first in line and until one of the desks becomes available. Then, the random generator determines service time for a passenger at the desk, based on the probability density function of the check-in service $f_{sckin}(x)$. The set of passenger variables (6) will be provided with queueing times tqc_j and service time tc_j. Once the service time is completed, the passenger vacates the desk but is still present in Module 2. Then, the queueing time before transition to Module 3 is randomly selected according to the probability density function of queueing time in the general access zone $f_l(x)$. After this time, the passenger is transferred to Module 3. A simplified algorithm of passenger transition to Module 2 is presented in Fig. 4.

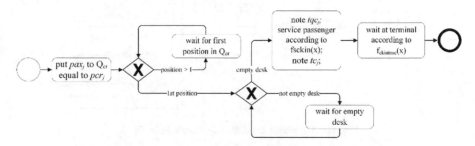

Fig. 4. Algorithm used to implement Module 2.

2.5 Module 3: Security Control Service

The structure of the security control system features one queue Q, where all streams of passengers from Module 2 meet. In addition, for this queue, passengers are directly generated from Module 1; the passengers are characterised by the parameter $pckin_j = 0$. The number of available lines L is variable in time and corresponds to the input data set scR (4). The structure of the system is shown in Fig. 5.

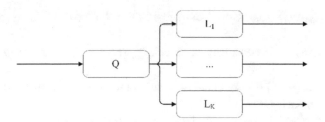

Fig. 5. Structure of the security control queueing system.

The principle of the process is in accordance with the FIFO strategy. The passenger is positioned at the end of the Q queue. The queueing passenger is waiting to be the first pin line and until one of lanes becomes available. Then, the random generator determines service time in the lanes, based on the probability density function of security control $f_{ssc}(x)$. The set of passenger variables (6) will be provided with queueing times $tqsc_j$ and service time tsc_j. Once the service time is completed, the passenger vacates the desk and is transferred to Module 4. A simplified algorithm of passenger transition to Module 3 is presented in Fig. 6.

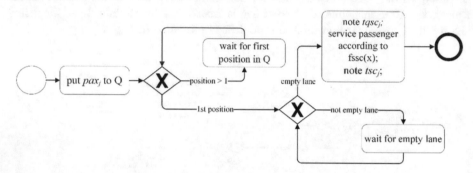

Fig. 6. Algorithm used to implement Module 3.

2.6 Module 4: Level of Service Estimation

For each passenger, Module 4 collects data relating to completion times of the processes and queueing times. Based on the times, a set of partial assessments LoS_j (9) is allocated to each passenger. Assessment values are allocated based on the regression functions developed in accordance with Sect. 2.2.

$$LoS_j = \langle LoS_{cw_j}; LoS_{cs_j}; LoS_{scw_j}; LoS_{scs_j}; LoS_{T_j} \rangle \tag{9}$$

where

- $LoS_{cw_j} = f_{cw}(tqc_j)$ – assessment of queueing time before check-in as given by the j-th passenger;
- $LoS_{cs_j} = f_{cs}(tc_j)$ – assessment of check-in service time as given by the j-th passenger;
- $LoS_{scw_j} = f_{scw}(tqsc_j)$ – assessment of queueing time before security control as given by the j-th passenger;
- $LoS_{scs_j} = f_{scs}(tsc_j)$ – assessment of security control service time as given by the j-th passenger;
- LoS_{T_j} – total assessment of the system as given by the j-th passenger.

In order to determine the total assessment of the system as given by each passenger, the multiple regression method will be applied. For passengers who completed full service in the landside zone (check-in and security control), the total assessment will be determined with the following formula (10). The passengers going only through security control provide their assessment according to the formula (11).

$$LoS_{T_j} = w_0 + w_1 \cdot LoS_{cw_j} + w_2 \cdot LoS_{cs_j} + w_3 \cdot LoS_{scw_j} + w_4 \cdot LoS_{scs_j} \tag{10}$$

$$LoS_{T_j} = w_{00} + w_5 \cdot LoS_{scw_j} + w_6 \cdot LoS_{scs_j} \tag{11}$$

where

- w_0, w_{00} – free terms;
- w_1, w_2, ..., w_6 – regression coefficients.

Free terms and regression coefficients can be adjusted, e.g. by means of the least squares method. Studies on the possible application of multiple regression in the analysis of queueing process at the airport are shown in [15].

The output assessment LoS_e of the subsequent simulation experiment e will be given as the arithmetic mean of the assessments from all passengers during the experiment (e.g. the same operational day at the airport):

$$LoS_e = \frac{1}{k} \sum_{j=1}^{k} LoS_{T_j} \tag{12}$$

The final result of *LoS* assessment will be estimated using a simulation with the Monte Carlo method. Multiple repetitions of the experiment will provide the expected value of the level of service indicator (13).

$$LoS = \frac{1}{r} \sum_{e=1}^{r} LoS_e \tag{13}$$

3   Summary

The presented approach is aimed at leading the passenger service management process to the management oriented towards increasing revenues on the non-aviation activity of the airport. The model takes into account a series structure of the passenger service system in the general access zone, and hence the relationship between subsequent sub-processes performed in this system. Therefore, appropriate management of the subsequent sub-processes can be used to predict the total assessment of the system.

Further research work will consist of the verification and validation of the presented model using data acquired from an actual system (Wroclaw Airport) to be able to present possible implementation of the proposed model in the actual system. The model will also be extended by subsequent processes of the passenger service performed at the airport terminal in the restricted zone of the airport.

References

1. Regulation (Eu) No. 1303/2013 of The European Parliament and of the Council of 17 December 2013
2. Neufville, R.: Building the next generation of airport systems. Transp. Infrastruct. **38**(2), 41–46 (2008)
3. Restel, F.J.: Measures of reliability and safety of rail transportation system. In: European Safety and Reliability Conference (ESREL) 2011 Advances in Safety, Reliability and Risk Management, pp. 2714–2719 (2012). WOS:000392426504029
4. Restel, F.J.: Defining states in reliability and safety modelling. In: 10th International Conference on Dependability and Complex Systems (DepCoS-RELCOMEX) 2015, Theory and Engineering of Complex Systems and Dependability Book Series: Advances in Intelligent Systems and Computing, vol. 365, pp. 413-423, Springer (2015). WOS:000365127100039
5. Werbinska, S.: The availability model of logistic support system with time redundancy. Maint. Reliability **3**, 23–29 (2007)
6. Tubis, A.: Process assessment of risks in the production company with the use of linguistic variables and the FMEA analysis. In: Proceedings of the First International Conference on Intelligent Systems in Production Engineering and Maintenance, ISPEM 2017, pp. 368–379. Springer (2018)
7. Zajac, M., Swieboda, J.: An unloading work model at an intermodal terminal. In: Theory and Engineering of Complex Systems and Dependability, pp. 573–582. Springer (2015)
8. Walkowiak, T., Mazurkiewicz, J.: Soft computing approach to discrete transport system management. Lecture Notes in Computer Science, Lecture Notes in Artificial Intelligence, vol. 6114, pp. 675–682 (2010) https://doi.org/10.1007/978-3-642-13232-2_83
9. Bruno, G., Genovese, A.: A mathematical model for the optimization of the airport check-in service problem. Electron. Notes Discrete Math. **36**, 703–710 (2010)
10. Walkowiak, T., Mazurkiewicz, J.: Analysis of critical situations in discrete transport systems. In: Proceedings of DepCoS - RELCOMEX 2009, Brunów, Poland, 30 June–02 July 2009, pp. 364–371. IEEE (2009). https://doi.org/10.1109/depcos-relcomex.2009.39
11. Solak, S., Clarke, J.-P.B., Johnson, E.L.: Airport terminal capacity planning. Transp. Res. Part B Methodol. **43**(6), 659–676 (2009). https://doi.org/10.1016/j.trb.2009.01.002
12. Restel, F.J.: The Markov reliability and safety model of the railway transportation system, In: Proceedings of the European Safety and Reliability Conference (Esrel), Safety and Reliability: Methodology and Applications, pp. 303–311 (2015). WOS:000380543400042
13. Skorupski, J., Uchroński, P.: Managing the process of passenger security control at an airport using the fuzzy inference system. Expert Syst. Appl. **54**, 284–293 (2016)
14. Kierzkowski, A., Kisiel, T.: Simulation model of security control system functioning: a case study of the Wroclaw Airport terminal. J. Air Transp. Manag. **64**(Part B), 173–185 (2017). https://doi.org/10.1016/j.jairtraman.2016.09.008
15. Anderson, R., Correiaa, A.R., de Wirasinghe, S.C., Barros, A.G.: Overall level of service measures for airport passenger terminals. Transp. Res. Part A Policy Pract. **42**, 330–346 (2008)
16. Wilson, D., Roe, E.K., So, S.A.: Security checkpoint optimizer (SCO): an application for simulating the operations of airport security checkpoints. In: Winter Simulation Conference, pp. 529–535 (2006)
17. Van Dijk, M., Van der Sluis, E.: Check-in computation and optimization by simulation and IP in combination. Eur. J. Oper. Res. **171**, 1152–1168 (2006)

18. van Boekhold, J., Faghri, A., Li, M.: Evaluating security screening checkpoints for domestic flights using a general microscopic simulation model. J. Transp. Secur. **7**, 45–67 (2014)
19. Kierzkowski, A., Kisiel, T.: A model of check-in system management to reduce the security checkpoint variability. Simul. Model. Pract. Theor. **74**, 80–98 (2017). https://doi.org/10.1016/j.simpat.2017.03.002
20. DKMA: Four steps to a great passenger experience (without rebuilding the terminal). DKMA (2014)
21. Bernoulli, J.: Ars Conjectandi opus posthumum. Basileae, Impensis Thurnisiorum, Fratrum (1713)

Consistency-Driven Pairwise Comparisons Approach to Software Product Management and Quality Measurement

Waldemar W. Koczkodaj[1], Paweł Dymora[2], Mirosław Mazurek[2], and Dominik Strzałka[2(✉)]

[1] Computer Science, Laurentian University, Sudbury, Canada
[2] Faculty of Electrical and Computer Engineering,
Rzeszów University of Technology, al. Powstańców Warszawy 12,
35-959 Rzeszów, Poland
strzalka@prz.edu.pl

Abstract. In this study, the software product quality measurement, based on the consistency-drive pairwise comparisons (PC) is proposed as a new way of approaching this complicated problem. The assessment of software quality (SQ) is a complex process. It is usually done by experts who use their knowledge and experience. Their subjective assessments certainly involve inaccuracy (which is difficult to control) and consistency of assessments (which can be measured and may influence accuracy). The inconsistency analysis, which is proposed in this approach, is used to improve assessments. Weights, reflecting the relative importance of the attributes, are computed as opposed to the commonly practiced arbitrary assignment. The PC method allows to define a consistency measure and use it as a validation technique. A consistency-driven knowledge acquisition, supported by a properly designed software, contributes to the improvement of quality of knowledge-based systems.

Keywords: Software quality · Pairwise comparison · Complex system Concluder

1 Introduction

Weinberg's Second Law:
"If builders built buildings the way programmers wrote programs, then the first woodpecker that came along would destroy civilization" (cited in [1, p. ix]).

Unlike a distance measure with one meter as a standard unit, there is no commonly acceptable unit of measurement for software product quality. This work proposes necessary basic preliminaries for a method of creating such a measure by pairwise comparisons (PC) of software quality (SQ) attributes. According to [2, p. 388], "In the context of software engineering, software quality refers to two related but distinct notions that exist wherever quality is defined in a business context:

- software functional quality reflects how well it complies with or conforms to a given design, based on functional requirements or specifications. That attribute can

© Springer International Publishing AG, part of Springer Nature 2019
W. Zamojski et al. (Eds.): DepCoS-RELCOMEX 2018, AISC 761, pp. 292–305, 2019.
https://doi.org/10.1007/978-3-319-91446-6_28

also be described as the fitness for purpose of a piece of software or how it compares to competitors in the marketplace as a worthwhile product.
• software structural quality refers to how it meets non-functional requirements that support the delivery of the functional requirements, such as robustness or maintainability, the degree to which the software was produced correctly."

Modern software has become so complex that older methods are no longer relevant. In particular, it does make sense, in our humble opinion, to compare software with a monograph. The International Standard Bibliographic Description (ISBD) is applicable when it comes to identification as well as classification of the printed matter, however it is inappropriate for its quality assessment. Our approach may be one of very few models for improving the measurement of SQ [3–5] and it is worth trying. We have decided to provide the PC interpretation of IEEE standard ISO-9126 because this standard is the most recognized by both software development practitioners and researchers. The use of PCs allows us to compute weights assumed by the standard to be all equal (hence 1 by each PC). Undoubtedly, it is possible to compare individual books in pairs what is also applicable to software. The question is however, how much the method in question will be useful in practice?

Necessary preliminaries that allow a construction of SQ measurement based on PC have been presented in this work. Such approach is already used in medical science where a rating scale is used for diagnosing mental health problems, such as depression or burnout [6]. Rating scale is a set of categories designed for data acquisition about a quantitative or a qualitative entity (e.g., SQ). Common examples of rating scales are the Likert response scale and 1–10 rating scales when a user selects the number to reflect the perceived intensity (both positive and negative) of an attribute [7].

A software engineer elicits knowledge about software product quality from experts and software developers, refines it with them and represents it in the PC matrices. The elicitation of knowledge from an expert can be done manually or with the aid of computers. The main purpose of computerized support to an expert is to reduce or eliminate the potential problems of assessments inconsistencies. Often, it dominates the gathering of data for the initial model and the interactive refinements of our knowledge about SQ. Visual modelling techniques are very important in constructing the initial domain model. The objective of this approach is to give the user the ability to visualize real-world problems and to manipulate elements of it through the use of graphics. The expert's knowledge is expressed by assessing the preferences, relevant criteria or factors, or possible alternatives. When devising methods for formulating and assessing preferences, a knowledge engineer has to take human capability limitations into account when making such an endeavor. One possible technique of extracting the expert's knowledge and preferences is based on the PC method.

The whole paper is divided into seven parts. Section 2, which follows an Introduction, shows an outline of necessary preliminaries presenting pairwise methodology. The problem of assessment inconsistencies measured by *ii* indicator is presented in Sect. 3, whereas Sect. 4 deals with an analytical approach which allows elimination of triads inconsistency. Section 5 gives a review of software measures with attributes on the basis of literature data. Last but not least part (Sect. 6) reveals a PC model that refers to original Boehm's approach and ISO-9126 standard for SQ measurement. The paper summary is in Sect. 7.

2   Pairwise Comparisons Preliminaries

Ramon Llull, the 13th-century mystic and philosopher is regarded as the first who used PC [8]. Thurstone applied PC in the form of "the law of comparative judgement" in [9]. A variation of this law is known as the BTL (Bradley-Terry-Luce) model [10]. A number of controversial customized PC followed in numerous studies. The authors of this paper do not intend to favor any particular customization here, however the seminar work [11] had a considerable impact on the PC research and we feel that it should be acknowledged despite serious controversies surrounding it and recently published in [12]. Yet, the most convincing evidence is in [13], none of which disqualifies the PC, but its ineffectual use in the past.

The PC method takes advantage of the statements about the expert's preferences and assessments. They are expressed by the examination of pairs of criteria or objectives. The presented methodology utilizes mapping of inconsistent evaluations by an expert into a numerical scale (see Table 1) that closely approximates his or her assessments. Ordinal numbers are used to express relative preferences. In particular, the numbers do not represent "absolute" measure of the mapped criteria as such may simply not exist (for example, it is hard to define a global measure of public safety but it is still practical to compare it, in relative terms, with the degree of environmental pollution).

Table 1. Example of comparison scale.

Intensity	Definition	Explanation
1	Equal importance	Equal contribution
2	Weak importance of one over another	Slightly favor one criterion over another
3	Essential or strong importance	Strongly favor one criterion over another
4	Demonstrated importance	Strong dominance
5	Absolute importance	The highest preference
1.2, 2.3,…, etc.	Intermediate values	When compromise is needed

The traditional matrix representation of PC is generated by using a matrix M of the following format:

$$M = \begin{bmatrix} 1 & m_{1,2} & \cdots & m_{1,n} \\ \frac{1}{m_{1,2}} & 1 & \cdots & m_{2,n} \\ \vdots & \vdots & \vdots & \vdots \\ \frac{1}{m_{1,n}} & \frac{1}{m_{2,n}} & \cdots & 1 \end{bmatrix}$$

PC matrix elements represent the intensities of an expert's preference between individual pairs of entities (or criteria) expressed as ratios chosen from an assumed scale for subjective data and transformed by the recently published formula in [13]. Note the criteria $E_1, E_2, \ldots, E_n$ (n is the number of criteria to be compared). The entry m_{ij} in i-th row and j-th column of the PC matrix M, denotes the relative importance of entity (or criterion) E_i compared with objective E_j, as expressed by an expert. This PC matrix M has all positive elements and the following reciprocal property:

$$m_{ij} = \frac{1}{m_{ji}} \tag{1}$$

PC matrix M is called consistent if $m_{ij} \cdot m_{jk} = m_{ik}$ for all $i < j < k$. Vectors consisting of three values in the consistency condition: (m_{ij}, m_{ik}, m_{jk}) are called "triads". Because of reciprocity condition, triads are mirrored in the lower triangle (these are also triads), however only the upper triangle is usually analyzed.

Let w_i denote the unknown weight of the criterion i. The vector $w = [w_1, w_2, \ldots, w_n]$ is estimated on the basis of the PC matrix M by the geometric means method. For consistent assessments, it is:

$$a_{ij} = \frac{w_i}{w_j} \tag{2}$$

The following heuristic:

$$w_i = \left(\prod_{j=1}^{n} m_{ij} \right)^{1/n} \tag{3}$$

was proposed in [14] for finding vector w for inconsistent PC matrices. In fact, it also works trivially for consistent PC matrices. Equation (3) leads to the important meaning of the cardinal consistency. The assessments A_i, A_j and A_k are said to be consistent if $a_{ij} \cdot a_{jk} = a_{ik}$. In this case, the triad (A_i, A_j, A_k) is said to be consistent. PC matrix M is consistent if all possible triads are consistent, i.e., Eq. (4) holds for any $i, j, k = 1, 2, \ldots, n$:

$$a_{ij} = \frac{a_{ik}}{a_{jk}} \tag{4}$$

In this paper we take into account a different definition of consistency than was proposed by Saaty in [11]. We follow Koczkodaj's approach proposed in [15], which in contradiction to Saaty's approach allows to locate the most inconsistent assessments and re-examine them. New and more consistent assessments may be expressed in an interactive way. They may contribute to the overall reduction of the inconsistency.

3 Inconsistency in Pairwise Comparisons

For a single triad (x, y, z), the inconsistency indicator ii is given by the following formula:

$$ii = 1 - min(y/x/z, x * z/y). \tag{5}$$

This new definition was proposed in [15], formally generalized to the entire matrix M by the use of the *max* function for all triads (defined by the consistency condition), and simplified in [16]. Making comparative assessments of intangible criteria (e.g., the degree of an environmental hazard or pollution factors) not only involves imprecise or inexact knowledge, but also inconsistency in our own assessments. The improvement of knowledge elicitation by controlling the inconsistency of experts' assessments is not only desirable but absolutely necessary.

Equation (5) is defined as the relative distance to the nearest consistent reciprocal matrix represented by one of these three vectors for a given metric. Checking consistency in the PC method could be compared to checking that the divisor is not equal to 0. The proposed solution of the PC method is based on the assumption that the given reciprocal matrix is consistent. However, expecting that all subjective assessments are consistent is not realistic, especially, if they are subjective. We know that most assessments are subjective, inaccurate, and nearly always contain some kind of bias, therefore, the total consistency is not expected. In practice, we can set a threshold for ii values where we accept the existing inconsistency.

We must have at least three compared criteria in order to have inconsistent assessments. Consequently we may assume, that all indexes i, j, k must be pairwise different. Having $n \geq 3$ criteria, we may calculate inconsistencies only for triads with indexes holding the property $0 < i < j < k \leq n$.

The ii indicator of a PC matrix is the indicator of the quality of knowledge, which begins with computing the inconsistency of the assessments. The triad with the largest inconsistency can be displayed for the experts to have an opportunity to revise their preferences.

4 Making a Single Triad Consistent

Denote the vector of arithmetic means of rows of a matrix A by $AM(A)$ and the vector of geometric means by $GM(A)$. For a consistent matrix, the vector of rows geometric means especially important since it can be used as a vector of weights [17].

For a matrix A with positive coordinates, we can define the matrix $B = \lambda(A)$ such that $b_{ij} = \ln(a_{ij})$. Reversely, for a matrix B define the matrix $A = \mu(B)$ such that $a_{ij} = \exp(b_{ij})$. Obviously,

$$\ln(GM(A)) = AM(\lambda(A)) \tag{6}$$

and

$$\exp(AM(B)) = GM(\mu(B)) \tag{7}$$

If A is consistent, then elements of $B = \lambda(A)$ satisfy:

$$b_{ij} + b_{jk} = b_{ik}$$

for every $i, j, k = 1, 2, \ldots, n$. We call such a matrix *additively consistent*.

Let's consider an additive (after the *ln* transform) triad (x, y, z), which is consistent if $y = x + z$. It is also equivalent to the scalar vector multiplication of $v = (x, y, z)$ by the vector $e = (1, -1, 1)$ giving 0. This means that v and e are perpendicular vectors. If a triad is inconsistent, an orthogonal projection onto subspace perpendicular to vector $e = (1, -1, 1)$ in the space $\mathbb{R}^3$ makes it consistent. These projections can be expressed as:

$$\tilde{v} = v - \frac{v \circ e}{e \circ e} e, \tag{8}$$

where $u_1 \circ u_2$ is the scalar product of vectors u_1 and u_2 in $\mathbb{R}^3$. It holds:

$$e \circ e = 3, v \circ e = x - y + z,$$

hence $\tilde{v} = (\tilde{x}, \tilde{y}, \tilde{z})$ where:

$$\tilde{x} = \frac{2}{3}x + \frac{1}{3}y - \frac{1}{3}z, \tag{9}$$

$$\tilde{y} = \frac{1}{3}x + \frac{2}{3}y + \frac{1}{3}z, \tag{10}$$

$$\tilde{z} = -\frac{1}{3}x + \frac{1}{3}y + \frac{2}{3}z. \tag{11}$$

Remark. For a given inconsistent triad (x, y, z) the transformation into a consistent one can be done by the following three ways:

$$(x, x + z, z), (y - z, y, z), (x, y, y - z)$$

The averages of the above three triads give a triad $\tilde{v} = (\tilde{x}, \tilde{y}, \tilde{z})$.

For a multiplicatively inconsistent triad (a_{ik}, a_{ij}, a_{kj}) its transformation to the consistent triad $(\tilde{a}_{ik}, \tilde{a}_{ij}, \tilde{a}_{kj})$ is given by:

$$
\begin{aligned}
\tilde{a}_{ik} &= a_{ik}^{2/3} a_{ij}^{1/3} a_{kj}^{-1/3}, \\
\tilde{a}_{ij} &= a_{ik}^{2/3} a_{ij}^{2/3} a_{kj}^{1/3}, \\
\tilde{a}_{kj} &= a_{ik}^{-1/3} a_{ij}^{1/3} a_{kj}^{2/3}.
\end{aligned}
\tag{12}
$$

The proof of convergence was provided (independently on [18]) in [19].

5 Software Quality Attributes

Most departments (e.g. Defense, Treasury, Energy) as well as the US Government agencies require their Automated Information Systems (AIS), which process sensitive or classified information, to be certified and credited [20]. The importance of SQ is unquestionable. An extensive research has been done focusing on developing a set of attributes which can be used to measure the quality of a software system [21]. Taxonomies have been proposed to cover the problem of quality issues. Two of the most prominent SQ models are those introduced in [22] and in [23] which defines 11 and 15 quality attributes, respectively. These two models are considered to be the predecessors of today's quality models and in this paper we will focus on Boehm's model [23]. However, many other models have been introduced with different attributes. For example, usability, reliability, performance, supportability, and other (serving as general category) are suggested in [24]. Others have used acronyms that capture their perspective as in FURPS (functionality, usability, reliability, performance and supportability) which then has been further developed into FLURPS+ [2]. In most quality models, taxonomies covering larger number of elements have been proposed to ensure that the defined attributes address all areas of the systems. In [22, 25, 26] authors came up with quality attributes as illustrated in Table 2. Similarities between the terms used to describe quality attributes indicate that there is a consensus in the industry for describing the problem of quality landscape.

Those rather broad taxonomies are usually broken into several schemes what illustrates that no single stakeholder can adequately speak of all factors of quality for a product and this is even more so when it comes to prioritizing the quality attributes. See, for example [26] that it is based on stakeholder's interests or [22] where the list is broken down into the product's operational characteristics, its ability to undergo change, and its adaptability to new environment.

It is critical that all stakeholders are involved in the discussion. As a result, engineers working on the system development can achieve the quality plan. It should also include a definition of the quality assessment process. This needs to be agreed upon the way of assessing whether some quality (e.g., maintainability or robustness) must be present in the product and its importance. The trade-offs between quality attributes for which good methods, techniques and tools have been provided by research are required [27].

Table 2. Software product attributes as proposed in [22, 23, 25, 26].

McCall et al.	RADC	Wiegers	Boehm et al.
Correctness	Correctness		
Reliability	Reliability	Reliability	Reliability
Usability	Usability	Usability	Usability
Integrity	Integrity	Integrity	Integrity
Efficiency	Efficiency	Efficiency	Efficiency
Portability	Portability	Portability	Portability
Reusability	Reusability	Reusability	Reusability
Interoperability	Interoperability	Interoperability	
Maintainability	Maintainability	Maintainability	
Flexibility	Flexibility	Flexibility	
Testability	Verifiability	Testability	Testability
	Survivability	Robustness	Robustness
	Expandability		
		Installability	
		Safety	Safety
		Availability	
			Security
			Resilience
			Understandability
			Adaptability
			Modularity
			Complexity
			Learnability

The software engineer usually has to cope with quite a large number of criteria during the product development process. A simple and very popular model proposed by Summerville in [28] is considered. It is an extraction of the quality attributes defined in [23, 29] reflecting the progress in software development (Table 2 column Boehm et al.)

6 Software Quality via Pairwise Comparisons

According to [23], the prime characteristic of quality is "general utility". This means that first and foremost, a software system must be useful to be considered as a system of high quality [30]. General utility comprises as-is utility (concerning how well a software system can be used as-is), maintainability (concerning how easy a software system can be maintained), and portability (concerning whether a software system can still be used if environment is changed). Furthermore, quality factors have been defined to address high-level characteristics of the model. Those quality factors together represent the qualities expected from a software system. For example, reliability (as-is utility characteristics) is defined as follows: a code has a reliability characteristic to the extent that it can be expected to perform its intended functions satisfactorily [29]. At the lowest level of the original Bohem's model, there are 15 primitive characteristics which provide the foundation for defining qualities metrics [23].

We have subdivided them (Table 3) into groups according to the original Boehm's proposal (see [23], Fig. 1, p. 595). Due to the fact that there is no possibility in PC to use the method when the analyzed structure is not hierarchical (whereas the original Boehm's proposal has this feature) we have proposed some modifications: if in an original proposal one final (low level) attribute is related to two or more upper level attributes its occurrence is repeated. This can be done, because in the original Boehm's proposal some of low level attributes have complex nature and they, in fact, represent more than one property. For example, Robustness, is related to Reliability and Human Engineering and when it's analyzed, the final results are expressed in terms of reliability and engineering [23].

Table 3. Software product attributes.

Upper level attribute	Lower level attribute	Acronym
Portability	Device Independence	(Dev-In)
	Self Containedness	(S-C_port)
Reliability	Accuracy	(Acc)
	Completeness	(Com)
	Robustness	(Rob/Int_rel)
	Self Containedness	(S-C_rel)
	Consistency	(Cons_rel)
Human Engineering	Robustness	(Rob/Int_hum-e)
	Accessibility	(Acc_hum-e)
	Communicativeness	(Com_hum-e)
Efficiency	Accountability	(Acc_eff)
	Device Efficiency	(Dev-Eff)
	Accessibility	(Acce_eff)
Testability	Accountability	(Acc_test)
	Accessibility	(Acce_test)
	Communicativeness	(Com_test)
	Self-Descriptiveness	(S-D)
	Structuredness	(Stru_test)
Understandability	Self-Descriptiveness	(S-D_und)
	Consistency	(Cons_und)
	Structuredness	(Stru_und)
	Conciseness	(Con)
	Legibility	(Leg)
Modifiability	Structuredness	(Stru_mod)
	Augmentability	(Aug)

Fig. 1. Boehm's model for SQ in the context of PC.

Figure 1 demonstrates how this model can be implemented in "Concluder" tool (see [31]). Visualization is facilitated by [32]. It shows a preliminary case when all of 15 low-level Boehm's attributes have exactly the same final influence on SQ. However, in reality this is not the right case pointing out that SQ is pretty difficult to quantify. The trend that is seen in the industry comes less from traditional metrics and attributes derived from the code and more from achievability of the project to which it is attached; a lot of this comes out of the Agile development movement, which at this point in time is de-facto a standard way that software is currently developed in Industry [33].

Given the Agile methodology, a number of tasks is set out that need to be accomplished in a number of iterations (called sprints). Quality then tends to be a measure of how many of those sprint items were achieved within the stated timelines, and how many need to be re-examined in further iterations. For instance, if we have 3 tasks, A, B, C and two sprints, then if we set out to achieve tasks A and B in sprint 1, and C in sprint 2, but end up having to also re-develop task B in sprint 2 this information can be used to point to a problem in software quality. Usually that kind of measurement is used with a few added metrics. One of which are redeployments. A piece of software that is redeployed multiple times outside of a sprint wrap-up (planned deployment of a finished iteration) points to issues in either customer satisfaction with the product or some problem with the items that were set to be developed in the sprint. The inference then is that the software is of poor quality [34]. This is linked with the fact that prediction of software defects, improvement of SQ and testing efficiency is done by the construction of different predictive classification models from code attributes to enable a timely identification of fault-prone modules; for example in [35] there is a large-scale empirical comparison of 22 classifiers over 10 public domain data sets from the NASA Metrics Data repository. But in fact, it has been noted that no significant performance differences were detected among the top 17 classifiers.

The SQ, as a scientific problem, has a long history. As it was written, many different approaches were proposed to determine the structure, classification and terminology of attributes, and metrics applicable to SQ. The most important are outlined by the ISO 9126 and ISO 25023:2016. ISO/IEC 9126 standard consists of: (i) internal metrics given by 9126-3:2003 for measuring attributes of six external quality characteristics, which are defined (ii) in ISO/IEC 9126-1, (iii) ISO/IEC 9126-2 that defines external metrics, and (iv) ISO/IEC 9126-4, which defines quality in use metrics, for

Fig. 2. Extended ISO-9126 model in Conluder tool with equal node weights.

measurement of the characteristics or the subcharacteristics. However, it is also assumed that metrics shown in 9126-3:2003 are not intended to be an exhaustive set. Moreover, in 2016 ISO/IEC 9126-3 standard has been revised by ISO/IEC 25023:2016 (note that ISO/IEC 9126 was for the first time proposed in 1991, then changed in 2001 (as 9126:2001), and was finally replaced by ISO/IEC 25010:2011) that provides internal metrics for measuring attributes of six external quality characteristics defined in ISO/IEC 9126-1 (see [36, 37] for details). The ISO 9126 model has six major characteristics with attributes: Functionality, Reliability, Usability, Efficiency, Portability, Maintainability. A proposal of this model in Concluder system is given in Fig. 2.

Based on the models listed above the Consortium for IT Software Quality (CISQ) has defined five major desirable structural characteristics needed for a piece of software to provide business value: Reliability, Efficiency, Security, Maintainability and (adequate) Size. Yet this another example of SQ measure only confirms that depending on the standardization organization we can have different measures without any guaranty that results obtained are comparable and convincing.

7   Conclusion

Software controls such devices as cars and medical equipment where malfunctioning may be dangerous for our lives (e.g., sudden acceleration out of driver's control and recall of cars by Toyota). In the past, such systems were required to be supervised by a professional engineer (PE) who was personally responsible and essentially he/she was betting his/her livelihood by overlooking malfunctioning of software. There was a name and registration number on the key drawings and documents. Such process takes no place for the software control component. Software developers claim that "*it is impossible for a single person to take responsibility for the entire code.*" Clearly, if it was possible for the CN Tower and other skyscrapers to have a PE in charge, then it was possible for software engineers to have a similar role in the structure of organizational responsibility.

It is a long way to go but the aim may be fulfilled if the direction is well-established what the authors hope has been achieved by this study. The idea of comparing two SQ attributes cannot be wrong since comparing all of them at once has not turned to be so fruitful otherwise. The SQ measurement would not be a problem since the model (proposed by Boehm in the late 1970's) has survived, but the quality of software measurement still remains a problem.

SQ, according to Feigenbaum [38], is a customer determination, not an engineer's determination, neither a marketing determination, nor a general management determination. It is based on the customer's actual experience with the product or service, measured against his or her requirements – stated or unstated, conscious or merely sensed, technically operational or entirely subjective – and always representing a moving target in a competitive market [38]. We have implemented our model with such vision.

Our approach allows the user to compute weights each of these dimensions. It is a step forward since the SQ model, according to [36, 37] (see Fig. 2 for example), assumes (by an arbitrary decision) to what extent a software rates on each of these dimensions. Frequently, all dimensions are assumed to have equal importance giving the most trivial case for our PC matrix: with all entries equal to 1's. When applied to SQ model, PC method can be regarded as a paradigm shift. However, more research is needed. For this model there is a need to conduct a set of tests that will check its validity. We hope that our study may make SQ a less elusive target than it was signaled in [39].

References

1. Chemuturi, M.: Mastering Software Quality Assurance: Best Practices. Tools and Technique for Software Developers. J. Ross Publishing, Fort Lauderdale (2010)
2. Pressman, S.: Software Engineering: A Practitioner's Approach, 6th edn. McGraw-Hill Education Pressman, Boston (2005)
3. Zhang, X.M., Teng, X.L., Pham, H.: Considering fault removal efficiency in software reliability assessment. IEEE Trans. Syst. Man Cybern. Syst. Hum. 33(1), 114–119 (2003)
4. Khoshgoftaar, T.M., Bhattacharyya, B.B., Richardson, G.D.: Predicting software errors, during development, using nonlinear regression models: a comparative study. IEEE Trans. Reliab. 41(3), 390–395 (1992)
5. Kenny, G.Q.: Estimating defects in commercial software during operational use. IEEE Trans. Reliab. 42(1), 107–115 (1993)
6. Koczkodaj, W.W.: Testing the accuracy enhancement of pairwise comparisons by a Monte Carlo experiment. J. Stat. Plan. Infer. 69(1), 21–31 (1998)
7. Likert, R.: A technique for the measurement of attitudes. Arch. Psychol. 140, 1–55 (1932)
8. Faliszewski, P., Hemaspaandra, E., Hemaspaandra, L.A., Rothe, J.: Llull and Copeland voting computationally resist bribery and constructive control. J. Artif. Intell. Res. 35, 275–341 (2009). Conference: 2nd International Workshop on Computational Social Choice Location: Liverpool, England
9. Thurstone, L.L.: A law of comparative assessments. Psychol. Rev. 34, 273–286 (1927)
10. Colonius, H.: Representation and uniqueness of the Bradley-Terry-Luce model for pair comparisons. Br. J. Math. Stat. Psychol. 33, 99–103 (1980)

11. Saaty, T.L.: A scaling methods for priorities in hierarchical structure. J. Math. Psychol. **15**, 234–281 (1977)
12. Koczkodaj, W.W., Mikhailov, L., Redlarski, G., Soltys, M., Szybowski, J., Tamazian, G., Wajch, E., Yuen, K.K.F.: Important facts and observations about pairwise comparisons. Fundamenta Informaticae **144**, 1–17 (2016)
13. Koczkodaj, W.W.: Pairwise comparisons rating scale paradox. In: Transactions on Computational Collective Intelligence XXII. Lecture Notes in Computer Science, vol. 9655, pp. 1–9 (2016)
14. Williams, C., Crawford, G.: Analysis of subjective judgement matrices. The Rand Corporation Report R-2572-AF, pp. 1–59 (1980)
15. Koczkodaj, W.W.: Statistically accurate evidence of improved error rate by pairwise comparisons. Percept. Mot. Skills **82**(1), 43–48 (1996)
16. Koczkodaj, W.W., Szwarc, R.: Pairwise comparisons simplified. Appl. Math. Comput. **253**, 387–394 (2015)
17. Jensen, R.: An alternative scaling method for priorities in hierarchical structures. J. Math. Psychol. **28**, 317–332 (1984)
18. Bauschke, H.H., Borwein, J.M.: On projection algorithms for solving convex feasibility problems. SIAM Rev. **38**(3), 367–426 (1996)
19. Bozoki, S., Fulop, J., Koczkodaj, W.W.: An LP-based inconsistency monitoring of pairwise comparison matrices. Math. Compute Model. **54**(1–2), 789–793 (2011)
20. Khoshgoftaar, T.M., Allen, E.B., Bullard, L.A., Halstead, R., Trio, G.P.: A tree-based classification model for analysis of a military software system. In: Proceedings of IEEE High-Assurance Systems Engineering Workshop, pp. 244–251 (1997)
21. IEEE Standard 1061-1992: Standard for a Software Quality Metrics Methodology. Institute of Electrical and Electronics Engineers, New York (1992)
22. McCall, J., Richards, P., Walters, G.: Factors in Software Quality, NTIS (1977)
23. Boehm, B.W., Brown, J.R., Lipow, M.: Quantitative evaluation of software quality. In: Proceedings of the 2nd International Conference on Software Engineering. IEEE Computer Society Press (1976)
24. Leffingwell, D., Widrig, D.: Managing Software Requirements: A Use Case Approach, 2nd edn. Addison Wesley, Boston (2003)
25. The Rome Air Development Center is now the Rome Laboratory, as of 1991
26. Wiegers, K.: Software Requirements, 2nd edn. Microsoft Press, Redmond (2003)
27. Wohlin, C., Lundberg, L., Mattsson, M.: Special issue: trade-off analysis of software quality attributes. Softw. Qual. J. **13**, 327–328 (2005)
28. Sommerville, I.: Software Engineering, 9th edn. Addison-Wesley Longman Publishing Co., Inc., Boston (2006)
29. Boehm, B.W., Brown, J.R., Kaspar, H., Lipow, M., Macleod, G., Merrit, M.: Characteristics of Software Quality. North-Holland, Amsterdam (1978)
30. Pfleeger, S.L., Atlee, J.M.: Software Engineering: Theory and Practice. Prentice Hall PTR, Upper Saddle River (2009)
31. https://sourceforge.net/projects/concluder/. Accessed 10 Feb 2017
32. Heer, J., Card, S.K., Landay, J.A.: Prefuse: a toolkit for interactive information visualization. In: Proceedings of the SIGCHI Conference on Human Factors in Computing Systems, Portland, Oregon, USA, pp. 421–430. ACM (2005)
33. The Agile Movement. http://agilemethodology.org/, see also W. W. Royce, Managing the development of large software systems: concepts and techniques, Proceeding ICSE 1987, pp. 328–338 (1987)
34. Duncan, G.O.: Private communication. Accessed 02 Mar 2017

35. Lessmann, S., Baesens, B., Mues, C., Pietsch, S.: Benchmarking classification models for software defect prediction: a proposed framework and novel findings. IEEE Trans. Softw. Eng. **34**(4), 485–496 (2008)
36. https://www.iso.org/standard/35733.html. Accessed 10 Feb 2017
37. ISO/IEC 25010:2011, Systems and software engineering - Systems and software Quality Requirements and Evaluation (SQuaRE) - System and software quality models (2011)
38. Feigenbaum, A.V.: Total Quality Control. McGraw-Hill, New York (1983)
39. Kitchenham, B., Pfleeger, S.L.: Software quality: the elusive target. IEEE Softw. **13**(1), 12–21 (1996)

Platform for Software Quality and Dependability Data Analysis

Rafał Kozik[1(✉)], Michał Choraś[1,2], Damian Puchalski[2], and Rafał Renk[2]

[1] Institute of Telecommunications and Computer Science,
UTP University of Science and Technology in Bydgoszcz, Bydgoszcz, Poland
{rafal.kozik,chorasm}@utp.edu.pl
[2] ITTI Sp. z o.o., Poznań, Poland
{damian.puchalski,rafal.renk}@itti.com.pl

Abstract. In this paper, we present the innovative advanced data analysis dedicated for product owners in software development teams. The goal of our work is to provide advanced data analysis to improve software quality, management and dependability. The major contribution of this work is the methodology and the tool that gathers the input raw data from the tools such as GitLab or SonarQube, and processes the data further (e.g. using Apache Kafka, Kibana and Spark) to calculate and visualize more advanced metrics and to find correlations between them. We present results and helpful charts and visualization supporting product owners.

Keywords: Data analysis · Code quality · Software dependability
Software quality requirements

1 Introduction

In the current IT ecosystems, that are highly interconnected and relying on software components, challenges such as optimization of the software code development process, minimization of the risk of software failures and code testing/debugging are critical for business, service providers, and societies.

Software flaws and bugs can impact not only its usability, functional value and user experience [1], but also security of users, due to the fact that bugs in the design or implementation phase can be exploited by cyber criminals [2,3].

One of the mechanisms implemented in the software engineering process is software testing with the objective to detect bugs and flaws in the code and then to address them before product deployment. However, the costs of quality assurance and testing in IT are growing from year to year. Currently, IT organizations spent approximately 1/3 of their budgets on quality assurance with the trend of raising this value to approx. 40% in next three years [4,5]. Although the process of debugging software during its design phase costs 4 to 5 times less than fixing bugs after its release [6] it is a non-trivial task that consumes a significant part of budgets and companies' effort. It could be less impactful for a

© Springer International Publishing AG, part of Springer Nature 2019
W. Zamojski et al. (Eds.): DepCoS-RELCOMEX 2018, AISC 761, pp. 306–315, 2019.
https://doi.org/10.1007/978-3-319-91446-6_29

big companies and software houses, however SMEs operating often with limited budgets and resources [7] are becoming more and more focused on techniques allowing automation and adequacy of the testing process, to be competitive in relation to big players in the market in terms of software quality, optimization of the development cost and time to market factor.

In the H2020 Q-Rapids project [8], the concept of quality-aware decision making based on key strategic indicators is proposed. The overall goal of the project is to support strategic decision-making processes by providing strategic indicators in the context of quality requirements in agile and rapid software development. For the purposes of the project, a strategic indicator is defined as a specific aspect that a software development company has to consider as crucial for the decision making process during the software development. Aspects such as e.g. time-to-market, maintenance cost, customer satisfaction, etc. can be considered as strategic indicators depending on the context. These strategic indicators are built on top of the measurements and factors calculated on the basis of the software development related data, stored in the management tools such as GitLab or SonarQube. In our previous Q-Rapids related paper [9] we showed the presented results of software related data analysis and correlation. Hereby in this paper, we present more advanced solutions that allow for synchronization and inspection of data related to software development gathered in GitLab and SonarQube tools. Hereby, we focus on improving cognition of the product owners and decision support by innovative advanced software-related data analysis.

This paper is structured as follows. First, we introduce the general architecture of the proposed system. In particular, we introduce implementation details of our two proposed data miners collecting the data from GitLab and SonarQube respectively. Afterwards, we demonstrate the capabilities of the presented solution. First, we demonstrate the visualization capabilities that are relevant from the project manager or project owner point of view. Finally, we demonstrate data mining capabilities that allow the operator to find potentially significant correlations in the collected data. The paper is concluded with final remarks and plans for the future work.

2   System Architecture

In this section, the architecture of the proposed system is presented. First, we demonstrate the general approach to system architecture, information flow used for the data analysis and the information data sources. Afterwards we present the subsequent steps of data collection, processing and presentation to provide Key Strategic Indicators for software engineering decision makers.

The conceptual architecture of the system is presented in Fig. 1. We use several data sources to measure the statistics (we call those metrics) related to the project. Currently, these metrics are mainly retrieved from GitLab and SonarQube project management tools.

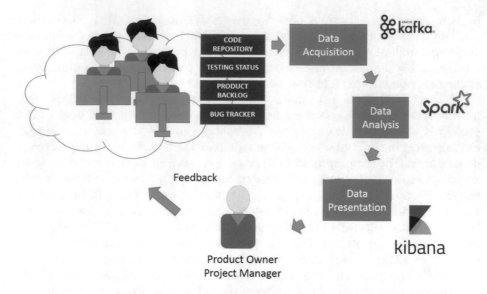

Fig. 1. The architecture of the proposed solution.

First, the data is acquired via HTTP Restful APIs by mean of Apache Kafka connect framework. The purpose of this tool is to simplify the process of integrating various solutions in a scalable manner.

Next, the data is processed, analyzed, and the results are published to ElasticSearch platform. Currently we use a mixture of scripts and Java applications to analyze the raw data and to produce intermediate results. However, we plan to have this part fully implemented in scalable Apache Spark framework.

Finally, we use capabilities of ElasticSearch and Kibana tools to build data presentation layer that is composed of various figures related to project management aspects.

2.1   Mining Project Management-Related Data

In the considered uses case scenario we have focused on software solution development using agile methodology in the context of rapid software development.

In general, the overall development process follows the iterative framework where consecutive tasks of requirements definition, system design, coding and testing are executed. The whole development process is monitored and tracked by means of GitLab project management system.

The progress of the project in the GitLab tool is tracked with so called issues. The key stages can be identified as:

– Conceptualisation – actions defining the general concept of the final product,
– Designing – actions defining functionalities and requirements,
– Implementation – actual development of the product,

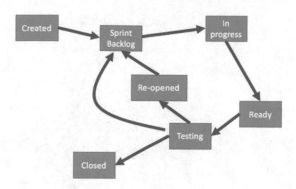

Fig. 2. Example of an issue lifetime. The 'opened' and 'closed' boxes indicates initial and terminal nodes

- Testing – executing test of the developed product,
- Deployment.

These stages reflect different types of issues that will be maintained in the Redmine system by particular users. We define the following type of issues:

- Story – general concept of the specific part of the product,
- Feature – single specific functionality of the developed system,
- Task – specific development activity,
- Bug – an error (flaw) in the developed software.

During the project lifetime, the state of an issue could change. It can evolve from conceptual definition in order to be materialized as an artifact such as piece of software serving specific functionality.

As it is shown in Fig. 2, each issue starts in a 'created' state, where the details of a functionality and acceptance criteria are formulated. Afterwards, an issue may be put into the backlog. This state is indicated as 'Sprint Backlog'. When a developer is assigned to a specific issue, the state is changed to 'In progress'. When the issue is ready, it has assigned 'Ready' state. From that moment, the result of the developer work can be tested by the Q&A engineer and hence the issue can be moved to the 'Testing' state. When all the functional tests are successfully passed, the issue is moved to the terminal state indicated as 'Closed'. However, when a test fails, the issue is either moved to re-open state or (it depends on the project) it goes directly to the backlog.

The number of issue-states-diagrams we have to track is equal to the number of issues that have been defined in GitLab. It would be overwhelming from the point of view of a user to analyze all of them. Therefore, we calculate project-wise aggregated metrics for each daily snapshot.

On the basis of such data, we can track the course of the given project in terms of work intensification and then combine and/or compare these data with the code-related metrics (as demanded by team leader and product owners).

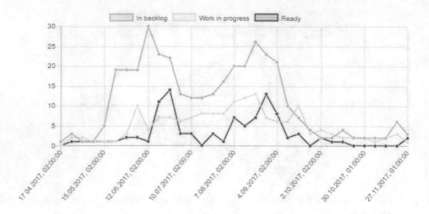

Fig. 3. Example of metrics related to sprints workload.

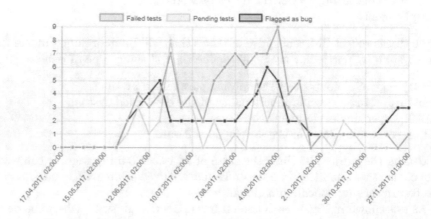

Fig. 4. Example of the calculated metrics assessing the testing performance.

Using that approach we can plot a time-series indicating how the specific metric changed over time. In example, as it is shown below, we can visualize the per-sprint average number of tasks being in backlog, 'in progress' or finished (ready) states.

Currently, our data mainer works in a pull mode. It means that the GitLab project management tool is queried periodically to fetch the data. The tool itself exposes fully-functional RESTful API. In principle, the API allows us to identify (and eventually download) only these elements (project related artifacts) which have been modified since previous pull request. Afterwards, the data collected locally is analyzed in order to collect the relevant measurements.

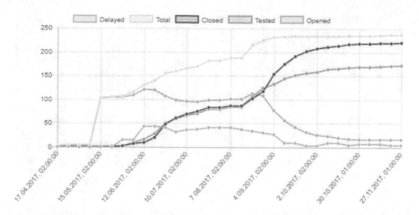

Fig. 5. Example of the calculated metrics assessing the overall project progress and performance.

2.2 Mining Code Quality

In the considered use-case scenario the code-quality assessing infrastructure is be based on SonarQube framework. It works in a distributed environment. The deployment of this tool is compliant with typical SonarQube configuration and it is as follows:

1. Developers code in their IDEs and use static analysis tools (e.g. SonarLint) to run local analysis,
2. Developers upload their code into the Git repository,
3. The Continuous Integration Server triggers an automatic build, and the execution of the SonarQube Scanner required to run the SonarQube analysis,
4. The analysis report is sent to the SonarQube Server for processing,
5. SonarQube Server processes and stores the analysis report results in the SonarQube Database, and displays the results in the UI,
6. Developers review, comment, and challenge their Issues to manage through the SonarQube UI,
7. Managers receive Reports from the analysis.

The SonarQube data mainer works in a pull mode as well. The code related metrics are downloaded periodically. Unfortunately, SonarQube does not provide historical data (only current snapshot).

Therefore, we store the data locally in our computing infrastructure.

3 Experiments and Tentative Results

The goal of our experiments was to validate the correctness of the architectural assumptions, to assess the usefulness of provided functionalities and the approach to visualize the calculated and aggregated data (Fig. 6).

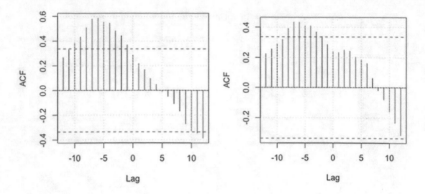

Fig. 6. The Correlation (ACF) between the backlog size and the cognitive complexity for varying lag (left) and the backlog size and the number of duplicated lines (right). The dashed lines represent an approximate confidence interval (95%).

3.1 Functional Capabilities

The results have been presented at the validation session to product owners involved in the development project in companies. Our primary goals were to collect feedback and understand the needs of the potential users of our system.

In result, by using our approach we were able to extend the basic functionalities of GitLab Community edition tool. Thanks to ElasticSearch and Kibana capabilities we have been able to measure and visualize such characteristics as:

- sprints workload (Fig. 3) using time series visualizing amount o tasks in progress, planned in backlog and waiting to be tested.
- testing performance (Fig. 4) indicating a number of failed test, number of reported bugs and amount of pending test.
- overall project performance (Fig. 5) using a number of deleted tasks, number of opened and pending tasks, growing number of new tasks, and the velocity of delivering new features.

The positive feedback we get during the evaluation session let us believe that our concept and results (e.g. provided visualization and figures) are of added value for a product owner or project manager.

Nonetheless, we also believe that such kind of collected and calculated data can be successfully used to drive the future decisions in the software development process. Discussion of those aspects is provided in the next section.

3.2 Searching for Correlation Between Metrics

Having the data for several commercial software development projects, we hypothesize that there might exist a certain level of correlation between specific metrics.

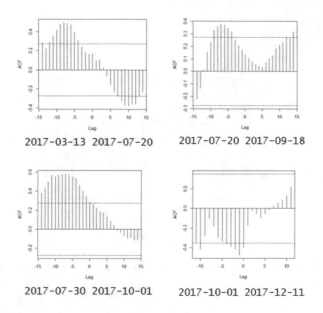

2017-03-13 2017-07-20   2017-07-20 2017-09-18

2017-07-30 2017-10-01   2017-10-01 2017-12-11

Fig. 7. The Correlation (ACF) between the backlog size and the amount of delayed tasks for a various period of project lifetime.

To asses the correlation between two metrics y_1 and y_2 we use following formula:

$$p = \frac{E[(y_1 - \mu_1)(y_2 - \mu_2)]}{\sigma_1 \sigma_2} = \frac{Cov(y_1, y_2)}{\sigma_1 \sigma_2} \tag{1}$$

where μ_1 and μ_2 are the means and σ_1 and σ_2 are the standard deviations for y_1 and y_2 respectively.

In one of the projects, we were able to show that overloading the sprint backlog highly correlates with deteriorated code quality, which in result increases the number of defects identified. That was the trend that held for this project during the entire project lifetime.

Obviously, some correlations will gradually vanish as the project advances. For example, in one of the analyzed project, the size of the backlog correlated with the number of delayed tasks until 70% lifetime. After that certain phase, the correlation disappeared. This is shown in Fig. 7.

In contrary, some correlation will also emerge as the project advances. For example, for the same project, we considered before the average complexity of files starts to correlate with the number of detected defects. It is shown in Fig. 8. This observation can be explained. As the project gets more complicated, new features introduced to that product will be more prone to errors.

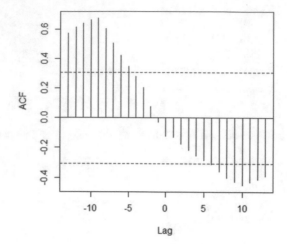

Fig. 8. The Correlation (ACF) between the average file complexity and the number of detected defects.

4   Conclusions

In this paper, we presented the framework architecture for advanced data analysis in order to improve product owners decision making capabilities and facilitate their cognitive processes.

We presented the concept, architecture, practical realization and the initial results achieved after the practical deployment for several projects.

Moreover, we presented the fundamentals for the task to find correlations between the metrics and the valid results for such innovative approach.

Acknowledgments. This work is funded under BSM 80/2017 and Q-Rapids project, which has received funding from the European Union's Horizon 2020 research and innovation programme under grant agreement No. 732253.

References

1. Kozik, R., Choraś, M., Flizikowski, A., Theocharidou, M., Rosato, V., Rome, E.: Advanced services for critical infrastructures protection. J. Ambient Intell. Humanized Comput. **6**(6), 783–795 (2015)
2. Choraś, M., Kozik, R., Puchalski, D., Holubowicz, W.: Correlation approach for SQL injection attacks detection. In: Herrero, A., et al. (eds.) Advances in Intelligent and Soft Computing, vol. 189, pp. 177–186. Springer (2012)
3. Choraś, M., Kozik, R.: Network event correlation and semantic reasoning for federated networks protection system. In: Chaki, N., et al. (eds.) Computer Information Systems Analysis and Technologies. CCIS, pp. 48–54. Springer (2011)
4. Jorgensen, P.C.: Software Testing: A Craftsman's Approach. CRC Press, Hoboken (2016)

5. Capgemini: World Quality Report 2016–17, 8th edn. https://www.capgemini.com/world-quality-report-2016-17/. Accessed 9 Oct 2017
6. Jones, C., Bonsignour, O.: The Economics of Software Quality. Addison-Wesley Professional, Reading (2011)
7. Felderer, M., Ramler, R.: Risk orientation in software testing processes of small and medium enterprises: an exploratory and comparative study. Softw. Qual. J. **24**(3), 519–548 (2016)
8. H2020 project Q-Rapids. http://www.q-rapids.eu/. Accessed 9 Oct 2017
9. Kozik, R., Choraś, M., Puchalski, D., Renk, R.: Data analysis tool supporting software analysis process. In: Proceedings of 14th IEEE International Scientific Conference on Informatics, Poprad, Slovakia, 14–16 November 2017, pp. 179–184 (2017)

Collaborative Filtering Recommender Systems Based on k-means Multi-clustering

Urszula Kużelewska[✉]

Bialystok University of Technology, Wiejska 45a, 15-351 Bialystok, Poland
u.kuzelewska@pb.edu.pl

Abstract. Recommender systems support users in searching on the internet by finding the items, which are interesting for them. This items may be information as well as products, in general. They base on personal needs and tastes of users. The most popular are collaborative filtering methods that take into account users' interactions with the electronic system. Their main challenge is generating on-line recommendations in reasonable time coping with large size of data.

Clustering algorithms support recommender systems in increasing time efficiency. Commonly, it involves decreasing of prediction accuracy of final recommendations. This article presents an approach based on multi-clustered data, which prevents the negative consequences, keeping reasonable time efficiency. An input data are clusters of similar items or users, however one of them may belong to more than one cluster. When recommendations are generated, the best cluster for the user or item is selected. The best means that the user or item is the most similar to the center of the cluster. As a result, the final accuracy is not decreased.

Keywords: Recommender systems · Multi-clustering
Collaborative filtering

1 Introduction

Recommender systems (RS) are electronic applications with the aim to generate for a user a limited list of items from a large item set. The list is constructed basing on the active user's and other users' past behaviour. People interact with recommender systems by visiting web sites, listening to the music, rating the items, doing shopping, reading items description, selecting links from search results. This behaviour is registered as access log files from web servers, or values in databases: direct ratings for items, the numbers of song plays, content of shopping basket, etc. After each action users can see different, adapted to them, recommendation lists depending on their tastes [6].

Recommender systems are used for many purposes in various areas. Multimedia services, such as Netflix or Spotify, are places, where recommendations are

© Springer International Publishing AG, part of Springer Nature 2019
W. Zamojski et al. (Eds.): DepCoS-RELCOMEX 2018, AISC 761, pp. 316–325, 2019.
https://doi.org/10.1007/978-3-319-91446-6_30

extremely helpful. Music recommender is described in [9] and a method applied in MoveLens system in [10]. They also offer great opportunities for business, government, education, e-commerce, leisure activities and other domains, with successful developments in commercial applications [7].

Scalability and performance are key metrics for deploying a recommender system in real environment [1]. Collaborative filtering techniques, although precise, calculate items for suggestion by searching similar users or items in the whole archived data. They deal with large amount of dynamic data, however the time of results generation should be reasonable to apply them in real-time applications. A user reading news expects to see next offer for him/her in seconds, whereas millions of archived news have to be analysed.

Clustering algorithms can be used to increase neighbour searching efficiency and thus decreasing time of recommendations generation. A drawback is that quality of predictions is usually slightly reduced in comparison to original neighbours identification strategy (original way is to apply k-Nearest Neighbours method) [12].

The explanation is in the way how clustering algorithms work. A typical approach bases on one partitioning scheme, which is generated once and then not updated significantly. The neighbourhood of data located on borders of clusters is not modelled precisely.

To improve quality of the neighbourhood modelling one can use multiple clustering schemes and select the most appropriate one to the particular data object. As a result, multi-clustering approach eliminates inconvenience of decreased quality of predictions keeping high time effectiveness. Figure 1 presents two different clustering results for the same dataset. For a particular data object it can be selected the scheme with this object located closer to the cluster centre, thus having more neighbours around.

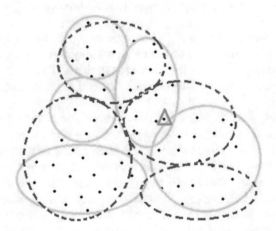

Fig. 1. Various neighbourhood modelling for particular data in case of k-means multi-clustering

This paper contains results of experiments on a collaborative filtering recommender system, which based on similarities among items identified a priori as multi-clusters. The set of clustering schemes was generated by k-means algorithm with the same values of their input parameters at every time. During searching the most similar items, every cluster is examined, and is selected the one, in which the appropriate items are the most similar to the centre.

The rest of the paper is organised as follows: Sect. 2 describes general aspects in recommendation systems, including problems in this domain and a role of clustering and multi-clustering algorithms. The following section, Sect. 3 describes the proposed approach, whereas Sect. 4 contains results of performed experiments. The last section concludes the paper.

2   Application of Multi-clustering Model of Data in Collaborative Filtering

Recommender systems face many challenges and problems. The vital, in the point of view on-line recommendations, is scalability. They deal with large amount of dynamic data, however the time of results generation should be reasonable to apply them in real-time applications. A user reading news expects to see next offer for him/her in seconds, whereas millions of archived news have to be analysed. The most effective recommender systems are hybrid approaches, which combine at least two different methods solving problems in each of them by strengths of the other one. Clustering algorithms are good support for determining neighbourhood before proper recommendation process positively influencing its scalability.

Multi-clustering is variously defined in literature. The major type is alternative clustering, which tries to find partitioning schemes on the same data that are different from each other. Bailey [2] provided a thorough survey on alternative clustering methods.

There are methods called multi-view clustering [8] that cope with different views of data. It is assumed that different views may represent distinct aspects of the same data leading to the same cluster structure. The multi-view clustering techniques take into account the mutual link information among the various views [4].

Another example is COALA [3] that search for alternative clusterings of better quality and dissimilarity with respect to the given clustering. It starts by treating each object as a single cluster and then iteratively merges a pair of the most similar clusters. The idea bases on $cannot - link$ constraints for guide the generation of a new, dissimilar clustering. Another example is MSC algorithm (for Multiple Stable Clustering) [5] that generates stable multiple clusterings. Advantage of this method is that it does not require to specify the number of clusters and provides users a feature subspace to understand each clustering solution.

The advantages of multi-clustering methods can be beneficial to the recommender systems domain. The better quality of the neighbourhood modelling lead to high quality of predictions keeping high time effectiveness provided by clustering methods. Although, there are few publications describing application multi-clustering methods in recommendations.

The method described in [11] combines content-based and collaborative filtering approach to recommendations. The system uses multi-clustering, however it is interpreted as clustering a single scheme on both techniques. It groups the items and the users based on their content-based then use the result, which is represented by the fuzzy set, to create an item group-rating matrix and a user group-rating matrix. As a clustering algorithm it uses k-means combined with fuzzy set theory to represent the level of membership an object to the cluster. Then finally prediction rating matrix is calculated to represent the whole dataset. In the last step of pre-recommendation process k-means is used again on the new rating matrix to find a group of similar users. The groups represent neighbourhood of users to limit a search space for collaborative filtering method.

Another solution is presented in [13]. The authors observed, that users might have different interests over topics, thus might share similar preferences with different groups of users over different sets of items. The method CCCF (Co-Clustering For Collaborative Filtering) first clusters users and items into several subgroups, where each subgroup includes a set of like-minded users and a set of items in which these users share their interests. The groups are analysed by collaborative filtering methods and the result recommendations are aggregated over all the subgroups.

3   Description of the Multi-clustering Algorithm

There are two main steps in the algorithm proposed in this article. The first of them (performed in off-line mode) prepares neighbourhood of the most similar users in the form of a set of many clustering schemes. The method k-means is run several times with the same values of k parameter. The results are stored as an input for the recommendation process.

Then, the second step concerning generating recommendations is performed. The main idea is to match users with the best clusters as their neighbourhood. While calculating items for recommendation for a particular user, the most appropriate neighbourhood is selected for searching for the candidates. The level of adequacy is calculated as a value of similarity between the particular user and a cluster centre. A similarity measure is estimated in the same way like in the recommendation process. Then candidates are searched by collaborative filtering item-based technique, but only within the cluster of neighbourhood.

The recommendation steps of the algorithm are described in Algorithm 1. The input set contains data of n users, who rated a subset of items - $A = \{a_1, \ldots, a_k\}$. The set of possible ratings - V - contains values $v_1, \ldots, v_c$. The input data are clustered ncs times into nc clusters every time giving as a result a set of clustering schemes CS. The algorithm generates a list of recommendations R_{x_a} for the active user.

Algorithm 1. A general algorithm of a recommender system based on multi-clustering used in the experiments

Data:
- $U = (X, A, V)$ - matrix of clustered data, where $X = \{x_1, \ldots, x_n\}$ is a set of users, $A = \{a_1, \ldots, a_k\}$ is a set of items and $V = \{v_1, \ldots, v_c\}$ is a set of ratings values,
- $\delta : v \in V$ - a similarity function,
- $nc \in [2, n]$ - a number of clusters,
- $ncs \in [2, \infty]$ - a number of clustering schemes,
- $CS = \{CS_1, \ldots, CS_{ncs}\}$ - a set of clustering schemes,
- $CS_i = \{C_1, \ldots, CS_{nc}\}$ - a set of clusters for a particular clustering scheme,
- $CS_r = \{c_{r,1}, \ldots, c_{r,nc \cdot ncs}\}$ - the set of cluster centres,

Result:

- R_{x_a} - a list of recommended items for an active user x_a,

begin
  $\delta_1 .. \delta_{ncs} \longleftarrow$ `calculateSimilarity`(x_a, CS_r, δ);
  $C \longleftarrow$ `findTheBestCluster`$(\delta_1 .. \delta_{ncs}, CS)$;
  $R_{x_a} \longleftarrow$ `recommend`(x_a, C, δ);

---

4   Experiments

Results of the experiments concern multi-clustering recommender system with respect to quality of recommendations and time effectiveness. There were examined various similarity measures in generation recommendations phase as well as different distance metrics during clustering process. The results were compared with memory-based recommender system and model-based one with clusters obtained from k-means as neighbourhood of objects in the recommendation process.

Quality of recommendations was calculated with RMSE measure (a classical error measure - Root Mean Squared Error) in the following way. For every users from the input set their ratings were divided into training (70%) and testing parts. The values from testing parts were removed and estimated with the selected recommender system. Difference between original and estimated number is taken for calculations. Time effectiveness is measured as average time of generating recommendations list composed of 5 elements for every of 100 users.

The clustering algorithm as well as the recommendation system were created using Apache Mahout library (http://mahout.apache.org/). The methods were tested with various similarity measures implemented in Apache Mahout: LogLikehood (*LogLike*), cosine coefficient (*Cosine*), Pearson correlation (*Pearson*), Euclidean distance (*Euclid*), CityBlock metrics (*CityBl*) and Tanimoto (*Tanimoto*) similarity. The clustering were obtained with Euclidean and cosine-based distance measures.

Recommendations were executed on benchmark LastFM music data [14]. The whole set contains over 16 millions ratings: 345 652 users who rated 158 697 songs. The data was split into several smaller sets and the clustering-based method as well as the multi-clustering approach were tested on the subset that contained 2 032 users who gave 99 998 ratings of 22 174 items.

Tables 1 and 2 contain results of RMSE values and time (in s) of execution for traditional collaborative filtering item-based system.

Table 1. RMSE of item based collaborative filtering recommendations

Similarity measure					
LogLike	Cosine	Pearson	Euclid	CityBl	Tanimoto
0.58	0.58	0.61	0.48	0.58	0.58

Table 2. Time [s] of item based collaborative filtering recommendations

Similarity measure					
LogLike	Cosine	Pearson	Euclid	CityBl	Tanimoto
0.090	0.125	0.127	0.121	0.071	0.077

The following experiments concerned comparison between the previous results and recommender system with modelling of neighbourhood by k-means clusters from a single clustering scheme. Tables 3 and 4 contain results of RMSE values and time (in s) of execution on clusters generated with Euclidean distance, whereas Tables 5 and 6 contain the same statistics but the distance measure while generating clusters was cosine-based distance. It can be noticed that in every cases of the second experiment RMSE is greater despite of similarity type measure, however time of generation the propositions is a few hundred times lower than in the first experiment.

Table 3. RMSE of item based collaborative filtering recommendations with neighbourhood determined by k-means with Euclidean distance measure

Number of clusters	Similarity measure					
	LogLike	Cosine	Pearson	Euclid	CityBl	Tanimoto
20	0.65	0.65	0.64	0.64	0.65	0.66
50	0.67	0.67	0.67	0.66	0.67	0.68
200	0.68	0.67	0.67	0.66	0.68	0.67
1000	0.66	0.64	0.65	0.62	0.65	0.66

Table 4. Time [s] of item based collaborative filtering recommendations with neighbourhood determined by k-means with Euclidean distance measure

Number of clusters	Similarity measure					
	LogLike	Cosine	Pearson	Euclid	CityBl	Tanimoto
20	0.017	0.019	0.018	0.019	0.015	0.016
50	0.026	0.027	0.027	0.027	0.024	0.024
200	0.011	0.012	0.011	0.011	0.010	0.010
1000	0.010	0.02	0.020	0.020	0.010	0.010

Table 5. RMSE of item based collaborative filtering recommendations with neighbourhood determined by k-means with cosine distance measure

Number of clusters	Similarity measure					
	LogLike	Cosine	Pearson	Euclid	CityBl	Tanimoto
20	0.89	0.88	0.58	0.81	0.89	0.90
50	0.84	1.01	0.80	0.91	1.03	1.23
200	1.11	1.11	0.62	1.00	1.12	1.12
1000	1.14	1.11	0.76	1.02	1.11	1.11

Table 6. Time [s] of item based collaborative filtering recommendations with neighbourhood determined by k-means with cosine distance measure

Number of clusters	Similarity measure					
	LogLike	Cosine	Pearson	Euclid	CityBl	Tanimoto
20	0.0034	0.0032	0.0024	0.0030	0.0027	0.0024
50	0.0010	0.0012	0.0009	0.00090	0.0011	0.0014
200	0.0003	0.0003	0.0003	0.0003	0.0003	0.0004
1000	0.0001	0.0001	0.0002	0.0002	0.0001	0.0002

Table 7. RMSE of item based collaborative filtering recommendations with neighbourhood determined by multi-clustering k-means with Euclidean distance measure

Number of clusters	Similarity measure					
	LogLike	Cosine	Pearson	Eucl	CityBlock	Tanimoto
20	0.15	0.15	-	0.11	0.15	0.15
50	0.15	0.15	-	0.10	0.16	0.16
200	0.20	0.22	-	0.16	0.22	0.21
1000	0.34	0.35	0.00	0.31	0.34	0.34

Table 8. Time [s] of item based collaborative filtering recommendations with neighbourhood determined by multi-clustering k-means with Euclidean distance measure

Number of clusters	Similarity measure					
	LogLike	Cosine	Pearson	Eucl	CityBlock	Tanimoto
20	2.21	2.65	-	2.62	2.91	2.30
50	3.11	3.41	-	3.21	3.26	3.14
200	13.50	15.70	-	14.26	16.28	13.15
1000	40.62	46.65	43.68	39.95	40.50	46.00

The following experiment concerned generating recommendation basing on a set of 3 clustering schemes (multi-clustering) generated separately for every 20, 50, 200 and 1000 clusters. All schemes for particular number of clusters were taken for recommendations process. The dataset in this experiment was the same as in the previous one - contained 99 998 ratings. Tables 7 and 8 contain RMSE and time of recommender system executed separately for every scheme. In this experiment clusters were created basing on Euclidean distance measure. Tables 9 and 10 contain results of the same experiment, however clusters were obtained with distance based on cosine coefficient.

The time of recommendations generation is greater than in the experiments where neighbourhood was modelled by a single clustering scheme, however the value of RMSE is tremendously lower. It is particularly visible when clusters are created basing on Euclidean distance. The experiment improved, that application multi-clustering in recommender systems and dynamic selection the most suitable clusters is very valuable.

Table 9. RMSE of item based collaborative filtering recommendations with neighbourhood determined by multi-clustering k-means with distance measure based on cosine (* - it was not possible to generate recommendations in one case, ** - it was not possible to generate recommendations in 2 cases, etc.)

Number of clusters	Similarity measure					
	LogLike	Cosine	Pearson	Eucl	CityBlock	Tanimoto
20	0.19**	0.19**	0.58	0.18**	0.19**	0.19**
50	0.34*	0.44*	0.09***	0.06***	0.47*	0.44*
200	0.19	0.26	0.00*	0.24	0.27	0.27
1000	0.32	0.32	0.00*	0.29	0.33	0.32

Table 10. Time [s] of item based collaborative filtering recommendations with neighbourhood determined by multi-clustering k-means with distance measure based on cosine

Number of clusters	Similarity measure					
	LogLike	Cosine	Pearson	Eucl	CityBlock	Tanimoto
20	2.68	2.31	0.01	2.02	2.20	1.49
50	2.38	3.05	2.57	3.12	3.11	2.43
200	13.86	15.70	16.13	14.26	16.28	13.15
1000	44.00	61.00	58.15	51.00	53.00	47.00

5   Conclusions

This article presented an approach to recommender systems based on multi-clustered data. Application of clustering algorithms based on a single scheme is characterised by high time efficiency and scalability. However, this benefit usually involves decreased accuracy of prediction feature. It results from inaccurate modelling of object neighbourhood in case of data located on borders of clusters.

Multi-clustering approach eliminates inconvenience of decreased quality of predictions reasonable time efficiency for low number of clusters (e.g. 20). Neighbourhood is modelled by multiple clustering schemes and for recommendations is selected the most appropriate one to the particular data object. The results confirmed significant reduction of RMSE with time efficiency permitting on-line application for the case of 20 clusters.

Acknowledgment. The present study was supported by a grant S/WI/1/2018 from Bialystok University of Technology and founded from the resources for research by Ministry of Science and Higher Education.

References

1. Bobadilla, J., Ortega, F., Hernando, A., Gutiérrez, A.: Recommender systems survey. Knowl.-Based Syst. **46**, 109–132 (2013)
2. Bailey, J.: Alternative clustering analysis: a review. In: Aggarwal, C.C., Reddy, C.K. (eds.) Intelligent Decision TechnologieData Clustering: Algorithms and Applications, pp. 522–548. Chapman and Hall/CRC (2014)
3. Bae, E., Bailey, J.: COALA: a novel approach for the extraction of an alternate clustering of high quality and high dissimilarity. In: Proceedings of the IEEE International Conference on Data Mining, pp. 53–62 (2006)
4. Guang-Yu, Z., Chang-Dong, W., Dong, H., Wei-Shi, Z.: Multi-view collaborative locally adaptive clustering with Minkowski metric. Expert Syst. Appl. **86**, 307–320 (2017)
5. Hu, J., Qian, Q., Pei, J., Jin, R., Zhu, S.: Finding multiple stable clusterings. Knowl. Inf. Syst. **51**, 991–1021 (2017)
6. Jannach, D.: Recommender Systems: An Introduction. Cambridge University Press, Cambridge (2010)

7. Jie, L., Dianshuang, W., Mingsong, M., Wei, W., Guangquan, Z.: Recommender system application developments: a survey. Decis. Support Syst. **74**, 12–32 (2015)
8. Mitra, S., Banka, H., Pedrycz, W.: Rough-fuzzy collaborative clustering. IEEE Trans. Syst. Man Cybern. Part B (Cybern.) **36**(4), 795–805 (2006)
9. Nanopoulos, A., Rafailidis, D., Symeonidis, P., Manolopoulos, Y.: Musicbox: personalized music recommendation based on cubic analysis of social tags. IEEE Trans. Audio Speech Lang. Process. **18**, 407–412 (2010)
10. Recio-Garcia, J.A., Jimenez-Diaz, G., Sanchez-Ruiz, A.A., Diaz-Agudo, B.: Personality aware recommendations to groups. In: Proceedings of the Third ACM Conference on Recommender Systems, pp. 325–328 (2009)
11. Puntheeranurak, S., Tsuji, H.: A multi-clustering hybrid recommender system. In: Proceedings of the 7th IEEE International Conference on Computer and Information Technology, pp. 223–238 (2007)
12. Sarwar, B.: Recommender systems for large-scale e-commerce: scalable neighborhood formation using clustering. In: Proceedings of the 5th International Conference on Computer and Information Technology (2002)
13. Wu, Y., Liu, X., Xie, M., Ester, M., Yang, Q.: CCCF: improving collaborative filtering via scalable user-item co-clustering. In: Proceedings of the Ninth ACM International Conference on Web Search and Data Mining, pp. 73–82 (2016)
14. A Million Song Dataset. https://labrosa.ee.columbia.edu/millionsong/lastfm/. Accessed 5 Jan 2018

Models and Scheduling Algorithms
for a Software Testing System Over Cloud

Paweł Lampe[(✉)] and Jarosław Rudy

Department of Computer Engineering, Wrocław University of Science and
Technology, Wybrzeże Wyspiańskiego 27, 50-370 Wrocław, Poland
{pawel.lampe,jaroslaw.rudy}@pwr.edu.pl

Abstract. In this paper a Testing-as-a-Service cloud system is considered. The problem of distributing software tests over cloud is modeled
as a variation of the online scheduling problem with the goal of minimizing the average flowtime of jobs. History of over 20 million tests
is used to develop statistical and empirical models of system activity.
Developed models are used to generate a number of problem instances.
The instances are then used to test the effectiveness of several scheduling
algorithms. Results indicate that SJLOF algorithm is superior to other
tested algorithms, especially when the cloud resources are limited.

Keywords: Discrete optimization · Task scheduling
Queueing systems · Modeling and simulation · Software testing
Testing-as-a-Service

1 Introduction

Software testing is an important phase of software development, requiring a lot
of resources like time and equipment. This is especially true for large companies
that employ many software developers and use constant regression testing. Thus,
there is a need to automate the testing process and optimize it in order to reduce
the resources needed for testing.

Software testing systems are sometimes implemented using cloud computing.
Such systems are often referred to as Testing-as-a-Service (or TaaS) and are
an extension of the Software-as-a-Service (or SaaS) cloud computing model. TaaS
systems can be very complex: one example of a complete architecture of a TaaS
system can be found in paper by Yu *et al.* [11]. Software testing over cloud is
a relatively new concept, gaining popularity over the last decade. This resulted in
a number of papers, topics ranging from discussion and challenges of TaaS [2] to
practical considerations [7]. The idea of TaaS systems is still expanding and new
possibilities are discovered and researched. For example, Tsai *et al.* suggested
that thanks to multi-machine cloud environment TaaS systems can be effectively
employed for combinatorial testing and presented a TaaS design based on Test
Algebra and Adaptive Reasoning [10]. For a more thorough reading on general
software testing over cloud please consider paper by Inçki *et al.* [3].

W. Zamojski et al. (Eds.): DepCoS-RELCOMEX 2018, AISC 761, pp. 326–337, 2019.
https://doi.org/10.1007/978-3-319-91446-6_31

One of the components of TaaS system is distribution of the workload (software test) over the available cloud resources, which is understood as a form of cloud scheduling (or load balancing). Workload distribution in cloud environments is an older and more mature field of research with many approaches including metaheuristic methods like Ant Colony Optimization [6] or fuzzy sets [9]. For further reading on various approaches to scheduling and load balancing in cloud environments please refer to [1,4].

In this paper we consider a TaaS system described in paper [5]. This system uses proprietary off-premises Amazon-compatible cloud service based on Eucalyptus software and served by an external vendor. The cloud is enhanced by on-premises software testing toolset. Because of the complexity of the system, we focus on the problem of distribution of workload among the cloud machines to minimize the time required for testing (flowtime). The first aim of this paper is developing a mathematical model for describing and predicting the activity of the system. Once such model has been constructed, it could be used to generate problem instances of arbitrary length. Sufficient number of such instances would then allow to reliably compare effectiveness of constructive algorithms, which is the second aim of this paper.

The remainder of this paper is structured as follows. Section 2 contains mathematical formulation of the problem as well a number of parametric and emprical distributions for modeling system parameters. Section 3 briefly describes a problem instance generator based on developed models. Section 4 describes constructive algorithms consdiered in this paper. Section 5 contains the description of computer experiment and results of the comparison of aforementioned constructive algorithms. Finally, Sect. 6 contains conclusions.

2 Mathematical Models

In the first part of this section we will present the mathematical model of the considered TaaS system in general. Such description can be written from several points of view, including (1) queueing theory, (2) load balancing and (3) scheduling. Due to certain properties of the system (like job weights, setup times and different numbers of operations per jobs) we have decided to present the considered problem as a variation of the online scheduling problem. In the second and third part of the section we will present some statistical and empirical models using the data accumulated from the work of the real-life system.

2.1 Problem Formulation

There is a set $\mathcal{J} = \{1, 2, \ldots, n\}$ of n jobs (which represent test suites – groups of software tests cases to be tested). Each job is made up of operations (which represent individual test cases). Let $\mathcal{O}_j = \{1, 2, \ldots, o_j\}$ be set of operations of job j. The size of this set is o_j. Next, let p_j^i, t_j^i, r_j^i be the processing time, type and arrival (ready) time of i-th operation of j-th job respectively.

If two operations are from the same job, they have the same arrival time. Moreover, if two operations have the same type, they have the same processing time (this represent the same piece of code being scheduled for testing twice). In other words if a and b are operations from jobs j and k respectively then:

$$j = k \implies r_j^a = r_k^b, \tag{1}$$

$$t_j^a = t_k^b \implies p_j^a = p_k^b. \tag{2}$$

Next, let $\mathcal{M} = \{1, 2, \ldots, m\}$ be a set of m identical machines. Each operation has to be assigned to one machine. Any machine can process any operation. However, one operation can be processed by at most one machine at a time and one machine can process at most one operation at a time. The processing of operations cannot be interrupted (no preemption). Moreover, processing of each operation has to be preceded by setup time. This represents the time needed to transfer necessary files to the given machine. Setup time is 30 s (this was measured to be enough to transfer the needed file to the target machine) if the machine has not processed any operations yet or if the previously processed operation was of different job. Otherwise the setup time is 0 s (*e.g.* all files required for the next operation are already present on the target machine).

Let s_j^i be the starting time of processing of i-th operation of job j. Then the completion time c_j^i of that operation is given as:

$$c_j^i = s_j^i + p_j^i. \tag{3}$$

The goal is to choose such starting times for all operations as to minimize the average flowtime of all jobs $\bar{f}$, given as:

$$\bar{f} = \frac{1}{n} \sum_{j=1}^{n} \max_{i \in \mathcal{O}_j} c_j^i - r_j^i. \tag{4}$$

It is important to notice that the considered problem is online. This means that the information about jobs and operations become available only when they arrive. Especially, number of jobs is generally unknown (in fact, jobs are expected to continue to arrive indefinitely). Moreover, processing time p_j^i is only known if another operation of the same type have already finished processing. Otherwise, p_j^i is unknown even after operation arrival. Let us also notice, that operations do not have to be assigned to machine immediately after their arrival and can be kept in a queue. This, and the half-known processing times, distinguish this problem from simple load balancing.

2.2   Job Arrival Time Estimation

For the purpose of modeling job arrival time distribution, data from 5 month Monday–Friday work of a real-life TaaS system was gathered. This data contained 177 167 jobs. The aggregated (each point represents sum of 600 s) histogram of in-day arrival time of jobs is shown in Fig. 1. Two distinct bell-like

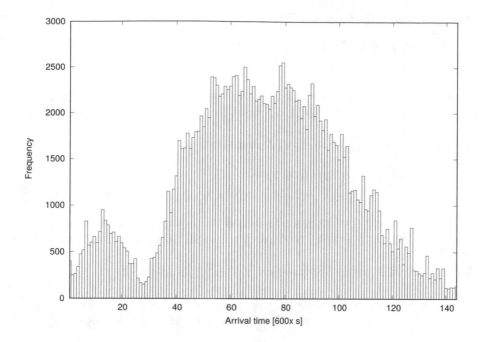

Fig. 1. Histogram of in-week job arrival time

peaks are clearly visible. Our hypothesis is that this is caused by two large distinct teams working in different timezones. There is also a third team, but it is so small that it is negligible. It is also important to note that due to the specific properties of the system and its automation it is very difficult to determine which job was run by which team. This forces us to treat the distribution of arrival times as mixture distribution (multimodal distribution), instead of estimating each peak separately.

We considered parametric models from 5 distribution families:

1. Bimodal normal distribution (5 parameters):

$$N(x \mid \mu_1, \sigma_1, \mu_2, \sigma_2, \alpha) = \alpha \mathcal{N}(x \mid \mu_1, \sigma_1^2) + (1 - \alpha)\mathcal{N}(x \mid \mu_2, \sigma_2^2), \qquad (5)$$

where $\mathcal{N}(x \mid \mu, \sigma^2)$ is normal distribution with location parameter μ and scale parameter σ at point x.

2. Bimodal Weibull distribution (7 parameters):

$$W(x \mid k_1, \lambda_1, o_1, k_2, \lambda_2, o_2, \alpha)$$
$$= \alpha \mathcal{W}(x - o_1 \mid k_1, \lambda_1) + (1 - \alpha)\mathcal{W}(x - o_2 \mid k_2, \lambda_2), \ (6)$$

where $\mathcal{W}(x \mid k, \lambda)$ is Weibull distribution with shape parameter k and scale parameter λ at point x. The offset parameters o_1 and o_2 are required to let Weibull distributions start later than at $x = 0$.

3. Uni-shaped bimodal Weibull distribution (6 parameters):

$$W_S(x \mid k, \lambda_1, o_1, \lambda_2, o_2, \alpha)$$
$$= \alpha \mathcal{W}(x - o_1 \mid k, \lambda_1) + (1 - \alpha)\mathcal{W}(x - o_2 \mid k, \lambda_2). \quad (7)$$

This is similar to the previous model, except $k_1 = k_2 = k$. This model assumes that the underlying process is the same for both teams, hence the same shape for both modes.

4. Bimodal Gamma distribution (7 parameters):

$$\Gamma(x \mid k_1, \theta_1, o_1, k_2, \theta_2, o_2, \alpha)$$
$$= \alpha \Gamma(x - o_1 \mid k_1, \theta_1) + (1 - \alpha)\Gamma(x - o_2 \mid k_2, \theta_2), \quad (8)$$

where $\Gamma(x \mid k, \theta)$ is Gamma distribution with shape parameter k and scale parameter θ at point x. The offset parameters have the same purpose as for Weibull model.

5. Uni-shaped bimodal Gamma distribution (6 parameters):

$$\Gamma_S(x \mid k, \theta_1, o_1, \theta_2, o_2, \alpha)$$
$$= \alpha \Gamma(x - o_1 \mid k, \theta_1) + (1 - \alpha)\Gamma(x - o_2 \mid k, \theta_2). \quad (9)$$

This is similar to the previous model, except $k_1 = k_2 = k$.

The parameters were estimated using a simple Monte Carlo method. Starting parameter ranges were set empirically, after which the method worked in stages. At each stage parameters were sampled and resulting distributions were evaluated. Each stage had tens to hundred thousands of samples. After each stage the range for each parameter was reduced, based on the best parameters found in that stage. The goodness of fit of the models was evaluated using the Kolmogorov-Smirnov (K-S) statistic. The results are shown in Table 1.

Table 1. Results of parameters estimation

K-S statistic	Model
0.013476	$N(x \mid 43420, 16462, 4543, 3173, 0.9644)$
0.009828	$W(x \mid 2.2616, 37138, 12967, 1.8046, 8077, 812, 0.9179)$
0.009700	$W_S(x \mid 2.2903, 37672, 12387, 8698, -536, 0.9208)$
0.017715	$\Gamma(x \mid 5.9917, 6226, 10095, 4.7869, 3215, -1242, 0.8692)$
0.052010	$\Gamma_S(x \mid 4.655, 6852, 13796, 2838, -4604, 0.9332)$

We see that both Weibull models yielded the best results (K-S statistic below 0.01), while Gamma distributions had the worst results. Let us also note that the use of K-S statistic is strictly for comparison of models, as all of the proposed models were rejected by the Kolmogorov-Smirnov test: for 177 167 observations (jobs), the critical value was 0.003226327, which is 3 times smaller than the value obtained from the W_S model. We suspect that this is caused by insufficient data – while there are over 177 167 data points, there are also 86 400 possible values for each observation, yielding slightly over 2 observations per value on average.

Another hypothesis was made, stating that a better fit could be achieved by modeling lunch breaks of employees – we believed those lunch breaks were responsible for visible drop of activity at the top of the bigger peak in Fig. 1. However, attempts to fit the data to 4-modal distribution (2 modes for big peak and 2 for small peak) did not yield better result, suggesting the underlying process is more complex (*i.e.* more non-lunch breaks). Unfortunately, the available company data does not contain information that could help model the breaks (especially for the second team).

Non-parametric statistic models were also used and achieved a good fit (passing Kolmogorov-Smirnov test) for 23-modal normal distribution (68 parameters in total). However, the number of modes was different for big and small peak. Moreover, the resulting model was sensitive to any changes resulting from adding more observations. This, coupled with large number of parameters, made us disregard that model.

To summarize, we choose W_S (a 6-parameter uni-shaped bimodal Weibull distribution) as a model of system activity (job arrival times) based on the currently possessed TaaS system data.

2.3 Empirical Distributions

Aside from arrival times, other system properties have to be modeled. This includes: (1) operation processing time, (2) operation type and (3) number of

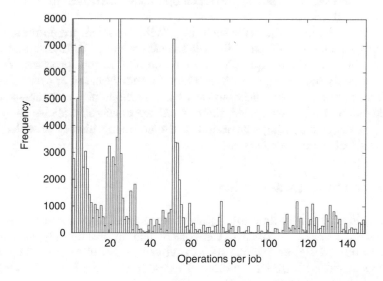

Fig. 2. Histogram of operations per job

operations per job. However, those properties are less regular and thus difficult to model. For example, histogram of number of operations per job (clipped for readability) is shown in Fig. 2. Because of this at present time those properties are modeled through empirical distributions calculated directly from accumulated system data. This time the data set contained 182 592 jobs and 21 987 210 operations.

3   Instance Generator

In this short section we will describe the generator of problem instances developed from the models shown in Sect. 2 and the benefits it grants over using data gathered directly from the system.

The generator accepts two input parameters: (1) number of jobs to generate N and (2) number of jobs per day D. If $N \equiv k \pmod{D}$, then the generated instance will have $\lfloor \frac{N}{D} \rfloor$ days of D jobs and one day with k jobs. For $k = 0$ this simplifies to $\frac{N}{D}$ days, each with D jobs. On full days (e.g. number of jobs equal to D) the arrival time of each job is drawn from the W_S distribution presented earlier with no restraints. Thus, job has 0.9208 and 0.0792 chance to be drawn from big and small peak respectively. The resulting times (in-day seconds) are modified by offsets and then ensured to fit into $[1; 86400]$ interval. In the case of non-full day (number of jobs $k < D$) the same procedure applies, but each arrival time has to fit into $[1; \lfloor \frac{86400k}{D} \rfloor]$ interval. This mechanism ensures that generation of non-full days will have expected result. For example with $N = 1000$ and $D = 2000$ all resulting jobs will have arrival times of 43200 or less, simulating the first 12 h of a daily workload. Finally, for each job, its number of operations is drawn from empirical distribution constructed from the collected system data. Similarly, processing time and type of each operation are drawn from appropriate empirical distributions.

Ideally, the number of jobs for each day would be drawn from another distribution, but such a distribution is difficult to estimate from the system data. This is because the company is expanding and the number of jobs processed each day has been steadily increasing over time. Thus, further research or system analysis is required to estimate the distribution of the number of jobs processed daily. It should be noted, however, that the current approach still allows to generate problem instances matching any period in the company history provided that a correct value of parameter D is chosen.

4   Algorithms Description

In this section we briefly describe algorithms we considered for scheduling software tests on the cloud. Because of the system size (approximately 240 000 operations scheduled daily) and short time available for scheduling (over 5 operations arriving each second during top activity), applying more advanced metaheuristic algorithms is somewhat difficult. Because of that simple priority-rule based constructive algorithms seem to be attractive alternative due to their low time complexity. Such algorithms were previously tested for a multi-criteria version of TaaS system, though tested problem instances were very limited in size [8].

All implemented algorithms work based on the concept of queue of operations – algorithms have to order this queue to decide which operations should be scheduled on the cloud first. Below is a brief description of implemented algorithms:

1. First In First Out (FIFO) – no rescheduling happens: jobs are scheduled in the order of their arrival and operations in jobs are scheduled in the order they appear in the job. Thus, this algorithm is the quickest with computational complexity of $O(1)$.
2. Shortest Operation First (SOF) – operations in the waiting queue are sorted in the order of their processing times, with shorter operations being scheduled first. Job to which operation belongs does not matter.
3. Longest Operation First (LOF) – similar to SOF, but the sorting order is reversed, so longer operations are scheduled first.
4. Smallest Job Longest Operation First (SJLOF) – operations in the waiting queue are sorted, first according to the size of the job they belong to with shorter jobs going first and then according to processing times of operations inside equal jobs (with longer operations going first). Size of a job is understood as the total time of all unstarted operations of that job.
5. Largest Job Shortest Operation First (LJSOF) – similar to SJLOF, but sorting rules for jobs and operations are reversed.
6. Max – this algorithm first selects one of the jobs from the waiting queue (*i.e.* one of the jobs that has unstarted operations remaining). The jobs are selected cyclically in deterministic order in a round-robin fashion. An unstarted operation with the longest processing time of the selected job is then scheduled. This algorithm is based on the one currently employed in the real-life TaaS system considered in this paper.
7. Min – similar to Max, but shortest unstarted operation of a job is scheduled each time a job is selected.
8. Random – operations in the queue are ordered randomly. This is the only non-deterministic algorithm.

It should be noted that some of the algorithms have their counterparts not listed above. This includes SJSOF and LJLOF algorithms. Moreover, Earliest Longest Operation First (ELOF) and Earliest Shortest Operation First (ESOF) are two algorithms similar to FIFO. Those four algorithms were performing similarly to their counterparts listed above, thus the results presented in Sect. 5 only include the better or faster of them (meaning FIFO, SJLOF and LJSOF algorithms).

5   Computer Experiment

A computer experiment was performed to test the effectiveness of the algorithms described in Sect. 4. Such tests are usually difficult for two reasons. First, considered instances are large. With 2 000 jobs per day, during rush hours (9 a.m. to 15 p.m) 2.5 jobs arrive each minute on average. 10-min instance would contain around 3 000 operations. Computing optimal values for such instances is already difficult. Thus, it makes more sense to compare algorithms to each other by normalizing them to the best amongst them for a given instance.

Second problem is the number of possible problem instances and their variety – algorithms may perform very differently for highly unusual instances. However, in the case of the considered TaaS system, we know that the problem instances follow certain probability distributions and deviations are unlikely to happen. Even if data deviating from those distributions were too appear they will affect the system only temporary, letting it get back too normal during low-activity hours. In result, the algorithm comparison for the considered system can be done more reliably, using instances created from instance generator.

For testing purposes, a number of problem instances was generated using the generator described in Sect. 3. Considered problem sizes were 333 (simulating early morning, 4 h), 1 000 (half a day), 2 000 (a full day) and 4 000 (two full days) jobs. For each problem size 50 instances were generated, for 200 instances in total. For each instance four different number of machines were considered: 100, 200, 400 and 800. This resulted in 800 actual problem instances. For each such instance 8 algorithms were run (for Random algorithm the presented results are average of 10 runs), resulting in 8 numbers. Let $R_A(i, n, m)$ be the result of running algorithm A on i-th instance with n jobs and m machines. Also let $R_{\text{best}}(i, n, m)$ be the best (*i.e.* lowest average job flowtime) algorithm for the i-th instance of n jobs and m machines. Then:

$$R'_A(i, n, m) = \frac{R_A(i, n, m)}{R_{\text{best}}(i, n, m)}, \tag{10}$$

is normalized value of $R_A(i, n, m)$. The results—averages of values of R' for various instance categories—are shown in Table 2. The most important values are highlighted in bold.

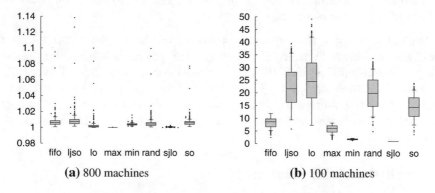

(a) 800 machines   (b) 100 machines

Fig. 3. Boxplots for all instances with a given number of machines

Table 2. Overall results of algorithm comparison. (n, m) means the average was for all instances with n jobs and m machines. Star means that instances with all possible values were used

(n, m)	FIFO	LJSOF	LOF	MAX	MIN	RANDOM	SJLOF	SO
$(*, *)$	3.34	9.03	10.66	2.45	1.28	7.41	**1.0015**	5.60
$(*, 100)$	8.09	22.91	26.18	5.71	1.81	19.93	**1.00**	14.57
$(*, 200)$	3.18	11.04	14.34	2.04	1.29	7.61	**1.0004**	5.73
$(*, 400)$	**1.086**	1.147	1.114	**1.0015**	1.026	**1.010**	**1.0049**	1.086
$(*, 800)$	**1.008**	1.010	1.003	**1.00**	1.004	**1.006**	**1.0008**	1.007
$(333, *)$	2.25	4.68	5.18	1.64	1.16	4.11	**1.0017**	3.19
$(1000, *)$	3.04	6.93	7.84	2.24	1.26	6.05	**1.0014**	4.62
$(2000, *)$	3.93	10.90	13.01	2.88	1.33	8.90	**1.0016**	6.70
$(4000, *)$	4.14	13.59	16.61	3.04	1.37	10.59	**1.0015**	7.89

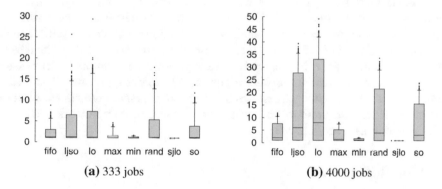

(a) 333 jobs   (b) 4000 jobs

Fig. 4. Boxplots for all instances with a given number of jobs

First thing to note is that there are only two values of 1.0 in the table. This means that in general no algorithm is superior in all cases – almost always there is at least one instance for which a given algorithm will fail to outmatch all others. When all instances are considered (first row, marked as $(*, *)$), SJLOF algorithm performs the best, being over 25% better than the second best algorithm overall – MIN. When the number of machines is increased, all algorithms perform better and the differences between them decrease. At 400 and 800 machines the MAX algorithm outperforms SJLOF. However, the difference is very small (around 0.6–0.8%) and by this time all algorithms have become similar. This is visible by the fact that for 400 and 800 machines FIFO and RANDOM algorithms are 0.8% away from the best algorithm on average as well. Let us note this is the only situation where average of any algorithm outperformed SJLOF. For lower number of machines SJLOF starts to outperform other tested algorithms greatly. When the number of jobs is considered, SJLOF algorithm outperforms other algorithms, being 16% to 37% better than the second best MIN algorithm.

That advantage only increases with the number of jobs. We conclude that, when the average results are concerned, SJLOF performs the best and is the only algorithm that stays near the top in all situations.

Averaged results are not fully conclusive, thus we also present boxplots in Figs. 3 and 4. Each box represents the interquartile range, with the line inside the box representing the median. The whiskers span from the 5th to the 95th percentile. Values outside the whiskers are indicated with dots. When the number of machines is high enough (Fig. 3a), all algorithms are rarely more than 3 or 4% away from the best one (MAX in this case). Still, MAX and SJLOF algorithms clearly outperform the rest. More interesting is the case when the number of machines is small (Fig. 3b). We see that all algorithms besides SJLOF and MIN can perform from around 5 to even 50 times worse than the best algorithm in some cases. Moreover, those algorithms very rarely manage to get closer than 3 times away from the best algorithm. In short, SJLOF algorithm decidedly outperforms other algorithms when number of machines is low.

When the number of jobs is low or the system activity is low (Fig. 4a), SJLOF algorithm is still the best, while the other algorithms perform worse (from 1 to over 20 in some cases). However, the medians of all algorithms are similar to the performance of SJLOF. When the number of jobs increases (Fig. 4b), the effect is increased as well – the other algorithms now performing even 50 times worse at times. However, the median of other algorithms is now from around 2 to 8 times higher than that of SJLOF algorithm. These results further confirm that SJLOF is the best algorithm in almost all situations. Its closest competitors – MIN and MAX algorithms – are less reliable.

6   Conclusions

In this paper we presented a model for the problem of scheduling software tests on Testing-as-a-Service cloud system. The problem was modeled as a variant of the online scheduling problem with the goal of minimizing flowtime of test suites. Uni-shaped bimodal Weibull distribution was used to model test suites arrival time in the system and other empirical distributions were used to model other parameters of the system (like test processing time). Instance generator that can be used to easily generate instances of the problem with desired size and suites per day density was also described.

A computer experiment for 800 instances with various job and machine size was used to test effectiveness of a number of constructive algorithms. Presented results indicate that SJLOF algorithm is the most effective in all situations except when the number of machine is high. However, for high number of machines all algorithms perform similarly and SJLOF algorithm still remains competitive. Thus, SJLOF algorithm could be used to reduce the cloud resources required for software testing (thus saving money or allowing the resources to be used for other tasks), while minimizing the increase in flowtime of test suites.

References

1. Al Nuaimi, K., Mohamed, N., Al Nuaimi, M., Al-Jaroodi, J.: A survey of load balancing in cloud computing: challenges and algorithms. In: 2012 Second Symposium on Network Cloud Computing and Applications (NCCA), pp. 137–142. IEEE (2012)
2. Gao, J., Bai, X., Tsai, W.T., Uehara, T.: Testing as a Service (TaaS) on clouds. In: Proceedings of the 2013 IEEE Seventh International Symposium on Service-Oriented System Engineering, SOSE 2013, pp. 212–223. IEEE Computer Society, Washington (2013). http://dx.doi.org/10.1109/SOSE.2013.66
3. Inçki, K., Arı, I., Sözer, H.: A survey of software testing in the cloud. In: 2012 IEEE Sixth International Conference on Software Security and Reliability Companion (SERE-C), pp. 18–23. IEEE (2012)
4. Kalra, M., Singh, S.: A review of metaheuristic scheduling techniques in cloud computing. Egypt. Inform. J. **16**(3), 275–295 (2015). https://doi.org/10.1016/j.eij.2015.07.001.   http://www.sciencedirect.com/science/article/pii/S11108665 15000353
5. Lampe, P., Rudy, J.: Job scheduling for TaaS platform: a case study. In: Innowacje w zarządzaniu i inżynierii produkcji, vol. 1, pp. 636–646. Oficyna Wydawnicza PTZP (2017)
6. Nishant, K., Sharma, P., Krishna, V., Gupta, C., Singh, K.P., Rastogi, R., et al.: Load balancing of nodes in cloud using ant colony optimization. In: 2012 UKSim 14th International Conference on Computer Modelling and Simulation (UKSim), pp. 3–8. IEEE (2012)
7. Riungu-Kalliosaari, L., Taipale, O., Smolander, K.: Testing in the cloud: exploring the practice. IEEE Softw. **29**(2), 46–51 (2012)
8. Rudy, J.: Online Multi-criteria Scheduling for Testing as a Service Cloud Platform. Springer International Publishing, Cham (in print)
9. Shojafar, M., Javanmardi, S., Abolfazli, S., Cordeschi, N.: Fuge: a joint meta-heuristic approach to cloud job scheduling algorithm using fuzzy theory and a genetic method. Cluster Comput. **18**(2), 829–844 (2015). https://doi.org/10.1007/s10586-014-0420-x
10. Tsai, W.T., Qi, G., Yu, L., Gao, J.: TaaS (Testing-as-a-Service) design for combinatorial testing. In: 2014 Eighth International Conference on Software Security and Reliability, pp. 127–136. IEEE (2014)
11. Yu, L., Tsai, W.T., Chen, X., Liu, L., Zhao, Y., Tang, L., Zhao, W.: Testing as a service over cloud. In: Proceedings of the 2010 Fifth IEEE International Symposium on Service Oriented System Engineering, SOSE 2010, pp. 181–188. IEEE Computer Society, Washington (2010). https://doi.org/10.1109/SOSE.2010.36

A Trustworthy and Privacy Preserving Model for Online Competence Evaluation System

Vasiliki Liagkou$^{(\boxtimes)}$ and Chrysostomos Stylios

Computer Engineering Department,
Technological Educational Institute of Epirus, Kostakioi, Arta, Greece
liagkou@kic.teiep.gr, stylios@teiep.gr

Abstract. Nowadays, all academic institutions exhibit and distribute their material over Internet. Moreover e-learning and e-evaluation products is one of the most rapidly expanding areas of education and training, with nearly 30% of U.S. college and university students now taking at least one online course. However, Internet increases the vulnerability of digital educational content exploitation since it is a possibly hostile environment for secure data management. Educational institutions have to deal with all the open security challenges that can cause huge data and financial losses, harm their reputation and strictly affect people's trust on them. In this paper we propose a trust preserving approach for utilizing online evaluation of acquired student competences and we present a trust preserving approach for handling the increasingly complex issues of designing, developing and e-competence evaluation systems suitable for educational and e-learning environments. The proposed architecture for an online competence evaluation system will offer to the authorized users procedures for evaluating competences and providing their feedback but they have to prove their eligibility to participate in the evaluation while, at the same time, it will preserve their privacy and it will ensure system's trustworthiness. The proposed model addresses a list of fundamental operational and security requirements and implements a small scale proof of concept of privacy enhancement technologies.

Keywords: Privacy preserving technologies · Cryptography · Trust
Educational data · Personal data

This work has been partially supported by the "Implementation of Software Engineering Competence Remote Evaluation for Master Program Graduates (ISECRET)" project No 2015-1-LVo1-KA203-013439 funded by ERASMUS+ of the European Commission and by the National funds allocated by the Greek General Secretarial of Research and Development project 2006SE01330025 as continuation of FP7-PEOPLE-IAPP-2009, Grant Agreement No. 251589, Acronym: SAIL.

W. Zamojski et al. (Eds.): DepCoS-RELCOMEX 2018, AISC 761, pp. 338–347, 2019.
https://doi.org/10.1007/978-3-319-91446-6_32

1   Introduction

The huge expansion of Internet has increased the vulnerability of electronic educational environments (see Rjaibi et al. in [18], Younis et al. in [28], Chen et al. in [7], Jain and Ngoh in [10] and Morrisson in [16]). E-learning portals would be characterized as hostile environment from secure data management perspective. In most cases, educational organizations have to deal with all the open security challenges that could cause huge data losses, harm their reputation and strictly affect people's trust on them. One of the main obstacles for the wide adoption of online evaluation and e-educational tools, is the reluctance of users to participate. This reluctance can be, partially, attributed to the, relatively, low penetration of technology among citizens. However, the main reason behind this reluctance is the lack of trust towards the educational online system, which stems from the fear of users that systems implementing online evaluation services and procedures may violate their privacy. Our point of view, is that trust in educational and on line evaluation systems should ensure and prove to users that the system respects their privacy. Special care has to be taken that the algorithms, services, applications and data uploaded to the educational portal concerning the teaching material, personal information and evaluating competencies or courses and in general the whole operation of the educational management system will remain secure and confidential to all users.

1.1   Our Contribution

In this paper we describe a privacy preserving approach for an online competence evaluation system that was developed within the context of the project ISECRET [14]. The users have to prove their eligibility to participate in the evaluation process and they are able to securely share their learning outcomes and their possessed competences. The proposed e-competence evaluation system includes the development of a privacy preserving approach based on the cryptographic primitives called Attribute Based Credentials (Rannenberg et al. in [17]). The online educational system will offer to the authorized users the procedures for evaluating their competences and providing their feedback. Moreover the users have to prove their eligibility to participate in the evaluation while, at the same time, the evaluation process preserves their privacy and ensures system's trustworthiness.

2   Online Evaluation and the Challenge of Security and Privacy

World Wide Web is a critical requirement for a series of administrative tasks such as how instructors assess student progress both formatively and summative, how they distribute evaluation activities and how they provide effective feedback. Any competence evaluation system want to market its evaluation services to enterprises, graduates and potential user. A competence evaluation system collects data to contact individuals to rally support for their cause and raise awareness. So, an e-competence evaluation system should protect user's privacy data.

2.1 Discussion on Existing Solutions and Comparison

Till now, assessment of student learning has attracted much attention during the past few years (see the relevant work of Jain and Ngoh in [10], Morrisson in [16], Yang and Lin in [27], Romansky et al. in [19], Huu Phuoc Dai et al. in [9] and Aljbori et al. in [1]). There are proposals to evaluate professional skills in literature. LeBelau et al. introduce an Integrated Design Engineering Assessment and Learning System in [12]. Moreover there are a lot of academic institutions that build systems for implementing assessment on the provided knowledge and learning outcomes (see [8,20,22]). The vast majority of related research work (see research work of Rjaibi et al. in [18], Younis et al. in [28], Chen et al. in [7], Kritzinger and Solms in [11], Mohd and Fan in [15], Saxena in [21], Yong in [23], Trek in [24], Reddy in [25], Whitson in [26] and Yand and Lin in [27]) propose a collection of security tools in order to solve certain security issues. All the relevant work has focused only on the traditional security requirements and do not focus on privacy oriented requirements in order to ensure the users' trust about the evaluation procedure. Our model addresses the special challenge that for competence evaluation results to be as accurate and impartial as possible, the privacy of the participants must be preserved while, at the same time, their eligibility should be confirmed. We, thus, require partial identification of the participants. Attribute Based Credentials (ABC) are able to handle this requirement by guaranteeing than no information is sent to the competence evaluation system which can later be used to identify the users who participated. At the same time, the competence evaluation system ensured that only eligible users can have access to the competence evaluation quiz.

3   Privacy and Privacy ABCs

In the online educational evaluation system we do not to use the commonly user authentication methods (e.g. PKI based) for controlling access to the online evaluation services in order to preserve users' *privacy* and increase system's trustworthiness. Online competence evaluation system don't need to reveal user's full identity profile in order to give access to educational services. In such types of applications there is, clearly, a need for a *partial*, and not complete, revelation of the user's identity thus we suggest the use of Attribute Based Credentials. Attribute Based Credentials (ABC) are a form of authentication mechanism that allows to flexibly and selectively authenticate different attributes about an entity without revealing additional information about the entity, for more details see Bichsel et al. in [2], Camenish in [3] and Camenish et al. in [4]. *Privacy Attribute-Based Credentials* or Privacy-ABCs, for short, is a technology that enables privacy preserving, partial authentication of users (see Camenish et al. in [5,6], Rannenberg et al. in [17], Zhang et al. in [29] and

Liagkou et al. in [13]). Privacy-ABCs are issued just like normal electronic credentials (e.g. PKI based) using a secret signature key owned by the credential issuer. However, and this is a key feature of this technology, the user is in position to transform the credentials into a new form, called *presentation token*, that reveals only the information about him which is really necessary in order to access a service. This new token can be easily verified with the issuer's public key. The main ABC entities are four: the *Issuer*, the *User* and the *Verifier*. In general, the Issuer issues credentials containing certified user attributes, thereby attesting the validity of the attributes. The Verifier, or *relying party*, on the other hand, offers a service with access limited only to those users for which it can verify the possession of certain attributes (or credentials).

4   The Operational Environment

The e-competence evaluation system will be used by users that want to certify their competences that they have acquire. Our system must be located in academic institution premises,where the competence evaluation system will be installed, operated, and monitored. Figure 1 shows the system and network infrastructure. Network security relies, partly, on a pair of firewalls which are connected to a high availability configuration (active-standby, without NAT, with automatic fail-over capability between them). The firewalls appear in between the border router and the internal network, inspecting incoming and outgoing traffic and ensuring protection against malicious attacks. For instance, these firewalls can block suspicious source IP addresses in the case of detected DoS attacks as well as traffic directed towards internal servers. However, they alone cannot block packets with malicious content (e.g. viruses) which are taken care of by other components of the security subsystem.

In addition, a *DMZ* subnet must exist in academic institution network infrastructure. A DMZ is a physical or logical subnetwork that encompasses and publicizes an organization's computing services to an external, untrusted network, most commonly the Internet. The DMZ offers an additional security layer to an organization's local area network and services since an attacker from the outside can only access the DMZ and not parts of the internal infrastructure of the organization.

Finally, at the perimeter of Academic Institution's network infrastructure must be placed a border router between the firewalls and the external network and performs some basic checks on incoming and outgoing network activity, such as *ingress* and *egress* filtering that may be helpful in blocking some Internet-based worms from reaching the firewall. In computer security terminology, *ingress filtering* refers to techniques which are employed to verify that incoming traffic actually comes the originators that the traffic packets claim to be from. Complementarily, *egress filtering* refers to techniques of monitoring and, possibly, restricting the type of outgoing traffic from one network to another. Most commonly, this outgoing traffic may contain information from private LANs which may be maliciously directed to the outside network (e.g. the Internet) and should

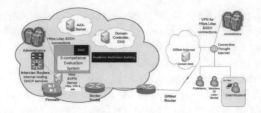

Fig. 1. E-competence evaluation system network infrastructure

be intercepted and, perhaps, blocked. This border router also implements some generic access list based control in order to increase the level of security and handle some types of attacks like DoS (Denial of Service) or DDoS (Distributed Dos).

5   Description of Our Architectural Model

The architecture of the privacy preserving course evaluation system is shown in Fig. 2. As it can be seen, the architecture is based on various components that have different functionalities and roles. In what follows, we will describe their properties and interactions within the pilot's context.

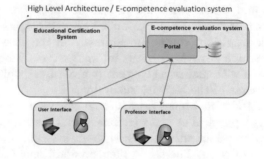

Fig. 2. Our architectural model

- *Portal*: This component is web base information portal. Through this portal,the users will get information about the e-competence evaluation system and functionality as well as information about its usage. The architecture of the portal course evaluation system is consisted on the following main components: (i) Link to the home page. (ii) Information of the provided academic educational program. (iii) Registration and login information. (iv) Help links about the use of the portal. (v) Professors' and Students' user interface (vi) The Software Engineering Competence Evaluation application.
- *Certification System*: This component issues credentials to the students. In particular:
  - An officer from an academic institution is authorized to insert student information in the database of the Certification System.

- The administrator can revoke a user credential.
- Users are issued credentials that certify that they are, indeed, members of the Academic institution eg. If they are undergraduate student or graduate. In a more general scenario the academic institution could offer its competence evaluation services to subscribed users, and the academic institution could certify if the user is a subscribed user for participating the e-competence evaluation.
- Users are able to browse their personal data that is stored academic institution database through their user interface.
- Users are able to manage, themselves, a limited subset of their personal information.
- Users are issued credentials that certify that they could participate to the e-competence evaluation system.

When a user requests a credential, through the e-competence evaluation portal then the e-competence evaluation system initiate the issuance protocol for the provided attribute based credentials.

- *E-competence Evaluation System:* This component implements the evaluation of the user's knowledge. The procedure is the following, a user could select a study subject through the academic educational program and run the competence evaluation application. Through the Learning Outcome link, he/she will see the list of the expected learning outcomes for the chosen Study Subject. When the obtaining competences phase has finished, then the user is able to evaluate any chosen competence. Whenever a user wants to evaluate a competence, he/she can access the Portal though his/her computer and complete the quizzes. After completing the evaluation, the student is informed about the evaluation result and his/her user profile will be updated automatically. The competence evaluation system performs access control to the e-competence Evaluation procedure. This access control is achieved by presenting a policy to the users stating what credentials they must possess in order to proceed. Only users who own the required credentials are given access to the evaluation. Potential users of this application are the users, the professors, and members of labor market. More precisely:
  - The professors can:
    - Create a template to describe the set of competences for the study subject of the academic educational program that are responsible for.
    - can rate or insert a weight factor for the competence that want to provide evaluation.
    - Create tests for the competence evaluation.
  - Users are able to anonymously evaluate the competences to which they are registered. A user could be registered because he is a member of academic institution or he is an subscribed user.
  - When the competence evaluation procedure is completed, professors and subscribed members of labor market could access the competence evaluation results.
- *User's Interface:* In order the internet browser to be recognize anonymous credentials any user must install in their pc a specific application which runs

through users interface. Here we don't use smartcards in order to store the anonymous credentials. The installed program helps user to perform operation on his credentials and initiate credential issuance and verification protocols through his user interface. A member of academic educational institution like a undergraduate or graduate student or a subscribed user thought his interface he will be able to: (i) View a Study Subject. (ii) View Announcements. (iii) View Rubrics. (iv) Access the evaluation area. (v) Participate in self-evaluation of a selected academic or professional competence. (vi) Submit evaluation. (vii) View evaluation results. (viii) View the set of professional and academic competence.

– *Professor's Interface:* A professor could log into the portal and access his interface in order to: (i) Edit/Insert data in a specific Study Subject. (ii) View a Study Subject. (iii) Add Announcements. (iv) Edit/View Rubrics. (v) Edit the contents on the evaluation area. (vi) Insert/edit e-Competence. (vii) Insert new academic competence.

6   High Level Description of the Online Educational Evaluation System

We will now take a closer look at the stages involved in the realization of the online competence evaluation. An authorized user has to collect credentials that proves, anonymously, that they are students or graduates of the educational institute and that they have registered to the study subject that provides the competence evaluation. Finally, they can anonymously participate the e-competence evaluation system using the E-competence evaluation system.

The Users interacted with the Educational Certification System in order to obtain their credentials, proving their studentship or their membership and their registration to the study subject. The Educational Certification System is the system which the students use in order to evaluate, anonymously, the competence that they have selected. Also, the students had to install an ABC User Client (User Service + GUI) on their computers in order their User Interface to be able to interact with system components. As far as security tokens are concerned, the use of a smart card is suggested. As a tamper proof device it offers security and it is the ideal hardware token for storing the User's device key. Additionally, it features a cryptographic processor which can be utilized for performing the cryptographic operations (exponentiations etc.) that are required during issuance or presentation. Finally, it makes a User who stores her personal data on it, more confident and trustful. When a user want to be registered to the online e-competence evaluation system he submits an application to the educational institute. The educational institute is responsible to check the submitted applications and to register a student, a graduate or an employee to the e-competence evaluation system. Then the secretary of the educational institute sends to the registered users an envelope containing a properly initialized smart card and the card's PIN and PUK values. We also give to each of them a contact smart card reader and a slip of paper containing a one-time-password for the initial logging in the Educational Certification System. The first step for the new users was to log in the Educational Certification System using their matriculation numbers as usernames and their one-time passwords. Then, they are able to register their smart cards so that the E-competence evaluation system could link their smart cards with the user' information residing in the system database. After a user has registered his smart card, he was able to obtain the evaluation credentials from the E-competence evaluation system. The evaluation credential proves that the user is registered to the study subject and he can access E-competence evaluation system.

It should be stressed here that each user is allowed to access the CE-competence evaluation system and provide his evaluation several times. However, only his last evaluation is taken into account due to the use of scope-exclusive pseudonyms. The E-competence evaluation system contains a database for storing eligibility policies and competence evaluation data for subsequent analysis.

7 Conclusions

Here we proposed a trust preserving approach for handling the increasingly complex issues of designing, developing and creating online competence evaluation systems suitable for educational and e-learning environments. Our future work is to use the e-competence evaluation system as a small scale proof of concept of privacy enhancement technologies in the e-learning environments in order to introduce, in the future, of these technologies to the educational communities of all levels in EU. These technologies will be the vehicle for supporting privacy preserving e-education activities in discussion groups whereby participants will provide their opinion anonymously but after proving that they are eligible to participate in group discussions.

References

1. Aljbori, M.A., Guirguis, S.K., Madbouly, M.M.: Adaptable mobile user interface for securing e-learning environment. Trans. Netw. Commun. **2**(4), 64–83 (2013)
2. Bichsel, P., Camenisch, J., Gro, T., Shoup, V.: Anonymous credentials on a standard java card. In: Proceedings of 16th ACM Conference on Computer and Communications Security, CCS, pp. 600–610 (2009)
3. Camenisch, J.: Protecting (anonymous) credentials with the trusted computing group's TPM, vol. 1, pp. 135–147 (2001)
4. Camenisch, J., Lysyanskaya, A.: An efficient system for non-transferable anonymous credentials with optional anonymity revocation. In: Proceedings of EUROCRYPT , pp. 93–118 (2001)
5. Camenisch, J., Lysyanskaya, A.: Dynamic accumulators and application to efficient revocation of anonymous credentials. In: Proceedings of EUROCRYPT, pp. 61–76 (2002)
6. Camenisch, J., Lysyanskaya, A.: Signature schemes and anonymous credentials from bilinear maps. In: Proceedings of EUROCRYPT, pp. 56–72 (2004)

7. Chen, Y., He, W.: Security risks and protection in online learning: a survey. Int. Rev. Res. Open. Distance Learn. **14**(5), 109–127 (2013)
8. Farias, G., Muñoz de la Peña, D., Gómez-Estern, F., De la Torre, L., Sánchez, C., Dormido, S.: Adding automatic evaluation to interactive virtual labs. Interact. Learn. Environ. **24**(7), 1456–1476 (2016)
9. Huu Phuoc Dai, N., Kerti, A., Rajnai, Z.: E-Learning security risks and its countermeasures. J. Emerg. Res. Solut. ICT. **1**(1), 17–25 (2016)
10. Jain, K., Ngoh, L.B.: Motivating factors in e-learning -a case study of UNITAR. Student Affairs Online, vol. 4(1) (2003). http://www.studentaffairs.com/ejournal/eLearning.html
11. Kritzinger, E., von Solms, S.H.: E-learning: incorporating information security governance. Issues Inf. Sci. Inf. Technol. **3**, 319–325 (2006)
12. LeBeau, J.E., McCormack, J., Beyerlein, S., Davis, D., Trevisan, M., Leiffer, P., Thompson, P., Davis, H., Howe, S., Brackin, P., Gerlick, R., Khan, M.J.: Alumni perspective on professional skills gained through integrated assessment and learning. Int. J. Eng. Educ. **30**(1), 48–59 (2014)
13. Liagkou, V., Metakides, G., Pyrgelis, A., Raptopoulos, C., Spirakis, P., Stamatiou, Y.: Privacy preserving course evaluations in greek higher education institutes: an e-participation case study with the empowerment of attribute based credentials. In: Proceedings of Privacy Technologies and Policy, Lecture Notes in Computer Science, vol. 8319, pp. 140–156. Springer Berlin Heidelberg (2014)
14. Liagkou, V., Stylios, C. : The case study of a software engineering e-competence evaluation portal. In: Proceedings of Actual Problem of Education (MIP-2017), pp. 21–23 (2017)
15. Mohd Alwi, N., Fan, I.: E-learning and information security management. Int. J. Digit. Soc. **1**, 148–156 (2010)
16. Morrison, D.: E-Learning Strategies. Wiley, Chichester (2005)
17. Rannenberg, K., Camenisch, J., Sabouri, A.: Attribute-Based Credentials for Trust: Identity in the Information Society. Springer Publishing Company, Incorporated, Cham (2014)
18. Rjaibi, N., Rabai, L.B.A., Aissa, A.B., Louadi, M.: Cyber security measurement in depth for e-learning systems. Int. J. Adv. Res. Comput. Sci. Softw. Eng. **2**(11), 1–15 (2012)
19. Romansky, R., Noninska, I.: Implementation of security and privacy principles in e-learning architecture. In: Proceedings of the 29th International Conference on Information Technologies (InfoTech-2015), St. Constantine and Elena, Bulgaria, pp. 66–77 (2015)
20. Ruano, M., Colomo, P.R., Gómez, J., García Crespo, A.: A mobile framework for competence evaluation: innovation assessment using mobile information systems. J. Technol. Manag. Innov. **2**(3), 49–57 (2007)
21. Saxena, R.: Security and online content management: balancing access and security, breaking boundaries: integration and interoperability. In: Proceedings of the 12th Biennial VALA Conference and Exhibition Victorian Association for Library Automation (2004)
22. Tsinakos, A., Kazanidis, I.: Identification of conflicting questions in the PARES system. Int. Rev. Res. Open Distrib. Learn. **13**(3), 298–314 (2012)
23. Yong, J.: Digital identity design and privacy preservation for e-learning. In: Proceeding of the 2007 11th International Conference on Computer Supported Cooperative Work in Design, pp. 858–863 (2007)
24. Treek, D.: An integral framework for information systems security management. Comput. Secur. **22**(4), 337–360 (2003)

25. Reddy, G.S.: Security issues and threats in educational clouds of e-learning: a review on security measures. Comput. Technol. Appl. **4**, 312–316 (2013)
26. Whitson, G.: Computer security: theory, process and management. J. Comput. Small Coll. **18**(6), 57–66 (2003)
27. Yang, C., Lin, F.O., Lin, H.: Policy based privacy and security management for collaborative e-education systems. In: Proceedings of the 5th IASTED International Multi-Conference Computers and Advanced Technology in Education (CATE 2002), pp. 501–505 (2002)
28. Younis Alsabawya, A., Cater-Steel, A., Soar, J.: IT infrastructure services as a requirement for elearning system success. Comput. Educ. **69**, 431–451 (2013)
29. Zhang, Z., Yang, K., Hu, X., Wang, Y.: Practical anonymous password authentication and TLS with anonymous client authentication. In: Proceedings of the 2016 ACM SIGSAC Conference on Computer and Communications Security (2016)

Effective Algorithm of Simulated Annealing for the Symmetric Traveling Salesman Problem

Mariusz Makuchowski[✉]

Department of Control Systems and Mechatronics, Wrocław University of Science and Technology, Wybrzeże Wyspiańskiego 27, 50-370 Wrocław, Poland
mariusz.makuchowski@pwr.edu.pl

Abstract. The paper deals with a symmetric traveling salesman problem (sTSP). An efficient algorithm, based on the simulated annealing (SA), is constructed. A heavy experimental evaluation of the proposed algorithm, that confirms its efficiency for known from literature instances of sTSP, is presented. The algorithm can find a solution both for the small and very large instances of the problem. It provides a good quality solution in very short time. Obviously, increasing time-limit of computations enables to improve quality of solutions found.

Keywords: Traveling salesman problem · Simulated annealing

1 Introduction

One of the most commonly studied combinatorial problems is a traveling salesman problem (see, e.g. [1]). This fact results from its simple formulation and the difficulty in determining an optimal solution. This problem belongs to class of *NP-hard* problems [2]. Many dedicated heuristics [3–8] to metaheuristics [9–12] have been constructed for this problem.

A general TSP problem can be defined as follows: There is a set $C = \{1, 2, \ldots, n\}$ given representing n cities and a square matrix d of a size $n \times n$. The value $d_{i,j} \in \mathbb{R}_+ \cup \{0\}$ defines the distance from i city to j city. Let π be the permutation of cities C. The set of all permutations will be denoted by Π. Permutation π is identified with a cyclic route of successively visited cities. For each route π we can determine its length denoted by $D(\pi)$;

$$D(\pi) = \sum_{i=1}^{n-1} d_{\pi(i),\pi(i+1)} + d_{\pi(n),\pi(0)}. \tag{1}$$

The problem relies in designating of permutation π^* minimizing the length of the cycle,

$$\pi^* \in \arg\min_{\pi \in \Pi} D(\pi). \tag{2}$$

© Springer International Publishing AG, part of Springer Nature 2019
W. Zamojski et al. (Eds.): DepCoS-RELCOMEX 2018, AISC 761, pp. 348–359, 2019.
https://doi.org/10.1007/978-3-319-91446-6_33

2   SA Algorithm

The presented algorithm hereinafter referred to as SA is based on the simulated annealing method. In the paper [13] the idea of the simulated annealing algorithm was presented for the first time. This is an improvement type of an algorithm, which means that every step of the way tries to "improve" a current solution moving to some neighbor solution. The neighbor solution being analyzed is accepted (i.e. it becomes the current solution for the next iteration) with a certain probability of P. The accepted solution can be either better and worse (in the sense of the value of the objective function)than the solution from previous iteration.

In the presented algorithm, the probability of accepting a specific example of the neighboring solution depends both on the Δ difference in the value of the objective function, (of current and accepted solutions) and the T parameter called temperature:

$$P(\Delta, T) = \begin{cases} 1, & \Delta \leq 0 \\ e^{-\Delta/T}, & \Delta > 0 \end{cases} \tag{3}$$

High temperature increases the probability of accepting weaker solutions, and during the algorithm operation is gradually reduced. A better solution than the current one is accepted unconditionally.

2.1   Neighborhood Solution

In the presented algorithm, the neighborhood of the current solution is based on five types of moves. Each type of the proposed moves is based on the so-called 'inverse'-type moves. The $inv(a, b)$ move inverts the solution fragment between a and b.

The algorithm in the initial phase creates the k list of the nearest neighbors for each city. Thanks to the use of the $kd - tree$ [14] structure this process runs very smoothly, i.e. for instance of size $n = 100.000$ and $k = 20$ takes 0.2 s. An analogous process with a traditional list of cities in the same example takes 40 s.

In all types of moves, the a position is always drawn, whereas the item b (or more precisely the $\pi(b)$ city number in the position b) is selected depending on the type of move:

- type 1: $\pi(b)$ is a city that belongs to the list of neighbors of the city $\pi(a)$, which minimizes the route for $inv(a, b)$ move, with restriction called *Radius Fixed*.
- type 2: the same as in type 1, but when finding a $inv(a, b)$ move that improves the route length, for further browsing, the *Radius Fixed* restriction no longer applies
- type 3: the same as in type 1, with the function of avoiding moves that do not change the length of the route,
- type 4: is the nearest city for the city $\pi(a)$,
- type 5: is a random city (the choice of the city is not limited to the list of neighbors).

In each iteration of the algorithm there is selected, randomly with equal probability, the type of move to be executed. Then, the algorithm generates random move of a given type and with the probability of P this move is executed.

Moves of type 1,2,3 work similarly as in a descending algorithm. Type 5 move is a typical random move used both in random algorithms and the simulated annealing type of an algorithm, while type 4 move has indirect characteristics of other moves. A set of different moves turned out to be decisive for the effectiveness of the algorithm. Where the presented algorithm provides solutions with a relative error of a fraction of a percent, an analogous algorithm using only one type of move provided solutions with a relative error of a few percent.

The *Radius Fixed* constraint significantly accelerates the algorithm, limiting the number tested neighbors. Avoiding the moves that do not change the length of the route, and using the moves that do not make the solutions worse instead is very beneficial in case of large instances, however, in case of small instances, it is unfavorable.

2.2   Starting Solutions

Simulated annealing type of algorithms, in the initial phase of operating have a high value of T called temperature. This parameter is directly responsible for the probability of acceptance of a worse solution resulting from the execution of a random move. Because in the initial phase the probability of accepting the aggravating move is large, the algorithm can easily leave local minima, as the result of which it generates poor quality solutions. In case of using a relatively high initial temperature, the starting solution is of little importance to the quality of the entire SA algorithm.

The algorithm can be accelerated by leaving the initial phase of calculations at high temperature. It is suggested that the SA algorithm should start with a lower initial temperature. In this case, the algorithm will not explore the entire solution space, and the search will be concentrated on a smaller area of this space. In this case, the quality of the initial solution has a great impact on quality of the whole algorithm and there should be provided solutions as good as possible.

As a result of the study, it was also observed that solutions obtained with the use of Nearest Neighbour (NN) algorithm, although they are of lower quality than the solutions obtained by the Farthest Insertion (FI) algorithm, were much easier to be improved. In addition, using data structures of $kd - tree$ [14] the operation time of the NN algorithm was much shorter than the FI algorithm. Ultimately, the NN algorithm was chosen to generate the initial solution.

2.3   Automatic Tuning of an Algorithm

The value of the initial temperature is very important for the quality of the algorithm. Its selection depends on the initial solution. For weaker solutions it should be larger so that the algorithm could have the chance to move to other areas of the solution space. For good starting solutions, the temperature should not be too high, so that the algorithm stays in the current good region of the solution space.

The algorithm works in two phases. In the first phase, it tunes the initial and final temperature values, and generates a starting solution for a second phase. The *iter* algorithm parameter specifies the total number of iterations performed. In the presented algorithm, there are 50% of all iterations belonging to each phase.

The method of determining the initial temperature in phase 1 is determined on the basis of average nearest neighbor distance:

$$T_{begin} = \frac{40}{n} \cdot \sum_{c \in C} \min_{k \in C, k \neq c} D(c, k).$$

The final temperature is responsible for optimization in the final stages of the algorithm. It should be small enough for the algorithm to accept only better solutions. However, too fast process of achieving such a state will result in the situation when an algorithm gets stuck in the local minimum and further optimization will be stopped. Due to the fact that for each problem instance, distance of the route is always an integer, the smallest possible solution improvement in one move is 1. This constant smallest difference in the length of two routes is the reason to adopt a fixed final temperature:

$$T_{end} = 2.$$

This temperature value means that the acceptance of the worse solution by 1 will be done with probability $P(\Delta = 1, T = 2) = e^{-1/2} \approx 0.6$

Geometric scheme was chosen out of different cooling schemes where the temperature geometrically decreases from T_{begin} to T_{end}. The algorithm has 100 steps and in one step algorithm performs 1% of scheduled iterations to be executed in this phase. The temperature parameter is changed after completing each step of the algorithm according to the formula:

$$T := T \cdot \alpha.$$

where $\alpha = (T_{begin}/T_{end})^{1/99}$. Value of parameter α is designated in the beginning of phase 1.

Then, during phase 1 of the algorithm, the initial and final temperature are tuned for the second phase. This is done in a way as described below. After each step of the phase 1 of the algorithm, the γ parameter is designated defining the relation of the number of accepted deteriorating solutions to the number of all deteriorating solutions. If the $\gamma > 0.2$ value is modified, the initial temperature is changed to the current temperature value; $T_{begin} = T$. If the $\gamma > 0.01$ value is modified, the final temperature is modified analogously. In this way, in the second phase in the initial steps, the algorithm accepts about 20% of worse solutions, whereas in the last steps only about 1% of these solutions are accepted.

3   Numerical Studies

Numerical studies were carried out on a MacBook Pro computer with a dual-core 2.5 GHz Intel i5 processor in OSX 10.9.5. The algorithm was written in C++ language and compiled with options optimizing efficiency.

3.1   Test Instances

In the [15] literature, instances of the TSP problem are available, in which data are the locations of cities on the plane and the distance is calculated according to the formula:

$$d_{i,j} = \left\lfloor \sqrt{|x_1 - y_1|^2 + |x_2 - y_2|^2} + \tfrac{1}{2} \right\rfloor. \tag{4}$$

The distance $d_{i,j}$ between the cities i, j can take values from a set of total non-negative numbers $d_{i,j} \in \mathbb{N} \cup \{0\}$ for $i,j \in C$. Note that the use of a rounding operator in determining the distance between cities results in the fact that the both the triangle condition and the identicality condition cease to apply. The commonly used in literature distance definition is therefore not a metric.

The test instance set contains all instances available in the TSPLIB [15] online library in which distance metric is defined by the formula 4 and additionally an instance of *monalisa* [16] with size of $n = 10^5$.

3.2   Relative Error of the Algorithm

For each instance, there are $tour^{ref}$ reference solutions given. These are the best known solutions. For all instances except the monalisa instance their optimality has been proven.

During tests relative error δ of the tested *tour* solution was computed in relation to reference $tour^*$ solution in a following manner:

$$\delta(tour) = D(tour)/D(tour^{ref}) \cdot 100\%. \tag{5}$$

The quality of the provided solutions was examined for the SA algorithm being run with a different number of set iterations. The first 'fast' version with an iteration number of 10^5 and average operation time for a single run at the level of 0.1s. The 'exact' version with an iteration number of 10^7 and average operation time for a single start at the level of 2.5 s. An intermediate version of the algorithm was also tested with an iteration number of 10^6.

Since the algorithm is non-deterministic and with every run can deliver different solutions, for each problem instance it was run 10 times. On the basis of the formula 5, 10 relative error values were determined and the total operating time of all runs was measured. On this basis, for each instance, for each algorithm there were determined:

- δ_{min} the smallest value of relative error,
- δ_{av} mean value of relative error,
- t_{sum} total operating time of all runs for an algorithm.

Table 1. Minimum and mean error of an algorithm and its computational time.

Instances	$SA\,iter = 10^5$			$SA\,iter = 10^6$			$SA\,iter = 10^7$		
	$\delta_{av}[\%]$	$\delta_{min}[\%]$	$t_{sum}[s]$	$\delta_{av}[\%]$	$\delta_{min}[\%]$	$t_{sum}[s]$	$\delta_{av}[\%]$	$\delta_{min}[\%]$	$t_{sum}[s]$
eil51	0.047	0	0.4	0	0	2.8	0	0	26.8
berlin52	0	0	0.4	0	0	2.6	0	0	25.2
st70	0.148	0	0.4	0	0	2.8	0	0	26.7
eil76	0.223	0	0.4	0	0	2.9	0	0	28.0
pr76	0.335	0	0.3	0	0	2.4	0	0	22.4
rat99	0.091	0	0.4	0.008	0	2.5	0	0	24.1
kroA100	0.295	0	0.4	0.044	0	2.4	0	0	22.6
kroB100	0.698	0	0.3	0.053	0	2.4	0	0	22.5
kroC100	0.797	0	0.3	0.030	0	2.4	0	0	22.4
kroD100	0.783	0.113	0.4	0.021	0	2.4	0	0	23.2
kroE100	0.467	0.172	0.3	0.228	0	2.5	0.112	0	23.0
rd100	0.851	0	0.4	0.061	0	2.5	0.023	0	24.2
eil101	0.064	0	0.4	0	0	3.1	0	0	30.2
lin105	0.204	0	0.4	0	0	2.5	0	0	23.7
pr107	0.081	0	0.4	0	0	2.5	0	0	25.4
pr124	0.338	0	0.3	0	0	2.3	0	0	21.7
bier127	0.645	0.026	0.4	0.182	0	2.6	0.016	0	25.1
ch130	1.322	0.164	0.4	0.293	0	2.6	0.049	0	24.3
pr136	1.044	0.101	0.4	0.158	0.033	2.3	0.058	0	21.6
pr144	1.284	0.091	0.4	0.122	0	2.4	0.007	0	22.4
ch150	0.426	0	0.4	0.348	0	2.4	0.095	0	23.1
kroA150	0.848	0.132	0.4	0.316	0.004	2.4	0.032	0	22.6
kroB150	0.817	0.065	0.4	0.111	0.023	2.4	0.034	0.008	22.7
pr152	0.817	0.185	0.3	0.177	0	2.3	0.108	0	22.0
u159	0.943	0.751	0.4	0.225	0	2.4	0.026	0	22.5
rat195	1.025	0.646	0.4	0.633	0.301	2.4	0.370	0	22.7
d198	0.210	0.089	0.4	0.125	0	2.8	0.022	0	26.5
kroA200	1.334	0.596	0.4	0.515	0.197	2.4	0.202	0	22.7
kroB200	1.306	0.333	0.4	0.227	0.007	2.4	0.017	0	22.7
ts225	0.956	0.136	0.4	0.087	0	2.3	0.035	0	21.2
tsp225	1.811	0	0.4	1.239	0.434	2.6	0.271	0	24.4
pr226	1.297	0.164	0.4	0.167	0	2.4	0.079	0	23.0
pr264	0.793	0	0.4	0.198	0	2.3	0	0	21.0
a280	1.039	0	0.4	0.636	0	2.5	0.019	0	23.7
pr299	1.408	0.299	0.4	0.172	0	2.4	0.121	0	22.6

(continued)

Table 1. *(continued)*

Instances	$SA\,iter = 10^5$			$SA\,iter = 10^6$			$SA\,iter = 10^7$		
	$\delta_{av}[\%]$	$\delta_{min}[\%]$	$t_{sum}[s]$	$\delta_{av}[\%]$	$\delta_{min}[\%]$	$t_{sum}[s]$	$\delta_{av}[\%]$	$\delta_{min}[\%]$	$t_{sum}[s]$
lin318	2.056	1.627	0.4	1.362	0.340	2.5	0.440	0.050	23.7
rd400	1.998	1.302	0.4	0.942	0.517	2.5	0.641	0.373	23.8
fl417	2.138	0.194	0.5	1.671	0.784	2.8	0.927	0.776	26.1
pr439	2.190	0.464	0.4	1.103	0.416	2.4	0.639	0.198	22.5
pcb442	1.939	0.957	0.4	0.886	0.331	2.7	0.427	0.114	25.2
d493	1.930	1.226	0.5	0.935	0.529	2.9	0.714	0.246	27.1
u574	2.610	1.607	0.4	1.676	1.070	2.6	0.975	0.534	23.7
rat575	2.339	1.432	0.5	1.543	0.856	2.7	0.905	0.576	24.8
p654	3.178	1.267	0.5	2.193	0.398	3.3	1.944	0.029	27.3
d657	2.571	1.452	0.5	1.211	0.940	2.6	0.817	0.570	23.7
u724	2.786	2.124	0.5	1.307	0.954	2.6	0.848	0.644	24.0
rat783	2.671	1.681	0.5	1.632	1.079	2.7	0.883	0.341	24.8
pr1002	3.884	2.886	0.6	1.933	1.402	2.6	1.331	0.924	23.5
u1060	3.453	2.689	0.6	1.776	1.275	2.8	0.743	0.327	24.6
vm1084	3.960	2.582	0.5	1.920	1.444	2.5	1.208	0.783	22.1
pcb1173	3.716	2.440	0.6	2.362	1.970	2.6	1.199	0.870	23.2
d1291	6.073	3.850	0.6	2.203	1.140	2.5	0.909	0.396	22.0
rl1304	6.360	4.309	0.6	2.964	2.040	2.4	0.787	0.204	21.2
rl1323	4.358	2.764	0.6	1.910	1.043	2.4	0.903	0.359	21.3
nrw1379	2.972	2.376	0.7	1.477	1.248	2.8	0.916	0.731	24.3
fl1400	5.416	2.767	0.8	2.536	0.576	3.5	1.265	0.422	30.7
u1432	3.180	2.191	0.8	1.314	0.894	3.2	0.754	0.478	28.1
fl1577	10.183	5.578	0.7	3.460	1.362	2.7	0.901	0.103	23.4
d1655	4.993	2.979	0.7	2.203	0.515	2.7	1.340	0.621	24.0
vm1748	5.163	3.894	0.7	2.239	1.156	2.7	1.289	0.908	22.8
u1817	4.670	3.631	0.7	2.470	1.830	2.8	1.008	0.848	24.5
rl1889	4.705	3.722	0.7	3.143	2.431	2.6	1.355	0.994	22.1
d2103	6.384	4.629	0.8	1.996	0.924	2.7	0.706	0.365	21.8
u2152	5.098	3.885	0.8	2.237	1.846	2.9	1.212	0.794	25.2
u2319	1.515	1.248	1.0	0.952	0.830	4.0	0.621	0.532	34.3
pr2392	5.357	4.430	0.8	2.585	2.049	2.9	1.564	1.181	23.9
pcb3038	5.057	4.164	1.0	2.457	2.108	3.1	1.436	1.224	25.4
fl3795	6.767	5.328	1.2	5.274	2.777	3.5	2.016	1.227	26.6
fnl4461	4.849	4.105	1.4	2.195	2.071	3.6	1.454	1.267	28.2
rl5915	7.001	5.497	1.8	3.662	3.387	3.6	1.980	1.340	23.9

(continued)

Table 1. *(continued)*

Instances	$SA\,iter = 10^5$			$SA\,iter = 10^6$			$SA\,iter = 10^7$		
	$\delta_{av}[\%]$	$\delta_{min}[\%]$	$t_{sum}[s]$	$\delta_{av}[\%]$	$\delta_{min}[\%]$	$t_{sum}[s]$	$\delta_{av}[\%]$	$\delta_{min}[\%]$	$t_{sum}[s]$
rl5934	7.617	6.471	1.8	4.329	3.861	3.6	2.242	1.725	23.8
rl11849	8.852	7.348	3.5	4.619	3.663	5.5	2.453	2.119	29.7
usa13509	9.418	8.378	4.0	4.382	4.220	6.5	2.279	1.985	30.0
brd14051	8.481	7.006	4.1	3.509	3.370	6.7	1.959	1.758	35.2
d15112	8.055	7.055	4.3	3.612	3.445	7.1	1.955	1.724	37.3
d18512	9.531	7.942	5.3	3.468	3.317	8.3	1.996	1.833	38.4
monalisa	12.493	11.954	34.8	3.948	3.616	60.7	1.531	1.511	217.6
all	3.022	2.045	95.0	1.376	0.922	284.8	0.691	0.442	2104.9

The results of the tests have been placed in the Table 1. The accuracy of the reported error is of 0.001%, whereas an error value of exactly zero is marked with symbol 0.

3.3   Start Solution

In the case of small instances, the impact of the starting solution is practically irrelevant. Even when starting from a random solution, the algorithm very quickly achieves the quality of solutions obtained with the use of construction algorithms. The situation changes in case of large instances. The SA algorithm then needs a lot of iterations to get quality solutions, e.g. the solution obtained by the NN algorithm. The results of errors and the operation time of the SA algorithm for different methods of generating a starting solution are presented in the Table 2. In addition, there is an increase in time of the SA algorithm for large instances starting from a random solution. This is due to the fact that the

Table 2. Minimum and mean error of an algorithm and its operating time of SA algorithm for different starting solutions and different sizes of instances

Algorithm	$n < 300$			$300 < n < 1000$			$n > 10000$		
	$\delta_{av}[\%]$	$\delta_{min}[\%]$	$t_{sum}[s]$	$\delta_{av}[\%]$	$\delta_{min}[\%]$	$t_{sum}[s]$	$\delta_{av}[\%]$	$\delta_{min}[\%]$	$t_{sum}[s]$
R	888	831	5.0	3668	3595	4.9	10575	10520	4.8
NN	24.79	18.55	5.0	24.85	22.60	5.4	23.18	22.38	7.1
R + SA(10^5)	0.77	0.10	13.9	6.15	4.14	29.7	582.58	575.66	183.4
NN + SA(10^5)	0.71	0.12	13.1	4.20	2.88	26.0	9.47	8.28	55.9
R + SA(10^6)	0.19	0.03	90.8	2.16	1.30	108.1	16.04	15.11	456.9
NN + SA(10^6)	0.18	0.03	87.0	2.11	1.34	102.9	3.92	3.61	94.9
R + SA(10^7)	0.05	0.01	862.4	1.14	0.64	904.8	2.07	1.87	847.7
NN + SA(10^7)	0.05	0.00	829.1	1.09	0.64	887.5	2.03	1.82	388.2

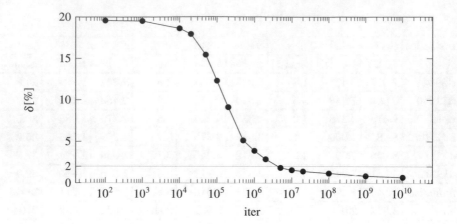

Fig. 1. δ error of SA algorithm depending on the number of iterations, *monalisa* instance

algorithm for the weak current solution more often generates accepted moves than in case of solutions close to the local minimum. Iteration with accepted move takes longer than when the move is rejected.

3.4   Study of the Convergence of the Algorithm

By the convergence of the algorithm, we mean generating solutions with smaller error, increasing the number of iterations to be done. For the largest instances, the δ error of the solutions obtained with the SA algorithm run with $iter = 10^7$ is at 2%. The current study is to check whether further increase in the number of iterations will continue to result in a significant improvement in the quality of solutions. Because this problem mainly concerns great instances, the study was made on the largest of them, the *monalisa* instance. The obtained solution error for the number of iterations from $iter = 10$ do $iter = 10^{10}$ was presented in the Fig. 1. From the analysis of the graph it follows that for iterations to the number $iter = 10^7$ the δ error quickly decreases to around 2%. With further increase of iteration, though not so fast, the δ error continues to decrease. For the largest number of iterations tested $iter = 10^{10}$ the algorithm's operating time was 6 h, and the obtained solution had a $\delta = 0.7\%$ error.

3.5   Comparison with Literature Algorithms

The following study aims at comparing the presented algorithm with the algorithms known from the literature. The most frequently tested examples is a set of 5 instances $KroA,...,KroE$ of the size of $n = 100$. In the Table 3 there are presented errors of the literature solutions of algorithms. The presented SA algorithm for each of the listed instances was run ten times with the number of iterations $iter = 10^6$. For each of the tested instances, among ten solutions there was an optimal solution found. The total time of all 10 runs for each instance is about 2.5 s.

Table 3. Relative errors of literature algorithms

Algorithm	KroA $\delta[\%]$	KroB $\delta[\%]$	KroC $\delta[\%]$	KroD $\delta[\%]$	KroE $\delta[\%]$
Presented SA ($10 \times iter = 10^6$, $t = 2.5$ s)	0	0	0	0	0
Meta-RaPS [18]	0	0.25	0	0	0.17
Christofides & 2opt, 3opt [3]	2.51	1.40	1.53	0.17	3.03
Convex Hull & 3opt [3]	0.37	1.46	1.06	0.04	2.46
Farthest Insertion & 3opt [3]	0.46	1.57	0.49	0.45	0.98
Nearest Neighbor & 2opt, 3opt [3]	0.14	1.46	1.06	0.73	2.46
LK (by Padberg and Rinaldi) [17]	0.26	0	0.70	0.17	0.16
mod LK(by Mak and Morton) [6]	0	0.17	0	0	0.21
GA by Chatterjee et al., Exp A [9]	0.70	N/A	1.78	1.45	N/A
Neural Net by Matsuyama [7]	1.81	3.37	N/A	N/A	N/A
Neural Net by Modares et al. [8]	0.31	1.43	N/A	N/A	N/A
Simulated Annealing by Malek et al. [12]	0.09	N/A	N/A	N/A	N/A
Tabu Search by Malek et al. [12]	0.33	N/A	N/A	N/A	N/A
Tabu Search & 3opt by Tsubakitani et al. [11]	1.37	N/A	N/A	N/A	N/A

Currently, the best algorithms are based on the idea of the Lin-Kernighan descending algorithm [4], for example the Chained Lin-Kernighan algorithm [19]. The above algorithm solves optimally in a reasonable time all instances of TSPLIB of size $n < 1000$. Unfortunately, the presented SA algorithm optimally resolves all instances from TSPLIB of size $n < 1000$. For the largest instances, the SA algorithm also turns out to be weaker than the Chained Lin-Kernighan version.

In addition, a package dedicated to the TSP problem called Concord by W. Cook, containing the described algorithm and other methods, including precise methods, optimally solved, with the guarantee of optimality, all of the instances discussed in this paper (except monalisa instance). For large $n > 10.000$ instances, the computation time on a 2.4 GHz clocked computer was several years for one instance. For the largest of them D18512, the best solution is presented in the paper [20] and proving its optimality with the use of the Concorde package took 57.5 CPU $\times$ years [19].

4   Summary

The SA algorithm presented in the paper is very easy to implement. It is suitable for solving both small instances (it finds quickly optimal solutions) as well as very large instances (it quickly finds solutions with an error of 1%). By slightly changing the parameters of the presented algorithm, it is possible to get an algorithm that deals even better with small or large instances.

The algorithm controlled in the way described in the paper during one minute almost always finds the optimal solution for the instance of up to $n = 300$, and for large instances, in a minute it finds a solution below 2%.

The presented algorithm is better than many other literature approaches (see Table 3). However, it works less efficient than currently best algorithms based on LKH heuristics proposed by S. Lin and B. Kernighan, in particular the Concorde package by W. Cook.

References

1. Cook, W.: In Pursuit of the Traveling Salesman Mathematics at the Limits of Computation. Princeton University Press, Princeton (2014)
2. Reinelt, G.: The Traveling Salesman: Computational Solutions for TSP Applications, Lectures Notes in Computer Science. Springer-Verlag, Berlin, Heidelberg (1994)
3. Lawler, E.L., Lenstra, J.K., Rinnooy Kan, A.H.G., Shmoys, D.B.: The Traveling Salesman Problem. Wiley, New York (1985)
4. Lin, S., Kernighan, B.: An effective heuristic algorithm for the traveling-salesman problem. Oper. Res. **21**(2), 498–516 (1973)
5. Helsgaun, K.: An effective implementation of the Lin-Kernighan traveling salesman heuristic. Eur. J. Oper. Res. **126**(1), 106–130 (2000)
6. Mak, K., Morton, A.: A modified Lin-Kernighan traveling salesman heuristic. ORSA J. Comput. **13**, 127–132 (1992)
7. Matsuyama, Y.: Self-organization neural networks and various Euclidean traveling salesman problems. Syst. Comput. Japan **23**(2), 101–112 (1992)
8. Modares, A., Somhom, S., Enkawa, T.: A self-organizing neural network approach for multiple traveling salesman and vehicle routing problems. Int. Trans. Oper. Res. **6**, 591–606 (1999)
9. Chatterjee, S., Carrera, C., Lynch, L.: Genetic algorithms and traveling salesman problems. Eur. J. Oper. Res. **93**, 490–510 (1996)
10. Grefenstette, J.J., Gopal, R., Rosmaita, B., Van Gucht, D.: Genetic algorithms for the traveling salesman problem. In: Grefenstette, J.J. (ed.) Proceedings of an International Conference on Genetic Algorithms and Their Applications, pp. 160–168. Carnegie-Mellon University (1985)
11. Tsubakitani, S., Evans, J.R.: An empirical study of a new metaheuristic for the traveling salesman problem. Eur. J. Oper. Res. **104**, 113–128 (1998)
12. Malek, M., Guruswamy, M., Pandya, M., Owens, H.: Serial and Parallel simulated annealing and Tabu search algorithms for the traveling salesman problem. Annal. Oper. Res. **21**, 59–84 (1989)
13. Kirkpatrick, S., Gelatt, C.D., Vecchi, M.P.: Optimisation by simulated annealing. Science **220**, 671–680 (1983)
14. Bentley, J.L.: Multidimensional binary search trees used for associative searching. Commun. ACM **18**(9), 509–517 (1975)
15. Reinhelt, G.: Set of TSP instances: TSPLIB. http://comopt.ifi.uni-heidelberg.de/software/TSPLIB95/tsp/
16. Bosch, R., Lisa, M.: TSP Challenge, February 2009. http://www.math.uwaterloo.ca/tsp/data/ml/monalisa.html
17. Padberg, M., Rinaldi, G.: A branch-and-cut algorithm for the solution of large-scale traveling salesman problems. SIAM Rev. **33**, 60–100 (1991)

18. DePuy, G.W., Whitehouse, G.E., Moraga, R.J.: Using the meta-raps approach to solve combinatorial problems. In: Proceedings of the 2002 Industrial Engineering Research Conference, Orlando, 19–21 May 2002
19. Applegate, D., Bixby, R., Chvatal, V., Cook, W.: The Traveling Salesman Problem: A Computational Study. Princeton Series in Applied Mathematics. Princeton University Press, Princeton (2006)
20. Tamaki, H.: Alternating cycle contribution: A tour-merging strategy for the travelling salesman problem, Max-Planck Institute Research Report, MPI-I-2003-1-007, Saarbucken, Germany (2003)

System for Child Speech Problems Identification

Jacek Mazurkiewicz[(⊠)]

Department of Computer Engineering, Faculty of Electronics,
Wroclaw University of Science and Technology,
ul. Wybrzeze Wyspianskiego 27, 50-370 Wroclaw, Poland
Jacek.Mazurkiewicz@pwr.edu.pl

Abstract. The main aim of this work is to develop and test the system that would automatically assess correctness of articulation of <r> consonant in child speech. The system should contain automatic speech recognition module and classification module. As there are many solutions connected with speech analysis with heavy load of algorithms based on complicated mathematical and statistical operations, motivation was to exchange them with simpler, but quite powerful softcomputing methods. The major assumption is to develop system, that would fulfill following requirements: speaker independence, recording device independence, high performance - immediate response, high accuracy. MLP used as classifier recognized properly more than 89% of sound probes. The systems is created for Polish language.

Keywords: MLP · Speech sound disorder · NLP · Consonant
Softcomputing

1 Introduction

Child speech problems are very common and quite reasonable, as it happens very rare that when child learns new words, they sounds properly from the very beginning. Some children develop speech sounds over time, but there are many of them which need help of speech therapist to make their articulation proper. It is important to cure speech disorders, as they have influence on further child's life. Untreated disorder may cause stress, which can transform even into aggressive behavior. It has also impact on child's self-confidence - he/she may become shy or stand-offish [1].

At the moment, if the parents want to diagnose their children, they have to visit a specialist in order to recognize type of disorder. There is no way to perform a test on their own to make such diagnose. There is a need to create a system that would aid parents with taking decision whether their child needs help with developing his/her speaking skills. Actually there are many systems available that provide auxiliary exercises, which may be applied during therapy. There is a group of them which speech therapist can use during the meetings with child. There are also some exercises available online in the Internet, which can be used by anyone, but they require a subscription. There is possibility to use exercises that help during therapy, however they require the knowledge about the type of child's disorder, so anyway the visit at

© Springer International Publishing AG, part of Springer Nature 2019
W. Zamojski et al. (Eds.): DepCoS-RELCOMEX 2018, AISC 761, pp. 360–372, 2019.
https://doi.org/10.1007/978-3-319-91446-6_34

specialist is necessary. Even if parents intend to perform the therapy at home, there is a need to know which exercises are suitable to the child's age. Many of online provided exercises do not have a description when they should be used. Using incorrect exercise (in mean of child's age or disorder) may become more harm than cure. There is a need to provide a solution to inform the parents whether the disorder exists or not, in reference to natural speech development process [8].

The main aim of this work is to develop and test the system that would automatically assess correctness of articulation of <r> consonant in child speech. The system should contain automatic speech recognition module and classification module. As there are many solutions connected with speech analysis with heavy load of algorithms based on complicated mathematical and statistical operations, motivation was to exchange them with simpler, but quite powerful softcomputing methods. The major assumption is to develop system, that would fulfill following requirements: speaker independence, recording device independence, high performance - immediate response, high accuracy.

Speaker independence requirement is obvious, as it is impossible to predict in advance whose speech will be tested. Moreover as the child might have an articulation disorder of <r> consonant there will be no possibility to train the system with both correct and incorrect samples of this particular child's voice [6]. System should be independent from recording device, as the motivation is that anyone should be able to use the program in usual environment, i.e. by family at home. On the other hand it would be also an advantage that system do not require special devices and conditions to effectively use it. Each user should be able to record sample on his/her personal computer using simple microphone connected to it. It would be also appreciate to get response immediately, as it would be used during an interview with child by speech therapist or parents as real-time application.

2 Proposed Approach

To recognize if there is a disorder in an articulation of <r> following steps have to be performed: load prepared recordings, preprocessing - cut off silence part - preemphasis, feature extraction, create features vector, put features vector as input for Artificial Neural Network - classifier, results [5, 9].

The solution will be adapted to disorder in articulation of <r> occurring in Polish language. Selected disorder has several varieties, however in most cases it is connected with substitute <r> consonant with <l> consonant. System will be prepared to differ this two consonants from each other. Speech therapist called such ability "phonemic hearing" [1, 8]. System will return an answer for question "Is the articulation of <r> proper or is it substituted by <l> consonant". There are two possible answers, which system may give - "articulation of <r> is proper" or "articulation of <r> is not proper and <r> is substituted by <l>". As an articulation of <r> should be developed between 4–5 year of life, the selected age group of children to record samples will be from 4 years old to 10 years old [3]. Although children should develop <r> sound at the latest to 5 year of life (if therapy was required it may be a little later), the system have also to be trained with consonants, which are said properly. The analyzed samples of speech has to be restricted to properly adjust parametrization. On the other hand it would be advantage to record such samples, which can be obtained during speech therapist diagnosis a child. Having this recommendations it approaches that recording separate words will be the best choice. Speech therapists uses picture questionnaire, which means they show picture to a child and he/she has to answer what is on the picture (usually it is single thing). The pictures are arranged in way that gives possibility to observe eventual disorders [1, 4, 8]. It is assumed that analyzed recordings will contain separate words which verify an articulation of <r>. The recordings will be gather in home-like environment on simple devices, available for everyone at reasonable price. The analyzed unit will not be the whole word, but crucial phoneme, which is <r>. This approach decreases load of analyzed information and do not require to learn neural network patterns of whole words, but patterns of single phonemes. It also gives possibility to easily add new patterns to learn - there is a limited number of them in Polish language. Number of words in every language is much greater than number of phonemes.

3   Consonants

According to articulatory phonetics, a consonant is a speech sound that is articulated with complete or partial closure of vocal tract. It is kind of phone that contains rustle elements and is opposite to vowel, which is a pure tone. Each spoken consonant can be distinguished by several phonetic features: the manner of articulation, the place of articulation, the phonation, the airstream mechanisms. The manner of articulation is a configuration and interaction on the articulators (i.e. tongue, lips and palate) and it describes how air escapes from vocal tract when the consonant sound is made. The place of articulation is the point of contact where an obstruction occurs in the vocal tract between an articulatory gesture - an active articulator (typically some part of the tongue) and a passive location (typically some part of the roof of the mouth). In table below there are examples of each articulator type. There are many potential combinations of them that give possibility to make particular sound [7, 9].

Selected consonants differs from each other in manner of articulation and airstream direction. Consonant <r> is trill, which means it is produced by directing air over the articulator so that it vibrates. Consonant <l> is approximant, which means that distance between tongue and palate is not enough to produce turbulent airstream. Place of articulation is for both the same – they are articulated with either the tip or the blade of the tongue at alveolar ridge. Both are voiced and oral consonants - the vocal cords vibrate during the articulation and air is allowed to escape through the mouth only. Next difference is airstream direction – alveolar trill is produced by directing the airstream along the center of the tongue, that is why it is central. Consonant <l> is lateral, which means that airstream is directed over the side of the tongue, rather than down the middle.

It is easy to understand that <r> consonant may cause some problem for a child to pronounce it. It require quite complicated work from tongue and specific tension of mouth muscles. Recognition of this sound disorder in proper stage of child development is crucial - it should be analyzed from 5 year of life – not earlier [3]. It is natural developmental process that child exchanges <r> consonant with <l> or <j> during his/her early years and trying to repair this may cause only unnecessary stress. It is not valuable to put such pressure on child until it reaches adequate age. There are many possible reasons from which speech disorders may results. There are authors that divide the speech problems reasons into two basic groups: extrinsic (also called environmental) and intrinsic [1, 8]. Extrinsic (environmental) reasons:

- incorrect speech models used by the people from the closest surrounding of the child (also include too infantile language),
- to rare engaging the child into conversations, dialogue and too poor stock of vocabulary, which he/she encounters,
- incorrect models, educational methods and related therewith so called parental attitudes that may adversely influence development of speech, i.e. overprotective attitude.

Intrinsic reasons:

- physical anomalies in the anatomical structure of speech organs and/or incorrect way of their functioning, i.e. occlusion defect, cleft palate,
- damage of the brain structures that are responsible for the speech transmission and its reception,
- hearing loss, genetic syndromes, i.e. Down syndrome,
- incorrect course of the emotional processes which may lead to serious disorders, i.e. stuttering or mutism,
- developmental disorders, i.e. autism.

4   Neural Networks for Speech Disorders

There are several scientific works connected with speech disorders. However, most of them do not deal with articulation difficulties, but with speech disorders that are connected with neurological issues, such as dysarthria [5] and apraxia [4] of speech. Considered problems with speech occurs when some parts of brain are damaged. There is very poor research in analysis of speech problems for Polish language. There were some tries, but they also concerned speech difficulties connected with neurological issues [8]. Unfortunately, from point of view of the work, mentioned research was not continued toward disorders direction. Polish scientists are more interested in such applications as speaker recognition or continuous speech translation into written text on computer than speech disorders. One of the most familiar projects concerning neural networks in speech analysis is yearbook published by the Polish Phonetic Association for almost 40 years [8]. Although there are few reports about usage of neural network in speech recognition, they do not deal with speech disorders, but speaker recognition and speech synthesis. There is some discussion about programs connected with speech disorders [1]. It is a fresh disputation and its result is not published yet. Besides neural networks, data mining techniques [6], such as Primary Component Analysis [7] are also popular in speech disorder recognition.

5   System

As stated in previous sections, there is a number of steps that need to be fulfilled in order to successful recognition of speech disorder in articulation of <r>. Figure 1 presents the sequence of those steps in the implemented solution.

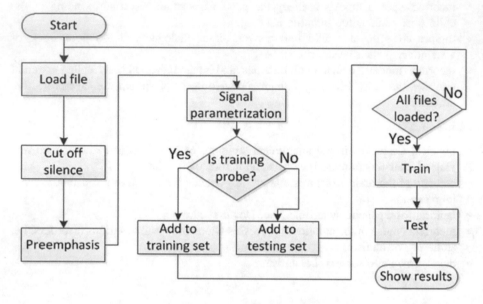

Fig. 1. Flowchart of system.

5.1   Samples

The recorded samples of child's speech pronunciation are separates words, which verify an articulation of <r>. Following words were selected from Polish language: "rower", "rak", "ryba". This words are simple and known by children. There are not many of them, but it is also an advantage, because children in selected age group do not concentrate long on one activity [8]. On the other hand opportunity to use microphone and computer makes children treat gathering sample recordings as fun and enjoyment, not a survey.

Phoneme Detection. In constrained it was stressed that analyzed speech signal will not be a whole word, but separate phoneme. Unfortunately, currently available algorithms which divide words into phonemes are vitiated by an error. Especially <r> and <l> consonants are not easy to analyze in reference to their features. That is why phonemes were cut off by hand using Audacity [2]. Research in area of separating speech into phonemes is still in progress. There are algorithms that are quite successful in separating vowels from continuous speech [9], but different works are applied to different languages. Each phoneme can have many physical realizations. There is a need to recognize separated phonemes in many languages, but because of that "physical realization" there are proposed variety of solutions for numerous of languages. The main idea may be similar, but phonemes in different language may have different physical features, so they have to be analyzed in some other way. What is also crucial from point of view of current thesis, many works are recognizing vowels only(which are the easiest case), not consonants that are much harder to cut them off from word. As the HMMs are still quite popular, there are more and more works stating that ANNs, if properly developed, may be more accurate than HMMs in separating phonemes also [9]. There are only few works that are quite successful in separating all phonemes given in particular language. The teams, which find out accurate solution (about 98% of accuracy [6]), consist of the group of scientist (three people or more) and their research is usually externally financed. There is no ready to use solution for Polish language and the subject of phoneme recognition is obviously too wide to cover it in current thesis. However, using such automation would give a real boost to applications connected with speech disorder recognition.

Features. Recorded samples contains both proper and improper pronunciation of <r> consonant. Separated by hand phonemes have the following features: duration from 100 ms to 220 ms, three possible sampling frequencies: 22 500 Hz, 44 000 Hz or 48 000 Hz. Duration may differ because of signal parametrization circumstances. Moreover, it may be affected by speaker individual habits or emotions while speaking if somebody is nervous, he/she may speak quietly or very quickly. On the other hand, as <l> and <r> are liquid consonants, they have slightly different shape and length when adjacent vowel changes. Samples were recorded on three different devices: dictaphone with sampling frequency 22 500 Hz, headphones with built-in microphone, frequency set in software (Audacity [2]), simple microphone (ambient), frequency set in software (Audacity [2]). Sampling frequency of dictaphone is fixed, but those samples which were recorded with computer usage have two possible frequencies, which were set in software.

5.2 Preprocessing

There are two steps done as preprocessing part: cut off silence part, preemphasis filter usage. Silence detection algorithm is based on volume threshold [9]. As far as the signal energy does not exceeds the threshold, it is assumed as silence part. Necessary steps to properly cut off silence parts: set length of analyzed frame, count number of frames in signal, check if there is a silence at beginning of signal – (a) calculate frame's energy – (b) if energy exceeds threshold, break and go to (d), else go to (c) – (c) get next frame from signal, go to (a) – (d) store number of frame, in which proper signal starts, check if there is a silence at the end of signal - do similar steps as (a) – (d), but in the opposite direction, only the part that lies between remembered number of frames is the proper signal. Cut off other parts. The preemphasis is done in classical way.

5.3 Feature Extraction

There are many parameters that can be obtained from signal analysis, cepstral coefficients are the most suitable, as they contain speaker independent features of speech. Observation of the shape of speech signal both for <r> and <l> consonants allow to notice that there are also two parameters that may be useful in analysis of this signal - autocorrelation and temporal centroid. They refer to envelope of signal, which differs between <r> and <l>. But if the single consonant is considered, the envelope does not change much for various of speakers (Figs. 2 and 3).

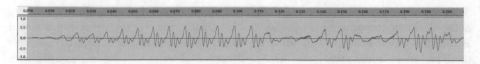

Fig. 2. Shape of <r> consonant signal (Audacity [2])

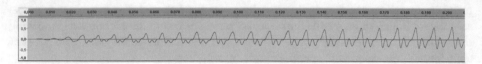

Fig. 3. Shape of <l> consonant signal (Audacity [2])

5.4   Disorder Recognition

The module, which realizes disorder recognition is MLP. Its input is feature vector that contains cepstral coefficients, autocorrelation, temporal centroid or some combination of them. It has binary output, which answers mean: 1 - articulation of <r> is proper, 0 - articulation of <r> is not proper and <r> is substituted by <l>. There are 12 cepstral coefficients calculated for each frame. As the probes may have different duration time

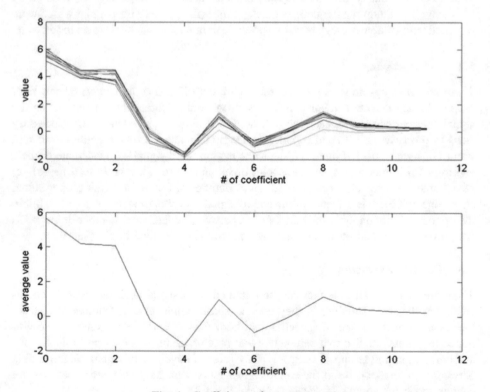

Fig. 4. Coefficients of <r> consonant

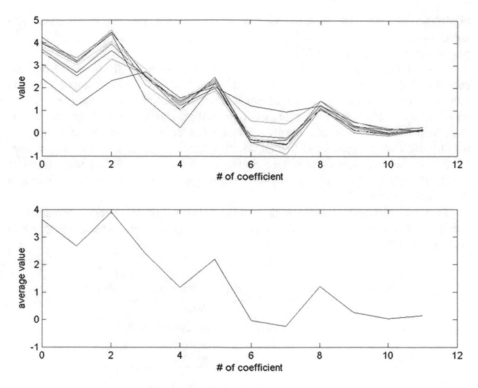

Fig. 5. Coefficients of <l> consonant

(from 100 ms to 220 ms) it is obvious, that they contain different number of frames to analyze. To unify the features vector for every consonant, independently of their duration time, not all cepstral coefficients are put into, but the mean value for each of them. It is fully justified as mean value is very similar to every of the coefficients, which is presented in Fig. 4 and in Fig. 5.

Each color on upper graph means cepstral coefficients of single frame. Number of frames depends on the length of recorded consonant. Lower graph presents mean value for every of the coefficients.

6 Results

6.1 Tests

Before it can be proceeded to disorder recognition, vector of features has to be prepared. It contains parameterized description of sample. In previous section, three parameters were pointed out: mean value of cepstral coefficients (vector, 12 values), maximum autocorrelation (scalar), temporal centroid (scalar). Mean values of cepstral coefficients are essential parameters. Two other are "helpers", which may improve disorder recognition. To test the utility of selected parameters, following combinations of them will be tested: cepstral coefficients only (CC), cepstral coefficients with autocorrelation (CC+A), cepstral coefficients with temporal centroid (CC+TC), all of possible parameters (CC+A+TC).

Each of the variants affects the input layer of MLP - the more parameters are considered, the more input neurons In it has (from 12 to 14 values). Output layer has only one neuron as the answer has binary nature. Number of hidden neurons is based on arithmetic mean value of number of input and number of output neurons. In this case, there will be 7 hidden neurons. Although this neural network is quite small, it may be still tested if number of hidden neurons affects its accuracy. For this purpose, MLP with different number of hidden neurons will be also used. Important thing of speech recognition in current work is to build system independent from recording device and speaker.

To check device independence the training set contains only samples recorded on ambient microphone and testing set contains only samples recorded on dictaphone and headphones with microphone (Table 1).

Table 1. Samples collection $Test_1$, for device independence test.

Features	Training set	Testing set
Recording device	Ambient microphone	Headphones with microphone Dictaphone
Quantity	21	39

Sampling frequency is one of the features that may vary on different recordings. It will be also tested if samples recorded with different frequencies can be recognized by system. To check it, training set contains only sound samples with frequency 44 000 Hz and testing set contains samples with other frequencies (Table 2).

Table 2. Samples collection $Test_2$, for sampling frequency independence test.

Features	Training set	Testing set
Frequency [Hz]	44 000	48 000 22 500
Quantity	21	39

To check speaker independence it is preserved that all samples that were recorded from one person are put into only one set of data (training or testing) and cannot appear in second set. Sound probes was taken from 20 people. As each one of them said 3 selected words, there are 60 sound samples. The training set contains 21 recording from first 7 people and testing set contains the rest of samples. To perform tests uniformly, all of the training sets contain 21 sound probes (Table 3).

Table 3. Samples collection *Test₃*, for speaker independence test.

Features	Training set	Testing set
Group of people	1–7	8–20
Quantity	21	39

Such division of data is a try to answer few questions:
Is it possible to use different devices, even if training set contains only samples from one device? Is it possible to analyze samples with different sampling frequency than in training set? Is it possible to properly recognize what was said, even if analyzed speaker was not known while training? To summarize, tests cover fields listed below: speaker independence, recording device independence, parameters utility, network topology.

6.2   Observations

Device Independence. Results of *Test₁* are presented in Table 4 below:

Table 4. Results of *Test₁*.

Parameters	Accuracy [%]		
Number of hidden neurons	4	7	10
CC	76.92	74.35	76.92
CC+A	74.35	76.92	87.17
CC+TC	74.35	79.48	74.35
CC+A+TC	76.92	79.48	82.05

Although system was trained by samples from single recording device, which was ambient microphone, it has high accuracy (about 87%) in recognizing samples recorded on other devices, such as dictaphone and headphones with built-in microphone. It is a big advantage, as these appliances are available in every supermarket and they are not an aggravating expense. It is likely that parents are able to buy such device.

Sampling Frequency Independence. Table 5 presents results of *Test₂*. The MLP accuracy is slightly higher than in previous *Test₁*. It appears that solution works quite well with changing frequencies of probes. As the sampling frequency is not obviously predetermined by device and it may be edited programmatically it is desirable score.

Table 5. Results of *Test₂*.

Parameters	Accuracy [%]		
Number of hidden neurons	4	7	10
CC	84.61	84.61	84.61
CC+A	87.17	89.74	89.74
CC+TC	79.48	71.79	71.79
CC+A+TC	82.05	82.05	82.05

Speaker Independence. Speaker independence is the crucial feature of designed system, as it is not possible to determine who will be examined. The neural network cannot be trained in advance. Table 6 presents accuracy of the solution.

Table 6. Results of $Test_2$.

Parameters	Accuracy [%]		
Number of hidden neurons	4	7	10
CC	71.79	84.61	84.61
CC+A	84.61	89.74	89.74
CC+TC	79.48	76.92	79.48
CC+A+TC	82.05	84.61	84.61

Received results shows that MLP recognized samples properly quite well. As in training and testing set there are samples from groups of people that are separable, it shows that solution is able to analyze entirely new sample speech taken from unknown child. It does not matter who is speaking - solution will manage with it.

Network Topology and Utility Parameters. The results (Table 7) are presented from point of view of network topology and parameters selection.

Table 7. Test results – MLP topology and utility parameters.

Parameters	Accuracy [%]		
Number of hidden neurons	4	7	10
CC	71.79	84.61	84.61
CC+A	84.61	89.74	89.74
CC+TC	79.48	76.92	79.48
CC+A+TC	82.05	84.61	84.61

It appears that mean value of cepstral coefficients connected with MLP gives really satisfying results. Adding maximum of signal autocorrelation to features vector slightly improves accuracy. The temporal centroid unexpectedly seems to be useless and makes results even worse. As the mean values of cepstral coefficients are calculated, system is immune to change of sound probes length. Such selection of parameters means size of MLP, which has good influence on its performance. Furthermore, even if neural network is not expanded, it is still successful.

7   Conclusions

Neural networks can be successfully used for speech disorder recognition in Polish language. As it was shown in this work, MLP with very simple structure may be effective in such task. The crucial point is to properly describe object (in this case - speech sample) using parameterization. There are many possible attributes that can be extracted from signal. It is important to select such group of parameters, which contains the most suitable information to resolve selected task. Solution presented in this work has several advantages. First of all, it is based on simple and clear conception. It may be easily expanded, as it requires to learn only new patterns of single phonemes in opposite to systems based on words patterns, where load of information is much bigger. As the mean values of cepstral coefficients concatenated with maximum of the auto-correlation are the only neural network inputs, it is really small and fast. Accuracy is also quite high, as MLP recognized properly more than 89% of sound probes.

On laboratory level this result is satisfactory, but even 89% of accuracy could be still too low to apply such solution into professional software (i.e. certified as medical). It is worth to try improve this result in future work to make solution fully reliable and trustworthy. One more important thing is automatic phoneme detection, but this is a good subject for another research and cannot be covered in current work. However, it could be really useful and convenient to separate phonemes automatically in this kind of program.

Such articulation disorder recognizer may be a part of an expert system. It may be help for parents to make decision, if a child needs a speech therapy. Besides part of speech recognition, it should contain helper questions upon child's speech development to get know in which from described periods he/she is. The system should be able to provide an answer, i.e. "Your child do not pronounce <r> properly, but in his/her age it is normal. He/she do not need a speech therapy yet". It may also contains some hints how to help child to expand his/her speech skills. There are also few possible technical improvements, which may significantly affect solution's effectiveness:

- Add new phonemes to recognize another articulation disorder, i.e. substituting <r> by <j>,
- Test, if another parameters obtained from signal may improve accuracy, i.e. Mel-Frequency Cepstral Coefficients (MFCC) or formants,
- Invent algorithm for automatic phoneme detection and implement it in already done solution.

Automatic detection is especially interesting, as such possibility to extract single phonemes expands widely area of interests. It gives a chance to analyze not only disorders in articulation of <r>, but any articulation problem.

References

1. American Speech-Language-Hearing Association, Speech sound disorders. https://www.asha.org/public/speech/disorders/SSDcauses.htm. Accessed 21 Mar 2018
2. Audacity Team, Audacity, Free, open source, cross-platform software for recording and editing sounds. http://audacity.sourceforge.net/. Accessed 08 Mar 2018

3. Bowen, C.: Ages and stages summary. http://www.speech-language-therapy.com. Accessed 08 Feb 2018
4. Hosom, J.P., Shriberg, L., Green, J.R.: Diagnostic assessment of childhood apraxia of speech using automatic speech recognition (ASR) methods. http://www.ncbi.nlm.nih.gov/pmc/articles/PMC1622919/. Accessed 13 Mar 2018
5. Su, H.-Y., Wu, C.-H., Tsai, P.-J.: Automatic assessment of articulation disorders using confident unit-based model adaptation. http://ieeexplore.ieee.org/xpl/login.jsp?tp=&arnumber=4518659&url=http%3A%2F%2Fieeexplore.ieee.org%2Fxpls%2Fabs_all.jsp%3Farnumber%3D4518659. Accessed 12 Mar 2018
6. Maheswari, N., Kabilan, A.P., Venkatesh, R.: Speaker independent phoneme recognition using neural networks. http://www.jatit.org/volumes/research-papers/Vol6No2/12Vol6No2.pdf. Accessed 20 Feb 2018
7. Matsumasa, H., Takiguchi, T., Ariki, Y.: PCA-based feature extraction for fluctuation in speaking style of articulation disorders. http://www.me.cs.scitec.kobe-u.ac.jp/publications/papers/2007/IS070653.pdf. Accessed 21 Feb 2018
8. Polish Phonetic Association, Speech and language technology, reports and technical notes. http://ptfon.pl/en/slt. Accessed 11 Mar 2018
9. Prica, B., Ilic, S.: Recognition of vowels in continuous speech by using formants. http://facta.junis.ni.ac.rs/eae/fu2k103/11prica.pdf. Accessed 11 Mar 2018

Development of Web Business Applications with the Use of Micro-services

Aneta Poniszewska-Marańda[(⊠)]

Institute of Information Technology,
Lodz University of Technology, Łódź, Poland
aneta.poniszewska-maranda@p.lodz.pl

Abstract. Nowadays, the applications support almost every activity we do, especially in companies to handling almost every business process. As the size of the business grows, the requirements for IT systems are also increasing. The answer to these problems is the use of the micro-service architecture.

The paper presents the evaluation of practical value of the micro-service architecture in relation to the creation of business web applications by designing and implementing an Internet application based on this architecture.

Keywords: Micro-services · Software-oriented architecture
Web business applications

1 Introduction

Currently, the applications support almost every activity we do. This is especially evident in companies where we can find a software dedicated to handling almost every business process that occurs in a company. As the size of the business grows, the requirements for IT systems are also increasing. These systems handle more and more processes, and as a result they are getting bigger and more complex. This causes the problems with maintenance of such software, problems with its upgrading, fixing the errors and the removal of failures, especially if the system is created as a whole in a classic, monolithic approach [1, 2].

The answer to these problems is the use of the micro-service architecture. The micro-service architecture, citing Martin Fowler is "an approach to creating a single application as a collection of small services, each operating in its own process and communicating with others through lightweight mechanisms; these services are built around business goals and are independently implemented by automated mechanisms" [2]. According to this definition, an application based on micro-service architecture consists of number of small, independent applications that need to work together to provide a functionality. The fact that each service can be independently created and implemented demonstrates the benefits of such approach, but also some difficulties that need to be addressed when creating such software, such as communication or the need to isolate the running services.

Micro-services are one of many already existing approaches, which define the software architecture. In this architecture, processes are divided on many individual and

© Springer International Publishing AG, part of Springer Nature 2019
W. Zamojski et al. (Eds.): DepCoS-RELCOMEX 2018, AISC 761, pp. 373–383, 2019.
https://doi.org/10.1007/978-3-319-91446-6_35

rather small components, which communicates with each other, forming one system. This means that, unlike the monolith architecture, all possible changes can be applied faster and easier to the system. The alteration in one of the components can have no influence on other existing elements in the system. Furthermore, development of such architecture model in the application gives an opportunity of usage of the created elements in another project [13].

The paper presents the evaluation of practical value of the micro-service architecture in relation to the creation of business web applications by designing and implementing an Internet application based on this architecture. The created application consists of two parts – server and client – supporting the business process. The server part was created as a set of micro-servers, working together to provide the required functionality. It provides an open API (Application Programming Interface) via HTTP (HyperText Transfer Protocol) to communicate with the client. The client component is a Single Page Application (SPA) web application that serves as the user interface.

The paper is structured as follows: Sect. 2 gives an outline of information about software architecture, SOA (Service-Oriented Architecture) and micro-services. Section 3 describes the solutions for management and monitoring of micro-services. Section 4 presents the created Internet application based on micro-service architecture.

2   Software Architecture

There is no definitive answer to a question what the software architecture is. Definitions that can be found cover a vast area of variance, but tend to orbit around a similar themes. *Software architecture* is a set of rules, which defines the organization of a software system, all its technical as well as operational requirements, information about not only relations between elements in an architecture but also about their interactions. One of the modern definitions [7] considers *software architecture* to be the structure of software elements together with their externally visible properties and relations connecting them [13].

2.1   SOA (Service-Oriented Architecture)

A *service oriented architecture* is based on the communication between services. Such services act as independent components, which cooperate to provide functionality of the application. Communication between services is described by different access protocol like for example REST (*Representational State Transfer*) or SOAP (*Simple Object Access Protocol*) [8, 9].

In most serious SOA implementations it is possible to find common components (Fig. 1), that contribute to the architecture's operation. SOA systems can be usually divided into three parts. First, there is the provider section. There live low-level, implementation-based services, that are more or less fine-grained. They are more often than not unsuitable for external use and they don't provide any meaningful context to business. It is also common to find some forms of service registries and managers [13].

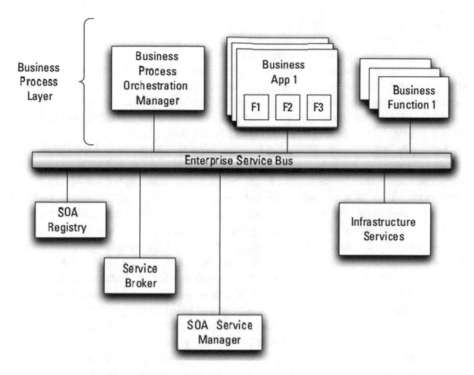

Fig. 1. Common elements of Service-Oriented Architecture [9].

2.2   Micro-Services Architecture

The base idea behind micro services can be described with canonical definition written by James Lewis and Martin Fowler: the *micro-service architectural style* is an approach to developing a single application as a suite of small services, each running in its own process and communicating with lightweight mechanisms, often an HTTP resource API. These services are built around business capabilities and independently deployable by fully automated deployment machinery. There is a bare minimum of centralized management of these services, which may be written in different programming languages and use different data storage technologies [1, 3].

Micro-services architecture causes *decentralization of governance*, which leaves more design possibilities and creates more options for the service teams. Each team might want to take a slightly different approach than others, maybe even use a different technology stack, that is more suited for their service's task. Rather than focusing on standardisation micro-services encourage writing tools, that can be improved over time and shared with other teams [4, 7]. *Decentralization* is a common theme when talking about micro-services. Another area it touches is *data management*. In case of monolith it is common to have a single database storing data for the application. It can go even further when enterprises decide to use a single data store integrating many applications and systems. Micro-services encourage teams to manage their own data storage dedicated for their service. It is worth noting that most of the time it is a good idea to allow services to access exclusively their storage and communicate with others using only their APIs (Application Programming Interfaces).

Another important aspect is *automation*. Advanced systems can consist of many services that need to be built, tested and deployed often. Doing it manually wastes time and is prone to human errors. Development infrastructure should be able to handle automated testing (unit and integration) as well as deployment to different environments. It is useful to implement methods such as *Continuous Deployment* and *Continuous Integration* to ease operational effort in the future [13].

Each micro-service is running in its own process and communicates with other services. This type of application split forces it to run concurrently. This means that many operations can be performed at the same time. This allows the application to work much faster than a single-threaded program. Concurrency is also possible in programs running on one process, but dividing the application into separate processes eliminates the difficulty of implementing such a solution [3, 6].

The greatest advantage of micro-services is their independence in the sense that they can be created separately. The programming language, platform, and development environment, but also the way the data is stored by the application, and even the repository of the generated code, may be different for different services. This is especially true when the creating application is large and many developers are working on it.

Speaking of communication, it is necessary to mention the basic difficulty that micro-services architecture is introduced into. Every part of the application needs to communicate with others, and to make it possible and effective, the consistent communication rules should be introduced. The micro-service application should use as seamless technology as possible for internal communication. Each service must specify exactly what features it offers and how it can be called, and what data and format it accepts and returns. This is the reason why it is important to create the documentation of interfaces of individual services, especially when many teams of developers are working on them. Once created interface should be modified as rarely as possible – this will avoid changes in the services that use it [5, 6].

3 Solutions for Management and Monitoring of Micro-Services

Independence of micro-servers, despite the advantages, can be problematic when running the entire application. Each part may require other environment variables, additional configuration, appropriate folder structure or another dependency version. The requirements of some micro-services may in some cases be mutually exclusive. This problem can be solved by running them on separate machines, but such solutions are often unprofitable and impractical during application development. The answer to this problem is the use of virtualization or containerization.

Virtualization and *containerization* are based on the same premise – providing separate, isolated environments for running programs, in this case single services. Both first and second option allows to create an easy to transfer and duplicate environment and run application in different environments on one machine. The basic difference between them is the scope to which they work. The virtual machine contains the application, all its dependencies, but also the entire operating system. This makes it less efficient and requires more resources. Access to hardware resources is done through a *hypervisor* [13].

Containerization is an advanced implementation of operating-system-level virtualization. Before going into details of containers, it is worth touching the subject of virtualization. In enterprise environment it is an idea of running multiple operation systems on a single server. Every operating system is separated from others by a layer of abstraction provided by a hypervisor. Virtualized operating system is called a virtual machine, due to the fact that it packs up virtual hardware, *kernel space* (a layer between hardware and applications consisting of API that applications can use via system calls) and *user space* (part of operating system where applications are running) [10, 11].

OS-level virtualization is different from a virtual machine, because it abstracts user space only making applications think that they are running on different OS-es. *Container* is simply a process or a branch of processes that are running in an abstracted environment under control of the same operating system.

Concept of containers isn't new, first implementations that could be classified as containers go as far back as to year 2000 with FreeBSD software called Jails and it was based on a simple file system abstraction. Later, around 2008, necessary features for containerization were embedded in the Linux kernel and a notion of LXC (Linux Containers) had begun. LXC is a set of hints, best practices and tools making creation of containers easier for operators of systems. It is taking advantage of two important kernel structures to provide necessary abstraction. First one is called *cgroups* (control groups), it limits and isolates system resources (such as CPU, memory, network) for processes. Second one are Linux namespaces, they are responsible for virtualization of e.g. process and user IDs, file systems, IPC (interprocess communication) and networking [12, 13].

4 Web Business Application Using Micro-services Architecture

To present the advantage of micro-services architecture and tools connected with it a simple Internet business application, named Booking Services, was created [6]. It is an application for managing the service reservations, customer data and supporting the process of booking a service by a customer. It has been designed to support a variety of services – managing them is possible through the application. can be used to streamline work in any company offering services to customers. The program was built using the micro-services architecture. This section describes the functional requirements of the application, the actors and roles in it, the documentation of created micro-services.

4.1 Functional Requirements

Services Booking application meets the following functional requirements [6]:

– Support the user accounts – log in the application using a login and password.
– Creating user accounts.

- View the list of all offered services.
- View the detailed information about each service.
- Possibility to create the service reservations for a given day.
- Possibility to list the reservations by the logged in user.
- User privilege system support.
- Creation of three administrator roles: user administrator (has the ability to change user account data including ability to change or delete the permissions), service administrator (has the ability to add new services, modify and delete existing ones) and the reservation administrator (has the ability to review all users' reservations and cancellations).
- User can play each role independently.

There are five types of actors in the application:

- Not logged-in user – any user who does not have an account in the application or is not logged in. It can only perform the basic operations such as browsing available services, creating an account and logging in.
- Logged-in user – any user who has logged in to its account on the system. It can perform basic operations on services, just as not logged in user, but can also create reservations and view reservations made by himself.
- Users Administrator – administrator of users accounts. It can perform all actions that a logged-in user performs, and is also authorized to modify user accounts, delete them and transmit and receive access rights.
- Services Administrator – administrator who can add, remove and change the existing services. It can also perform all actions that a logged-in user performs.
- Reservations Administrator – administrator authorized to view the list of all users and cancel each list item. In addition, it can perform all actions that a logged-in user performs.

The application implements a permission system and each actor has a dedicated role with the appropriate access level. The roles of administrators are not mutually exclusive, which means that the user can administer multiple aspects of the application simultaneously. The Users Administrator role is the supervisor – only he is entitled to change the access rights of other users.

4.2 Micro-services of Created Application

Services Booking application consists of micro-services, responsible for strictly defined functionalities groups. The micro-services are described below (Fig. 2) [6]:

Rest API Service
This is one of the most important services in Booking Micro-services application. It represents the implementation of API Gateway pattern. It provides API interface through HTTP. It communicates with other micro-servers through the event bus. Any communication coming from the outside to the server part of the application must access this service and only then it is forwarded. Because all inquiries from the user passed by this service, they are forwarded to *Authentication Micro-service* before being forwarded to execution by the other services.

Authentication Micro-service

This micro-service supports the user rights system in the application. It is responsible for validating the logging-in data and for generating access tokens. A token is created by encrypting a JSON document containing information about the user and its rights. Encryption is performed using DES (Data Encryption Algorithm) algorithm. This service also checks whether the user identifying the given token has the authority to perform the requested action. It communicates with other micro-servers through the event bus.

Users Management

It is a micro-service that supports functionality related to user accounts. It includes algorithms used to perform the actions on user accounts, including validating data. It does not contain the code responsible for communication with the database. It communicates with the assigned database via *Mongo Users Database Proxy* micro-service.

Mongo Users Database Proxy

This micro-service is responsible for communication with the database that stores user account information. It does not contain any business logic – all the algorithms are in other micro-servers, its only task is to communicate with the database. It is the implementation of proxy pattern.

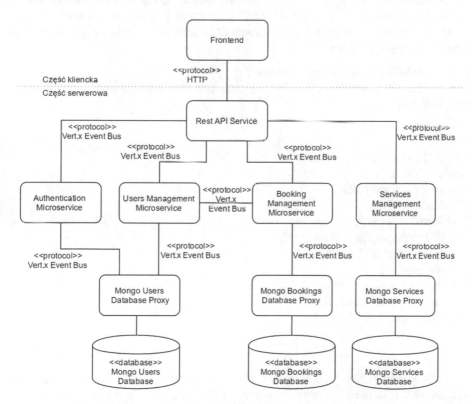

Fig. 2. Micro-services structure of Booking services application [6].

Booking Management
The task of this micro-service is to support the booking functionality. Like Users Management, it contains only business logic and the corresponding database communicates through Mongo Bookings Database Proxy service.

Mongo Bookings Database Proxy
It is a micro-service responsible for communication with the database where reservation information is stored. It is the implementation of proxy pattern.

Services Management
This micro-service supports service-related functionalities. Just like *Booking Management* and *Users Management*, it only contains business logic and does not communicate directly with the database.

Mongo Services Database Proxy
Mediator in communication with database that stores the information about services available in the system. It is implementation of proxy pattern.

4.3   Databases Operating in Micro-Services Application

In Booking Services application, data is stored in three separate databases. All of them use MongoDB database management system which means that the information is stored in documents in JSON format. Each database is in a separate, isolated Docker container, and each database communicates only with one micro-service. Databases used in the application are [6]:

Users database (Mongo Users Database)
This database has only one collection – Users. The data is stored in documents of the following format:

```
{
"_id" : "1234567",
"login" : "jsmith",
"password" : "jsmith123",
"firstName" : "John",
"lastName" : "Smith",
"email" : "john.smith@example.pl",
"createdDate" : "2017-12-02",
"permissions" : {
"canManageServices" : true,
"canManageUsers" : true,
"canManageBookings" : true}
}
```

Each document has a user ID field (*_id*) – it is added automatically by the database management system and is always unique. In addition, the saved documents contain information about login and password of the user, its first name, last name and email address, and the date when the account was created. In nested permissions document, the user rights data is stored. This collection has a created index in login field – it supports the queries to database, searching for data based on this field. In addition, the created index is unique, so there cannot be multiple users in the system identifying the same login.

Services database (Mongo Services Database)

This database has a collection called Services. It stores data in documents of the following format:

```
{
"_id" : "12345",
"name" : "service1",
"price" : 15,
"description" : "service1 description",
"createdDate" : "2017-11-15"
}
```

Each document contains automatically generated, unique id (_id), service price information, service description, the date it was saved in the system (createdDate) and its name. The name field also has a created index, and the service names are unique.

Reservations database (Mongo Reservation Database)

This database contains data in collection called Bookings. It stores data in documents of the following format:

```
{
"_id" : "123",
"name" : "service1",
"price" : 10,
"description" : "service1 description",
"createdDate" : "2017-11-15"
}
```

Each document contains an automatically generated id (_id), as well as user login information who has reserved the service (userLogin), name of reserved service, its description and date it was created (createdDate) and date of reservation (Date). There were created index for fields: userLogin, serviceName and date. In addition, each set of serviceName, date and userLogin values must be unique – this prevents the same service from being repeatedly booked by the same user on the same day.

5 Conclusions

When working on large application it is common to organize teams based on technology layer, which leads to teams responsible for the UI, server-logic teams, database administration etc. On the other hand, creating technologically mixed teams focused on a single function causes system to be constructed of a set of functional components. Micro-services similarly to SOA approaches application's division problem from the business side. It encourages team with a complete range of development skills, from UI design, through project management, all the way to database administration.

Service-Oriented Architecture encourages reuse and reducing redundancy. It is a common occurrence for large companies to create many slightly varied versions of the same software, which leads to the redundant or obsolete applications. More often than not when a department asks for something slightly different a whole new version of the program is created. Those variations cause the whole system to be difficult to understand and maintain. SOA encourages the business processes to become sealed containers with software specific to a single process – business. Then, services end up connecting with each other to realize more complex functions.

When creating web-based business applications, it is important to ensure data security and application reliability. The created exemplary application fulfilled the set of requirements. Thanks to the use of micro-services architecture, high fault tolerance and modification are achieved without disabling the entire application. The tests have been performed to prove the applicability of fault tolerance and the ability to turn off parts of micro-services without affecting others. They consisted of shutting off some micro-services and observing the behavior of the system. The results of these tests show that some of the functionality is working all the time, even after shutting down part of the system. This would be impossible with monolithic architecture.

The created application ensures data security and prevents unauthorized actions. This was achieved through the implementation of the token mechanism and API Gateway design pattern used in the micro-services architecture. This has enabled to provide an open API that is also usable in external applications (e.g. mobile applications). In addition, a dedicated micro-service corresponds to the token mechanism. This means that changing the algorithm that generates and checks tokens can be done at any time without having to change other parts of the application. Technologies used in exemplary application work well with each other. Thanks to the *Vert.x* platform, the asynchronous communication and a unified internal communication interface were introduced. All messages between micro-services are sent in JSON format, which in turn makes it possible to save data in MongoDB without having to change them to objects corresponding to the structure of relational database tables.

References

1. Richardson, C.: Micro-service architecture pattern, 15 March 2017. microservices.io
2. Fowler, M.: Micro-services, 20 March 2017. martinfowler.com
3. Newman, S.: Building Micro-Services – Designing Fine-Grained Systems. O'Reilly, Massachusetts (2015)

4. Wolff, E.: Micro-services: Flexible Software Architecture. Addison-Wesley, New Jersey (2016)
5. Posta, C.: Micro-services for Java Developers, A Hands-on Introduction to Frameworks and Containers. O'Reilly, Massachusetts (2016)
6. Bubel, P.: Creation of web based business applications using micro-services architecture, Eng. thesis. Lodz University of Technology (2017). (in Polish)
7. Bass, L., Clements, P.C., Kazma, R.: Software Architecture in Practice, 2nd edn. (2003)
8. Richards, M.: Micro-services vs. Service-Oriented Architecture (2016)
9. Rod, S.: Beginning Software Engineering (2015)
10. MSDN, Service-Oriented Architecture. https://msdn.microsoft.com/en-us/library/aa480021.aspx. Accessed 29 Jan 2017
11. Architecting Containers Part 2: Why, the User Space Matters. http://rhelblog.redhat.com/2015/09/17/architecting-containers-part-2-why-the-user-space-matters-2/. Accessed 30 Jan 2017
12. Architecting Containers Part 1: Why, Understanding User Space vs. Kernel Space Matters. http://rhelblog.redhat.com/2015/07/29/architecting-containers-part-1-user-space-vs-kernel-space. Accessed 30 Jan 2017
13. Mikicin, J.: Management and monitoring of applications based on micro-services, Eng. thesis, Lodz University of Technology (2017)

Investigation of Small-Consequence Undesirable Events in Terms of Railway Risk Assessment

Franciszek J. Restel$^{(\boxtimes)}$, Agnieszka Tubis, and Łukasz Wolniewicz

Wroclaw University of Science and Technology,
27 Wybrzeze Wyspianskiego Str., Wroclaw, Poland
{franciszek.restel,agnieszka.tubis,
lukasz.wolniewicz}@pwr.edu.pl

Abstract. Managing the risks in a railway transport company is related to many challenges. This is influenced by the characteristics of the transport process itself, which is carried out in a complex system. The transportation system consists of a very large number of elements and interactions between them. This situation makes, that it is impossible to diagnose all events, states and relations that occur during the transport process. In the case of rail transport, legal provisions are also an important obstacle. Directive 2004/49/EC - Railway Safety Directive, imposed by European Parliament, classifies the events, which are assessed in risk analysis. This classification significantly determines the form of historical data reporting. It is also focused on improving the transport system (macro scale) and not transport processes on the organisational level of the company (micro scale). In the article authors focus on the issues connected with risk analysis in rail transport processes performance. The aim of the article is to present a multi-criteria risk analysis for events that are classified as railway incidents. The presented approach to risk analysis is dedicated to improving transport processes, which are provided by the railway operator.

Keywords: Rail transport · Risk assessment · Consequences

1 Introduction

Risk management today is firmly established on the agenda of business. To qualify as legitimate and responsible, organizations must be accountable for their capacity to deal with risk by means of various tools and techniques [14, 21, 24, 25]. The pressure on organizations to account for how they manage risk has given rise to organizational responses [14, 15, 21] depending on how risks are understood, perceived and constituted. The increasing demand for risk management on the part of organizations, businesses and government authorities has been identified as a general societal trend emphasizing public accountability and responsibility [21, 22, 25]. Especially, we can observe this in the transport sector, which has not only business but also social roles.

It is really difficult to manage the risks in a transport company. This is influenced by the characteristics of the transport process itself, which is carried out in a complex system. The transportation system consists of a very large number of elements and

© Springer International Publishing AG, part of Springer Nature 2019
W. Zamojski et al. (Eds.): DepCoS-RELCOMEX 2018, AISC 761, pp. 384–399, 2019.
https://doi.org/10.1007/978-3-319-91446-6_36

interactions between them. This situation makes, that it is impossible to diagnose all events, states and relations that occur during the transport process. In the case of rail transport, legal provisions are also an important obstacle [35, 36, 41]. Directive 2004/49/EC - Railway Safety Directive, imposed by European Parliament, classifies the events, which are assessed in risk analysis. This classification significantly determines the form of historical data reporting. It is also focused on improving the transport system (macro scale) and not transport processes on the organisational level of the company (micro scale). However, as Gibson [12] points out, most managers want four things from their risk management information systems:

- calculate value at risk
- perform scenario analyses,
- measure current and future exposure to each counterparty,
- give the possibility to aggregate information across various groups of risks, product types, and across subsets of counterparties.

The authors focus on the issues connected with risk analysis in rail transport processes performance. The aim of the article is to show the small-consequence event influence on railway safety, and the importance of these events. Following, the main goal is to describe a method for small-consequence event handling during the operation process.

The presented approach to event analysis is dedicated to improving transport processes, which are provided by the railway operator. As a result, the authors focus on the presentation of the issues, which apply risk assessment for rail transport. There is provided a brief overview of the literature in area of risk assessment and consequence analysis. Then the authors presented a new approach to event handling for events that are qualified as railway incidents and safe events. An important element of the presented method is the proposed classification of events, according to their causes and consequences. In section four, the authors used the proposed analytical method to classify real events and how to deal with them. The classification of incidents was carried out for events registered in the selected railway line. The article concludes with a summary and guidelines, including directions for further research.

2   Risk Analysis in the Railway Industry – Literature Review

Every year we observe an increase in security and logistics requirements, which are reported by the participants of the transport process. This phenomenon means that transport companies are more and more interested in tools/concepts that will enable them to better manage basic processes and secure the high quality of services provided [29].

Currently, there is no one, unified and commonly used definition of risk term [5]. We can even state that the underlying concepts of risk are hard to define and even harder to assess [13]. In recent decades, we have observed this term being applied to many research areas, like decision theory, management, emergency planning, or critical structures operation, including transport systems performance [27]. The historical development trends of risk concept are discussed e.g. in [1, 5]. The risk perspectives review and discussion are given e.g. in [2–4, 6].

A literature review of the risk area [2] leads to the conclusion that the common part of all these definitions is that the concept of risk comprises events (initiating events, scenarios), consequences (outcomes) and probabilities. Uncertainties are expressed through probabilities (1). We can formalise this by writing [5]:

$$Risk = \{A, C, P\} \tag{1}$$

where A represents the events (initiating events, scenarios), C the consequences of A, and P the associated probabilities.

There are basically two ways of interpreting the probability P [7, 8]:

(a) A probability is interpreted as a relative frequency Pf: the relative fraction of times the event occur if the situation analysed were hypothetically "repeated" an infinite number of times. The underlying probability is unknown, and is estimated in the risk analysis. We refer to this as the relative frequency interpretation.

(b) Probability P is a measure of uncertainty about future events and consequences, seen through the eyes of the assessor and based on some background information and knowledge. Probability is a subjective measure of uncertainty, conditional on the background knowledge (the Bayesian perspective).

Consequence analysis is the second integral part of risk assessment [31]. It is one of four major steps in quantitative risk assessment, which includes [23]: hazard identification, consequence analysis, frequency assessment, and risk quantification. Consequence analysis is generally defined as an assessment of likely consequences if an accident happens. The complexity of a QRA depends on the scenario and the availability of data and consequence information [10, 37, 40, 42]. Risk assessment methods and supporting applications are also important for the obtained results. In [16] authors presented an interesting overview of risk assessment methods used in the railway industry. In the article the authors also presented applications based on the Monte Carlo method and Fuzzy Reasoning approach.

However, many researchers, like Burgman et al. [9] stated that the majority of applications of risk analysis consider the probability of events in detail, and pay almost no attention to the severity of consequences. Limited work has been done on quantifying consequences [31]. Zadakbar et al. proposes to use the function of loss in assessing the consequences (Fig. 1).

Fig. 1. The flowchart of consequence analysis using loss functions [31]

It should also be noted that in the risk analysis for the rail transport process, the concept of *risk* combines with the concept of *hazard*. This is confirmed by research conducted by Bohus [26]. He determined hazardous events, the hazard factors and Detailed Hazardous Factors for each type of accident in the proposed risk assessment model. This is logically linked to the cause-and-effect chain of accidents, where the primary link in the accident is the hazard (potential source of damage). The cause and effect chain of loss formation is shown in the Fig. 2.

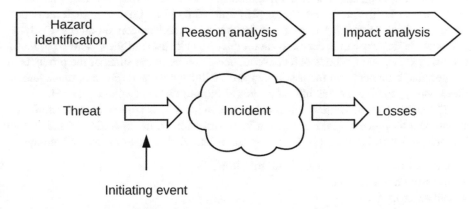

Fig. 2. The cause-effect chain in occurrence of losses [20]

The analysis of consequences for emerging events or scenarios is particularly important in risk assessment for the rail transport process. The analysis of the consequences of the event is the basis of its classification and undertaken operating procedures in accordance with the Instruction Ir-8 [18] (Fig. 3).

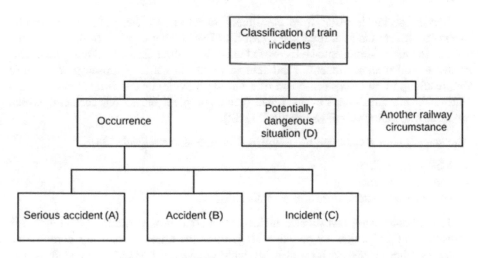

Fig. 3. Classification of undesirable train events

The basis for the qualification of the subject event is to determine the number of fatalities, injured people or the amount of costs. It has a significant impact on the event data collected and their use in the process of risk analysis for the railway system.

3 Preliminary Event Analysis Approach for Incidents, Potentially Dangerous Situations and Safe Events

In engineering contexts, risk is often linked to the expected loss, see e.g. Lirer et al. [17], Mandel [19], Verma and Verter [28] and Willis [30]. However, such an understanding of risk means that there is no distinction made between situations involving potential large consequences and associated small probabilities, and frequently occurring events with rather small consequences, as long as the sums of the products of the possible outcomes and the associated probabilities are equal. For risk management these two types of situations normally would require different approaches [5].

The railway transportation system is very complex. This applies to structure of the system and all processes taking place in it. Information on the operation of the system is stored in several places. The most important knowledge bases currently belongs to:

- national rail accident research commissions;
- infrastructure managers;
- railway carriers.

Each of these knowledge bases is characterized by a different profile, in accordance with the main business.

The main unwanted events investigated by the National rail accident research commissions (NRARC) are [32]:

- serious accidents;
- accidents;
- incidents.

Serious accidents result from derailment of a rail vehicle, collision or similar events, which results in at least one fatality, or five serious injuries, or damage to the system and the environment at a level of at least two million Euro. Accidents are events related to collision, derailment of rail vehicles, collisions at level crossings, or vehicle fire, resulting in no deaths (on board of a railway vehicle) or losses up to two million Euro. Incidents are all other events (other than an accident and serious accident), which are related to safety of railway traffic [11, 32].

An accident commission can be established at one of three levels [34]:

- local commission;
- company commission;
- national rail accident research commissions.

Local commission investigates incidents occurred during shunting, rolling stock collisions, hit pedestrian, or people who jumps into a train. Company commission examines other events. Each event being incident, accident or serious incident is also reported to the national commission for research of railway accidents. NRARC supervises proceedings, takes part in the proceedings, or take control of the proceedings. Each event studied in this procedure results in:

- circumstances of the incident;
- causes of the event;
- prevention requests.

Research institutions for railway accidents have the best competence, so they may develop the most accurate and insightful reports. Research carried out by these institutions can bring a lot to the railway transportation system. In practice, these studies are restricted to a small fraction of all events in the system. For a required region in Poland (from 2009 to 2011) less than 0.1% of all registered events have a characteristic such as to be notified by the national safety institutions [32, 33]. Of the fraction of reported incidents 1.4% was investigated directly by NRARC. To the European Railway Agency (ERA) only 0.18% was reported [32].

Accidents and serious accidents have the largest consequences, but in the same time the smallest occurrence probability. Therefore, the risk related to such events is small. The events are very well investigated.

For accident prevention, it will more important to learn from incidents, events may effect on safety and safe events. This is due to the high frequency of these events and their influence on accident rising. So, the risk assessment should also systematically include them.

In previous studies was built an accident occurring model, basing on Fault Tree Analysis [38, 39]. Using the model there have been identified four important barriers preventing the accidents. The barriers where shown in Fig. 4 in form of Reason's Swiss-cheese model.

The first barrier is to implement actions preventing decisions leading to incorrect system using. These activities are designed to train staff for a more effective solving of problems arising during operation. As a result, traffic crew have to solve emergency situations without changing to incorrect system operation.

Another barrier is the correct operation of the system, including using of all safety equipment and procedures. The third barrier is the undisrupted train traffic that prevents collisions of trains expedited by the traffic crew. On the other hand, that barrier protects only partially against train driver's failures. The last barrier is the correct assessment of the traffic situation by railway employees. The occurrence of a serious accident is possible if all barriers fail, only when there are at the same time the following factors:

- decision for incorrect system operation,
- incorrect system operation at a given time interval,
- traffic disruptions,
- human failures.

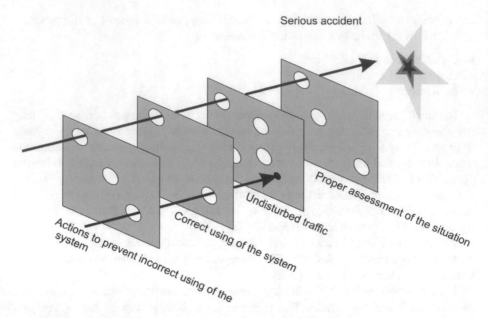

Fig. 4. Model of new identified barriers against serious railway accidents.

Using [14, 21] were identified causes of undesirable events and divided into classes C41–C67 basing on their causes. At first, the causes of the events were grouped. Finding the source of failure gives an easier possibility to analyse it [29]. All identified causes were split into four groups:

- related to environmental impact,
- infrastructure failures,
- human failures,
- rolling stock (vehicle) failures.

Furthermore, human failures were split into three subgroups:

- related to network staff,
- related to the train crew and passengers,
- related to man not belonging to the system (third party).

The results of such a categorisation due to the causes of undesired events was graphically shown in Table 1.

Events related to rolling stock or staff of the carrier can be easily analysed by the carrier. For the carrier, it is also possible to prepare improvements. Therefore, for that events, that risk analysis should be done mainly by the carrier. Other events can be most detailed investigated by the infrastructure manager.

Table 1. Classification of incidents by their sources

	Human			Vehicle	Infrastructure	Environment
	Network staff	Staff of the carrier	Third party			
C41–C43	•					
C44–C47		•				
C48	•					
C49–C50		•				
C51	•				•	
C52					•	
C53–C54				•		
C55–C56			•	•		
C57			•			
C58–C59				•		
C60		•				
C61			•			
C62						•
C63					•	
C65		•	•			
C64, C66, C67			•			

On the next stage, the event consequences have to be analysed. The authors propose a classification on five categories of consequences:

- financial losses,
- health losses,
- delay or lack of service,
- equipment repair,
- other.

The classification was done also in relation to the event types placed in [18]. The results were presented in Table 2.

Basing on the types of consequences, an integrated loss function can be estimated (2). At this stage of research, the cost function estimation will be subjective.

$$FL(C) = f(F, H, S, R, A) \qquad (2)$$

where: FL (C) – loss function for each class; F – financial losses; H – damage to health; S – delay or lack of service; R – rapair; A – another.

Finally, a method was proposed for handling undesirable events, are not accidents or serious accidents. That means, incidents, events with potential influence on the system safety, and safe events. It was presented in Fig. 5.

Table 2. Events with potential losses

	Financial losses	Health losses	Delay or lack of service	Repair	Other
C41, C43		•	•		
C42					•
C44		•			
C45					•
C46–C48	•	•			
C49	•				
C50, C52	•	•	•		
C51	•		•		
C53–C54		•	•	•	
C55–C57	•	•	•		
C58	•	•	•	•	
C59	•	•	•		
C60	•		•	•	
C61		•	•		
C62–C65	•	•	•	•	
C66–C67	•		•		

The first and most important question arises after occurring of an unwanted event, if such an event has happened in the past. It should be checked if some circumstances are the same as in the past. For example place, time period, train type, carrier and others. After a positive answer, such a repeatable situation can be classified as potential hazardous situation. Therefore, it should be analysed quantitative and/or qualitative. If a conclusion can be formulated, and any improvements can be done in the system, than the situation is classified as safe. Otherwise, more data must be collected, and the situation is classified as hazardous. This is due to the unknown phenomena.

Another decision branch will be followed if the event was not repeatable. Afterwards, it should be found what were the circumstances related to typical operation. So, looking for factors beyond operation routine. When such circumstances have occurred, the question arises if the system was operated correctly. That means if all safe guards were correctly in operation and procedures were followed. When the system was operated correctly, the causes and consequences of the event should be investigated and it should be stored in a database.

Finally, it has to be investigated what impact on the event occurring had human failures. Depending on that there should follow a large scale analyse, or a dedicated to the human failure.

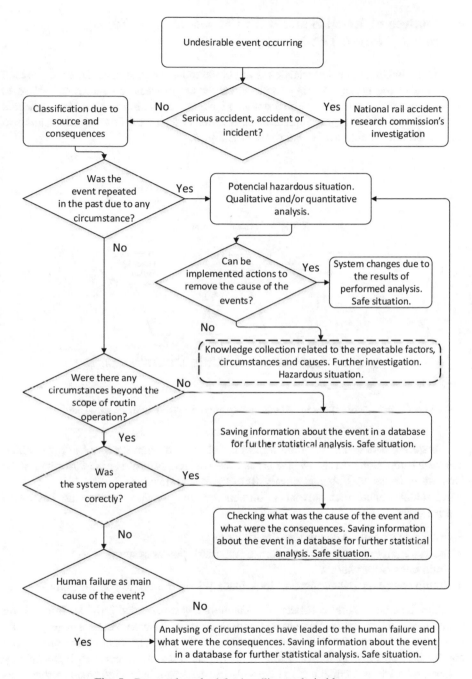

Fig. 5. Proposed method for handling undesirable events

4 Analyse of Event Sources and Consequences Basing on the Proposed Categories

The main challenge in event analysis is to get the most accurate data. There are several databases in the system. During traffic management process, operating data have to pass several instances of the infrastructure manager. In relation to that, data collected by the carrier during regular operation is a little. The general flow of information in case of an undesirable event was shown on Fig. 6.

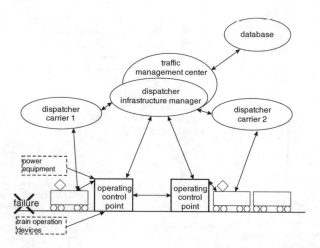

Fig. 6. Information flow in railway transportation system after an unwanted event

Operating data (related to train traffic) is collected by traffic control points. Which can be located along a railway line or as a centralized control centre. Information on location of trains are obtained mostly from track-side devices.

Random events which prevents or hinders system operation, are registered at a train control point via:

- trackside signalling equipment (interlocking),
- power equipment (traction network or auxiliary power system),
- train crew observations,
- traffic controller (when incident took place within the traffic control point).

A traffic control point connects to dispatching department and provides information about an event. The dispatching department is a part of infrastructure manager's traffic management centre. Parallel communication takes place between train crews and appropriate carrier dispatchers. As a result of the dynamic interaction between dispatchers of various companies it is possible to rise errors in the information that goes into the traffic database.

The main task of the dispatching department is a rapid reaction to an unwanted event. This process should allow to minimize secondary disruptions of failures for passengers, carriers, etc. In view of those obligations, carriers' dispatchers do not have complete data on total number of events that affects the reliability of railway system.

Finally data are archived by infrastructure manager traffic management centre. Data are available as traffic books and failure books, usually contained in one IT system.

For a punctual train or a small disrupted one data is collected only in relation to time when a control point is passed. However, after a disruption causing in delayed trains by more than five minutes (sometimes this limit has a different value), or a deviation from normal operation of the system, that do not result in delay, range of collected information will increase. A typical information collecting process after an unwanted event includes the following parameters:

- type of failure,
- occurrence date,
- repair time,
- place of the event (route/station, kilometre of the line, number of track),
- number of delayed passenger trains,
- summed delays of passenger trains,
- number of delayed freight trains,
- summed delays of freight trains,
- own delay of train (if cause of the event is that train).

The carrier has the most specialized database. It contains only information about carrier activities. Data on train traffic are secondary - originally derived from infrastructure manager.

At this level of information availability, it was only possible to group events due to the proposed cause and consequence categories. For an existing railway line were analysed about two thousand undesirable events, have occurred during one year of operation. During the observed operation, accidents and serious accidents have not occurred on the railway line. Only 0.4% of all events were classified as incidents. The remaining events were classified as safe ones or as potentially dangerous.

In Fig. 7 was shown the results for classification of events by sources and consequences. For the consequences estimation, if an event had more than one type of consequences, than that one were named which had the biggest impact. The main part of all undesirable events consists of rolling stock and infrastructure failures, with repairing actions or delays as consequences. Therefore, a hypothesis can be built, that analysing of the events can lead to improve the system and reduce risks occurring in it.

In Table 3 are shown results for the analysed event grouping. In terms of the algorithm shown in Fig. 5. Events have been grouped into six categories. Each one requires other actions. A new group of potential hazardous events was introduced. It is the fifth one. Its quota is about 0.091. A quiet high value for events, which were not investigated yet. The highest frequency (0.614) is related to events that have to registered for potential further analysis.

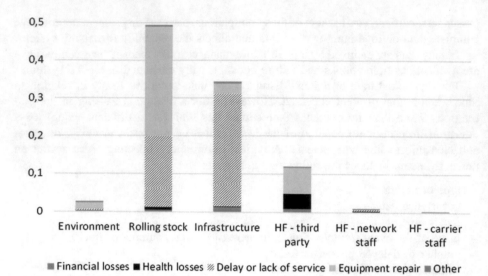

Fig. 7. The classification of incidents by source of hazards and potential effects for a selected railway line.

Table 3. Results for event grouping

	Analysing of circumstances of the human failure and the consequences. Saving information	Checking what was the cause and the consequences. Saving information about the event in a database	Saving information about the event in a database for further statistical analysis	System changes due to the results of performed analysis	Knowledge collection related to the repeatable factors, circumstances and causes	National rail accident research commission's investigation
Potential hazardous situation	No	No	No	No	Yes	Yes
Safety investigation required	No	No	No	Yes	Yes	Yes
Changes required	No	No	No	Yes	Yes	Yes
Part of events from the sample	0.097	0.032	0.614	0.165	0.091	less 0.001

5   Summary

In rail transport, it has been observed a large number of safe events and events with potential impact on safety, which are not analysed in risk context. There is also lack of methods to deal with such events.

Therefore, it was proposed to categorize undesirable events according to the sources of their occurrence. The elaborated categories base on the types of events used in the description of serious accidents, accidents and incidents. As a result, four main categories were obtained. One of the categories is related to human failures. It was split into sub-categories, depending on the human occurrence in the system.

Subsequently, the documents of the railway accident investigation and operational data were analysed. Five groups of consequences were developed. With the classification of events, a method for dealing with all undesirable events was developed. It gives the opportunity to control events and prepare preventive actions to reduce operation risks. Observation of real data has shown that the most events are related to infrastructure and vehicles. Therefore, it possible to influence the subsystems and increase safety.

The shown method for undesirable event handling allows to make decisions about possible steps of event investigation. It means, how to find relevant events in the set of safe events. On the other hand, it allows to minimize the number of investigated events due to the high occurrence frequency. After grouping of events into relevant and not relevant for safety issues, the first group has to be investigated by already known risk assessment methods.

For further research, it is necessary to build cooperation with the railway industry to improve the proposed method. The next stage have to include implementation in a IT environment for automatization of the method.

References

1. Anderson, E.L.: Scientific trends in risk assessment re-search. Toxicol. Ind. Health **5**(5), 777–790 (1989)
2. Aven, T., Krohn, B.S.: A new perspective on how to under-stand, assess and manage risk and the unforeseen. Reliab. Eng. Syst. Saf. **121**, 1–10 (2014)
3. Aven, T.: Practical implications of the new risk perspectives. Reliab. Eng. Syst. Saf. **115**, 136–145 (2013)
4. Aven, T.: A risk concept applicable for both probabilistic and non-probabilistic perspectives. Saf. Sci. **49**, 1080–1086 (2011)
5. Aven, T.: On how to define, understand and describe risk. Reliab. Eng. Syst. Saf. **95**, 623–631 (2010)
6. Aven, T.: Perspectives on risk in a decision-making context – review and discussion. Saf. Sci. **47**, 798–806 (2009)
7. Aven, T.: Foundations of Risk Analysis. Wiley, Chichester (2003)
8. Bedford, T., Cooke, R.: Probabilistic Risk Analysis. Cambridge University Press, Cambridge (2001)
9. Burgman, M., Franklin, J., Hayes, K.R., Hosack, G.R., Peters, G.W., Sisson, S.A.: Modeling extreme risks in ecology. Risk Anal. **32**(11), 1293–1308 (2012)

10. CCPS/AIChE. Guidelines for Chemical Process Quantitative Risk Analysis. Wiley, New York (2000)
11. European Parliament, Directive 2004/49/EC – Railway Safety Directive (2004)
12. Gibson, M.: Information systems for risk management. FRB International Finance Discussion Paper No. 585, July 1997. Available at SSRN, http://dx.doi.org/10.2139/ssrn. 231755. Accessed January 2017
13. Heckmann, I., Comes, T., Nickel, S.: A critical review on supply chain risk – definition, measure and modelling. Omega **52**, 119–132 (2015)
14. Hood, C., Rothstein, H., Baldwin, R.: The Government of Risk: Understanding Risk Regulation Regimes. Oxford University Press, Oxford (2001)
15. Hutter, B., Power, M.: Organizational Encounters with Risk. Cambridge University Press, Cambridge (2005)
16. Leitner, B.: A general model for railway systems risk assessment with the use of railway accident scenarios analysis. Procedia Eng. **187**, 150–159 (2017)
17. Lirer, L., Petrosino, P., Alberico, I.: Hazard assessment at volcanic fields: the Campi Flegrei case history. J. Volcanol. Geoth. Res. **112**, 53–73 (2001)
18. Lk, S.A.: Safety Office: Instruction about procedure in case of serious accidents, accidents and incidents on railway lines (Ir-8). PKP Polskie Linie Kolejowe S.A, Warsaw (2015)
19. Mandel, D.: Toward a concept of risk for effective military decision making. Defence R&D Canada—Toronto. Technical report. DRDC Toronto TR 2007-124 (2007)
20. Młyńczak, M., Nowakowski, T., Valis, D.: How to manage risks? the normative approach. Problemy Eksploatacji **1**, 137–147 (2011). (in Polish)
21. Power, M.: Organized Uncertainty: Designing a World of Risk Management. Oxford University Press, Oxford (2007)
22. Power, M., Scheytt, T., Soin, K., Saklin, K.: Reputational risk as a logic of organizing in late modernity. Organ. Stud. **30**(02–03), 301–324 (2009)
23. Pula, R., Khan, F., Veitch, B., Amyotte, P.: A grid based approach for fire and explosion consequence analysis. Trans. IChemE, Part B, Process Safety and Environmental Protection **84**(B2), 79–91 (2006)
24. Risk and safety in the transport sector RISIT – A state-of-the-art review of current knowledge. Research Report, The Research Council of Norway. http://www.forskningsradet. no/csstorage/vedlegg/english_report.pdf. Accessed January 2017
25. Rothstein, H., Huber, M., Gaskell, G.: A theory of risk colonization: the spiralling regulatory logics of societal and institutional risk. Econ. Soc. **35**(1), 91–112 (2006)
26. Sasidharan, M., Burrow, M.P.N., Ghataora, G.S., Torbaghan, M.E.: A review of risk management applications for railways. In: Railway Engineering (2017)
27. Tubis, A., Werbińska-Wojciechowska, S.: Operational risk assessment in road passenger transport companies performing at Polish market. European Safety and Reliability ESREL 2017, 18–22 June 2017, Portoroz, Slovenia (2017)
28. Verma, M., Verter, V.: Railroad transportation of dangerous goods: Population exposure to airborne toxins. Comput. Oper. Res. **34**, 1287–1303 (2007)
29. Wieteska, G.: Risk Management in Supply Chain on B2B Market. Difin, Warsaw (2011)
30. Willis, H.H.: Guiding resource allocations based on terrorism risk. Risk Anal. **27**(3), 597–606 (2007)
31. Zadakbar, O., Khan, F., Imtiaz, S.: Development of economic consequence methodology for process risk analysis. Risk Anal. **35**(4), 713–731 (2015)
32. Leśniowski, R., Ryś, T.: National Rail Accident Research Commission – Annual report 2011. Ministerstwo Transportu, Budownictwa i Gospodarki Morskiej, Warszawa (2012). (in Polish)

33. Polish railway infrastructure manager – PKP PLK S.A.: Data about failures on railway lines in a selected polish region for the years 2009–2011 (2012). (in Polish)
34. Polish Journal of Laws no. 89 Poz. 593: Rozporządzenie Ministra Transportu w sprawie poważnych wypadków, wypadków i incydentów na liniach kolejowych, Warszawa (2007)
35. Jodejko-Pietruczuk, A., Werbińska-Wojciechowska, S.: Block inspection policy for non-series technical objects. Safety, reliability and risk analysis: beyond the horizon. In: Proceedings of the 22nd Annual Conference on European Safety and Reliability (ESREL) 2013, Amsterdam, pp. 889–898 (2014)
36. Kierzkowski, A., Kisiel, T.: Simulation model of security control system functioning: A case study of the Wroclaw Airport terminal. J. Air Transp. Manag. http://dx.doi.org/10.1016/j.jairtraman.2016.09.008
37. Kierzkowski, A.: Method for management of an airport security control system. In: Proceedings of the Institution of Civil Engineers - Transport (2016). http://dx.doi.org/10.1680/jtran.16.00036
38. Restel, F.J.: Defining states in reliability and safety modelling. AISC, vol. 365, pp. 413–423 (2015). https://doi.org/10.1007/978-3-319-19216-1_39
39. Restel, F.J., Zajac, M.: Reliability model of the railway transportation system with respect to hazard states. In: 2015 IEEE International Conference on Industrial Engineering and Engineering Management (IEEM), pp. 1031–1036 (2015)
40. Walkowiak, T., Mazurkiewicz, J.: Soft computing approach to discrete transport system management. LNCS, vol. 6114, pp. 675–682 (2010). https://doi.org/10.1007/978-3-642-13232-2_83
41. Walkowiak, T., Mazurkiewicz, J.: Analysis of critical situations in discrete transport systems. In: Proceedings of DepCoS - RELCOMEX 2009, Brunów, Poland, 30 June–02 July, 2009, pp. 364–371. IEEE (2009). https://doi.org/10.1109/depcos-relcomex.2009.39
42. Zak, L., Kisiel, T., Valis, D.: Application of regression function - two areas for technical system operation assessment. In: CLC 2013: Carpathian Logistics Congress - Congress Proceedings, pp. 500–505 (2014). WOS:000363813400076

Identification of Information Technology Security Issues Specific to Industrial Control Systems

Dariusz Rogowski[1,2](✉) iD

[1] Faculty of Automatic Control, Electronics and Computer Science,
Institute of Informatics, Silesian University of Technology, Akademicka 16,
44-100 Gliwice, Poland
dariusz.rogowski@polsl.pl
[2] Institute of Innovative Technologies EMAG, Leopolda 31,
40-189 Katowice, Poland
dariusz.rogowski@ibemag.pl

Abstract. The paper presents current cybersecurity issues in industrial automation and control systems (IACS). It also reviews the state of the art in literature, standards and frameworks used to evaluate and certify industrial control devices. Nowadays the Common Criteria (CC) security assurance methodology is commonly used for the vast majority of information technology (IT) products but not for IACS components. The paper proposes a security evaluation method of IACS to be based on the CC approach. The CC standard has not been used in industry so far and this is why it became the main motivation of the author's doctoral research work in that field. The implementation of CC security requirements can enhance the "safety" of functional features in control devices by adding "security" measures typical of IT products. The paper delivers input information to the first stage of the author's research whose goal is the identification of design needs and requirements for building the security evaluation method. As a result, in the next stage, the evaluation method can be built according to the model of a control system and to the criteria taken from the CC standard adjusted to IACS needs. Coupling both "security" and "safety" for industrial control systems is a promising way of using the CC assurance methodology for a new kind of devices.

Keywords: Industrial control system · Security evaluation · Common criteria

1 Introduction

In recent years information technology (IT) has been widely used in Industrial Automation and Control Systems (IACS) – aka Industrial Control Systems (ICS) – which makes these systems more vulnerable to security attacks from the Internet leading to serious consequences. In most cases the structure of a typical ICS consists of a corporate network, supervision network and production network. These interconnected networks are interdependent and thus vulnerable to typical IT threats. These security threats can be mitigated in a similar way like in standard IT systems by using IT security evaluation standards. A promising approach is proposed by the Common

© Springer International Publishing AG, part of Springer Nature 2019
W. Zamojski et al. (Eds.): DepCoS-RELCOMEX 2018, AISC 761, pp. 400–408, 2019.
https://doi.org/10.1007/978-3-319-91446-6_37

Criteria (CC) standard [1–3] with Common Evaluation Methodology (CEM) [4, 5], which both constitute the basic set of stringent rules for IT security functionality development and assessment.

The CC standard could be applied to evaluate ICS functional features but some adaptation work is needed first. CC should be adjusted to the needs of industrial devices in order to combine the "safety" aspects of ICS with the "security" aspects of IT. The CC evaluation process is difficult, time-consuming and demands a deep knowledge of CC requirements. This is why the author's doctoral research goal is to work out a new evaluation method which could support developers and users in using CC for industrial devices. On the basis of a literature review, the paper demonstrates that little attention has been paid to using CC in the evaluation of IACS security. It proves the need and pertinence of the author's research work in that area.

A major current focus in cyber-security of IACS systems is to identify major threats, vulnerabilities and risks to industrial control devices and to find out security controls preventing these threats. There are institutions which identify ICS risks and publish special security reports like ICS-CERT (The US Industrial Control Systems Cyber Emergency Response Team) [6]. ICS-CERT publishes Annual Assessment Report every year. For instance they conducted 130 security assessments [7] in Fiscal Year (FY) 2016 report in which they summarized risks and recommended mitigation measures associated with six most prevalent weakness categories like: boundary protection, least functionality, identification and authentication, physical access control, audit review, analysis and reporting, authentication management.

Security issues in IACS became one of major concerns especially after the Stuxnet cyberattack in 2010 [8–10]. It revealed that firmware, control logic and services of Programmable Logic Controller (PLC) should be more protected than in the past because of a rising number of software vulnerabilities. Please note that firmware modification is one of the main types of attacks on firmware (cf. an attack on the Ukrainian power grid [11]). Another example of attacks relates to control logic and services. For instance, an attacker can alter or delete a PLC code [12] what can lead to freezing of outputs and destroying a whole control system.

Security problems in ICS have been worsening what stimulates researchers and vendors to find and implement methods for security evaluation of industrial control systems. Unfortunately, researchers focus mainly on applying standards and methods concerning the "safety" of ICS. They often diminish the importance of "security" enhancing methodologies. This is why in this paper the Common Criteria methodology is introduced to solve the "security" problems of ICS. This study investigates the use of the CC standard in industry to prove the novelty and originality of the author's doctoral research work.

The paper is organized as follows. The first part of Sect. 2 presents the state of the art in the field of cybersecurity in IACS concerning certification frameworks, general security issues, standards with a focus on Common Criteria application. The second part of Sect. 2 presents a short introduction to the Common Criteria methodology. Section 3 summarizes the results of literature review and Sect. 4 contains discussion and future work.

2 State of the Art

The paper presents a literature review of cyber security issues in ICS. In the review a systematic approach based on [13] was used to identify current organizations, standards, papers, and technical reports relevant to security issues in ICS with the consideration of the Common Criteria standard.

First, European publications dealing with the framework for evaluation and certification of ICS were reviewed. Five documents were chosen for the study.

Second, the databases of such publishers like: Elsevier, IEEE, Springer and Wiley were reviewed with the following keywords: "cybersecurity", "security", "evaluation", "industrial control systems", "Common Criteria", "ISO/IEC 15408". Several search strings were created with the combination of these keywords. Next, aggregative databases like: IEEE Xplore, Web of Science (WoS), Scopus were searched.

As a result 460 records were received from the IEEE Xplore database. That result was refined by using the "security evaluation" phrase which gave 27 records. After an in-depth review of these publications the most relevant 17 papers were chosen for further research.

Similar search in the WoS database (excluding results from IEEE) was made and 68 records were received. From these records 8 publications were chosen for a detailed analysis. The search in the Scopus database gave duplicates of papers found in WoS and IEEE. The duplicate papers were excluded from the results. Finally, 30 publications in total (5 about European initiatives, 17 from IEEE, 8 from WoS) were chosen for further analysis. Beneath there are main insights of the literature review.

2.1 Certification Frameworks for ICS

The European Reference Network for Critical Infrastructure Protection (ERNCIP) in the thematic area of industrial control systems and smart grids gives proposals for a European IACS components cybersecurity compliance and certification scheme. The ERNCIP Thematic Group (TG) treats Common Criteria as a possible candidate standard for testing and certification of IACS components [14], although the standard is recognized as difficult and costly. The group also indicates that some countries like Germany or England work on adaptations of CC to IACS environments.

The European Union Agency for Network and Information Security (ENISA) [15] helps EU member states in implementing relevant legislation and works to improve the resilience of critical information infrastructures and networks. It also monitors communication network dependencies [16] and delivers guides for CERTs (Computer Emergency Response Teams) [17] on how to deal with ICS security incidents.

In the report "Protecting Industrial Control Systems – Recommendations for Europe and Member States" [18] ENISA suggests to base a model of a security framework on Common Criteria.

2.2 Security Issues in ICS

There are hundreds of publications describing security problems in IACS as it was demonstrated by the number of records found during publishers' databases search.

Thus in this paper only several most relevant publications were chosen for further analysis. As it was mentioned in the introduction, many threats to industrial control devices have its source in the Internet. There are over one million ICS/SCADA systems connected to the Internet with unique IP addresses [19]. This is a potential area of sophisticated attacks. These can be memory overflow, malware scripts injected in the code of the client websites, SQL (Structured Query Language) injections.

In general, the vulnerabilities of ICS can be divided into the following categories [20]:

- Security management policies and procedures – vulnerabilities are related to: poor management, insufficient audit, lack of contingency plans.
- Hardware and software platform and applications – vulnerabilities are related to: outdated equipment and software, default settings, lack of backups, configuration loss, inadequate authentication control, flaws in software components, no controls to protect from malware, bad application design.
- Network configuration – vulnerabilities are related to: badly designed network architecture, poor authentication mechanisms, insecure physical ports, missing or badly configured firewall, lack of traffic monitoring, using standard protocols such as Telnet or FTP (File Transfer Protocol) without encryption means.
- Protocols – vulnerabilities are related to: lack of message authentication and encryption, Denial-of-Service (DoS) attacks, buffer overflow, Main-in-the-Middle (MitM) attacks.

All the above mentioned security issues of ICS are getting more attention of vendors and security evaluation bodies that is why a few security evaluation methodologies for industrial control devices were developed [21], for instance: Embedded Device Secure Assurance (EDSA) developed by ISA (International Society of Automation) or Achilles Platform for Industrial Cyber Security for Operational Technology (OT) developed by General Electric [22].

Several publications concerning security issues in ICS were reviewed but only one paper [21] mentioned the possible use of the Common Criteria approach. It seems from this short review that researchers focus rather on the "safety" issues of ICS and do not consider the CC methodology. Future work should broaden the literature review to get more reliable results.

The next section describes the most commonly known security standards for IACS.

2.3   Standards for ICS

Governmental bodies, standard organizations like ISO and IEC, and international agencies like NIST (National Institute of Standards and Technology) or ENISA developed general standards and guidelines for securing SCADA systems.

According to [23, 24] there are the following main standards relevant and dedicated to cyber security applicable to IACS: IEC 62443 (ISA-99), ISO/IEC 27019, NIST SP 800-82 Guide.

The IEC 62443 (ISA-99) technical standards are used worldwide and based on the ISA-99 (ISA–International Society of Automation) [25] series of technical reports. Currently they are further developed by the IEC (International Electrotechnical Commission). The standards include security models and technologies for IACS, security risk assessment, guidance for security management system implementation. They also address some privacy aspects related to IACS. The recent new program of work led by the ISA99 WG7 committee is coupling both safety and security issues together.

ISO/IEC 27019 Information technology – Security techniques – Information security management guidelines [26] based on ISO/IEC 27002 is dedicated to process control systems specific to the energy utility industry.

NIST (National Institute of Standards and Technology) SP 800-82 Guide to Industrial Control Systems Security [27] is used in the US and it specifies main threats sources, vulnerabilities, and incidents for IACS (see p. 137 in 800-82 guide). It also recommends security controls to mitigate risks. The NIST guide lists threat events that can directly endanger ICS components like PLCs, RTUs (Remote Terminal Units), DCS (Distributed Control Systems) or SCADA.

According to [28] security objectives for ICS are in contrast with the standard information security approach (e.g. Common Criteria), which focuses on such security attributes as: Confidentiality, Integrity, Availability (CIA). In IT systems priority is given to confidentiality of information, then comes integrity, and finally availability. It is quite contrary in ICS where priorities are reversed. It is important to note that fact when security measures are proposed for ICS devices.

After the review of standards it was found out that the author of only one publication [24] recommends CEM to be applied for ICS security evaluation.

In the next section the Common Criteria approach was shortly described.

2.4   Common Criteria for ICS Evaluation

ISO/IEC 15408 (Common Criteria – CC) is a 3-part standard describing criteria for security development and evaluation of IT products. Part 1 includes an introduction and general model, part 2 includes components to express security functional requirements (SFRs), and part 3 includes security assurance requirements (SARs). Every IT product may be used in many ways and in many types of environments, specifically in IACS environments.

In CC products are called TOE (Target of Evaluation). Evaluated security features of TOE have a proper level of assurance which is gained during independent assessment made by a certified laboratory ITSEF (IT Security Evaluation Facility). This assurance is measured by EALs (Evaluation Assurance Level) from EAL1 to EAL7. Evaluation results are verified in a certification process led by a Certification Body (CB). If verification is successful then a certificate is issued and published on the Common Criteria webpage [29]. Certificates are mutually recognized between country-members of either of two agreements: (1) SOG-IS MRA (Senior Officials Group Information Systems Security Mutual Recognition Agreement of Information Technology Security Evaluation Certificates with 14 members) [30]; (2) CCRA (Common Criteria Recognition Arrangement with 28 members) [31].

CC is based on the paradigm that security assurance depends on the rigor applied to the security development, operational environment, testing, verification, documenting, etc. This rigor is evaluated according to CEM which can be applied to many different categories of products, e.g.: integrated circuits, smart cards, databases, firewalls, operating system, etc. Till now 2333 products in total were certified (301 products in 2017) [32] but none of them was an ICS-typical device.

It is worth to note the CC methodology consists of three main processes [33–35]: (1) IT security development process; (2) TOE development process, and (3) security evaluation process which can be adapted to security evaluation of ICS products.

3   Results

As outlined in the introduction, the IT technology commonly used in ICS can be a source of threats and vulnerabilities for industrial control devices. This is why CC and CEM methodologies for IT security evaluation can be also applied for ICS devices to introduce proper IT security measures.

In general, the literature survey shows that there are some European initiatives to develop a common European framework for testing and certification of IACS systems. European institutions claim the framework could be based on the Common Criteria methodology.

From the same survey it is also evident that there are many publications describing security issues in ICS and standards relevant to cyber security applicable for IACS. However, very few of these publications pertain to the Common Criteria standard.

As mentioned in the introduction, two organizations in Europe ERNCIP [14] and ENISA [18] take into account the use of the CC methodology in the new European certification framework for IACS.

Although on the European level there is a trend to use CC for IACS, it is barely seen among studies about ICS security issues. Among several publications only one [21] mentioned the possible use of CC. This result is consistent with the review of standards for ICS. Among many publications about the standards for ICS only one paper [24] indicated directly the CEM methodology as adaptable for ICS devices.

The literature review demonstrates that security problems of ICS have become a common concern in industrial control and information technology. It also indicates there is almost no research on using the Common Criteria for industrial devices.

This supports the author's motivation to run research in that area on building an evaluation method based on CC and CEM dedicated for ICS. The results of this paper will be further used in the identification of requirements for the method.

4   Conclusions

The aim of this paper was literature review and building a necessary background for finding proper requirements for the ICS security evaluation method. The paper includes the review of security evaluation frameworks, standards, certification methods, and current security problems and challenges in industrial automation systems. The paper also describes some of the common vulnerabilities and types of attacks.

It has been demonstrated that little attention has been paid to using CC in the evaluation of ICS security, especially among publications which describe security issues of ICS. It was revealed that the issue of ICS security had been taken seriously by such European organizations like ERNCIP and ENISA. They anticipate the use of the Common Criteria methodology in the new European certification framework. All these factors prove the sense of running research work in that area. The author's doctoral research can provide a new CC-based approach to ICS security.

The results will be applied in the licensed security evaluation laboratory of IT products in Institute EMAG. The laboratory will be built within R&D project financed by The National Centre for Research and Development. The project aim is to develop and implement a national scheme for IT Security & Privacy evaluation and certification based on the Common Criteria approach.

The research work is very challenging because it tries to combine CC "security" requirements and "safety" requirements in one security evaluation method.

The author is currently working on developing a basic set of requirements needed to build a model of the security evaluation method based on Common Criteria.

Acknowledgment. I would like to thank Prof. Andrzej Bialas for his support in my research work regarding the security evaluation method for ICS and Ms Barbara Flisiuk for her insightful proofreading.

Research activities are financed within the "Implementation Doctorates" program of the Polish Ministry of Science and Higher Education.

References

1. Common Criteria for Information Technology Security Evaluation. Part 1: Introduction and general model. CCMB-2017-04-001, Version 3.1, Revision 5, April 2017
2. Common Criteria for Information Technology Security Evaluation. Part 2: Security functional components. CCMB-2017-04-002, Version 3.1, Revision 5, April 2017
3. Common Criteria for Information Technology Security Evaluation. Part 3: Security assurance components. CCMB-2017-04-003, Version 3.1, Revision 5, April 2017
4. Common Methodology for Information Technology Security Evaluation – Evaluation Methodology. CCMB-2017-04-004, Version 3.1, Revision 5, April 2017
5. ISO/IEC 18045:2008 – Information technology – Security techniques – Methodology for IT security evaluation
6. ICS-CERT Homepage. https://ics-cert.us-cert.gov/. Accessed 22 Oct 2018
7. ICS-CERT Annual Assessment Report FY 2016. US Department of Homeland Security, National Cybersecurity and Integration Center (NCCIC) (2016)
8. Yang, W., Zhao, Q.: Cyber security issues of critical components for industrial control system. In: Proceedings of 2014 IEEE Chinese Guidance, Navigation and Control Conference, pp. 2698–2703. IEEE, Yantai (2014)
9. Miyachi, T., Yamada, T.: Current issues and challenges on cyber security for industrial automation and control systems. In: 2014 Proceedings of the SICE Annual Conference (SICE), pp. 821–826. IEEE, Sapporo (2014)
10. Karnouskos, S.: Stuxnet worm impact on industrial cyber-physical system security. In: 37th Annual Conference on IEEE Industrial Electronics Society, IECON 2011, pp. 4490–4494. IEEE, Melbourne (2011)
11. Robert M. Lee, R.M., Assante, M.,J., Conway, T.: Analysis of the Cyber Attack on the Ukrainian Power Grid. E-ISAC, Washington (2016)

12. Valentine Jr., S.E.: PLC Code Vulnerabilities Through SCADA Systems. (Doctoral dissertation), University of South Carolina (2013)
13. Webster, J., Watson, R.T.: Analyzing the past to prepare for the future: writing a literature review. MIS Q. **26**(2), xiii–xxiii (2002)
14. Theron, P., Bologna, S.: Proposals from the ERNCIP Thematic Group, "Case Studies for the Cyber-security of Industrial Automation and Control Systems", for a European IACS Components Cyber-security Compliance and Certification Scheme. EUR – Scientific and Technical Research series (2014). ISSN 1831-9424, ISBN 978-92-79-45417-2
15. ENISA Homepage. https://www.enisa.europa.eu/. Accessed 03 Oct 2018
16. Communication network dependencies for ICS/SCADA Systems. ENISA (2016). ISBN 978-92-9204-192-2, https://doi.org/10.2824/397676, https://www.enisa.europa.eu/publications/ics-scada-dependencies
17. Dufkova, A., Budd, J., Homola, J., Marden, M.: Good practice guide for CERTs in the area of Industrial Control Systems. Computer Emergency Response Capabilities considerations for ICS. ENISA (2013)
18. Leszczyna, R., et al.: Protecting Industrial Control Systems. Recommendations for Europe and Member States. ENISA (2011). https://www.enisa.europa.eu/publications/protecting-industrial-control-systems.-recommendations-for-europe-and-member-states/
19. Babu, B., Ijyas, T., Muneer, P., Varghese, J.: Security issues in SCADA based industrial control systems. In: 2017 2nd International Conference on Anti-Cyber Crimes (ICACC). IEEE (2017)
20. Calvo, I., Etxeberria-Agiriano, I., Iñigo, M.A., González-Nalda, P.: Key vulnerabilities of industrial automation and control systems and actions to prevent cyber-attacks. IJOE (Int. J. Online Eng.) **12**(1), 9–16 (2016)
21. Xie, F., Peng, Y., Zhao, W., Gao, Y., Han, X.: Evaluating industrial control devices security: standards, technologies and challenges. In: Saeed, K., Snášel, V. (eds.) CISIM 2014. LNCS, vol. 8838, pp. 624–635. Springer, Heidelberg (2014). https://doi.org/10.1007/978-3-662-45237-0_57
22. General Electric. https://www.ge.com/digital/cyber-security. Accessed 07 Mar 2018
23. Piggin, R.S.H.: Development of industrial cyber security standards: IEC 62443 for SCADA and Industrial Control System security. In: IET Conference on Control and Automation 2013: Uniting Problems and Solutions, pp. 1–6. IEEE, Birmingham (2013)
24. Leszczyna, R.: Cybersecurity and privacy in standards for smart grids – a comprehensive survey. In: O'Connor, R., Schummy, H. (eds.) Computer Standards and Interfaces, vol. 56, pp. 62–73. Elsevier (2018)
25. ISA, ISA99: Industrial automation and control systems security. https://www.isa.org/isa99/. Accessed 24 Jan 2018
26. ISO/IEC, ISO/IEC TR 27019:2017 Information technology — Security techniques — Information security controls for the energy utility industry. https://www.iso.org/standard/68091.html. Accessed 21 Jan 2018
27. Stouffer, K., Pillitteri, V., Lightman, S., Abrams, M., Hahn, A.: NIST SP 800-82 Guide to Industrial Control Systems (ICS) Security Revision 2. NIST (2015)
28. Vavra, J., Hromada, M.: An evaluation of cyber threats to industrial control systems. In: International Conference on Military Technologies (ICMT), pp. 1–5. IEEE, Brno (2015)
29. The Common Criteria Homepage. http://www.commoncriteriaportal.org/. Accessed 03 Jan 2018
30. SOG-IS MRA – Mutual Recognition Agreement of Information Technology Security Evaluation Certificates, version 3.0. Final version January 8th, 2010. https://www.sogis.org/uk/mra_en.html. Accessed 23 Jan 2018

31. CCRA – Arrangement on the Recognition of Common Criteria Certificates. In the field of Information Technology Security, 2 July 2014. http://www.commoncriteriaportal.org/ccra/. Accessed 23 Jan 2018
32. CC certified products list. http://www.commoncriteriaportal.org/products/. Accessed 23 Jan 2018
33. Bialas, A.: Common criteria related security design patterns for intelligent sensors—knowledge engineering-based implementation. Sensors **11**, 8085–8114 (2011)
34. Bialas, A.: Computer-aided sensor development focused on security issues. Sensors **16**, 759 (2016)
35. Rogowski, D.: Software support for Common Criteria security development process on the example of a data diode. In: Zamojski, W., et al. (eds.) Proceedings of the Ninth International Conference on Dependability and Complex Systems DepCoS-RELCOMEX. Advances in Intelligent Systems and Computing, vol. 286, pp. 363–372. Springer (2014)

Time-Dependent Reliability Analysis Based on Structure Function and Logic Differential Calculus

Patrik Rusnak[⊠], Jan Rabcan, Miroslav Kvassay,
and Vitaly Levashenko

Faculty of Management Science and Informatics, University of Zilina,
Žilina, Slovakia
{patrik.rusnak,jan.rabcan,miroslav.kvassay,
vitaly.levashenko}@fri.uniza.sk

Abstract. Investigation of system reliability is a complex problem. Importance analysis of the system components is part of this investigation. Information about importance of the system components can also be used in system maintenance or in optimization of system reliability. In this paper we present how reliability function can be derived from the structure function of the system and how logic differential calculus allows us to investigate importance of the system components via time-dependent Birnbaum's and criticality importance measures.

Keywords: Reliability · Structure function · Time-dependent analysis
Birnbaum's importance · Criticality importance · Logic differential calculus

1 Introduction

Important step in reliability evaluation of system is the development of its mathematical representation. The most often used mathematical representation of a system in reliability analysis is a model that takes into account two important states of system: failure and working state. This mathematical model has been introduced as one of the first [2–4]. At the present time, there are many methods to evaluate the system reliability and failure based on this mathematical model. All these methods can be divided into four groups depending on the mathematical background [1, 3]: methods based on structure function, stochastic methods, Monte-Carlo simulation and methods based on universal generation function. All of these methods are used in reliability analysis and are alternative in system evaluation. The structure function based methods permit mathematically representing system of any structural complexity [2].

The structure function defines univalent correlation of a system performance level on components states and it is used to represent repairable system composed of n components in the stationary state [5]. The structure function can be viewed as a Boolean function [5, 6]. Such a mathematical representation is time-independent. Important advantage of this representation is possibility to use the well-developed and simple mathematical approach of Boolean algebra in reliability evaluation of the investigated system. Effective methods in reliability analysis were developed with

© Springer International Publishing AG, part of Springer Nature 2019
W. Zamojski et al. (Eds.): DepCoS-RELCOMEX 2018, AISC 761, pp. 409–419, 2019.
https://doi.org/10.1007/978-3-319-91446-6_38

application of Boolean algebra for minimal cut/path sets definition [7], frequency characteristics of system reliability [5], importance measures calculation [6]. The disadvantage of these methods is analysis of system in stationary state.

In this paper we develop methods of system reliability analysis based on Boolean algebra considering the dependence of components and system states on time. These methods allow analyzing system in non-stationary state too. Calculation of the reliability function is considered based on the structure function of the system. Although, reliability function is important for reliability analysis, it is not sufficient to give a complete picture about system reliability. Another necessary constituent of reliability evaluation is importance analysis. Methods for calculation of Importance Measures (IMs) based on application of logic differential calculus have been considered in [8, 9] for a system in stationary state. In this paper we continue in investigation from [8, 9] and we propose a method for system importance analysis depending on time and based on the application of logic differential calculus. The efficiency and usability of this method is shown by reliability estimation of a storage system.

2   Reliability Analysis Based on Structure Function

In this paper we are considering systems and its' components with two states that can be interpreted as functioning and failure. The structure function is used to describe the analyzed system [3].

2.1   Structure Function

The structure function is a mapping that defines value of system state for each combination of states of the system components. If we assume that the system is composed of n components, then this mapping is as follows:

$$\phi(x_1, \ldots, x_n) = \phi(\boldsymbol{x}) : \{0, 1\}^n \to \{0, 1\}, \tag{1}$$

where x_i is a variable defining state of component i $(i = 1, \ldots, n)$ and $\boldsymbol{x} = (x_1, \ldots, x_n)$ is a vector of states of the system components (state vector).

In this paper we will assume that the analyzed system is coherent, which means that structure function $\phi(\boldsymbol{x})$ is not decreasing in any of the variables and all the components are relevant for system operation [3, 9].

In reliability analysis it is also needed to compute the probability that the system will be functioning during its mission time (period of time during which the system is required to operate properly). This can be done by computing reliability R of the system that is defined as follows [3, 9]:

$$R = R(\boldsymbol{p}) = \Pr\{\phi(\boldsymbol{x}) = 1\}, \tag{2}$$

where p_i is the probability that component i will be functioning during the mission time (it agrees with reliability of the component), and $\boldsymbol{p} = (p_1, \ldots, p_n)$ is a vector of probabilities of components being functioning during the mission time. If the components of the system are mutually statistically independent, reliability R of the system can be obtained from the structure function using several transformation rules [9].

A complementary measure to system reliability is system unreliability, which agrees with the probability that the system will fail during the mission time [3, 9]:

$$F = F(\mathbf{q}) = \Pr\{\phi(\mathbf{x}) = 0\} = 1 - R(\mathbf{p}), \tag{3}$$

where $q_i = 1 - p_i$ is the probability of a failure of component i during the mission time (it agrees with unreliability of the component) and $\mathbf{q} = (q_1, \ldots, q_n)$ is a vector of unreliabilities of the components. It is worth noting that probabilities p_i and q_i are also known as state probabilities of component i because they define the probability that the component is in state 1 or 0 respectively.

System reliability (2) and unreliability (3) do not depend on time. The time-dependent functions $R(t)$ and $F(t)$ are defined based on the time-dependent state vector $\mathbf{x}(t) = (x_1(t), \ldots, x_n(t))$, where $x_i(t)$ is a state function of component i that defines state of the component at time t [9]. Similarly, vector $\mathbf{p}$ of reliabilities of the components has to be replaced by $\mathbf{P}(t) = (P_1(t), \ldots, P_n(t))$ and vector $\mathbf{q}$ of unreliabilities of the components by time-dependent vector $\mathbf{F}(t) = (F_1(t), \ldots, F_n(t))$ and:

$$P_i(t) = 1 - F_i(t). \tag{4}$$

Functions $R(t)$ and $F(t)$ are defined as follows:

$$R(t) = R(\mathbf{P}(t)) = \Pr\{\phi(\mathbf{x}(t)) = 1\}; F(t) = F(\mathbf{F}(t)) = \Pr\{\phi(\mathbf{x}(t)) = 0\}. \tag{5}$$

The procedure described above allows us to find reliability/unreliability (failure) function of the system if the structure function of the system is known, and we have information about lifetime distributions of all the system components. This proves that structure function, which is a static representation of the system can be used in time-dependent (dynamic) reliability analysis.

2.2 Logic Differential Calculus

Definition (1) of the structure function corresponds to the definition of Boolean function [8]. This means it is possible to use mathematical tools of Boolean algebra in reliability analysis based on structure function. One useful tool that can be used to analyze how failure of a component affects the system operation is logic differential calculus [10]. In order to analyze direction of component state change, a *Direct Partial Logic Derivative* (DPLD) is needed. This type of logic derivative can be used to analyze how a specific change of component state (from 0 to 1 or from 1 to 0) affects the system functionality (from 0 to 1 or from 1 to 0). This derivative is defined as follows [8]:

$$\frac{\partial\phi(1 \to 0)}{\partial x_i(1 \to 0)} = \frac{\partial\phi(0 \to 1)}{\partial x_i(0 \to 1)} = \overline{\phi(o_i, \mathbf{x})} \wedge \phi(1_i, \mathbf{x}), \tag{6}$$

where $\wedge$ denotes operation AND and $^{-}$ is a negation of the argument. In reliability analysis, DPLD can be used to compute importance measures [8].

3 Importance Measures

An important part of reliability analysis is an estimation of influence of a component or a group of components on system operation. Such estimation is implemented by IMs and can be used to optimize system reliability or to plan its maintenance. There are many IMs, and each of them takes into account different factors that make a system component more important than another. According to [9] IMs can be divided into three categories: structure, reliability, and lifetime IMs.

Structure IMs are used to calculate importance of components according to their placement in the system and do not take the reliability of components into account [9]. One of the most commonly known structure IMs is Structural Importance (SI) that is defined as a relative number of state vectors at which a failure of component i results in system failure. Using DPLDs, this measure can be computed as [8]:

$$\text{SI}_i = \text{TD}\left(\frac{\partial\phi(1 \to 0)}{\partial x_i(1 \to 0)}\right), \tag{7}$$

where TD(.) is the truth density of the argument and this value agrees with the relative number of vectors for which the argument takes a nonzero value.

Reliability IMs are mostly used when the structure function and the probabilities of the components functioning and failure are known [9]. Two reliability IMs are considered in the rest of the paper: *Birnbaum's Importance* (BI) and *Criticality Importance* (CI). BI takes into account system topology and the probabilities of the components functioning/failure. This measure is computed using DPLDs as [8]:

$$\text{BI}_i = \Pr\left\{\frac{\partial\phi(1 \to 0)}{\partial x_i(1 \to 0)} = 1\right\}, \tag{8}$$

and it agrees with the probability that a failure of component i results in system failure, i.e., with the probability that the component is critical for the system.

CI corresponds to the probability that system failure has been caused by a failure of component i given that the system has failed [8, 9]:

$$\text{CI}_i = \text{BI}_i\frac{q_i}{F}, \tag{9}$$

and it allows us to find components whose failures have resulted in system failure with the greatest probability when we know that the system has failed. It is typically used in system maintenance to identify components whose repairs will result in system repair with the greatest probability [3, 9].

Lifetime IMs can be calculated if the lifetime distributions of components probabilities $P_i(t)$ and $F_i(t)$ are known.

Birnbaum's time-dependent importance of component i at time t agrees with the probability that the system is in a state at time t in which component i is critical for the system, and it can be computed by partial differentiation of reliability function $R(t)$ according to $P_i(t)$ [9]:

$$\mathrm{BI}_i(t) = \frac{\partial R(t)}{\partial P_i(t)}. \tag{10}$$

Another possibility, which is presented in this paper, is to calculate firstly DPLD $\partial\phi(1 \to 0)/\partial x_i(1 \to 0)$ and then transform it into a probabilistic function similarly as structure function $\phi(x)$ can be transformed into reliability function $R(t)$ (5). This procedure is shown in an example considered below.

Time-dependent CI can be computed from $\mathrm{BI}_i(t)$ as [9]:

$$\mathrm{CI}_i(t) = \mathrm{BI}_i(t) \frac{F_i(t)}{F(t)}, \tag{11}$$

and it corresponds to the probability that component i has failed by time t and that component i is critical for the system at time t, given that the system has failed by time t.

4 Reliability Analysis of Storage System

In this section we consider the reliability evaluation of storage system (Fig. 1) by the calculation of reliability function of a system based on its structure function and time-dependent BI and CI measures with the use of DPLDs.

The analyzed storage system is composed of two storage units parallel connected in communication network. These units have two Hard Disk Drives (HDDs) organized in a specific structure called Redundant Array of Inexpensive (or Independent) Disks (RAID). The first unit, which is in the upper path, consists of two HDDs in RAID 1, namely SEAGATE ST6000DX000 and WDC WD60EFRX. In RAID 1, data are

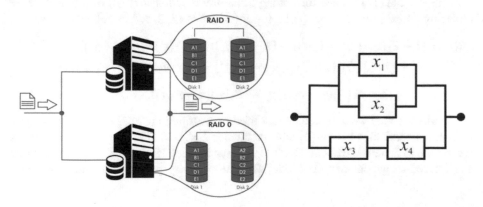

Fig. 1. Analyzed storage system (a) and its reliability block diagram (b)

written into two drives identically, which means that data can be read from any drive. Also RAID 1 will operate successfully as long as at least one drive is functioning [11]. The second unit, which is in the lower path, consists of two HDDs organized in configuration RAID 0, namely WDC WD30EZRX and HGST HDS5C3030ALA630. In RAID 0, the capacity of the unit is equal to the sum of capacities of the used drives, which implies no redundancy of data. Therefore, failure of one drive means that the entire RAID 0 is lost [11]. Each of both units has capacity 6 TB, and they are parallel connected to the network and used to store the same data. The HDDs WDC WD30EZRX and HGST HDS5C3030ALA are used in RAID 0, because they have lower read and write speeds than HDDs in RAID 1 and RAID 0 allows us to increase read and write speeds. Thanks to this the storage system is functioning (data can be stored to it or load from it) if at least one unit is working.

In reliability analysis of the storage system we will focus mostly on storage units, specifically their HDDs, and we will not take network reliability into account. According to the description of the storage system, the system can be in one of two possible states – state 0, which agrees with a situation in which data cannot be stored or retrieved, and state 1, in which it is possible to store data or retrieve them. The components are HDDs, and they can also be in one of two states: state 1 represents functioning HDD; state 0 represents HDD failure.

System topology expressed in the form of reliability block diagram can be viewed in Fig. 1, where x_1 denotes a variable defining state of HDD SEAGATE ST6000DX000 in RAID 1, x_2 is a state variable for HDD WDC WD60EFRX in RAID 1, x_3 represents a variable defining state of HDD WDC WD30EZRX in RAID 0, and x_4 agrees with state of HDD HGST HDS5C3030ALA in RAID 0. The reliability block diagram allows us to obtain the structure function of the system:

$$\phi(x_1, x_2, x_3, x_4) = (x_1 \lor x_2) \lor x_3 \land x_4. \tag{12}$$

The system reliability R (2) is calculated by the structure function (12) as:

$$(x_1 \lor x_2) \lor x_3 \land x_4 \rightarrow (p_1 + p_2 - p_1 p_2) + p_3 p_4 - (p_1 + p_2 - p_1 p_2) p_3 p_4. \tag{13}$$

The formula (13) is used for finding reliability function $R(t)$ (5) of this system by replacing probabilities p_i by $P_i(t)$ or $1 - F_i(t)$, for $i = 1, 2, 3, 4$ as follows:

$$\begin{aligned}
R(t) = {} & (1 - F_1(t)) + (1 - F_2(t)) - (1 - F_1(t))(1 - F_2(t)) + (1 - F_3(t))(1 - F_4(t)) \\
& - (1 - F_1(t))(1 - F_3(t))(1 - F_4(t)) - (1 - F_2(t))(1 - F_3(t))(1 - F_4(t)) \\
& + (1 - F_1(t))(1 - F_2(t))(1 - F_3(t))(1 - F_4(t)).
\end{aligned} \tag{14}$$

In this example we will assume that lifetime distributions $F_i(t)$ of the HDDs agree with exponential distribution [3] with parameter $\lambda = 1/\text{MTTF}$, where MTTF denotes mean time to failure. We obtained MTTF in days for each HDD from data published by Backblaze Storage company in 2016 [12]. The company publishes quarterly statistics about HDDs that they use in their storage solutions. Based on them, the company estimates Annualized Failure Rate (AFR) for individual models of HDDs. AFR is an estimation of the probability that a device (HDD) will fail during a full year of use [13], and its relation to MTTF in days is as follows [14]:

$$AFR = 1 - e^{\frac{-365.25}{MTTF}}. \tag{15}$$

AFRs of the HDDs estimated by the company based on data from April 2013 to December 2016 are shown in Table 1. MTTFs in days, which we need to obtain lifetime distributions $F_i(t)$ of the HDDs, were obtained from AFRs by transforming (15) into the following form:

$$MTTF = \frac{-365.25}{\ln(1 - AFR)}. \tag{16}$$

Based on MTTFs presented in Table 1, we can obtain the reliability function of the storage system. Its time course is presented by the blue dashed curve in Fig. 3. In this graph we can see how reliability of the storage system with increasing time gradually degrades.

Table 1. Properties of HDDs used in the storage system

Component	Component data			
	HDD name	Capacity [TB]	AFR	MTTF [days]
1	SEAGATE ST6000DX000	6	0.0143	25,359
2	WDC WD60EFRX	6	0.0568	6,246
3	WDC WD30EZRX	3	0.0738	4,764
4	HGST HDS5C3030ALA	3	0.0082	44,360

Reliability function $R(t)$ can also be used to find failure function $F(t)$ of the storage system. This function defines system unreliability, and it can be computed according to (3). Time course of this function for the storage system is presented as the red solid line in Fig. 2. Based on its shape, we can find the same results as in the case of the reliability function [15].

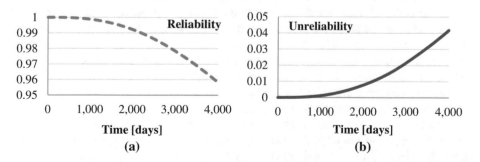

Fig. 2. Reliability (a) and unreliability (b) function of the storage system

Using functions $R(t)$ and $F(t)$, we can see how the system reliability and unreliability will change in time, but we are unable to conclude how the components are important for system operation. Therefore, we compute several importance measures for every component in the next step. All these computations can be done using structure function (12) and logic differential calculus.

In the first step, we compute BI of each component. Because the system is coherent, we can compute BI using DPLDs and by their transformation into a probabilistic form. For this purpose, let us firstly compute DPLD for the first HDD of the storage system. This can be done using formula (6) as:

$$\frac{\partial \phi(1 \rightarrow 0)}{\partial x_1(1 \rightarrow 0)} = \overline{((0 \vee x_2) \vee x_3 \wedge x_4))} \wedge ((1 \vee x_2) \vee x_3 \wedge x_4)) = \overline{x_2} \wedge (\overline{x_3} \vee \overline{x_4}) \quad (17)$$

This result implies that HDD 1 is critical for the system, i.e., its failure results in a failure of the system, if HDD 2 and at least one from HDDs 3 and 4 fail. Using the same procedure that was used to obtain the reliability from the structure function, we can transform Boolean formula (17) into probabilistic form, which agrees with the probability that HDD 1 is critical for the system, i.e., with the BI of HDD 1:

$$BI_1 = 1 - p_2 - p_3 p_4 + p_2 p_3 p_4. \quad (18)$$

BIs for other components are computed by the similar way. BI measures do not depend on time, and they allow us to compute and compare importance of the components only for given values of the state probabilities of the components. However, the probabilities of individual states of the system components change as time flows, what implies that the importance of the components also changes over the time. This changes are investigated based on the time-dependent versions of BIs that are computed by replacing probabilities p_i with time-dependent probability functions $P_i(t)$, for $i = 1, 2, 3, 4$, as we did above to find reliability function $R(t)$ of the storage system. For example, application of this procedure on BI_1 (18) results in the following formula for computation of time-dependent BI measure for HDD 1:

$$BI_1(t) = 1 - P_2(t) - P_3(t)P_4(t) + P_2(t)P_3(t)P_4(t). \quad (19)$$

Since probability $P_i(t)$ is defined as a complement of lifetime distribution $F_i(t)$ to value 1 ($i = 1, 2, 3, 4$). In the similar way, we can obtain time-dependent versions of BI measures of the rest of the components of the storage system. Time courses of all these measures are depicted in Fig. 3. In the figure we can see that all the components of the system have similar importance at the beginning of the system operation but at time 500+days, BI of HDD 1 is much greater than BI measures of the remaining HDDs, and its importance grows about four times faster than the importance of HDD 2. We can also conclude that the component with the least importance is HDD 4 and its BI does not change during the time. These results are quite reasonable because HDD 4 is connected in series with HDD 3, what means it influences system operation if HDD 3 is working. Since HDD 3 is unreliable (its MTTF presented in Table 1 is overall low), it is very likely this HDD is not functioning. Therefore, a failure of HDD 4 can result in system failure only with a small probability. Furthermore, since HDD 3 penalizes activity of HDD 4, which is the most reliable HDD according to MTTFs, importance of HDDs 1 and 2 grows. From these two HDDs, the most important is HDD 1 because if it fails, then it is very likely that the system fails. This results from the fact that a path composed of HDDs 3 and 4 (Fig. 1) is very unreliable since HDD 3 has little MTTF, and a path containing HDD 2 is less reliable than a path containing HDD 1 since MTTF of HDD 2 is about four times less than MTTF of HDD 1.

The BI measures presented in Fig. 3 show how the probabilities that a failure of individual components results in system failure changes over time. However, these measures do not consider reliability of components for which they are computed, i.e., BI of component i does not depend on lifetime distribution $F_i(t)$. To avoid this problem, static (without time) and dynamic (with time) version of CI measures is computed. Time-dependent CI can be computed for HDD 1 based on (12) as:

$$\mathrm{CI}_1(t) = \mathrm{BI}_1(t)\frac{F_1(t)}{F(t)} = \mathrm{BI}_1(t)\frac{F_1(t)}{1 - R(t)} = 1, \tag{20}$$

where $\mathrm{BI}_1(t)$ is time-dependent BI of HDD 1 computed in (19), $R(t)$ is reliability function (14) of the storage system, and $F_1(t)$ is lifetime distribution of component i defining the probability that the component fails throughout interval $0, t$ given that it worked at time 0. After substituting all functions by their formula, we obtain the result 1, which implies that CI of HDD 1 does not depend on time. This result agrees with our expectations because HDD 1 constitutes one path in reliability block diagram in Fig. 1 and, therefore, its repair surely results in system repair if we know that the system has failed. This corresponds to the meaning and usage of CI presented in Sect. 3 according which CI allows us to find components whose repair results in system repair with the greatest probability given that the system has failed.

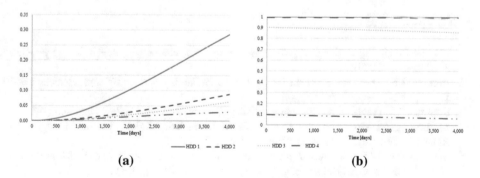

Fig. 3. Time-dependent BI (a) and CI (b) measures for HDDs of the storage system

The similar procedure can be used to find time-dependent CI measures for the rest of the components. Their time courses are shown in Fig. 3. From the graph, we can conclude that the most important components are HDDs 1 and 2, while a HDD with the least importance is HDD 4. This order of importance is similar to that obtained using the BI measures, but the values of CI measures are completely different from BI ones. This result from the fact that the CI does not compute the probability that a failure of a given component results in a failure of the system as BI, but instead it estimates the probability that a failure of the system has been caused by component i assuming that the system has failed.

5   Conclusion

In this paper, we presented application of structure function in time-dependent reliability analysis. Firstly, we showed how the structure function can be used to find reliability function $R(t)$ and failure function $F(t)$ of the system if lifetime distributions $F_i(t)$ of all the components of the system are known. In the second part, we showed how time-dependent or lifetime BI and CI measures can be computed using logic differential calculus. This generalizes results from [8, 9] where logic differential calculus was used only in computation of reliability (time-independent) IMs.

It is worth to notice that the main contributions of this paper are in showing that the structure function that defines system topology, which is independent of time, can also be used in time-dependent reliability analysis, and DPLDs can be used to find formulae for computation of most commonly known time-dependent IMs. This was illustrated in the last part of the paper, which deals with reliability analysis of the storage system composed of 4 HDDs. We showed how its reliability function can be derived from the structure function and how time-dependent importance analysis of such a system can be done using logic differential calculus.

Acknowledgment. This work was partly supported by the grants VEGA 1/0038/16 and 1/0354/17.

References

1. Zio, E.: Reliability engineering: old problems and new challenges. Reliab. Eng. Syst. Saf. **94**, 125–141 (2009)
2. Barlow, R.E., Proschan, F.: Statistical Theory of Reliability and Life Testing. Rinehart and Winston Holt, New York (1975)
3. Rausand, M., Høyland, A.: System Reliability Theory, 2nd edn. Wiley, Hoboken (2004)
4. Kaufmann, A., Grouchko, D., Cruon, R.: Mathematical Models for the Study of the Reliability of Systems. Academic Press, New York (1977)
5. Schneeweiss, W.G.: A short Boolean derivation of mean failure frequency for any (also non-coherent) system. Reliab. Eng. Syst. Saf. **94**, 1363–1367 (2009)
6. Armstrong, M.J.: Reliability-importance and dual failure-mode components. IEEE Trans. Reliab. **46**(2), 212–221 (1997)
7. Kvassay, M., Levashenko, V., Zaitseva, E.: Analysis of minimal cut and path sets based on direct partial Boolean derivatives. Proc. Inst. Mech. Eng. Part O: J. Risk Reliab. **230**(2), 147–161 (2016)
8. Zaitseva, E., Levashenko, V., Kostolny, J.: Importance analysis based on logical differential calculus and Binary Decision Diagram. Reliab. Eng. Syst. Saf. **138**, 135–144 (2015)

9. Kuo, W., Zhu, X.: Importance Measures in Reliability, Risk, and Optimization: Principles and Applications. Wiley, Hoboken (2012)
10. Steinbach, B., Posthoff, C.: Boolean Differential Calculus. In: Synthesis Lectures on Digital Circuits and Systems, vol. 12, no. 1, pp. 1–215 (2017)
11. Powell, G.: Beginning Database Design. Wiley, Hoboken (2006)
12. Klein, A.: Backblaze hard drive stats for 2016 (2016). https://www.backblaze.com/blog/hard-drive-benchmark-stats-2016/. Accessed 20 Dec 2017
13. Elerath, J.G.: AFR: Problems of definition, calculation and measurement in a commercial environment. In: Annual Reliability and Maintainability Symposium, pp. 71–76 (2000)
14. Szabados, D.: Diving into 'MTBF' and 'AFR': Storage reliability specs explained. https://web.archive.org/web/20100501151901/, http://enterprise.media.seagate.com/2010/04/inside-it-storage/diving-into-mtbf-and-afr-storage-reliability-specs-explained/. Accessed: 20 Dec 2017
15. Burgess, M.: Analytical Network and System Administration: Managing Human-Computer Systems. Wiley, Hoboken (2012)

Analysis of Information Transmission Security in the Digital Railway Radio Communication System

Mirosław Siergiejczyk$^{(\boxtimes)}$ and Adam Rosiński

Warsaw University of Technology, Warsaw, Poland
{msi,adro}@wt.pw.edu.pl

Abstract. GSM-R (Global System for Mobile Communications-Railways) is a European radio communication standard developed and used for the railway applications. is a telecommunication structure operating for the European Rail Traffic Management System (ERTMS) and providing digital radio communications between the rail vehicle and the Radio Control Centre, as well as between the personnel members responsible for traffic management (drivers, dispatchers, conductors, rolling stock security employees, etc.). Therefore, the transmission correctness of information determines the train operation safety. The article analyses the factors that constitute a potential source of danger for the information transmission in GSM-R. The possibilities of increasing the transmission security by monitoring the radio signal coverage ratio of the railway line as well as the analysis of the quality factor statistics of the air interface were presented. The system should be also complemented by additional encryption mechanisms of a radio link and signal channels, as well as additional infrastructure elements related to the security provision (e.g. firewall). It is necessary to create a formal security policy (including the records of active and withdrawn terminals, management rules for SIM cards, and *Disaster Recovery* scenarios) and to integrate the system with other areas, among others, with the switching subsystem monitoring. These activities are designed to provide comprehensive network protection, which contributes to maximum security of the transmitted information.

The abstract should summarize the contents of the paper in short terms, i.e. 150–250 words.

Keywords: GSM-R · Security · Radio coverage control

1 Introduction

GSM-R (Global System for Mobile Communications-Railways) is a European radio communication standard developed and used for the railway applications. It provides digital voice communication and digital data transmission. It offers the complex functionality of a GSM system [10, 11]. It is characterised by infrastructure located only near the railway lines. GSM-R is designed to support the systems being introduced in Europe: ERTMS (European Rail Traffic Management System) and ETCS (European Train Control System), which are designed to continuously collect and

© Springer International Publishing AG, part of Springer Nature 2019
W. Zamojski et al. (Eds.): DepCoS-RELCOMEX 2018, AISC 761, pp. 420–429, 2019.
https://doi.org/10.1007/978-3-319-91446-6_39

transmit data on a rail vehicle, such as speed or geographical location [2, 14]. GSM-R is a telecommunication structure operating for the European Rail Traffic Management System (ERTMS) and providing digital radio communications between the rail vehicle and the Radio Control Centre, as well as between the personnel members responsible for traffic management (drivers, dispatchers, conductors, rolling stock security employees, etc.) [18]. The transmitted data include, among others, the train speed and its current geographical location, which determine the decision on the content of further timetables (authorisation for movement at a specific speed and the lack of authorisation). Therefore, the transmission correctness of this information determines the train operation safety [8].

In the article, the factors that constitute a potential threat to correctness of the transmission of data sensitive to distortion will be analysed, and the concept of solutions, which have an impact on minimisation of the probability of the critical system failure occurrence, will be presented. It will be related to the issue of monitoring the coverage of the railway line and the statistics of air interface quality factors.

2   Threats Resulting from Technical Factors

The threats resulting from technical factors are the easiest to eliminate or minimise the frequency of their occurrence [7]. An effective method is to conduct regular inspections of the condition of the devices, as well as to install monitoring or automatic alarm systems (e.g. open-door alarm, no battery alarm, etc.) [9, 12, 13]. However, the most important issue is to provide the network redundancy and the effective system for switching the movement in case of a failure [17]. The network operation may be also affected by, e.g. damage to the antenna module of the mobile device. The tools for monitoring the statistics of its operation quality, which are used in mobile networks, allow to identify a single user and transmit information on the necessity of putting this IMEI number on a list of observed or blocked numbers (EIR list).

All the above-mentioned cases have a significant impact on the mobile network operation with radio access, and on information transmission security. The device that communicates with the transceiver station at incorrect frequencies or in an improper manner resulting from damage to the radio module of the equipment interferes with the radio access network [15].

In every system, a human being is the weakest link. It is a man that causes most situations that pose a real threat to information security. It is possible to divide them into unconscious and conscious ones.

The unconscious activities most often result from carelessness and distraction of the organization's employees, as well as the lack of knowledge of the current security policy or the failure to comply with it (lack of awareness of the threats resulting from such behaviour). The problem is also laziness being part of the human nature and resulting in a tendency to excessive automation of everyday duties without proper control of the created software operation, and over-reliance on used untested tools, etc. The resulting threats do not have to entail the consequences in the form of, e.g. thefts or data modification, because the user is not guided by bad intentions. However, these situations can be effectively used by people who have knowledge about the most common mistakes.

A much more serious threat to the security of information transmitted in the network is a deliberate destructive activity. It relates both to the company employees and outworkers. There are many ways of classification of attacks on information systems. Below, there are listed only several most common types of threats, which are the most important from the perspective of the subject of the article:

- information collection (Information Gathering);
- data interception (Passive Interception);
- transmission eavesdropping (Man-In-The-Middle);
- blocking access to services (Denial of Service);
- spoofing;
- information modification (Misconfiguration);
- faking;
- tracking;
- spam.

In the radio access networks, MITM (Man In The Middle) attacks are the most common, when another eavesdropping element, which the interlocutors do not know about, occurs between a sender and a receiver. The attack designed to obtain the information about a subscriber is IMSI catcher mechanism. The problem of networks based on 2G standard is the lack of their authorisation for the user. An attacker is able to create fake BTS that simulates an element of the home network. If the device emits a stronger signal than the home network, MS will detect it and try to log in, providing all of its identification data. The fake BTS, which is actually a device with a SIM card, will correctly log in to the network. At the same time, it will force the unencrypted transmission from the victim's mobile device, being, in a sense, a relay between MS and the network. Therefore, in the GSM-R network, it is necessary to create and strictly follow the security policy in the field of distribution of SIM cards, because by having the knowledge about the subscriber's IMSI number, it is possible to extort a lot of information from the network.

A type of MITM attack includes eavesdropping/interception of SMS messages, a voice call or the content of GPRS transmission – all these services are used in the GSM-R system. The call eavesdropping (regardless of the provided service) on the radio channel is very difficult due to the used encryption. Normally, in order to set up the call, the user must identify itself by means of keys stored in MSC and in VLR. In case of performing handover (which in case of GSM-R occurs very often) between VLRs, VLR that currently provide services to the VLR subscriber polls the previous set of keys in send Identification message. This transmission is not encrypted as standard. The attacker can intercept it, and therefore, decrypts the transmission.

Another type of the attack includes impersonating the subscriber (Faking type attacks). This is a method that results in the transmission eavesdropping, after prior use of the IMSI Catcher mechanism, consisting in faking the mobile device by an unauthorised change in information on the subscriber's location and rights. When the subscriber tries to perform the service, the network will redirect it to the attacker's device in accordance with the saved information.

An example of the attack aimed at blocking access to services is Denial of Service (DoS). In case of GSM-R, it can involve an unauthorised change of rights to perform the services (possibility of making voice calls or setting up GPRS sessions). Another example of DoS attack involves sending the mass amount of questions to the network elements, which results in its congestion and the lack of possibilities of using the resources by other users.

Another type of the attack involves tracking the subscriber, both in terms of its location and its provided services. By having the knowledge only about MSISDN number to the SIM card, there is the possibility of polling the centre about the subscriber's identification data (IMSI number and the number of serving VLR). With this information, it is possible to extort the actual location from the mobile device with an accuracy to the cell radius.

3 Concept of Solutions Affecting the Increase in Information Transmission Security in GSM-R System

3.1 Radio Coverage Control

The radio coverage control involves testing strength of a signal transmitted from transceivers, which is received by mobile devices. The recommendations regarding the GSM-R radio coverage testing have been described in detail by the UIC International Union of Railways in *Procurement & Implementation Guide V.1.0* [16] document and in SRS EIRENE specifications [5].

The requirements of these parameters for trains moving at a maximum speed of 220 km/h are as follows:

- received signal strength *Received Signal Level RxLev UL/DL* at the level ≥ -95 dBm for 95% of samples for lines equipped with ETCS of level 2 or 3;
- received signal strength *Received Signal Level RxLev UL/DL* at the level ≥ -98 dBm for 95% of samples for lines equipped with ETCS of level 1.

These values directly show that the so-called "coverage holes" – the areas, where the coverage requirements are not met, cannot be greater than 5 m per each 100 m (measurement sections along the railway line). It should be also noted that these values refer to the worst scenario resulting, e.g. from very bad weather conditions, a high level of external interference, etc.

The applied quality parameter is RxQual, indicating the quality of the signal received by the device before its decoding. It adopts integers from 0 to 7 [4]. In the GSM-R radio access network design process, the recommended RxQual at the level of 4 or lower for 95% of samples per each 100 m of the railway line is assumed [3], which corresponds to BER error rate in the range of 0.8% < BER < 1.6%. The interference occurrence on the air interface significantly affects the signal quality factor. In Fig. 1, the example distribution of RxQual parameter for the transceiver station in the diagram was presented.

The level of the radio signal interference also directly affects the correctness of setting up the calls, which in case of GSM-R should be >99% [5]. Figure 2 shows the

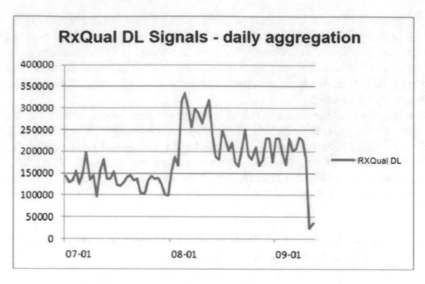

Fig. 1. Example of RxQual statistics (own development).

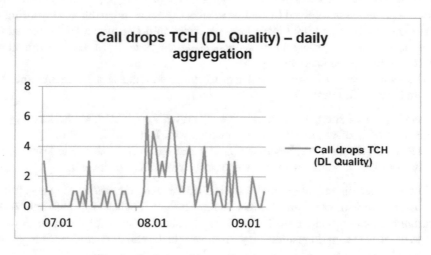

Fig. 2. Statistics of lost calls (own development).

distribution of the number of lost calls due to the insufficient signal quality for the same transceiver station. The correlation between the signal quality and the call loss is visible.

A decrease in the RxQual parameter value to the level of 5 results in ineffectiveness of information included in the GSM-R signal frame, due to a too high error rate and too much probability of information "distortion" [6]. Although the device is only an intermediary in communication between the SIM card and the services provided by the network, in case of GSM-R, the occurrence of the radio signal interference resulted from the mobile station malfunction, is unacceptable due to the real threat to the train movement safety.

In addition, it is proposed to use the monitoring of a signal received by individual mobile terminals of the GSM-R system, without the necessity of additional software. These measurements use the fact that the terminals continuously report the current *RxQual* and *RxLevel* parameters to the network. By using the appropriate commands entered on the BSC controller, the data can be saved. The disadvantage of this type of solution is the lack of information on the exact location of the device, because standard mobile terminals are not equipped with a GPS system. However, an undeniable advantage is that the measurement can be initiated by the personnel of the Network Monitoring Centre at any time, and due to the limited amount of data, and their analysis is not complicated.

Another method for controlling the coverage is to monitor the statistics of the average signal strength in the broadcast channel, in both transmission directions (Figs. 3 and 4).

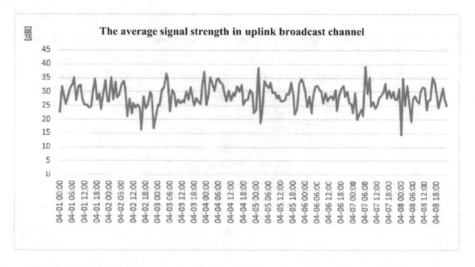

Fig. 3. The average signal strength in *uplink* broadcast channel (own development).

3.2 Radio Coverage Control

The second main parameter of the radio access network monitoring, along with the received signal strength, is its quality. According to the EIRENE SRS standardisation, the required *value of Received Signal Quality RxQual UL/DL* was determined at the level ≤ 4 for 95% of samples. This value directly affects the BER error rate and it is a basis for performing the service of transferring calls (*handover*).

3GPP standardisation clearly defines RxQual mapping on BER rate, which is presented in Table 1 [1].

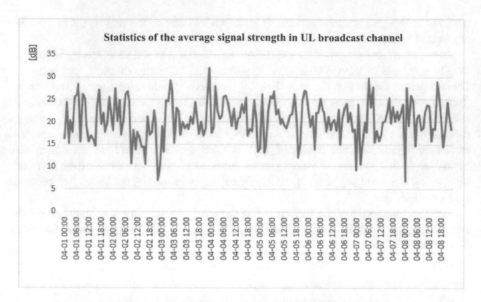

Fig. 4. Statistics of the average signal strength in UL broadcast channel (own development).

Tab. 1. RxQual mapping, and BER (development based on 3GPP TS 05.08).

RxQual	Bit Error Rate (BER) range
	BER < 0.2%
1	0.2% < BER < 0.4%
2	0.4% < BER < 0.8%
3	0.8% < BER < 1.6%
4	1.6% < BER < 3.2%
5	3.2% < BER < 6.4%
6	6.4% < BER < 12.8%
7	12.8% < BER

Due to a critical impact of the BER error rate on the reliability of the transmitted information and train movement operation security, in the proposed system, an alert threshold was determined, when 95% of samples in 15-min aggregation exceeds the value of 3.5.

Figures 5 and 6 present weekly statistics of the average value *RxQual* for the transmission from the station to mobile devices (DL) and the other way around (UL) for the example GSM-R station within the suburban area.

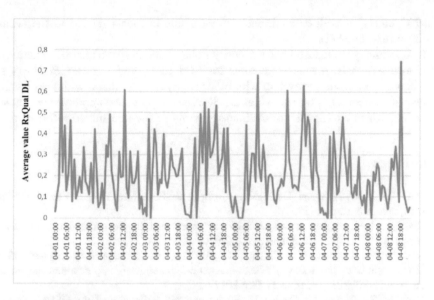

Fig. 5. Statistics of the average value RxQual DL (own development).

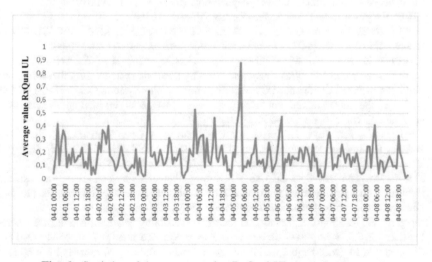

Fig. 6. Statistics of the average value RxQual UL (own development).

4 Conclusion

The article analyses the factors that constitute a potential source of danger for the information transmission in GSM-R. The possibilities of increasing the transmission security by monitoring the radio signal coverage ratio of the railway line as well as the analysis of the quality factor statistics of the air interface were presented. The system should be also complemented by additional encryption mechanisms of a radio link and signal channels, as well as additional infrastructure elements related to the security provision (e.g. firewall).

It is necessary to create a formal security policy (including the records of active and withdrawn terminals, management rules for SIM cards, and *Disaster Recovery* scenarios) and to integrate the system with other areas, among others, with the switching subsystem monitoring. These activities are designed to provide comprehensive network protection, which contributes to maximum security of the transmitted information.

References

1. 3GPP TS 05.08, Technical Specification Group GSM/EDGE Radio Access Network; Radio subsystem link control (1999)
2. Białoń, A.: Masterplan of the ERTMS implementation in the national and community perspective. Transport i Komunikacja 2 (2010)
3. Dudoyer, S., Deniau, V., Slimen, N., Adriano, R.: Susceptibility of the GSM-R Transmissions to the Railway Electromagnetic Environment. http://cdn.intechopen.com/pdfs-wm/34801.pdf. Accessed 11 Dec 2017
4. Dudoyer, S., Slimen, N., Deniau V., Berbineau, M.: Reliability of the GSM-R Communication. System against Railway Electromagnetic Interferences. http://www.railway-research.org/IMG/pdf/h03_dudoyer_stephen.pdf. Accessed 11 Dec 2017
5. EIRENE System Requirements Specification Version 16.0.0
6. GSM-R Radio Planning Guidelines, 02-2006, JERNBANEVERKET UTBYGGING Document number 3A-GSM-038. www.trv.jbv.no/tidligere-utgaver-T6009b00.pdf
7. Kasprzyk, Z., Rychlicki, M.: Analysis of phiysical layer model of WLAN 802.11 g data transmission protocol in wireles networks used by telematic systems. In: Proceedings of the Ninth International Conference Dependability and Complex Systems DepCoS-RELCOMEX, given as the monographic publishing series – "Advances in intelligent systems and computing", vol. 286, pp. 265–274. Springer (2014)
8. Lewiński, A., Perzyński, T., Toruń, A.: The Analysis of Open Transmission Standards in Railway Control and Management. In: Mikulski, J. (ed.) TST 2012. CCIS, vol. 329, pp. 10–17. Springer, Heidelberg (2012). https://doi.org/10.1007/978-3-642-34050-5_2
9. Paś, J., Rosinski, A.: Selected issues regarding the reliability-operational assessment of electronic transport systems with regard to electromagnetic interference. Eksploatacja i Niezawodnosc – Maintenance and Reliability 19(3), 375–381 (2017)
10. Siergiejczyk, M., Gago, S.: Safety and security, availability and certification of the GSM-R network for ETCS purposes. Archivess Transport Syst. Telematics 7(1), 45–49 (2014)
11. Siergiejczyk, M. (ed.): High Speed Railways in Poland. Railway Institute Publishing House, Warsaw (2015)
12. Siergiejczyk, M., Krzykowska, K., Rosiński, A., Grieco, L.A.: Reliability and viewpoints of selected ITS system. In: Proceedings 25th International Conference on Systems Engineering, ICSEng 2017, pp. 141–146. IEEE, Conference Publishing Services (CPS) (2017)
13. Siergiejczyk, M., Paś, J., Rosiński, A.: Issue of reliability–exploitation evaluation of electronic transport systems used in the railway environment with consideration of electromagnetic interference. IET Intel. Transp. Syst. 10(9), 587–593 (2016)
14. Siergiejczyk, M., Pawlik, M., Gago, S.: Safety of the new control command European system. In: Proceeding of the European Safety and Reliability Conference, pp. 635–642. CRC Press/Balkema (2014)

15. Sumiła, M., Miszkiewicz, A.: Analysis of the problem of interference of the public network operators to GSM-R. In: Mikulski, J. (ed.) TST 2015. CCIS, vol. 531, pp. 253–263. Springer, Cham (2015). https://doi.org/10.1007/978-3-319-24577-5_25
16. UIC, GSM-R Procurement & Implementation Guide. http://www.uic.org/IMG/pdf/2009gsm-r_guide.pdf. Accessed 11 Dec 2017
17. Wielemborek, R., Laskowski, D., Łubkowski, P.: Safety and reliability of data transmission over public networks. In: Proceedings of the European Safety and Reliability Conference, ESREL 2015, pp. 399–407. CRT Press/Balkema (2015)
18. Winter, P.: International Union of Railways, compendium on ERTMS. Eurail Press, Hamburg (2009)

Modeling Transport Process Using Time Couplings Methods

Emilia Skupień[✉] and Zdzisław Hejducki

Wroclaw University of Science and Technology,
27 Wybrzeze Wyspianskiego Str., 50-370 Wroclaw, Poland
{emilia.skupien, zdzislaw.hejducki}@pwr.edu.pl

Abstract. The paper concerns operational research and in particular problems of scheduling transport projects (ships traffic synchronization on inland waterways). The optimization issue presented by authors concerns in particular: the maximization of use the limiting locks (bottlenecks), i.e. scheduling ships to be in front of the lock when it is ready for lockage (shown on basis of Gliwice Canal). The basic problem is the synchronization of two-way vessel traffic on a limited section. The research, for calculation model, developed the Time Couplings Methods (TCM). Due to the computational complexity a deterministic approach to research was adopted.

Time Couplings Methods TCM base on internal time connections between transport processes and sectors (waterway sections, locks), taking into account resource constraints (e.g. types of ships) and technical (capacity of locks).

Scheduling of transport processes (ship traffic on canalized waterway) with the use of time couplings, consists in synchronizing the cruise, i.e. determining the start and end of the cruise on each section, ensuring lack of collisions of units moving bi-directionally. An example of ships schedule on canalized waterway was presented.

Keywords: Time couplings methods · Transport planning · Inland navigation

1 Introduction

The paper concerns operational research and in particular problems of scheduling transport projects (ships traffic synchronization on inland waterways). In current literature on linear planning issues, introduces are for example: linear scheduling (LSM - Linear Scheduling Model), linear diagrams (LOB - Line of Balance) and CPM/PERT network planning, with consideration a number of objective functions: the smallest cost, time, limited resources, work priorities etc., both in deterministic and probabilistic terms. These problems are solved by simulation methods and others. In this article the basic concepts and assumptions on the basis of the Time Couplings Methods [1, 2] were adopted to establish time parameters in scheduling transport processes of ships on inland, canalized waterways.

Scheduling continuous processes occurring in many areas, i.e. in: road transport, railway transport, building and others, has been the subject of many studies in particular with regard to the constraints and the minimal-time objective function [3, 4].

© Springer International Publishing AG, part of Springer Nature 2019
W. Zamojski et al. (Eds.): DepCoS-RELCOMEX 2018, AISC 761, pp. 430–439, 2019.
https://doi.org/10.1007/978-3-319-91446-6_40

Linear Scheduling Method (LSM), equivalent to TCM, but taking into account a different approach, was developed mainly in Canada and the USA. O'Brien in 1969 introduced the term "Line of Balance", this topic was also addressed by Carr and Meyer in 1974, Halpin and Woodhead in 1976, Peer in 1974 and Selinger in 1980, Handa and Barciain 1986, Chrzanowski and Johnston in 1986.

Research on the "Construction Planning Technique" were developed by: O'Brien in 1975, Barrie and Paulson in 1978 (Vertical Production Method); Birrell in 1980 (Time-Location Matrix Model); Johnston in 1981, Stradal and Cacha in 1982 (Time Space Scheduling Method); Whiteman and Irwig in 1988 (Disturbance Scheduling), Thabet and Beliveau in 1994 (Horizontal and Vertical Logic Scheduling for Multistory Projects), Arditi and Albulak in 1979.

Melin and Whiteaker in 1981 developed tools to optimize cyclographs. The works of El-Rayes and Moselhi (1998, 2001) concern the optimization of resource demand. Harris and Ioannou (1998) deal with rhythmic cyclographs taking into account technical limitations. Adeli and Karim (1997) worked on the issues of modelling using neural networks to optimize project costs [5–8].

The issue presented by authors concerns in particular: the maximization of use the limiting locks (bottlenecks), i.e. scheduling ships to be in front of the lock when it is ready for lockage (shown on basis of Gliwice Canal). The basic problem is the synchronization of two-way vessel traffic on a limited section. The research, for calculation model, developed the Time Couplings Methods (TCM) [1, 2, 9, 10]. Due to the computational complexity a deterministic approach to research was adopted.

Time Couplings Methods base on internal time connections between transport processes and sectors (waterway sections, locks), taking into account resource constraints (e.g. types of ships) and technical (capacity of locks).

The main problem in scheduling inland waterway transport is to avoid crossing in a lock (practically it is not possible, and putting it in a schedule would cause delays). By now scheduling was made only by plain calculation, with double checking each stage. Introducing TCM into scheduling inland waterway transport makes the process easier, faster and safer.

Scheduling of transport processes (ship traffic on canalized waterway) with the use of time couplings, consists in synchronizing the cruise, i.e. determining the start and end of the cruise on each section, ensuring lack of collisions of units moving bi-directionally. Cyclographs, i.e. biaxial diagrams, can be a way to map such transport processes.

The paper consists of: introducing Time Couplings Methods, customizing in into inland navigation planning, case study - using TCM to inland waterway transport modeling and conclusions.

2   Basic Model Assumptions for TCM

Time Couplings Methods (TCM) differ from the other scheduling methods in that initially the division into sectors (sections) and resource needs (e.g. availability of transport means). It is assumed: in the 1st method (TCM I) - no downtime of resources used, in the 2nd method (TCM II) - no downtimes in sectors, and in the 3rd method

(TCM III) - the minimum duration of processes, demurrage in resources and sectors is possible. Methods TCM IV, TCM V and TCM VI, take into account - minimum time and additional restrictions.

The TCM I is used when the priority is to ensure the continuity of processes $P1$ ... Pm (lockage and navigation of ships), in sectors $S1$ ... Sn (sections of waterway). In particular, this is related to ensuring the continuity of navigating (e.g. without stoppages in front of the lock). The same situation is with the leading processes (limiting lock), i.e. allowing all dependents to start (other locks or ships navigating).

For the calculations, one establishes known sectors $S1$ ... Sn (segments), transport processes $P1$ ... Pm arranged in the order given and their execution times $T1.1$ ... $Tn.m$ (time of going through one sector) like shown on Fig. 1. Therefore, one should assume a sequence of sectors and as it illustrates the waterway, the sequence cannot be changed. As a result of TCM I calculations one will get the total time of the TT undertaking (time of whole process). However, $LT1.1$ ... $LTn.m$ - time distances between processes in progress and subsequent ones occurring after them, are always equal to zero - that is how a continuity of $P1$ ... Pm processes is guaranteed.

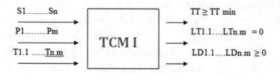

Fig. 1. Schematic diagram of calculations and initial conditions of TCM I [2, 10]

Figure 2 shows the priority relations, i.e. those for which LT must always be equal to zero and constitute the basis for scheduling.

$LT1.1$ ... $LTn.m = 0$

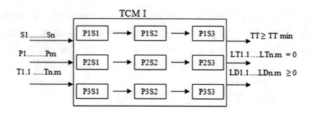

Fig. 2. Priority relationships in theTCM I [2, 10]

3   Matrix Model of the Issue

Matrix models can be used to perform calculations. They enable linking of transport processes running on sections (in sectors) in time and space.

Time matrix of process on segments (sectors) $T = [tij]$ $i = 1, 2, ..., n$ $j = 1, 2, ..., m$, has the dimensions $n \times m$.

Elements of the matrix can be times of transport processes determined on the basis of normative data or after the development of statistical survey results.

$$tij = f\,(Pij,\ Bij) \tag{1}$$

where: tij - element of the matrix,
Pij - the speed i,
Bij - the distance traveled.

When creating a time matrix for a transport process, one should create a collection of segments $d_i = \{d_i\} = \{d_1,\ d_2,\ ...,\ d_n\}$ and a collection of transport means (ships available) $P_j = \{p_j\} = \{p_1,\ p_2,\ ...,\ p_m\}$. When starting to build a matrix, it is most convenient to divide this activity into stages. In the first stage, one defines a set of column vectors corresponding to particular transport processes. The coordinates of these vectors are the times of transport processes.

$$P_1 = \begin{bmatrix} t_{11} \\ t_{21} \\ \cdots \\ t_{i1} \\ \cdots \\ t_{n1} \end{bmatrix} ;..,\ P_j = \begin{bmatrix} t_{1j} \\ t_{2j} \\ \cdots \\ t_{ij} \\ \cdots \\ t_{nj} \end{bmatrix} ;...P_m = \begin{bmatrix} t_{1m} \\ t_{2m} \\ \cdots \\ t_{im} \\ \cdots \\ t_{nm} \end{bmatrix} \tag{2}$$

In the second stage, one builds a matrix of work completion times (T), grouping column vectors in a fixed order, resulting from the order taken.

Generalized matrix model of transport processes on sections (sectors):

$$T = \begin{bmatrix} t_{11},\ ...,\ t_{ij},\ ...,\ t_{1m} \\ t_{21},\ ...,\ t_{2j},\ ...,\ t_{2m} \\ \cdots\cdots\cdots\cdots\cdots\cdots \\ t_{i1},\ ...,\ t_{ij},\ ...,\ t_{im} \\ \cdots\cdots\cdots\cdots\cdots\cdots \\ t_{n1},\ ...,\ t_{nj},\ ...,\ t_{nm} \end{bmatrix} \tag{3}$$

4   Modified Calculation Method

Adopting TCM to transportation processes determines a modified calculation method of time parameters of scheduling, because it needs to take into account the collision freeness of river vessels moving bi-directionally. As an objective function it was assumed to minimize the waiting times in places of units crossing (locks of the channel).

This method, allowing the planning of a continuous transport process, allows determining the minimum downtime in the area of the so-called "bottleneck", they can be eliminated only when modelling rhythmic processes.

The total duration of the transport process complex with the continuous use of implementation means:

$$T = \sum_{j=2}^{m} t_j^r + \sum_{i=1}^{n} t_{im} \tag{4}$$

where t_j^r is the duration of the development of the next partial stream (movement of an individual ship over the entire section) j, i.e. the difference between the start times j and $(j-1)$ of the part stream.

The second component of this sum, i.e.: $\sum_{i=1}^{n} t_{im}$ does not depend on the order of work on particular fronts. Therefore, the search for the shortest time T of the process can be limited to minimizing the first component, i.e.: $T1 = \sum_{j=2}^{m} t_j^r$

Development time t_j^r depends on the mutual synchronization of two neighbouring transport processes:

$$t_j^r = \max_{1 \leq k \leq n} \left[\sum_{i=1}^{k} t_{i,j-1} - \sum_{i=1}^{k-1} t_{i,j} \right] k = 1, 2, \ldots, n \tag{5}$$

Based on tables published in [2], the data for inland waterway transportation planning, using TCM are presented in Tables 1 and 2. Table 1 includes coordinates of points (route) to create a cyclograph and Table 2 includes duration of processes (navigation and lockage of ships). To include the fact, that ships navigate with different speed going up and down-stream, the downstream navigation parameters were marked using " ' ".

Table 1. Coordinates of points (route) to create a cyclograph

	Value on the axis y for x_0 to x_n
Value of y for process m	Y_{0m}
Value of y for process (m–1)	$Y_{0(m-1)}$
...	...
Value of y for process 2	Y_{02}
Value of y for process 1	Y_{01}

Table 2. Duration of processes (navigation and lockage of ships)

	Starting time of the process	Duration of process 1		Duration of process (n–1)	Duration of process n
Process m	t_{0m}	t_{1m}	...	$t_{(n-1)m}$	t_{nm}
Process (m–1)	$t_{0(m-1)}$	$t_{1(m-1)}$	...	$t_{(n-1)(m-1)}$	$t_{n(m-1)}$
...	...	...	...	...	...
Process 2	t_{02}	t_{12}	...	$t_{(n-1)2}$	t_{n2}
Process 1	t_{01}	t_{11}	...	$t_{(n-1)1}$	t_{n1}
	Starting time of the process	Duration of process 1'		Duration of process (n'–1)	Duration of process n'
Process 1	$t_{0'1}$	$t_{1'1}$	...	$t_{(n'-1)1}$	$t_{n'1}$
Process 2	$t_{0'2}$	$t_{1'2}$	...	$t_{(n'-1)2}$	$t_{n'2}$
...	...	...	...	...	...
Process (m–1)	$t_{0('m-1)}$	$t_{1'(m-1)}$	...	$t_{(n'-1)(m-1)}$	$t_{n'(m-1)}$
Process m	$t_{0'm}$	$t_{1'm}$	...	$t_{(n'-1)m}$	$t_{n'm}$

5 Case Study

TCM are mostly used in civil engineering to construction field works scheduling. However TCM as a scheduling method can be adapted to be used in inland waterway transportation planning.

Economic development goes together with growth of demand for transportation services. In near future in Poland we will probably experience the full use of rail and road capacity near sea ports [11]. That is why one can observe a development of revitalisation of inland waterways plans [12].

As inland waterway navigation is being developed to take bigger part in cargo transportation, it is good to establish what would be the maximum capacity of those links. Such an analysis has been done in the past, but the described model [13] was built without automation of calculations.

TCM Methods can be useful to establish a procedure to calculate the maximum capacity of canalized waterways, based on maximum usage of locks (which are bottle necks). It can be obtained by calculating minimum time between every two ships navigating in the same direction and making a schedule of ships navigating on the examined section, including ships going in opposite direction.

A basic difference from classical TCM methods, is that in inland waterway transportation planning the scheduling must come from two sides in the same time. To achieve a maximum use of a limiting lock (the lock on witch lockage last the longest on examined section). After one ship leaves a lock in one direction, the other should come from the opposite side. To build a matrix describing such a situation it has to be divided into two parts (one for ships going upstream, second for ships going downstream).

To simplify the case study it was adopted that each ship navigating on the same section in the same direction will navigate with the same speed (the duration time of processes *0 to n* will be the same, and duration time of processes *0' to n'* will be the same). It is in accordance with the limitations on Gliwicki Canal (ships are allow to navigate with a speed of 7 km/h downstream and 9 km/h upstream [14], which was used as an example of usage of TCM method to inland waterway transportation scheduling.

Graphical presentation of calculation results requires a table with coordinates. Particular points were calculated using measurement data and an Eq. (6).

$$Y0k = t0k = \text{MAX} \begin{cases} t_{0(k-1)} \\ t_{0(k-1)} + t_{1(k-1)} \\ \left(t_{0(k-1)} + t_{1(k-1)} + t_{2(k-1)}\right) - t_{1k} \\ (\cdots) \\ \left(t_{0(k-1)} + t_{1(k-1)} + t_{2(k-1)} + \ldots + t_{(n-1)(k-1)} + t_{n(k-1)}\right) - \left(t_{1k} + t_{2k} + \ldots + t_{(n-1)k}\right) \end{cases} \tag{6}$$

Table 3. Sample data for cyclograph coordinates

	Value on the axis y for x_0	Value on the axis y for x_1	...	Value on the axis y for x_9	Value on the axis y for x_{10}
Value of y for process 1	0	0,44	...	4,53	4,83
Value of y for process 2	0,6	1,04	...	5,13	5,43
Value of y for process 3	1,2	1,64	...	5,73	6,03
Value of y for process 4	1,8	2,24	...	6,33	6,63
Value of y for process 6'	10,58	10,02	...	4,53	4,23
Value of y for process 7'	11,18	10,62	...	5,13	4,83
Value of y for process 8'	11,78	11,22	...	5,73	5,43
Value of y for process 9'	12,38	11,82	...	6,33	6,03
Value of y for process 10'	12,98	12,42	...	6,93	6,63

Using Tables 1 and 2, the Table 3 including the sample data was built.

The duration time of processes, presented in Table 3 were calculated (navigation time was calculated based on allowed speed and distance between locks) and measured (time of lockage in each lock).

In Table 3 one can see an example of time table for ships navigating on Gliwicki Canal. For ships going upstream, time between x_1/x_2, x_3/x_4, x_5/x_6, x_7/x_8, x_9/x_{10}, x_{11}/x_{12} are the duration of lockage.

Time between processes $1 \dots m$ and $1' \dots m'$ is doubled time of lockage on limiting lock. That distance guarantee constant use of the limiting lock.

The limiting lock was one between x_9/x_{10} (a measured time of lockage on it was the longest on the Canal) and the full usage of this lock is shown in Table 3. Cells with the same colours shows when one ship going upstream leaves the lock, the other from the opposite side comes in, and so on.

Eg. the ship from process 7' comes to the lock $x_{9/10}$ at the time of 4,83 and leaves at the time of 5,13, and after that the ship from process 2 comes from the opposite direction to that lock $x_{9/10}$ at 5,13, leaves at 5,43, and then the ship 8' comes in and so on.

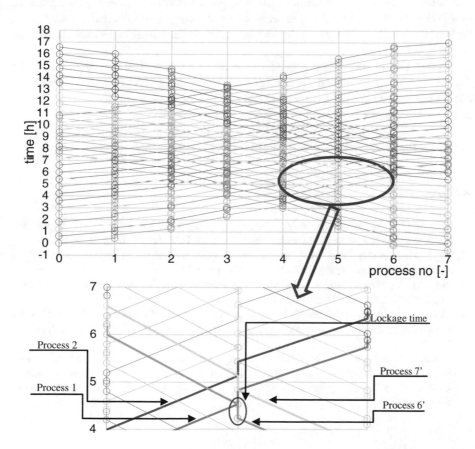

Fig. 3. Cyclograph and it's excerpt: example of ships navigating on Gliwicki Canal

Graphic interpretation of Table 3 is shown on Fig. 3. The cyclograph shows ships location on Gliwicki Canal and how much time does the navigation between locks takes. The excerpt shows the data from Table 3.

Having those information one can see, that it takes a least 6,22 h to go upstream, 8,11 h to go downstream. To calculate the time of cycle for one ship, the time between coming to the port and going in the opposite direction need to be added. In this case it is 14,78 h without taking into account the time of unloading.

The cycle length it required to estimate the maximum capacity of a waterway. And is also required to determine a timetable. The approach is shown e.g. in [13].

Using TCM procedure shown in paragraph 4, the length of transportation cycle may be calculated. The Fig. 3 is shown only to visualize the solution. The main problem in scheduling inland waterway transport, with the use of TCM can be solved easier than before.

6   Conclusion

The paper presents an application of TCM to provide the maximum exploitation time of a limiting lock on the examined section on the example of Gliwicki Canal. On the basis of shown algorithm the cycle time of a vehicle on inland waterway can also be determined.

Using TCM in scheduling ships allow to determine a critical path with the use of Theory of Constrain. It can further lead to determine a timetable and calculate a maximum capacity for the examined section and allow to plan cargo transportation on canalized waterways.

In presented research the authors adopted a deterministic approach. The next step is to include the randomness of the transportation phenomenon.

References

1. Afanasev, V.A., Afanasev, A.V.: Stream scheduling of works in civil engineering (Поточная организация работ в строительстве), St. Petersburg (2000). (in Russian)
2. Hejducki, Z., Rogalska, M.: Harmonogramowanie procesów budowlanych metodami sprzężeń czasowych, Monografie. 246 s., Lublin (2017)
3. Wen Fei, W., Won Young, Y.: Scheduling for inland container truck and train transportation. Int. J. Prod. Econ. **143**(2), 349–356 (2013)
4. Chen, X., Grossmann, I., Zheng, L.: A comparative study of continuous-time models for scheduling of crude oil operations in inland refineries. Comput. Chem. Eng. **44**, 141–167 (2012)
5. Arditi, D., Tokdemir, O.B., Suh, K.: Scheduling system for repetitive unit construction using line-of-balance technology. Department of Civil and Architectural Engineering, Illinois Institute of Technology, Department of Civil Engineering, Honan University, Honam, South Korea (2001)

6. Lucko, G.: Flexible modeling of linear schedules for integrated mathematical analysis. In: Henderson, S.G., Biller, B., Hsieh, M.-H., Shortle, J., Tew, J.D., Barton, R.R. (eds.) Proceedings of the 39th Winter Simulation Conference, Washington, District of Columbia, 9–12 December 2007, pp. 2159–2167. Institute of Electrical and Electronics Engineers, Piscataway (2007)
7. Mattila, K.G., Acse, A.M., Park, A.: Comparison of Linear Scheduling Model and Repetitive Scheduling Method. J. Constr. Eng. Manage. **129**(1), 56–64 (2003)
8. Wang, C., Huang, Y.: Controling activity interval times in LOB scheduling. Constr. Manage. Econ. **16**, 5–16 (1998)
9. Hejducki, Z.: Scheduling model of constructions activity with time couplings. J. Civ. Eng. Manage. **IX**(4), 284–291 (2003)
10. Rogalska, M., Bozejko, W., Hejducki, Z.: Time/cost optimization using hybrid evolutionary algorithm in construction project scheduling. Autom. Constr. **18**, 24–31 (2008)
11. Pluciński M.: Możliwości wykorzystania transportu wodnego śródlądowego w obsłudze zespołu portowego Szczecin – Świnoujście, Polskie Towarzystwo Ekonomiczne, Szczecin (2016)
12. Kulczyk, J., Skupień, E.: Śródlądowy transport wodny w Polsce: stan obecny i perspektywy rozwoju. Problemy Transportu i Logistyki. 2016, nr 4, s. 59–70 (2016)
13. Skupień, E.: Coal transportation model in Odra Waterway. Contemporary transportation systems: selected theoretical and practical problems: modelling of change in transportation subsystems. In: Janecki, R., Krawiec, S., (eds.) Wydawnictwo Politechniki Śląskiej, 2011. s. 123–132. Gliwice (2011)
14. Zarządzenie Dyrektora Urzędu Żeglugi Śródlądowej w Kędzierzynie – Koźlu z dnia 17 września 2004r. w sprawie szczegółowych warunków bezpieczeństwa ruchu i postoju statków na śródlądowych drogach wodnych. Kędzierzyn Koźle (2004)

Decision Model of Wind Turbines Maintenance Planning

Robert Adam Sobolewski[1(✉)], Guglielmo D'Amico[2],
and Filippo Petroni[3]

[1] Faculty of Electrical Engineering, Bialystok University of Technology,
Wiejska 45D, 15-351 Bialystok, Poland
r.sobolewski@pb.edu.pl
[2] Department of Pharmacy, University "G. d'annunzio" of Chieti-Pescara,
via dei Vestini 31, 66100 Chieti, Italy
g.damico@unich.it
[3] Department of Business and Economics, University of Cagliari,
V.le S. Ignazio 17, 09123 Cagliari, Italy
fpetroni@unica.it

Abstract. Performing a maintenance of wind energy system components under good wind conditions may lead to energy not served and finally – to financial losses. The best starting time of preventive maintenance will be, that reduces the energy not served in most. To find this time, a decision model is desired, where many circumstances should be taken into account, i.e. (i) the number and the order of components to be maintained, (ii) component maintenance duration, and (iii) wind turbine(s) output power prediction. Usually, preventive maintenance is planned a few days or weeks in advance. One of the decision problem representations can be influence diagram that enables choosing a decision alternative that has the lowest expected utility (energy not served). The paper presents an decision model that can support decisions-making on starting time of preventive maintenance and maintenance order of wind energy system components. The model relies on influence diagram. The conditional probability distribution of a chance nodes of the diagram are obtained relying on Bayesian networks (BN), whereas the utilities of value node in the diagram are calculated thanks to the second order semi-Markov chains (SMC). The example shows the application of the model in real case of two wind turbines located in Poland. Both the parameters of Bayesian network nodes and semi-Markov chain are derived from real data recorded by SCADA system of the both turbines and weather forecast.

Keywords: Decision model · Influence diagram · Wind turbine maintenance
Bayesian network · Semi-Markov chains

1 Introduction

The management of complex technical system contributes to higher both competitiveness and performance at lower costs. In system operation conditions the management addresses, among others, the maintenance activity (both preventive and corrective

© Springer International Publishing AG, part of Springer Nature 2019
W. Zamojski et al. (Eds.): DepCoS-RELCOMEX 2018, AISC 761, pp. 440–450, 2019.
https://doi.org/10.1007/978-3-319-91446-6_41

ones) intended to retain the system, or restore it to, a state of expected performance and effectiveness. In this way, the relevance of decision making process in the maintenance field increased. Nowadays, one of the major problems in this process is addressing the system modeling in relation to the increasing of its complexity and impact of extraneous conditions of its operation. This modeling task underlines issues referring the quantification of the model parameters and the representation, propagation and quantification of the uncertainty in the system behavior.

Concerning wind energy systems, maintenance activity involves wind turbines and other technical infrastructure that are responsible for power generation and its transfer to a power system. It is quite common that the components should be out of service within the maintenance (mainly because of safety reason). Both the number of components to be on outage simultaneously and their outage durations depend on technical, organizational and financial facilities. Performing a maintenance of components in question under good wind conditions may lead to energy not served and finally – to financial losses. In case of preventive maintenance (dedicated to reducing the probability of failures or degradation of the components performance) the starting time of maintenance can be quite flexible one. On the one hand, this flexibility should be restricted because of the technical requirements, guaranty conditions, service life and other circumstances. On the other hand, it must be subject to expected energy not served because of outage one or more components. Thus, the best starting time of maintenance and maintenance order will be, that reduces the energy not served in most. To find this time, a decision model is desired, where many circumstances should be taken into account, i.e. (i) the number and the order (if more than one component – all simultaneously or one by one) of components to be maintained, (ii) each component maintenance duration and (iii) wind turbine(s) output power prediction. Usually, preventive maintenance is planned a few days or weeks in advance. One of the decision problem representations can be influence diagram that enables choosing a decision alternative that has the highest expected utility.

The research in the field of wind turbines' maintenance is focused mainly on estimation of the effects of different maintenance strategies and their optimization, to limit the maintenance cost and production losses due to turbine's outage [1–4].

This paper presents an decision model that can support decision maker on starting time of preventive maintenance of wind energy system components. The model relies on influence diagram that is used to estimating the expected utility of the a few decisions involving: uncertain maintenance duration, the order of components to be maintained and staring time of maintenance. The conditional probability distribution of a chance nodes of the diagram are obtained relying on Bayesian networks (BN), whereas the utilities of value node in the diagram are calculated thanks to the second order semi-Markov chains (SMC). The example shows the application of the model in the real conditions of two wind turbines (E-48 and E-53 by Enercon) located in northern part of Poland. Both parameters of parent nodes of Bayesian network and kernel of semi-Markov chain are derived from real data recorded by SCADA of the both turbines and weather forecast.

The model relies on the ideas introduced in [5] where the influence diagram presented refers to one wind turbine only and deterministic duration of the maintenance activity. Here, the model takes the advantages of the previous one and includes additional circumstances mentioned in the paragraph above.

2 Influence Diagram

The influence diagram is an effective modelling framework for representation and analysis of the decision making process under uncertainty caused by stochastic nature of both the wind turbine energy yield and other circumstances.

The influence diagram is considered as a Bayesian network augmented with decision variables, utility functions defining the preferences of the decision maker, and precedence ordering that specifies the order of decisions and observations [6]. The objective of decision analysis is to identify the decision option that produces the highest expected utility. To identify this decision option one needs to compute the utility of each decision alternative. The expected utility of action a_i is following

$$\mathrm{EU}(a_i) = \sum_j \mathrm{U}(a_i, h_j) \cdot \mathrm{P}(h_j|\varepsilon) \tag{1}$$

where a_i is ith option of decision variable A $(i = 1, 2, \ldots, m)$, h_j is the jth state of hypothesis H $(j = 1, 2, \ldots, n)$, ε is a set of observations in the form of evidence, $\mathrm{P}(\cdot)$ represents the believe in H given ε and utility function $\mathrm{U}(\cdot)$ encodes the preferences of the decision maker on a numerical scale.

Decision maker should choose the alternative with highest expected utility, i.e. the option a^* such as

$$a^* = arg \max_{a_i \in A} \mathrm{EU}(a_i). \tag{2}$$

The influence diagram of two wind turbines maintenance scheduling decision problem is presented in Fig. 1. It contains three types of nodes: decision, chance and value.

There are four decision nodes (drawn as rectangles) that represent decision variables, i.e. 'WT1 maintenance duration', 'WT2 maintenance duration', 'WT1 starting time of maintenance' and 'WT2 starting time of maintenance'. These variables are under control of the decision maker and model available decision alternatives. For example, the possible states of decision nodes are as follows:

- 'WT1 Maintenance duration': <WT1_6 h>, <WT1_7 h>, <WT1_8 h>, <WT1_No>,
- 'WT2 Maintenance duration': <WT2_6 h>, <WT2_7 h>, <WT2_8 h>, <WT2_No>,
- 'Starting time of WT1 maintenance': <In_72 h>, <In_96 h>, <In_120 h>,
- 'Starting time of WT2 maintenance': <In_72 h>, <In_96 h> ,<In_120 h>.

Both nodes 'WT1 Maintenance duration' and 'WT2 Maintenance duration' have four discrete alternatives: 6 h, 7 h, 8 h (one shift for maintenance activity is assumed) and lack of maintenance. The alternatives can be chosen independently one another,

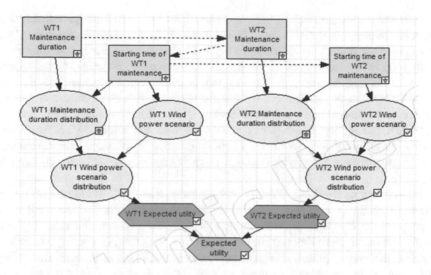

Fig. 1. Influence diagram of two wind turbines maintenance scheduling decision problem.

e.g. 6 h for the first turbine and 7 h for the second one. Two letter nodes 'WT1 Starting time of maintenance' and 'WT2 starting time of maintenance' have four discrete alternatives, i.e. 72 h, 96 h, 120 h and lack of maintenance. The number of hours is the time accounted from the time the decision is will be made (time 0) till the starting time of the maintenance. It is assumed that all the data needed for wind turbines output power prediction are known for time 0 (the historical data of wind turbines output power and wind conditions forecast). The maintenance order decision can be made relying on both letter nodes. For example, if we are going to assume WT1 maintenance then WT2 maintenance (starting time of WT1 maintenance is one or two days before starting time of WT2 maintenance), we should decide to WT1 starting time of maintenance (e.g. in 72 h) and WT2 starting time of maintenance (e.g. 96 h or 120 h). Assuming the simultaneous maintenance of both turbines means that we should decide to the same time of starting the maintenance for both turbines. The wind turbine maintenance durations and starting time of turbine maintenances can have the values as required in real application of decision model.

In general, the discrete alternatives of all decision nodes selection depend on a few main features, i.e.: expert knowledge of decision maker or management board, formal procedures of maintenance planning, maintenance requirements provided by turbines' producer, to name a few. The alternatives can be either the same for all turbines in questions or can be different for each turbine individually.

There are 6 chance nodes (drawn as ovals), i.e. 'WT1 Maintenance duration distribution', 'WT2 Maintenance duration distribution', 'WT1 Wind power scenario', 'WT2 Wind power scenario', 'WT1 Wind power scenario distribution' and 'WT2 Wind power scenario distribution'. They are discrete random variables with finite sets of mutually exclusive states and represent uncertain quantities that are relevant to the decision problem. They are quantified by conditional probability distributions. Each node represents random variable with finite set of mutually exclusive states.

Two first nodes are random variables quantified by conditional probability distribution of maintenance duration given the alternatives of predecessors (decision nodes). For example the probability distribution of 'WT1 maintenance duration distribution' is conditioned by alternatives of the nodes 'WT1 Maintenance duration' and 'Starting time of WT1 maintenance'. There are ten possible states of chance node 'WT1 maintenance duration distribution' – nine states of maintenance duration and starting time combinations and one state that refers to the lack of WT1 maintenance. In fact, the conditional probability tables (CPT) associated with the nodes 'WT1 maintenance duration distribution' and 'WT2 maintenance duration distribution' are deterministic ones, i.e. they are functionally determined by their parents and each state probability is either 0 or 1.

Two chance nodes 'WT1 Wind power scenario' and 'WT2 Wind power scenario' represent hypothesis with states determined by the combinations of the wind turbine output power states (total number of states is s, for each turbine individually) within two discrete instances of time: (i) starting time of the maintenance and (ii) one unit time preceding the starting time. The s states represent mutually exclusive partitions of wind turbine output power within the power range $0 \ldots P_R$, where P_R is rated power of wind turbine in question. The probability distribution of wind turbine output power (the probabilities of output power states) can be calculated relying on BN model (see Subsect. 3.1). Both nodes represent random variables quantified by conditional probability distributions obtained thanks to given alternative of decision node 'Starting time of maintenance'. The possible states of chance node 'WT1 Wind power scenario' are as follows: <WT1_1>, ..., <WT1_N> . For example, if the number of states $s = 5$ the total number of state combinations of the wind turbine output power states is $N = 25$.

Two latter chance nodes 'WT1 Wind power scenario distribution' and 'WT2 Wind power scenario distribution' are the child nodes of other chance nodes, i.e. 'WT1 Maintenance duration distribution' and 'WT1 Wind power scenario', and 'WT2 Maintenance duration distribution' and 'WT2 Wind power scenario', respectively. The conditional probability tables (CPT) associated with the nodes in question are deterministic ones. For example the CPT of node 'WT1 Wind power scenario distribution' given 10 alternatives of the node 'WT1 Maintenance duration' and 25 options of node 'WT1 Wind power scenario' has got 226 states, i.e. 225 states as a combinations of maintenance durations and wind power scenarios, and one state that refers to the lack of WT1 maintenance.

There are three value nodes (drawn as hexagons): 'WT1 Expected utility', 'WT2 Expected utility' and 'Expected utility'. Two first ordinary utilities can be calculated relying on second-order semi-Markov predictive models (see Subsect. 3.2) of wind turbine output powers – WT1 and WT2 respectively. Each of these nodes involves utility of one wind turbine only. It is usually easier for a decision maker to elicit utility functions over each of the attributes in separation and then combine them in a single multi-attribute utility function. Thus, the value node 'Expected utility' includes special case of Additive Linear Utility (ALU) functions. This node combines the parent utility nodes using the following linear function with weight 1.

$$\text{Expected utility} = 1 \cdot \text{WT1 Expected Utility} + 1 \cdot \text{WT2 Expected utility} \quad (3)$$

Although, a decision maker should choose the alternative with highest expected utility (see formula (2)) as it is the best option, in wind turbines maintenance planning analysis the best option is the one of the lowest "Expected utility'. The reason is that we look for the decision that would ensure the minimum energy to be lost because of the wind energy system components outage.

Each arc in Fig. 1 denotes influence, i.e. the fact that the node at the tail of the arc influences the value (or the probability distribution over possible values) of the node at the head of the arc. Some arcs have a causal meaning, e.g. the path from a decision node to a chance node. It means that the decision will impact the chance node in the sense of changing the conditional probability distribution over its outcomes.

3 Modeling of Wind Turbine Output Power Prediction

3.1 Bayesian Network Model

BN approach is used for both obtaining probability distribution of wind turbine output power and pointing out the power states as an input to predict amount of wind turbine energy to be produced (relying on second order semi-Markov chain model). As mentioned in Sect. 2 the probability distribution must be obtained for two discrete time instances: (i) starting time of maintenance $t = l$ and (ii) one unit time preceding the starting time $t = l - 1$. Number of states s is the same for each time instance and basically should be optimized regarding the satisfied error of the results. BN model consists of three states, i.e. two root nodes 'Wind speed' (WS) and 'Wind direction' (WD) and one child node 'Wind power' (WP) [1]. The latter node represents the wind turbine output power. Each node represents a random variable with a finite set of mutually exclusive states. Learning CPTs amounts essentially to counting data records for different conditions encoded in the network. It means that prior probability distribution of WS, WD and WP can be obtained from relative counts of various outcomes in those data records that meet the conditions described by a combination of the outcomes of the parent variables. The data records in question (wind speed and direction measured by anemometer installed at the top of turbine nacelle and output power) can be gathered from real data acquired by SCADA system of wind turbine (or the whole farm). Since BN has a built-in computational architecture for computing the effects of evidence on the states of variables in the model, it allows for updating probabilities of the variable states, on learning new evidence. It means that one can calculate conditional probability function of WP given forecasts of wind speed and direction within instances of time $t = l - 1$ and $t = l$. The error of probability distribution of WP depends on the errors of forecasting wind speed and direction (usually gathered from meteorological models), the number of states of the nodes in model and number of data used for learning parameters of BN.

Wind power scenario (WPS) probability in $t = l$ can be calculated from the formula

$$p_g^{WPS}(l - 1, l) = p_{u,l-1} \cdot p_{u,l} \tag{4}$$

where $p_{u,l-1}$ and $p_{u,l}$ probability of the output power state calculated relying BN model within instance of time $l - 1$ and l respectively, $u = 1, 2, \ldots, s$ and $g = 1, 2, \ldots, s^2$.

All details about BN model one can find in [1].

3.2   Second Order Semi-Markov Chain

We use a second order semi-Markov chain in state and duration as proposed in [6–8]. The model is applied for predicting wind turbine output power and energy given real data of output power within two instances of time $t = l$ and $t = l - 1$, and duration of output power value at $t = l - 1$. The time t = 0 refers to the last unit of time when the output power is recorded by SCADA system. All details about second order semi-Markov chain model one can find in [1].

4   Application

Decision model of wind turbines maintenance planning involves two turbines, i.e. E-48 and E-53 by Enercon, located in northern part of Poland. The distance between turbines is around 1 km. Both of them are of 800 kW rated power. The first is in operation from the last quarter of 2011 whereas the letter one – from the last quarter of 2014. Total number of records (1 h averages of wind speed and direction, and output power) taken as an input data to both BN and SMC models is 55530 and 29230 respectively.

The influence diagram is depicted in Fig. 1. Let introduce the index p for distinguishing the number of wind turbine in question. Let assume that WT1 ($p = 1$) represents the turbine E-48 and WT2 ($p = 2$) – the turbine E-53. The nodes explanation provided in Sect. 2 is valid in the example.

To learn parameters of BN models all the variables (WS_p – wind speed at the site of the pth turbine, WD_p – wind direction at the site of pth turbine and WP_p – output power of pth turbine) are discretized following two assumptions: (i) the method of discretization – uniform counts and (ii) number of bins (states of the node) – 10 for WS_p, 10 for WD_p and 5 for WP_p. Having learnt parameters of the BN model the validation of the results has been performed. The K-crossvalidation has been taken for it given fold count K = 10. As a results of validation the accuracy, confusion matrix, ROC Curve and Calibration have been obtained. The states of variables WS_p, WD_p and WP_p and their boundaries are provided in Table 1.

Let assume that decision problem of maintenance planning concerns the staring time of maintenance in 72 h, 96 h and 120 h, and the turbines maintenance duration can be of 6 h, 7 h an 8 h. Moreover, the turbines maintenance order can be considered as follows: both turbines simultaneously (within one day), WT1 (the first day) then WT2 (the second day), and WT2 (the first day) and WT1 (the second day).

The wind speed and direction forecast and wind turbines output power probability distributions given t are provided in Table 2.

Table 1. States of WS_p, WD_p and WP_p, and their boundaries of partitions represented by BN node states

State	WS_1/WS_2 [m/s]		WD_1/WD_2 [°]		WP_1/WP_2 [kW]	
	From	To	From	To	From	To
S1	0.0/0.0	3.2/2.7	0/0	71/81	0/0	20/16
S2	3.2/2.7	4.0/3.7	71/81	134/117	20/16	67/77
S3	4.0/3.7	4.7/4.5	134/117	165/136	67/77	139/178
S4	4.7/4.5	5.2/5.1	165/136	189/163	139/178	282/352
S5	5.2/5.1	5.7/5.7	189/163	214/182	282/352	835/828
S6	5.7/5.7	6.2/6.4	214/182	237/200	—	—
S7	6.2/6.4	6.9/7.1	237/200	263/222	—	—
S8	6.9/7.1	7.6/7.9	263/222	287/253	—	—
S9	7.6/7.9	8.8/9.0	287/253	316/285	—	—
S10	8.8/9.0	18.0/19.3	316/285	360/360	—	—

Table 2. The wind speed and direction forecast at the site both wind turbines are located and WP_p distributions given t (starting time in 72 h)

Time [h] →		$t = -5$	$t = -4$	$t = -3$	$t = -2$	$t = -1$	$t = 0$
Wind speed forecast [m/s] →		8.4	8.6	8.5	8.0	7.7	7.5
Wind direction forecast [°] →		190	191	190	188	186	186
State →	S1	0.049/0.014	0.049/0.014	0.049/0.014	0.024/0.014	0.024/0.003	0.006/0.003
	S2	0.009/0.002	0.009/0.002	0.009/0.002	0.002/0.002	0.002/0.001	0.003/0.001
	S3	0.003/0.009	0.003/0.009	0.003/0.009	0.006/0.009	0.006/0.006	0.012/0.006
	S4	0.082/0.016	0.082/0.016	0.082/0.016	0.074/0.016	0.074/0.865	0.919/0.865
	S5	0.864/0.958	0.864/0.958	0.864/0.958	0.894/0.958	0.894/0.124	0.060/0.124

5   Discussion of the Results

The 'Expected utility' results of the influence diagram (Fig. 1) for decision problem of two wind turbines maintenance planning given the maintenance durations, maintenance order and starting time of maintenances, are depicted in Fig. 2, 3 and 4. The numbers 1–9 represent the combinations of both WT1 and WT2 turbines maintenance durations, i.e. 1 (6 h of WT1 and 6 h of WT2), 2 (6 h of WT1 and 7 h of WT2), 3 (6 h of WT1 and 8 h of WT2), 4 (7 h of WT1 and 6 h of WT2), 5 (7 h of WT1 and 7 h of WT2), 6 (7 h of WT1 and 8 h of WT2), 7 (8 h of WT1 and 6 h of WT2), 8 (8 h of WT1 and 7 h of WT2), and 9 (8 h of WT1 and 8 h of WT2).

According to Fig. 2 the best decision that minimizes the energy not supplied because of wind turbine outages given the starting time of maintenance in 72 h is to start the maintenance of WT1 in 72 h (the first day) and of WT2 in 96 h (the second day). It is valid for each combination (from 1 up to 9) of wind turbines maintenance duration. The worst option would be performing the maintenance of both turbines in one day in 72 h. If the WT1 and WT2 maintenance duration are 8 h and 6 h (the combination 7) respectively the 'Expected utility' is almost the same for the options 'WT1 then WT2' and 'WT2 then WT1'.

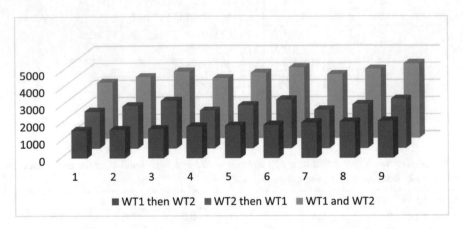

Fig. 2. The 'Expected utility' results (in MWh) given 9 combinations of WT1 and WT2 maintenance duration, 3 combinations of maintenance order, and starting time of maintenance in 72 h.

Other options to be considered in the decision problem are the starting times of WT1 and WT2 maintenance in 96 h and 120 h. The 'Expected utility' results given starting time in 96 h are depicted in Fig. 3. The best decision given each combination of maintenance durations is to perform the maintenance of both turbines simultaneously. If there are any restrictions in maintaining the turbines simultaneously (e.g. there is one maintenance crew only), the preferred decision can be to start the maintenance of WT2 in 96 h (the first day) and of WT1 in 120 h (the second day). It is valid for each combination of wind turbines maintenance duration, excluding combinations 4 and 7 (the 'Expected utility' is comparable so the maintenance order is not important at all).

The 'Expected utility' results given starting time of maintenance in 120 h are depicted in Fig. 4. There is one maintenance order, i.e. the simultaneous maintenance of both turbines. It is obvious, that the best decision is performing the maintenance under conditions of shortest duration of both turbines maintenance (6 h, combination 1 of maintenance durations). Another conclusion is there are not so essential differences among 'Expected utility' under assumption of some maintenance duration combinations, e.g. combination 2(6) − 6(7) h of WT1 maintenance and 7(8) h of WT2 maintenance, and combination 4(8) − 7(8) h of WT1 maintenance and 6(7) h of WT2 maintenance.

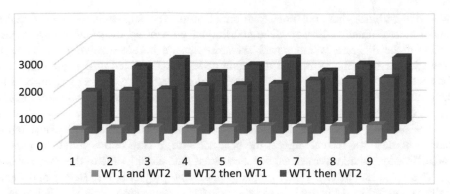

Fig. 3. The 'Expected utility' results (in MWh) given 9 combinations of WT1 and WT2 maintenance duration, 3 combinations of maintenance order, and starting time of maintenance in 96 h.

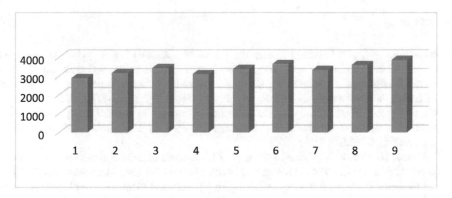

Fig. 4. The 'Expected utility' results (in MWh) given 9 combinations of WT1 and WT2 maintenance duration, maintenance order – WT1 and WT2, and starting time in 120 h'.

Having investigated the 'Expected utility' results depicted in Fig. 2, 3 and 4 one can conclude that the best decision is to perform the maintenance of both turbines simultaneously in 96 h, independently on the combination of turbine maintenance durations.

6   Conclusion

In the paper we present the approach to support making decision problem in the area of wind turbines maintenance activity. The problem concerns the choosing the best time in the future (usually in a few days or couple of weeks) for performing the maintenance of the turbines in question, i.e. the time when unserved energy will be of lowest amount. Although, the approach can affect the maintenance of many turbines (e.g. all turbines included in wind farm), the paper demonstrates a decision model involving two wind turbines that constitutes one small farm. The approach relies on decision model as an influence diagram. The conditional probability distribution of a chance nodes of the diagram 'Wind power scenario ...' are obtained relying on Bayesian networks, whereas the utilities of the value nodes '... Expected utility' are calculated thanks to both Bayesian networks and semi-Markov chains. To derive these probability distributions and utilities the real data of wind turbine output powers and wind forecast are needed.

The example shows that the decision model can have important practical implications because the results suggest the best decision, taking into account the maintenance duration, wind turbines maintenance order, and starting time of the maintenance.

This work has been partially prepared under the project S/WE/3/2018 and financially supported by Ministry of Science and Higher Education, Poland. All Bayesian network models in this paper were created and tested using GeNIe, a development environment for graphical probabilistic models, available for the research at https://www.bayesfusion.com/.

References

1. Pattison, D., Segovia Garcia, M., Xie, W., Revie, M., Whitfield, R.I., Irvine, I.: Intelligent integrated maintenance for wind power generation. Wind Energ. **19**, 547–562 (2016)
2. Kerres, B., Fisher, K., Madlener, R.: Economic evaluation of maintenance strategies for wind turbines: a stochastic analysis. IET Renew. Power Gener. **9**(7), 766–774 (2015)
3. Sperstad, I.B., McAuliffe, F.D., Kolstad, M., Sjomark, S.: Investigating key decision problems to optimize the operation and maintenance strategy of offshore wind farms. Energ. Procedia **94**, 261–268 (2016)
4. D'Amico, G., Petroni, F., Sobolewski, R.A.: Maintenance of wind turbine scheduling based on output power data and wind forecast. In: Zamojski, W. et al. (eds.) Dependability Problems and Complex Systems, Advances in Intelligent Systems and Computing, vol. 582, pp. 106–117. Springer International Publishing AG (2018)
5. Kjaerulff, U.B., Madsen, A.L.: Bayesian Networks and Influence Diagrams: A Guide to Construction and Analysis. Springer, New York (2008)
6. D'Amico, G., Petroni, F., Prattico, F.: First and second order semi-Markov chains for wind speed modeling. Phys. A **392**(5), 1194–1201 (2013)
7. D'Amico, G., Petroni, F., Prattico, F.: Reliability measures of second order semi-Markov chain with application to wind energy production. J. Renew. Energ. **2013**, 6 (2013)
8. D'Amico, G., Petroni, F., Prattico, F.: Performance analysis of second order semi-Markov chains: an application to wind energy production. Methodol. Comput. Appl. Probab. **17**, 781–794 (2015)

Meteorological Data Acquisition from MERRA-2 Reanalysis for Wind Energy Systems Modeling Support

Adrian Sołowiej[1](✉), Teodora Dimitrova-Grekow[1],
and Robert Adam Sobolewski[2]

[1] Faculty of Computer Science, Bialystok University of Technology,
Wiejska 45A, 15-351 Bialystok, Poland
adrian.solowiej94@gmail.com, t.grekow@pb.edu.pl
[2] Faculty of Electrical Engineering, Bialystok University of Technology,
Wiejska 45D, 15-351 Bialystok, Poland
r.sobolewski@pb.edu.pl

Abstract. Modeling of power systems with interconnected wind energy sources requires high resolution time series of wind turbine (or wind farm) output powers, as their variable and unpredictable nature, weather-dependent, with complex correlations over space and time, poses increasing challenges for power systems planning and operation. These time series are often difficult to acquire or simulate accurately. Conversely, meteorological data (wind speed and direction, air pressure, temperature, humidity, to name a few) is often freely-available. One of such data is reanalysis, e.g. MERRA-2. To acquire the particular data from MERRA-2 reanalysis to be used in wind energy systems modeling, the significant time and knowledge is required to explore huge amount of data. There are two reasons of it, i.e. the file format NC4 and huge size of the file that consists of the data involving one day. The developed software supports downloading and initial retrieving the complex meteorological data in NC4 format. Moreover, the database of NC4 files is examined, with a high degree of dependability, burden just a minimal computer resources. After a short overview of the actual technical state, a general presentation of the project is provided. Moreover, the implemented technics insuring the dependability of the whole system is shown. Practical demonstration and test are also presented.

Keywords: MERRA-2 · Reanalysis · Wind power · Software dependability

1 Introduction

Modeling of power systems with interconnected wind energy sources requires high resolution time series of wind turbine (or wind farm) output powers, as their variable and unpredictable nature, weather-dependent, with complex correlations over space and time, poses increasing challenges for power systems planning and operation. In other words, such power system models require some input data to represent the contribution from wind power generation as it cannot be controlled. Moreover, the

© Springer International Publishing AG, part of Springer Nature 2019
W. Zamojski et al. (Eds.): DepCoS-RELCOMEX 2018, AISC 761, pp. 451–460, 2019.
https://doi.org/10.1007/978-3-319-91446-6_42

profiles of wind power output need to be understood at regional or national scale. These data are often difficult to acquire or simulate accurately. Historic data can be lack due to commercial confidentially or huge costs of purchasing. Conversely, meteorological data (wind speed and direction, air pressure, temperature, humidity, to name a few) are often freely-available. One of such data is reanalysis that is product of an atmospheric model set to match historic weather observations, and estimated weather parameters on a regular grid, often with global coverage, spanning many years. More recently, reanalysis datasets have been explored as a means of modeling and simulating wind power generation and wind energy yield [1–3]. A key benefit of reanalysis is that it can infer meteorological data, for which there are no observations, e.g. wind speeds at 50 m (met masts are usually only 10 m tall), in locations that are either remote or out to sea (crucial for off-shore wind power modeling). Several reanalysis products are available [3], but only a few provide wind data at a height of 50 m above ground closer to those used by wind turbines. One of them is MERRA-2 [4].

Since 1980 MERRA-2 (Modern-Era Retrospective analysis for Research and Applications, Version 2) has been offering data about the complex climate on the Earth with a spatial resolution about 50 km in the latitudinal direction. It includes NASA's ozone profile observations, the GEOS model and the GSI assimilation system [4, 5]. Thanks to the online data, shared through the NASA Goddard Earth Sciences Data and Information Services Center (GES DISC), a quasi-real-time climate analysis are available [6]. All data, organized into a tree-structure, is accessible in NC4 file format [5]. It is a kind of HDF5 (Hierarchical Data Format) [7] - a multidimensional data format from the NetCDF library. It is used to store huge amounts of data and to enable quick access to selected parts. The format has been designed especially for meteorology, where a lot of them are recorded.

To acquire the particular data from MERRA-2 reanalysis to be used in wind energy systems modeling, the significant time and knowledge is required to explore huge amount of data. For example, one of the such most interesting MERRA-2 collection is M2T1NXSLV that consists of the data involving 47 variables, e.g. air pressure and temperature, humidity, eastward and northward wind, to name a few. The variables are of: 1 h time resolution (time averaged), full resolutions of the grid $0.625° \times 0.5°$ (576 points in the longitudinal direction and 361 points in the latitudinal direction). The granule size of the file is around 393 MB that refers to 1 day only. In wind energy systems modeling it is most common to use the historic meteorological data that represent long period of time (thousands of days). It means that obtaining some particular data relying on download an each file individually and extraction the data collected in the file of NC4 format is very inconvenient and time-consuming. Thus, the computer software is expected alternative in the process of meteorological data acquisition.

MERRA-2 file collection is equipped with a kind of table of contents that includes information about the file structure and the name of columns or variable headings, what allows a specific data extraction. The root of the tree structure is a data set (file), and subsequent branches are variables and parameters. In implementing a search algorithm of a file tree with an unknown structure, especially attention has to be paid to the references located in individual nodes. Otherwise it could crashed into an endless loop. All files inherit the structure of the predecessor, what is very convenient for the software developer.

The paper presents the computer software designed for: downloading MERRA-2 reanalysis files, data acquisition from them and producing an output XLSX file that consists of the data obtained, given collection, geographical coordinates, variables, and time span. Once designing software the two additional technical tasks have been taken into account: (i) insuring a trust data transfer and (ii) use of the optimal computer resources. We propose a solution, which performs both tasks, giving simultaneously a bunch of useful functions for data collecting. The computer program presented in this paper supports downloading and initial retrieving the complex meteorological data that are collected in the files of NC4 format, as it is in MERRA-2. Moreover the databases of NC4 files is examined, with a high degree of dependability, burden just a minimal computer resources. General information about the software and user interface details are provided as well.

2   Computer Software

There were several different attempts to achieve an optimal data collecting system. After a short but reasonable corresponding discussion with NASA, within several months of development we have obtained an well-equipped software, written in. NET. The WPF graphics engine has been used as well. Microsoft's technology supports the NC4 file format by a library for processing scientific data (HDF5DotNet version 1.8.9).

2.1   General Information

It should be mentioned that solutions such as *crawler*, downloading files and session storage have been used. Due to the sophisticated mechanisms of application operation, such as searching URLs and downloading files (due to download status updates, etc.) we present only two interaction diagrams that illustrate the most interesting functionalities.

In Fig. 1 the mechanism of finding the URLs of individual directories is presented. A complete sequence that begins from login into the system, through status update and all specifics of relations between the User, UI, Crawler and database (MERRA-2), and finally ends with a successful listing of the URLs, can be easy followed. On the other hand Fig. 2 presents adding and rounding off the coordinates for which the user data is request. The relations between the User, UI, Coordinates Addition and Coordinate Helper are considered as a sequence of ten steps. The process starts from the new coordinates user request and successful screening of the rounded off coordinates from the Coordinate Helper coordinates data. Moreover, a map of all coordinates (a dictionary of all indexes and coordinates) is being created. The mechanism of data rounding is forced by the MERRA-2 data content, what orders such a procedure. The end effect is an optimal for both the user and system data serving.

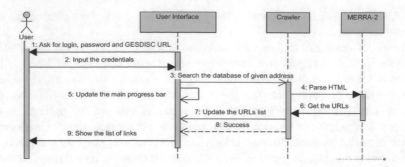

Fig. 1. Diagram of interaction of the mechanism finding the URLs of individual directories

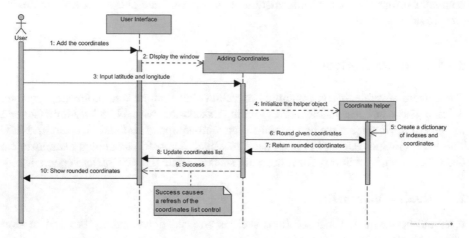

Fig. 2. Diagram of interaction of the coordinate adding mechanism

2.2 User Interface

The UI (*user interface*) is intuitive and user-friendly (Fig. 3). It was crucial to preserve the responsiveness [8]. In addition, UX (*user experiene*) [9] should be continuously update. This can be achieved by using *multiple threads* technics (see Sect. 3). The WPF ensured a smart appearance of the both: software and UX. The facilities include:

- selection of coordinates for which data extraction are being performed (inside the frame titled 'List of coordinates'),
- selection of headings (variables, parameters) in the file that refers to data are being collected (appears while pushing the key button 'Prepare the list of files'),
- list of files to be downloaded (the bottom part of window, where the columns of 'Date', 'Name', '%', 'State' and 'Address' are shown),
- download progress-bars - separated for files and designed to know how many files have already been downloaded (above the frame titled 'Data extraction'),

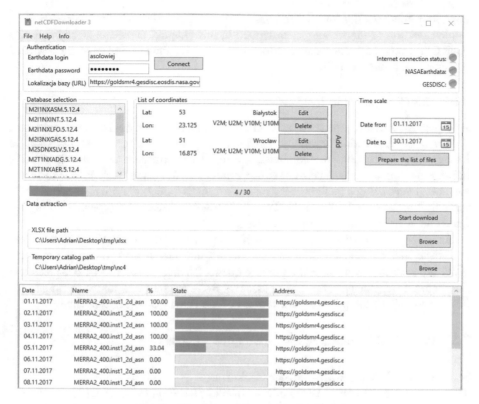

Fig. 3. The main window of the software.

- buttons for selecting file storage paths (inside the frame 'XLSX file path' and 'Temporary catalog path'),
- connection status with the servers required to carry out the download process (three dots in the right top corner of the window).

It was important to provide the information about the processes taking place at the stage of downloading the files. For this purpose, each item in the file list is equipped with its own progress bar and a separate column with a numerical interpretation of progress (in %). While downloading each of the files, the dedicated progress bar is supposed to update itself on a regular basis, whereas the global progress bar (in the middle of the window) updates itself only after a successful operation of file downloading.

3   Dependability of the Solution

In software engineering, dependability is the ability to provide services that can defensibly be trusted within a time-period [11]. To justify this mechanisms designed to increase and maintain the dependability of the system are presented. We achieved both the basic functionality of the software and its high stability and dependability. Implementation of the techniques described in this chapter, ensures a steady software operation. Some of them were mentioned in previous chapter and now they will be presented with more details.

3.1   Multi Threads and Responsiveness

A responsive interface was achieved thanks to the implemented *multi thread* technique. A *freeze* or application inactivity, often occurs due to starting one or a set of long-running instructions on the thread, serving the interface. In fact, it is intolerable because one thread cannot perform two operations at the same time. In our software, each new event being a method of the main class, calls a method from an external class, running it on a separate thread. In short, unnoticeable freeze that appeared during the operation is simply started on another thread.

In order to update the state of the download progress-bars mechanisms of *delegates* and *dispatcher* are applied. In short, this serves the inter thread-communication and gives a possibility to manage the priority queue. Thanks to the use of delegates and dispatcher, a responsive interface is created. Both mechanisms allow to update the information about the progress and success of the downloading in real time.

3.2   Dependability of the System – Adopted Methods

Scanning the Folders with Crawler Technique

There was a need to collect URLs of a set of files placed on the GES DISC server. For this purpose *crawler* technique was implemented. A complete file localizations data was collected: the site structure and the URLs. Such a tasks includes collecting information about individual nodes and mapping the tree with file addresses, as it is on MERRA-2. Moreover, this should be done in a limited time interval. A simple algorithm to search the tree was written, omitting fixed page elements (also being links) by opening a black list of unwanted elements.

We established a completed parsing the content of Web sites and many other operations on the code using *Html Agility Pack* library. It enables searching for tags by ID or other attributes. Therefore, the implementation of a simple crawler was sufficient. The mechanism is used to search all directories with data under the main URL, considering MERRA-2 data structure. This is important for the dependability of the system, because each directory contains a different type of data.

HTTPS Connection, Data Parsing and Preparation of a Dependable Result File

Files in the MERRA-2 database are available via the *HTTPS* protocol. Thus an Earthdata account must be created and used to access the data [10]. To authenticate the user and simulate a browser connection, the built-in .NET Framework classes *HttpWebRequest*, *HttpWebResponse* are used. This allows to get the total length of the stream and hence - the size of the file, as well as to download the file block by block. During the collection of subsequent data blocks, a delegate is started. It transmits the percentage of data that has been collected so far. HTTP requests are created for subsequent URLs that have been collected to the list by the previously discussed crawler. The process is illustrated in Fig. 2.

Although the variable names in all files are standardized, every file collection may consists of different parameter set. This requires data parsing to be adjusted to the aims of the project.

The result file refers to the Microsoft Excel standard. For this aim we used the *EPPlus* library. It gives a wide range of possibilities to create a new or modify an existing *XLSX*-file. When the first file is downloaded, headers for the chosen Earth – points and data parameter are created in the worksheet. In this way dates in the first column and the first two rows are filled. The first column contains the time sequence according the convention *day.month.year hour*, e.g.: *01.12.2017 06:00*. The recorded data is grouped in column sets. Each set refers to a geographical coordinates and contains the parameters chosen by the user. The used library allows to fill a whole series of spreadsheet cells. This is fundamentally different from simply appending to the end of a text file and makes much easier an implementation of columns modifying. When a new data is added to the output file, a record is made. Thanks to this, even if due to random reasons (e.g. no connection to the Internet) the application suddenly stops downloading data, the temporal progress will not be lost. The prepared information is ready to be read and further processed by Microsoft Excel and any program or procedure able to read *XLSX* spreadsheets.

4   Software Testing

To exam the effectiveness of this solution several performance tests were conducted. The final results prove that presented software is very frugal for the processor and safe for memory - does not cause, so-called *memory leak*. The machine we performed the tests on was: ACPI architecture computer, processor x64 Intel Core i7-2600 K, timing: 3.40 GHz, Windows 10 Education 64-bit, 8 GB RAM (2 × 4 GB).

To perform reliable tests *Performance profiler* was used. It is a dedicated tool, available from the Visual Studio 2015 and supports an observation the *CPU load* and *memory usage* during the application running.

CPU Usage Test

First, the processor load analysis was started. Next, an step by step regular work flow was executed, i.e.:

- connection to Earthdata via the software (during it the directory addresses automatically are being downloaded),
- coordinates and variables are selected,
- date range is determined,
- URL list for the selected time period is downloaded,
- data files download is started.

In Fig. 4 the observation of software operation for about twenty minutes (1160 s) is presented. During this time two whole files and part of the third one were downloaded. The third was not added to the result file, which was correct. During the operation, the CPU usage did not exceed 15% (see Fig. 4).

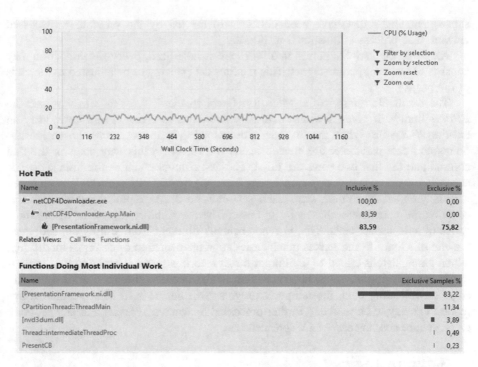

Fig. 4. Using the processor while downloading files from the MERRA-2 database

In Fig. 5 the *data call tree* of the software is presented. The largest load is generated by downloading files from the *Core* module. We do not take into account the presentation layer, because there we have hardly any influence.

Function Name	Inclusive Sa... ▼	Exclusive Sa...	Inclusive Samples %	Exclusive Samples...	Module Name
◢ netCDF4Downloader.exe	1 055 938	0	100,00	0,00	
◢ netCDF4Downloader.App.Main	882 673	0	83,59	0,00	netCDF4Downloader.exe
◢ [PresentationFramework.ni.dll]	882 673	800 591	83,59	75,82	PresentationFramework.ni.dll
▷ netCDF4Downloader.Core.MERRA2Downlo;	78 221	24	7,41	0,00	netCDF4Downloader.Core.dll
▷ [kernel32.dll]	1 389	657	0,13	0,06	kernel32.dll
▷ netCDF4Downloader.MainWindow+<Butto	1 103	0	0,10	0,00	netCDF4Downloader.exe
▷ netCDF4Downloader.MainWindow.ButtonA	971	0	0,09	0,00	netCDF4Downloader.exe
▷ netCDF4Downloader.MainWindow.ButtonS	303	0	0,03	0,00	netCDF4Downloader.exe
netCDF4Downloader.Core.MERRA2Crawler.	30	30	0,00	0,00	netCDF4Downloader.Core.dll
netCDF4Downloader.Core.MERRA2Crawler.	20	20	0,00	0,00	netCDF4Downloader.Core.dll
netCDF4Downloader.Core.MERRA2Crawler.	16	16	0,00	0,00	netCDF4Downloader.Core.dll
▷ netCDF4Downloader.Core.ExcelSheetBuilde	11	0	0,00	0,00	netCDF4Downloader.Core.dll
netCDF4Downloader.Core.MERRA2Crawler.	7	7	0,00	0,00	netCDF4Downloader.Core.dll
▷ HDF5DotNet.H5D.open	5	0	0,00	0,00	HDF5DotNet.dll
netCDF4Downloader.MainWindow.Downlo	4	4	0,00	0,00	netCDF4Downloader.exe
MS.Internal.SystemXmlExtension.IsXmlNod	1	1	0,00	0,00	PresentationFramework-SystemXml.
▷ netCDF4Downloader.MainWindow.<ResetP	1	0	0,00	0,00	netCDF4Downloader.exe
▷ [kernel32.dll]	173 258	1	16,41	0,00	kernel32.dll
▷ [nvd3dum.dll]	2	0	0,00	0,00	nvd3dum.dll
▷ [nvSCPAPI.dll]	1	0	0,00	0,00	nvSCPAPI.dll
▷ [mdnsNSP.dll]	1	0	0,00	0,00	mdnsNSP.dll
▷ [nvspcap.dll]	1	0	0,00	0,00	nvspcap.dll
[nvldumd.dll]	1	1	0,00	0,00	nvldumd.dll
▷ [nvapi.dll]	1	0	0,00	0,00	nvapi.dll

Fig. 5. Using the processor while downloading files from the MERRA-2 database

Memory Usage Test

The memory usage test was conducted in the same way as the CPU usage test. In Fig. 6 is visualized a RAM usage during nearly 30 min application work. The first increase of used memory is almost imperceptible: point 1 – 54 MB at 56.3 MB. At this point a connection is made and the list of directory URLs is downloaded to the memory. Then before point 2, a list of file URLs from a given time interval is downloaded. Point 2 indicates the moment when instruction *to start the download* was issued. Between point 2 and point 3, the memory usage increased from 71 MB to 277 MB. In point 4, the next file has been downloaded from the list and there has been another increase in usage at 485 MB.

Another interesting phenomenon is the drop located between points 5 and 6. It shows how of the garbage collector frees the memory, occupied by an already processed file. When a data file is completely looked through and is no longer used it automatically, it is cleaned from the memory. During this test, the program did not take up more than 490 MB of RAM.

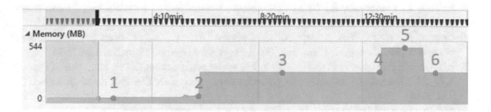

Fig. 6. Using the processor while downloading files from MERRA-2 database

The results of conducted tests, prove that there are no accidental memory leaks and the maximal memory usage is fully acceptable for a modern computer.

5 Conclusions

Wind energy systems modeling require meteorological data to derive power output or energy yield in regional or national scale. One of the resources of such data can be reanalysis. MERRA-2 reanalysis data offers a few meteorological data (wind speed and direction, air pressure, temperature, humidity, to name a few) useful in systems modeling. All the data, organized into a tree-structure, is accessible in NC4 file format. Moreover, the granule size of the file is usually huge. One file refers to 1 day only.

In wind energy systems modeling it is most common to use the historic meteorological data that represent long time period (thousands of days). It means that obtaining some particular data relying on download each file individually and extraction the data stored in NC4 files is very inconvenient and time-consuming. Thus, we have designed and implemented computer software that can support the collection and processing of meteorological data stored in the files of NC4 format and accessible to download from MERRA-2 collections, shared through the NASA Goddard Earth Sciences Data and

Information Services Center (GES DISC). The processing relies on copying selected data from the NC4 file to another, with a more user-friendly format, e.g. XLSX.

Acknowledgement. The MERRA-2 reanalysis data was provided by the Global Modeling and Assimilation Office (GMAO) at NASA Goddard Space Flight Center through the NASA GES DISC online archive.

This work has been prepared under the project S/WE/3/2018 and financially supported by Ministry of Science and Higher Education, Poland.

References

1. Cannon, D.J., Baryshaw, D.J., Methven, J., Coker, P.J., Lenaghan, D.: Using reanalysis data to quantify extreme wind power generation statistics: a 33 year case study in Great Britain. Renew. Energy **75**, 767–778 (2015)
2. Olanson, J., Bergkvist, M.: Modelling the Swedish wind power production using MERRA reanalysis data. Renew. Energy **76**, 717–725 (2015)
3. Staffell, I., Pfenninger, S.: Using bias-corrected reanalysis to simulate current and future wind power output. Energy **114**, 1224–1239 (2016)
4. Gelaro, R., et. al.: The modern-era retrospective analysis for research and applications, version 2 (MERRA-2). J. Clim. July 2017. https://doi.org/10.1175/jcli-d-16-0758.1
5. MERRA-2: File specification. Global Modeling and Assimilation Office. Note No. 9 (Version 1.1). 21 March 2016. https://gmao.gsfc.nasa.gov/reanalysis/MERRA-2/docs/. Accessed 27 Jan 2018
6. NASA.: General information. https://www.nasa.gov/. Accessed 14 Jan 2018
7. Brennan, J., Lee, J.H., Yang, M.Q., Folk, M., Pourmal, E.: Working with NASA's HDF and HDF-EOS earth science data formats. Earth Obs. **25**, 16–20 (2013)
8. Cleary, S.: Async Programming: Patterns for Asynchronous. MSDN Magazine 29/3 (2014). https://msdn.microsoft.com/en-us/magazine/dn605875.aspx. Accessed 5 Feb 2018
9. Careerfoundry.: 5 Big Differences Between UX And UI Design. https://careerfoundry.com/en/blog/ux-design/5-big-differences-between-ux-and-ui-design/. Accessed 14 Jan 2018
10. Earthdata. https://urs.earthdata.nasa.gov/. Accessed 14 Jan 2018
11. Kumar, R., Khan, S.A., Khan, R.A.: Revisiting software security: durability perspective. Int. J. Hybrid Inf. Technol. (SERSC) **8**, 311–322 (2015)

Spartan FPGA Devices in Implementations of AES, BLAKE and Keccak Cryptographic Functions

Jarosław Sugier[✉]

Faculty of Electronics, Wrocław University of Science and Technology,
Janiszewskiego St. 11/17, 50-372 Wrocław, Poland
jaroslaw.sugier@pwr.edu.pl

Abstract. The aim of this paper was to analyze efficiency of the Sparatn-7 platform in implementation of three cryptographic algorithms: the AES-128 symmetric block cipher, the BLAKE-256 hash function and Keccak-f[1600], a compression function selected for the SHA-3 hash standard. Each cipher was implemented in 5 variants: in the basic iterative architecture and two high speed loop unrolled organizations, with and without pipelining. The results identified potential benefits which the new FPGA family can bring to implementations of these ciphers with regard to both absolute parameters of the designs and efficiency of the loop unrolling mechanism. The paper also includes comparison of the findings with the corresponding results of analogous evaluations of the algorithms in the previous generations of Spartan-6 and 3 devices.

Keywords: Implementation efficiency · Pipelining · Loop unrolling
Hash function · Block cipher

1 Introduction

Contemporary ciphers are one of the most difficult kind of circuits for implementation in FPGA devices. Their large sizes, irregular internal processing and random data distribution resulting in chaotic yet widely spread routing create a lot of problems for implementation tools and stretch capabilities of the arrays to their limits. The aim of this work was to evaluate implementation efficiency of the three ciphers: AES-128, BLAKE-256 and Keccak-f[1600] in the new Spartan-7 FPGA devices and compare the results with analogous effects known for the Spartan-6 and Spartan-3 families. Each cipher was tested in five architectures which were implemented by the tools in the same S7 device. In addition, each group of cipher architectures was analyzed with regard to efficiency of the loop unrolling mechanism.

The paper is organized as follows. The algorithms and their architectures are presented briefly in the second section. In the third section parameters obtained after their implementation are discussed: they are analyzed in order to evaluate implementation efficiency of the ciphers on the new Spartan-7 platform and then also to compare efficiency of the loop unrolling mechanism within each cipher group. Conclusions in

© Springer International Publishing AG, part of Springer Nature 2019
W. Zamojski et al. (Eds.): DepCoS-RELCOMEX 2018, AISC 761, pp. 461–470, 2019.
https://doi.org/10.1007/978-3-319-91446-6_43

Sect. 4 summarize benefits which the new FPGA family has brought for implementations of each algorithm.

2   The Algorithms

The Advanced Encryption Standard (AES-128, [8]) is a symmetric block cipher which operates as a substitution-permutation network processing *the state* – a 4×4 matrix of 8-bit words – in a series of $n_r = 10$ almost identical rounds. Each round uses its own key which is generated by a separate key expansion routine from another set of four 32b words. Data encoding and key expansion share the same group of elementary transformations operating on the 8b words and constitute two 128b processing paths, hence the total data width in the hardware implementation is 256b.

The BLAKE-256 algorithm [1] is a hash function generating 256b output, internally handling a 512b state in a form of sixteen 32b words. The heart of the method is a compression function and its implementation is the subject of this study. The compression is done in $n_r = 14$ identical rounds with each round transforming the state twice by so called *G* function. The set of elementary operations comprises bitwise xor, arithmetic additions and rotations of the state words.

The KECCAK-f[1600] permutation function [2] is the essential part of so called *sponge construction* [3] which calculates hash values in the SHA-3 standard. The function operates on a state of 5×5 64b words (to the total state size of 1600b) in a series of $n_r = 24$ identical rounds: each round is a sequence of 5 transformations operating on individual bits (rather than words) of the state. Of the three ciphers considered in this comparison the KECCAK's operations are the most fine-grained ones.

In the literature there are many proposals of hardware implementations of these particular algorithms [5–7, 9] which are usually optimized for minimum size or maximum performance. For the purposes of this work a distinct approach was developed: to obtain results which are comparable between different ciphers, their architectures and the hardware platforms, all three algorithms were realized in the same set of generic architectures, applying the same implementation methodology.

The set of architectures every algorithm was tested in was based on the standard iterative one: a module executing transformations of one cipher round was instantiated in hardware once and to compute the result the state bits were passed through it n_r times, i.e. in n_r clock cycles. After universal taxonomy of architectures feasible for round-based cryptographic algorithms proposed in [4] this case will be denoted as "x1". Then the remaining high-speed architectures were derived by loop unrolling and pipelining [4]: the xk organizations instantiated a combinational cascade of k round modules with computations completed in n_r/k clock cycles and the PPLk cases were created by adding pipeline registers after each round. For AES and BLAKE $k = 2$ and 5 (i.e. their complete sets comprised of x1, x2, x5, PPL2 and PPL5 implementations) while for KECCAK, because of $n_r = 24$, as the most unrolled architectures x4 and PPL4 cases were chosen ($k = 4$ instead of 5).

3 Implementation Results

To create actual hardware implementations of the 3 algorithms a complete practical designs were prepared: in each one the main cipher unit was equipped with some basic input/output buffers providing means for iterative loading the plaintext and unloading the results. The buffers were necessary because number of input and output bits of the cipher units exceeded 250 available pins even in the largest package of the Spartan-7 chip. Nevertheless, the buffers consumed only flip-flops (and not combinational resources) so they interfered very little with the actual cipher units which were heavily logic-oriented.

The designs were automatically synthesized and implemented by Xilinx Vivado software for the Spartan-7 XC7S50 FGGA484-2 device [14]. The chip was sufficiently large to accommodate even the largest organizations: with 32600 LUT elements available, the most sized BLAKE designs took at most 67% of them while KECCAK and AES units fit within no more than, respectively, 36 and 16%.

3.1 Parameters of the Implementations

As explained above, each cipher was implemented in five architectures (the basic iterative one plus two loop unrolled and two unrolled pipelined) and the obtained parameters of the 15 cases are presented in Table 1. The first four columns describe utilization of various resources i.e. the size of the implemented designs while the last four ones provide characteristics related to their performance. The first two parameters give design size in number of occupied Look-Up Table elements (LUTs) and slices while the next two ones – utilization of F7 and F8 multiplexers which are located in every slice and are used in combining multiple LUT elements when Boolean functions of more than 6 inputs are synthesized [13]. The fifth column is the fundamental evaluation of circuit performance – the minimum clock period T_{clk} as it was estimated after static timing analysis of the final, fully routed design. Finally, the last three columns provide parameters describing the longest combinational path in the design (which determined T_{clk}): its number of logic levels in total, the dedicated carry logic elements among them, and the percentage of the propagation delay incurred by the routing resources.

For a better visualization the LUT numbers of all designs are presented graphically in the left-hand part of Fig. 1. The reported sizes are – in general – in accordance with expectations based upon internal sizes of the ciphers, but some interesting variations should be emphasized.

The AES has been known for problematic implementations in the older generations of FPGA chips where their smaller, 4-input LUT tables struggled in realizations of 8-bit substitution boxes. In Spartan-7 there are no symptoms of these problems and the AES turns out to be the smallest unit in all organizations. As the table proves this is accomplished with extensive use of F7/F8 multiplexers which were crucial in realization of the SBoxes – if the algorithm had used a substitution scheme for words any longer than 8b the problems would probably return but the configuration of LUT/F7/F8 in every slice of the Spartan-7 array handled its implementation quite efficiently.

BLAKE is the other cipher which uses F7/F8 multiplexers equally intensively. Looking at raw LUT and slice numbers we can see that this cipher led to the largest implementations of all the three algorithms – a rather unexpected record because a 512-bit BLAKE's state is about 3 times smaller than that of KECCAK (1600). Further analyses of the combinational paths of this cipher in the last paragraph of this section will help in explanation why the resource utilization is so above the expected numbers. The unrolled architectures of this cipher – the x5 and PPL5 cases – turned out to be the largest designs of all the cases, surpassing remarkably their KECCAK counterparts.

Table 1. Size and speed parameters of the ciphers implemented in the Spartan-7 device.

Architecture	LUTs	Slices	F7 Mux	F8 Mux	T_{clk} [ns]	Levels of logic	Carry elements	% route delay
AES								
x1	1479	499	144	40	5,61	6	0	80,9
x2	2464	799	384	160	8,79	11	0	78,4
x5	5309	1489	984	412	19,04	24	0	79,7
PPL2	2594	869	320	160	4,97	5	0	78,6
PPL5	5360	1503	1280	640	5,13	6	0	77,3
BLAKE								
x1	6048	1663	620	40	21,7	50	33	50,6
x2	9114	2434	1984	992	41,3	92	59	49,0
x5	21510	5575	5120	2560	100,0	222	148	49,8
PPL2	9485	2522	1984	992	22,0	49	32	49,5
PPL5	21736	5783	5120	2560	23,3	53	34	51,0
KECCAK								
x1	4526	1401	0	0	3,90	2	0	82,7
x2	7691	2183	0	0	7,92	6	0	82,2
x4	11649	3679	0	0	18,32	13	0	90,1
PPL2	6664	1891	0	0	4,17	2	0	85,5
PPL4	10392	2924	0	0	5,80	3	0	85,1

Implementations of the SHA-3 core cipher, on the other hand, produce very stable and predictable sizes in all architectural alternatives. This stands in bright contrast to our previous evaluations of this algorithm in Spartan-3 and Spartan-6 devices [11] where implementations of the full-sized KECCAK were either difficult to complete due to routing congestion or led to oversized designs. New capabilities of the Spartan-7 array offer substantial improvements here especially when compared to the Spartan-6 abilities. It should be also noted that KECCAK implementations are not as large as its raw size would suggest: with 1600 bits of state it is approx. 6 times bigger than AES (which holds a total of 256 bit of state in both data and key expansion paths) while the their ratios in the LUT and slice utilization are significantly lower – they vary over all the architectures from 2.2 to 3.1.

Turning now to the second part of the table – the performance characteristics, with the minimum clock period presented graphically in the right-hand part of Fig. 1 – one look at the graph reveals an interesting observation: the fastest designs comes from Keccak and not from (much smaller) AES. Even if the Keccak cases are 2 to 3 times larger than the AES ones, number of logic levels in their longest paths are noticeable lower (even 2 vs 6 in the x1 cases) so, with the logic packed much more efficiently in the LUTs, the overall delay is shorter. It also means that, still when comparing to AES, its 2x to 3x size increase affected very little the length of the propagation tracks; instead, the increase was absorbed in the width of the data paths. Such a regular, almost ideal scaling of logic with an enlarged size is difficult to achieve in FPGA implementations and is even more distinctive if presents itself in realizations of a very dense and involved cipher algorithm – the biggest one in our test suite.

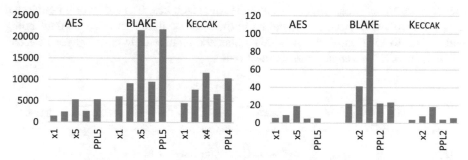

Fig. 1. Comparison of size and performance of the algorithms after implementation in the Spartan-7 device: number of occupied LUT elements (left) and minimum clock period (in ns, right).

The performance results are not so spectacular for the BLAKE algorithm. Parameters of the longest propagation paths reveal its peculiarity: this cipher (as the only one of the tested set) relies on utilization of the dedicated carry logic resources and this impairs efficient application of the LUT generators which cannot be utilized up to their extended Spartan-7 abilities. Because the BLAKE transformations are a mixture of xors, rotations *and* 32-bit additions – and the adder circuits must remain outside the LUT tables because they are implemented with dedicated carry logic – data paths in this cipher are implemented as relatively uncomplicated Boolean fragments (included in LUTs) running between the adders. This explains very high numbers of levels of logic in this cipher: 49 (x1) ÷ 222 (x5) compared to 2 ÷ 13 in the much bigger Keccak. Even if another but this time positive consequence of this feature is exceptionally low routing part in the delay of the longest path (approx. 50% vs 70 ÷ 80% in the other ciphers) this cannot compensate multiple extra propagation costs induced by the very large number of elements along the path.

3.2   The Basic Iterative Architectures on Various Platforms

In this point we will compare the Spartan-7 implementation of the basic iterative architecture of each cipher with its realizations on the previous platforms: Spartan-3 and Spartan-6. The data for comparisons were taken from our previous works published in [10–12] where the same approach was taken in construction of the code and evaluation of results. The analysis will be limited to the two most important parameters of the designs: number of utilized LUT elements (size) and minimum clock period (speed) – and these are presented in Fig. 2.

The left-hand graph generally confirms equivalent capabilities of LUT resources in Spartan-7 and 6 generations: sizes of all the three ciphers in these two devices are comparable within ±6 ÷ 10% margin while the reduced Spartan-3 LUT elements had to be utilized in much larger quantity. In AES this is nearly 6 times more while in BLAKE and KECCAK – approx. by 50%. The AES disproportion confirms that the biggest challenge in FPGA implementation of this particular cipher is realization of 8-bit substitution boxes which in Spartan-7 or -6 arrays need only 4 LUTs per output (plus F7 and F8 multiplexers of each slice) while in Spartan-3 – 16 LUTs for the 256 bits of function definition plus extra multiplexing resources. It is also worth noting that the most involved internal data handling in the KECCAK algorithm is not so demanding with regard to LUT capabilities: in this cipher the relative difference between S7&S6 utilizations vs S3 is actually the smallest one.

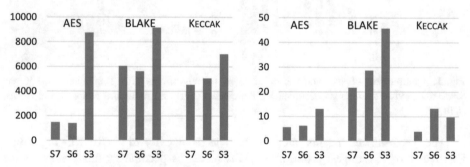

Fig. 2. Size and performance of the basic iterative architectures in the Spartan-7, Spartan-6 and Spartan-3 devices: number of occupied LUT elements (left) and the minimum clock period (in ns, right).

The results are even more interesting regarding the minimum clock cycle (the right-hand part of Fig. 2). In the AES case the big size difference on the Spartan-3 platform is not as evident in the clock cycle: it is longer but, in relation to S7&S6 results, more by a factor of 2 rather than 6 – which indicates that the logic of substitution boxes did utilize a lot of LUT resources in Spartan-3 but the implementation tools were able to keep their layout and propagation paths under control. The BLAKE case remains the slowest design across all the platforms but here the performance improvements of the newer Spartan-7 architecture begin to show the advantage in routing efficiency over the Spartan-6 predecessor. This advantage becomes critical in the KECCAK cipher: while its Spartan-6 implementation was unexpectedly slower than the Spartan-3 one (on both graphs of Fig. 2 this is *the only* case when a S6 or S7 result is inferior to the S3 one) such an abnormal situation returns to proper proportions on the Spartan-7 platform: this implementation is evidently the fastest one, obliterating the inappropriate Spartan-6 result.

3.3 Efficiency of the Loop Unrolling

Scalability of the loop unrolled architectures (in our case: the xk and PPLk cases, $k = 2$, 5 or 4) is the ability to keep the size and the minimum clock cycle in proportion to the number of rounds instantiated in hardware. As the previous studies have shown e.g. in [12] some hardware platforms may exhibit significant weaknesses in keeping the expected minimum clock period, mainly due to difficulties in reproduction of involved and irregular internal routing of a round when their long cascade must co-exist in the FPGA array.

Like in the previous subsection, the evaluation of loop unrolling efficiency observed in the Spartan-7 device will be extended with comparison to analogous results obtained in our previous works for the same ciphers and architectures on Spartan-6 and Spartan-3 platforms, albeit with one notable exception: implementation of any loop unrolled architecture of the Keccak cipher was impossible in the S6 array (although the implementations on the older Spartan-3 platform completed successfully) due to routing congestion which was reported by the tools. The only available results for Keccak on Spartan-6 were obtained for the smallest x1 case so this analysis of loop unrolling efficiency had to omit this case of cipher/platform combination.

The loop unrolling efficiency was investigated in a basic comparison within each group of 5 cipher implementations on a specific platform: taking the x1 design as a point of reference, its LUT size and T_{clk} period were used to compute the estimated size and speed of the derived architectures and then the estimations were compared against actual parameters. The estimates were computed assuming ideal efficiency of round replication during loop unrolling: size in the unrolled (xk) and pipelined (PPLk) organizations should increase linearly with k, while the clock period should remain constant in the PPLk cases (the data propagate in each clock cycle still through one round) and should increase in proportion to k in the unrolled xk ones.

Figure 3 presents results of such an analysis applied to sizes of the designs (number of LUTs), showing the ratios *actual_size*: *estimate* for all the derived architectures in

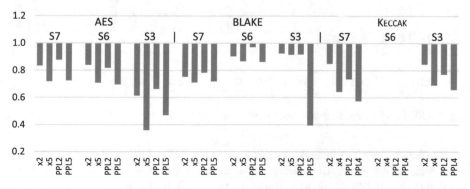

Fig. 3. Efficiency of the loop unrolling mechanism: scaling of size expressed in LUT elements.

each cipher group. The lower the ratio, the smaller (lower number of LUT) was the actual implementation in comparison to what could be expected from the relevant x1 case. The value of 1.0 is the threshold separating "better than" (*actual-_size* smaller than *estimate*) from "worse than" expected (*actual_size* bigger than *estimate*).

As the chart proves, all the unrolled architectures were implemented in Spartan-7 with amount of LUT which was smaller than simple multiplication by k would indicate but there are some differences in comparison with the older platforms. The S7 AES cases are the closest to the S6 results but the S3 numbers of this cipher are even lower – being the lowest across all cipher/platform combinations. The size optimization techniques of the implementation tools were especially efficient in the unrolled AES architectures on the Spartan-3 platform – i.e. exactly in the cases when the 8-bit substitution boxes made the overall design size so large. A different relation is in the BLAKE cipher where the S7 reductions are on average twice bigger than the S6 ones *and*, not counting the singularity of the PPL5 case, even bigger than the S3 ones. In KECCAK, finally, the reductions are only marginally better in S7 than they were in S3 so advantages of much more capable LUT resources are hardly seen in efficiency of size scaling in this cipher.

Again, the situation is more irregular (and thus more interesting) in Fig. 4 which visualizes scaling of the T_{clk} parameter. In this comparison only the unrolled AES architectures can operate consistently faster than expected – with the only exception of the pipelined organizations on the Spartan-3 platform which exceeded the x1 clock period but only by $3 \div 4\%$ – while the other two ciphers tend to make even bigger excesses a rule rather than an exception. In the BLAKE cases the worst situation is on the Spartan-6 platform: switching there from the older S3 device actually impairs unrolling efficiency and the newest S7 platform on average only returns to the results which were possible beforehand. Nevertheless, the BLAKE results were not the worst ones.

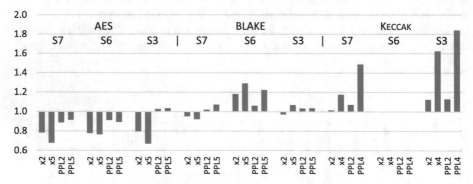

Fig. 4. Efficiency of the loop unrolling mechanism: scaling of the minimum clock period T_{clk}.

Scaling of the clock period was the most problematic in the KECCAK implementations and, unlike in BLAKE, these problems were shown already on the Spartan-3 platform were the PPL4 architecture was by a record 84% slower than expected and the x4 one – by 63%. The pipelining did not help to solve the difficulties brought by unrolling of this cipher into 4 rounds. While implementations on the Spartan-6 platform ended in failures, the new Spartan-7 device did offer substantial progress: not only the implementations were successful but excesses in the T_{clk} scaling were reduced to 18 and 49%.

This completes the picture of the benefits which the new S7 platform brought to the KECCAK implementations: with its x1 organization being even faster than the corresponding much smaller AES design and with very efficiently size implementations sizes, the problems in realization of this cipher remain only in the scaling of clock period in the unrolled architectures. Those problems are not so significant, though, and the overall improvements over the previous S6 platform remain undisputable.

4 Conclusions

In this work we have evaluated implementation efficiency of the three ciphers: AES-128, BLAKE and KECCAK-f[1600] compression function (the core of the new SHA-3 standard) in the new Spartan-7 devices, in 5 high speed architectural variants.

The oldest cipher in the test suite, AES, benefits to the least degree from the new capabilities of the S7 architecture. Implementations of this algorithm attained their optimal efficiency already on the S6 platform where the wide, 8-bit substitution boxes (which were so troublesome for the S3 resources) found an adequate realization with 6-input look-up tables. Neither AES size nor complexity of its internal data distribution created any challenge that would call for new capabilities of the S7 architecture. The situation is different, though, for the new ciphers developed for the SHA-3 contest.

The BLAKE algorithm still remains difficult in FPGA implementation due to its extensive use of the 32-bit adders which are realized with dedicated carry resources and cannot benefit from extended capabilities of the LUT elements. This factor leads to large implemented sizes which also on the S7 platform exceed proportions of the actual dimensions of this cipher based e.g. on size of its state words. Nevertheless, what the new platform can really offer in this situation are extended routing capacities which alleviate problems in scaling of the minimum clock period of the unrolled architectures demonstrated on the Spartan-6 platform.

It is the KECCAK compression function which benefits most from the new Spartan-7 potentials in this comparison. Despite the largest internal size, KECCAK implementations were consistently smaller than those of BLAKE and not as much larger than the AES ones as the difference in the raw data sizes would suggest. Details of the propagation paths confirmed very effective compression of the cipher transformations in S7 LUT elements. This high size efficiency was accompanied also by good performance characteristics: despite their size, the KECCAK S7 implementations were on average faster than their AES peers while the S6 and S3 cases did not obey this rule.

References

1. Aumasson, J.P., Henzen, L., Meier, W., Phan, R.C.W.: Sha-3 proposal BLAKE, version 1.3. https://www.131002.net/blake/blake.pdf. Accessed Mar 2018
2. Bertoni, G., Daemen, J., Peeters, M., Van Assche G.: The Keccak reference. http://keccak. noekeon.org. Accessed Mar 2018
3. Bertoni, G., Daemen, J., Peeters, M., Van Assche G.: The Keccak sponge function family. http://keccak.noekeon.org. Accessed Mar 2018
4. Gaj, K., Homsirikamol, E., Rogawski, M., Shahid, R., Sharif, M.U.: Comprehensive evaluation of high-speed and medium-speed implementations of five SHA-3 finalists using Xilinx and Altera FPGAs. In: The Third SHA-3 Candidate Conference (2012). Available: IACR Cryptology ePrint Archive, 2012, 368
5. Gaj, K., Kaps, J.P., Amirineni, V., Rogawski, M., Homsirikamol, E., Brewster, B.Y.: ATHENa – automated tool for hardware evaluatioN: toward fair and comprehensive benchmarking of cryptographic hardware using FPGAs. In: 20th International Conference on Field Programmable Logic and Applications, Milano, Italy, (2010)
6. Junkg, B., Apfelbeck, J.: Area-efficient FPGA implementations of the SHA-3 finalists. In: 2011 International Conference on Reconfigurable Computing and FPGAs (ReConFig), pp. 235–241. IEEE (2011)
7. Liberatori, M., Otero, F., Bonadero, J.C., Castineira, J.: AES-128 cipher. high speed, low cost FPGA implementation. In: Proceedings of the Third Southern Conference on Programmable Logic, Mar del Plata, Argentina. IEEE Computer Society Press (2007)
8. National Institute of Standards and Technology: Specification for the ADVANCED ENCRYPTION STANDARD (AES). Federal Information Processing Standards Publication 197. http://csrc.nist.gov/publications/PubsFIPS.html. Accessed Mar 2018
9. Strömbergson, J.: Implementation of the Keccak hash function in FPGA devices. http://www.strombergson.com/files/Keccak_in_FPGAs.pdf. Accessed Mar 2018
10. Sugier, J.: Implementation efficiency of BLAKE and other contemporary hash algorithms in popular FPGA devices. In: Zamojski, W., Mazurkiewicz, J., Sugier, J., Walkowiak, T., Kacprzyk, J. (eds.) Dependability Engineering and Complex Systems, Proceedings of the 11th International Conference on Dependability and Complex Systems DepCoS-RELCOMEX. AISC, vol. 479, pp. 457–467. Springer, Heidelberg (2016)
11. Sugier, J.: Low cost FPGA devices in high speed implementations of Keccak-f hash algorithm. In: Zamojski, W., Mazurkiewicz, J., Sugier, J., Walkowiak, T., Kacprzyk, J. (eds.) Proceedings of the 9th International Conference on Dependability and Complex Systems DepCoS-RELCOMEX. AISC, vol. 286, pp. 433–441. Springer, Heidelberg (2014)
12. Sugier, J.: Popular FPGA device families in implementation of cryptographic algorithms. In: Zamojski, W., Mazurkiewicz, J., Sugier, J., Walkowiak, T., Kacprzyk, J. (eds.) Theory and Engineering of Complex Systems and Dependability, Proceedings of the 11th International Conference on Dependability and Complex Systems DepCoS-RELCOMEX. AISC, vol. 365, pp. 485–495. Springer, Heidelberg (2015)
13. Xilinx, Inc.: 7 Series FPGAs Configurable Logic Block. UG474.PDF. www.xilinx.com. Accessed Mar 2018
14. Xilinx, Inc.: 7 Series FPGAs Data Sheet: Overview. DS180.PDF. www.xilinx.com. Accessed Mar 2018

Signal Feature Analysis for Dynamic Anomaly Detection of Components in Embedded Control Systems

Xin Tao[1], DeJiu Chen[1(✉)], and Juan Sagarduy[2]

[1] Mechatronics, Machine Design, School of ITM, KTH Royal Institute
of Technology, 100 44 Stockholm, Sweden
{taoxin, chendj}@kth.se
[2] Application Engineering Group, MathWorks AB, 164 21 Kista, Sweden
juan.sagarduy@mathworks.com

Abstract. Embedded Control Systems (ECS) are getting increasingly complex for the realization of Cyber-Physical Systems (CPS) with advanced autonomy (e.g. autonomous driving of cars). This compromises system dependability, especially when components developed separately are integrated. Under the circumstance, dynamic anomaly detection and risk management often become a necessary means for compensating the insufficiencies of conventional verification and validation, and architectural solutions (e.g. hardware redundancy). The aim of this work is to support the design of embedded software services for dynamic anomaly detection of components in ECS, through probabilistic inference methods (e.g. Hidden Markov Model - HMM). In particular, the work provides a method for classifying the signal features of operational sensors and thereby applies Monte-Carlo sensitivity analysis for eliciting the probabilistic properties for error estimation. Such approach, based upon a physical model, reduces the dependency on empirical data for bringing about confidence on newly developed components.

Keywords: Embedded Control Systems · Cyber-Physical Systems
Feature analysis · Anomaly · Hidden Markov Model
Monte-Carlo sensitivity analysis

1 Introduction

Increasing complexity of cyber physical systems (CPS) targeting advanced autonomy (e.g. autonomous driving) results in numerous unknowns and uncertainties linked to operational contexts, functional decisions and technical compositionality. The lack of *a priori* knowledge about open environments, system operations and other relevant factors, can lead to unacceptable hazards and accidents [1]. This makes it more difficult to guarantee overall system dependability, especially when components developed separately are also being integrated. Under the circumstances, dynamic anomaly detection and risk management become a necessary means for compensating for insufficiencies of conventional verification and validation, and architectural solutions (e.g. hardware redundancy) [2]. The support allows a system to infer its likely

© Springer International Publishing AG, part of Springer Nature 2019
W. Zamojski et al. (Eds.): DepCoS-RELCOMEX 2018, AISC 761, pp. 471–481, 2019.
https://doi.org/10.1007/978-3-319-91446-6_44

operational errors based on the observations of some data and signal emissions and thereby to avoid or mitigate unacceptable behaviors.

In this paper, we propose an approach to signal feature analysis and thus empowering embedded software services for dynamic anomaly detection in ECS components through probabilistic inference methods (e.g. HMM - Hidden Markov Model) [2]. The work provides a classification method for basic signal features of sensors and thereby applies a simulation technique for eliciting necessary probabilistic properties for error estimation. Such a model-based approach is motivated by the lack of empirical knowledge in regard to newly developed components. The work is supported by MathWorks technology and expertise, in particular Simscape for multi-domain system modeling and Simulink Design Optimization [3] for Monte-Carlo sensitivity analyses. The paper is organized as follows: in Sect. 2, the problem and methodology are introduced; Sect. 3 presents the proposed method for signal feature classification; Sect. 4 presents the case study of an MEMS (Micro-Electro-Mechanical Systems) accelerometer in automotive systems; Sect. 5 gives a conclusion of the work and a conception of future work.

2   State of the Art

Comprehensive research has been done in the areas of safety and reliability engineering for anomaly detection of embedded systems. A common method is building a fault model given by a mapping between the known failure modes in a system and the available symptoms, often captured in a Dependency Matrix (D-MATRIX) [4]. This type of model is at the core of an Integrated Vehicle Health Management (IVHM) system introduced in [5]. However, for systems with complex and integrated structures, it is difficult to give an accurate detection and diagnosis of the failure modes according to symptoms. The cause is the complexity in isolating the impact by disturbances [1] and conditions [6] such as environment temperature and noise. In [7], a board-level functional fault diagnosis has been conducted using various methods including artificial neural networks (ANN), support vector machines (SVM) and weighted-majority voting (WMV). These methods rely heavily on samplings of experience knowledge and field data.

In regard to ECS or CPS, uncertainties can be classified into two major categories: aleatory and epistemic [8]. Aleatory uncertainty, also known as statistical uncertainty, is the inherent randomness associated with the physical system or environment. Epistemic uncertainty is mainly derived from the lack of knowledge or information (domain unfamiliarity, limited experience). In [1], a conceptual uncertainty model has been proposed, considering explicitly three architectural levels: application, infrastructure and integration. Methods for uncertainty propagation are described in detail in [9], with different analysis methods including probability bound analysis, imprecise probability, evidence theory and possibility theory. In [10], a fault diagnostic tool was proposed with consideration of uncertainty according to available field data. A simulation-based approach has been introduced in [11], using Discrete Time Markov Chains and probabilistic model accommodating a diverse set of parameter range distributions in software architecture evaluation. In [12], a HMM algorithm has been developed for fault diagnosis with partial and imperfect tests, which can give a sequence of uncertain test outcomes over time.

3   Signal Feature Classification for Anomaly Detection

3.1   Problem Statement

In the development of ECS, it is essential to conduct quality management on the basis of a systematic verification and validation (V&V). The operational behaviors and conditions should be carefully analyzed in order to assess whether all related system requirements are satisfied and all the components are in conformity with their contextual assumptions and guarantees. However, when the V&V effort can only cover the overall system behaviors and conditions partially, which is typically the case when newly developed components are integrated as black-boxes, embedded services for dynamic anomaly detection and treatment are seen as necessary [2].

For newly developed components, error files or prior fault knowledge are often not accessible or documented. Thus, empirical anomaly detection with machine learning methods, such as ANN [13] and SVM [14] is severely hindered. A pragmatic solution is to firstly obtain the basic knowledge of potential anomaly by models and simulations and thereafter enrich such knowledge by operation monitoring. In [2], an approach to self-assessment by monitoring and inferring the operational conditions of component and system, referred to Monitoring and Assessment Service (MAS), has been proposed. The approach adopts HMM for the state estimation, which relies in turn on the specifications of conditional probabilities relating to the dynamic states and observed emissions. It assumes that such specifications would be predefined or derived by models. There are however many practical challenges in practical situations. For example, it could be difficult to define the operation characteristics to be observed due to the gaps between abstraction levels. Some expected features or patterns for the state inference may not be available because of instrumentation restrictions. Some characteristics of observations may not necessarily be caused by specific component, but by likely influences of external conditions. All these phenomena could undermine the trustworthiness and accuracy. As a consequence, an explicit classification of operation characteristics will facilitate the understanding of the system properties to be observed and assessed and thereby more effective decisions in anomaly detection and state estimation.

3.2   Anomaly Detection Method with Uncertainty Analysis

In order to tackle the issues mentioned above, a general scheme for the design of service for component anomaly detection is shown in Fig. 1. The main procedures involving are defined as follows. *Step 1: System modeling*. For newly developed components, detailed know-how on anomalies is yet to be discovered. Analysis models and simulation are used to gain insight and elicit potential anomalies. The descriptions listed below, which can be given collectively as contract [2], provide the information for creating the analysis models:

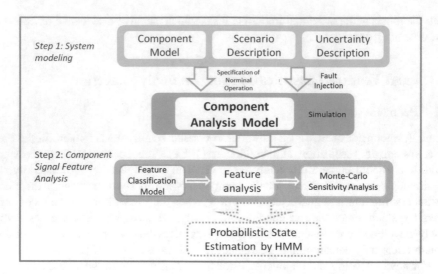

Fig. 1. General scheme for the design of service for component anomaly detection in ECS.

(1) *Component Description.* A component model specifies fundamental functional and technical design properties through physical & control parameters, logical transfer functions, and their relative configuration.

(2) *Scenario Description.* Given a component model, the scenario description captures the expected operational situations, including the possible inputs and outputs, their pre-&post-conditions. An extension of the description with faults through a systematic investigation of all possible pre-&post-conditions enables effective fault-injection and extraction of anomalous data by simulation.

(3) *Uncertainty Description.* The goal here is to support the quantification of the stochastic properties of a component, including the conditional probabilities from its inputs to its outputs. It is assumed that all the component faults under consideration are operational, i.e. relating to the inherent randomness of system operations. Such faults can be caused by noise patterns, temperature variation, etc. and lead to uncertainties in the overall system behavior.

Step 2: Component Signal Feature Analysis. The purpose of this analysis is to define and extract the signal features of a component. It is based on simulation results of a component analysis model. Input stimuli, as well as faults to be injected, are derived based on the corresponding scenario and uncertainty description. The system faults and states are defined according to combinations of input and output conditions. After each simulation, output signals are recorded and analyzed for anomaly identification. To this end, a generic definition of typical signal features is provided to support the reasoning (see Sect. 4.4). In order to quantify the emissions of various signal features given certain system states, an estimation of the conditional probabilities from such system states to the signal features becomes necessary. This is supported by Monte-Carlo sensitivity analysis through simulation. The definition of such operational features and their probabilistic properties constitute the basis for enabling HMM-based state estimation.

The tool used for this task is Simulink Design Optimization provided by Math-Works [3]. The tool automatically generates a large number of input data or stimuli according to probability distributions of the inputs. With some pre-defined output signal features as evaluation requirements, the statistical relation between the inputs and such signal features are calculated numerically based on Monte-Carlo simulations. For these features, a D-matrix can also be established for their mappings to different inputs or faults. During this process, uncertainty analysis will be performed with consideration of various disturbances that will influence the estimation accuracy.

4 Case Study of an MEMS Accelerometer

To validate the proposed method, a case study of a MEMS accelerometer is conducted. For advanced CPS or ECS such as autonomous cars, similar sensing devices are essential for the operation perception and thereby correct system behaviors. Depending on the failure modes of relevance, the provision of services for dynamic anomaly detection constitutes a vital mechanism for the overall system dependability.

4.1 Component Description

The component description is shown in Fig. 2. It captures the most basic configurations of mechanical and electrical properties based on a combination of a mass-spring-damp system and a resistance-induction-capacity system. Before performing fault injection and Monte-Carlo simulation, it is necessary to check the validity of the accelerometer model to assure that it can define the functionality correctly. The simulation result of the model validity checking is displayed in Fig. 3, in which the input and output almost overlap. This model is calibrated by adjusting the parameters of the model and observing the inputs and the outputs simultaneously. A normally functioning system can export a linear output when giving a linear input.

4.2 Scenario Description

The accelerometer of this case study is used in automotive vehicles. The expected operational situations are summed up in three driving conditions:

(1) *Braking*. When the car is braking, a sudden force is applied to the accelerometer and could be presented by a step signal;
(2) *Accelerating* or *decelerating*. When the car is accelerating or decelerating, a gradually increasing force is applied to the accelerometer, which can be presented by a ramp signal;
(3) *Bumping*. When the road suffers from bumps, a periodical force leading to vibration is generated and described as a sinusoidal signal.

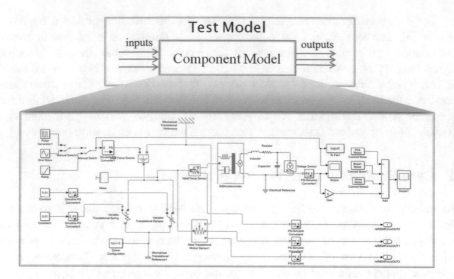

Fig. 2. Component analysis model for an MEMS accelerometer in Simscape, represented as a mass-spring-damper (coupled to vehicle) exciting an RLC electrical circuit.

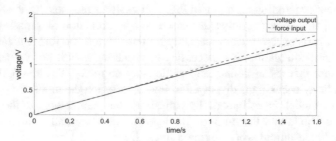

Fig. 3. Simulation result of model validity checking, showing that this model can measure the accelerated velocity correctly with the adjusted parameters of the component analysis model

4.3   Uncertainty Description

For the accelerometer, two operational conditions would have most influence on the component behavior: *temperature* and *noise*. Study has shown that the temperature around the electronical control unit (ECU) is typically between *300 K* and *400 K*, and negatively skewed with a *370 K* mode [11]. The failure rate (λ) is a function of its ambient Temperature T (in *Kelvin*) such as $\lambda = 4 \cdot 10^{-6} \times (T + 100)$. The PDF (*Probability Density Function*) of λ of *ECU X* is then given as a *Shifted-Beta* distribution [11]:

$$\lambda = (400 \times 4 \cdot 10^{-6}, 500 \times 4 \cdot 10^{-6}, 10, 2)$$

where the parameters refer to the minimum, maximum, α and β of the beta distribution respectively. By this way, the uncertainty due to temperature is converted into the failure rate of the component and modeled as an output offset.

In regard to the noise, there are mainly three categories: *White, Brown* and *Pink Noise. White Noise* is a random signal noise with constant power spectral density. *Brown Noise* is the signal noise that decreases in power by 20 dB per decade, while the pink noise is the signal noise with the power density falling off at 10 dB per decade. In practical situations, the observed output signals of a component or system can become obscured because of these signal noises.

4.4 Signal Feature Analysis

With a force input chosen from the scenarios described above (Sect. 4.2), four kinds of component operational faults are possible, relating to specific configurations of *spring stiffness* and *damping coefficient, inductance* and *capacitance*. Each type of component fault is reasoned by certain coefficient increase and decrease. By injecting the operational faults of interest into the analysis model and its simulation, the responses with different signal feature emissions can be observed. Some of the simulation results are described in Table 1. It is shown that different fault types and operational inputs will

Table 1. Signal input and output feature emissions under different scenarios with the dotted lines for the force inputs and the solid lines for the voltage outputs

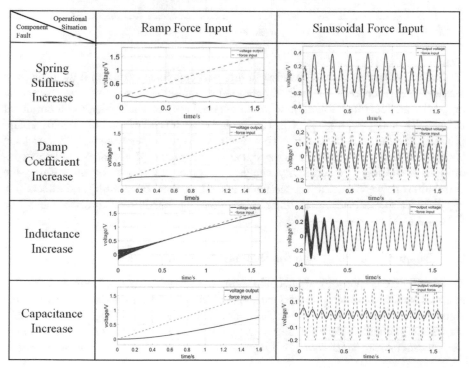

result in different outputs in terms of signal feature emissions with mathematical properties. For example, with a sinusoidal operational force input, the spring stiffness increase will result in a periodical output with multiple dominant frequencies. With a step operational force input, the damping coefficient increase will result in a logarithmic decrease of the output. With a ramp operational force input, the inductance increase will result in an output with sharp convergence. And with a sinusoidal operational force input, the capacitance increase will result in an output with amplitude amplification.

Table 2. D-Matrix mapping component failure to mathematical properties of features

Mathematical properties of features ＼ Component failure	Spring stiffness increase	Damping coefficient increase	Inductance average increase	Inductance sharp increase	Capacitance increase
Logarithmic/ Exponential growth		■●			■●
Sinusoidal	▲■	▲	▲●	▲■	▲
Multi dominant frequencies	▲				
Damping concussion			●■		

● Step force input   ■ Ramp force input   ▲ Sinusoidal force input

Table 2 presents the relation between different component faults and the respective mathematical properties of their signal feature emission. Four main mathematical properties are identified: logarithmic or exponential growth, sinusoidal, multi dominant frequencies and convergence. Clearly, complexity arises when one signal feature property is influenced by multiple component faults and operational conditions. One component fault may lead to similar signal features under different operational conditions. Accordingly, a sensitivity analysis becomes necessary for quantifying the correlations between causes (i.e. the faults and operational conditions) and effects (i.e. signal features) according to their corresponding conditional probabilities. See Sect. 4.5.

4.5   Monte-Carlo Simulation for Dynamic State Inference

To quantify the compositional effects in terms of probabilistic conditions, a sensitivity analysis with focus on component is carried out through iterative simulations (two output signal features; three system inputs). The overall results are given in Fig. 4, which clearly shows the numerical dependencies between the system inputs (i.e. *I – Induction, ab – accelerometer bandwidth, d – damping, g – noise gain*) and the evaluation criteria given by two output signal feature properties (i.e. *output signal bound, output signal variance*). Figure 5 shows the statistical analysis results quantifying such numerical dependencies, classified as correlation and standardized regression. As the configuration of parameters represents potential operational faults and states, such statistical results constitute the basis for defining the probabilistic conditions between the component states. For example, it is shown in Fig. 5 that the probability that g (the noise gain) has a strong correlation to the *output signal variance* is 90%, while the result is nearly zero when it comes to the correlation between g and the output signal bound. For further details of the tool support, please refer to [3].

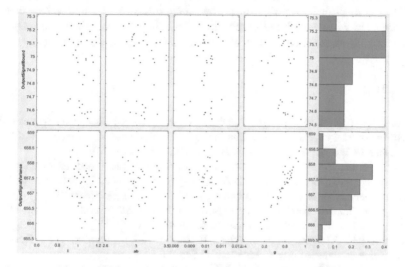

Fig. 4. Scatter plot of the results by Monte Carlo Simulation

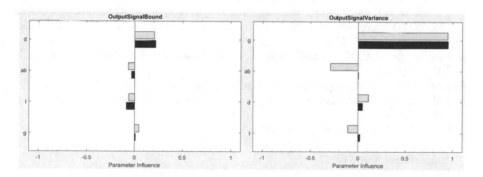

Fig. 5. Statistical results by Monte Carlo Simulation with the yellow bar for correlation and blue bar for standardized regression

5   Conclusion and Future Work

In this paper, we have proposed a dynamic anomaly detection method with uncertainty analysis. Signal processing methods are used to extract different features and establish a D-matrix to distinguish different fault types in the system. By modeling various hypothetical faults with explicitly defined uncertainties and performing Monte-Carlo simulation, a probabilistic estimation of the effects are given. A case study with MEMS accelerometer, which has a vital significance in automated driving system, has been presented to check the effectiveness of this approach. The approach contributes to the design of dynamic monitoring and assessment services with HMM by supporting the identification of prior knowledge for probabilistic inference. Further enhancements of current method will be explored in the future regarding the combinatorial features and failures, as well as the modeling of uncertainties for specific components.

References

1. Zhang, M., Selic, B., Ali, S., Yue, T., Okariz, O., Norgren, R.: Understanding uncertainty in cyber-physical systems: a conceptual model. In: European Conference on Modelling Foundations and Applications, pp. 247–264. Springer, Cham (2016)
2. Chen, D., Lu, Z.: A methodological framework for model-based self-management of services and components in dependable cyber-physical systems. In: Advances in Dependability Engineering of Complex Systems, pp. 97–105. Springer, Cham (2017)
3. Design Optimization Toolbox of Matlab Simulink, Matlab, Mathworks. https://se.mathworks.com/products/sl-design-optimization.html. Accessed 17 Mar 2018
4. Lanigan, P.E., Kavulya, S., Narasimhan, P., Fuhrman, T.E., Salman, M.A.: Diagnosis in automotive systems: a survey. Accessed 10 Sep 2011
5. Jennions, I.K. (ed.): Integrated Vehicle Health Management: Perspectives on an Emerging Field. SAE International, Warrendale (2011)
6. Kapur, K.C., Pecht, M.: Reliability Engineering. Wiley, Hoboken (2014)
7. Ye, F., et al.: Board-level functional fault diagnosis using artificial neural networks, support-vector machines, and weighted-majority voting. IEEE Trans. Comput. Aided Des. Integr. Circuits Syst. 32(5), 723–736 (2013)
8. der Kiureghian, A., Ditlevsen, O.: Aleatory or epistemic? Does it matter? Struct. Saf. 31(2), 105–112 (2009)
9. Aven, T., Baraldi, P., Flage, R., Zio, E.: Uncertainty in Risk Assessment: The Representation and Treatment of Uncertainties by Probabilistic and Non-probabilistic Methods. Wiley, West Sussex (2013)
10. Cannarile, F., Compare, M., Zio, E.: A fault diagnostic tool based on a first principle model simulator. In: International Symposium on Model-Based Safety and Assessment, pp. 179–193. Springer, Cham (2017)
11. Meedeniya, I., Moser, I., Aleti, A., Grunske, L.: Evaluating probabilistic models with uncertain model parameters. Softw. Syst. Model. 13(4), 1395–1415 (2014)
12. Ying, J., Kirubarajan, T., Pattipati, K.R., Patterson-hine, A.: A hidden Markov model-based algorithm for fault diagnosis with partial and imperfect tests. IEEE Trans. Syst. Man Cybern. Part C (Appl. Rev.) 30(4), 463–473 (2000)

13. Bangalore, P., Letzgus, S., Karlsson, D., Patriksson, M.: An artificial neural network-based condition monitoring method for wind turbines, with application to the monitoring of the gearbox. Wind Energy **20**(8), 1421–1438 (2017)
14. Zhang, X., Gu, C., Lin, J.: Support vector machines for anomaly detection. In: The Sixth World Congress on Intelligent Control and Automation, WCICA 2006, pp. 2594–2598. IEEE (2006)

Dependable Slot Selection Algorithms for Distributed Computing

Victor Toporkov$^{(\boxtimes)}$ and Dmitry Yemelyanov

National Research University "MPEI",
ul. Krasnokazarmennaya, 14, Moscow 111250, Russia
{ToporkovVV, YemelyanovDM}@mpei.ru

Abstract. In this work, we introduce slot selection and co-allocation algorithms for parallel jobs in distributed computing with non-dedicated and heterogeneous resources. A single slot is a time span that can be assigned to a task, which is a part of a parallel job. The job launch requires a co-allocation of a specified number of slots starting and finishing synchronously. Some existing resource co-allocation algorithms assign a job to the first set of slots matching the resource request without any optimization (the first fit type), while other algorithms are based on an exhaustive search. In this paper, algorithms for effective and dependable slot selection are studied and compared with known approaches. The novelty of the proposed approach is in a general algorithm efficiently selecting a set of slots according to the specified criterion.

Keywords: Distributed computing · Grid · Economic scheduling
Resource management · Slot · Job · Allocation · Optimization

1 Introduction

Modern high-performance distributed computing systems (HPCS), including Grid, cloud and hybrid infrastructures provide access to large amounts of resources [1, 2]. These resources are typically required to execute parallel jobs submitted by HPCS users and include computing nodes, data storages, network channels, software, etc. The actual requirements for resources amount and types needed to execute a job are defined in resource requests and specifications provided by users.

HPCS organization and support bring certain economical expenses: purchase and installation of machinery equipment, power supplies, user support, etc. As a rule, HPCS users and service providers interact in economic terms and the resources are provided for a certain payment. Thus, as total user job execution budget is usually limited, we elaborate an actual task to optimize suitable resources selection in accordance with a job specification and a restriction to a total resources cost.

Economic mechanisms are used to solve problems like resource management and scheduling of jobs in a transparent and efficient way in distributed environments such as cloud computing and utility Grid. The significant and important feature for well-known scheduling solutions for distributed environments is the fact that the scheduling strategy is formed on a basis of efficiency criteria [1–5]. The metascheduler [5, 6] implements the economic policy of a VO based on local resource schedules.

© Springer International Publishing AG, part of Springer Nature 2019
W. Zamojski et al. (Eds.): DepCoS-RELCOMEX 2018, AISC 761, pp. 482–491, 2019.
https://doi.org/10.1007/978-3-319-91446-6_45

The schedules are defined as sets of slots coming from resource managers or schedulers in the resource domains, i.e. time intervals when individual nodes are available to perform a part of a parallel job. In order to implement such scheduling schemes and policies, first of all, one needs an algorithm for finding sets of simultaneously available slots required for each job execution. Further we shall call such set of simultaneously available slots with the same start and finish times as execution window.

In this paper we study algorithms for optimal or near-optimal resources selection by a given criterion with the restriction to a total cost. Additionally we consider practical implementations for a dependable resources allocation problem.

2   Related Works

The scheduling problem in Grid is NP-hard due to its combinatorial nature and many heuristic-based solutions have been proposed. In [4] heuristic algorithms for slot selection, based on user-defined utility functions, are introduced. NWIRE system [4] performs a slot window allocation based on the user defined efficiency criterion under the maximum total execution cost constraint. However, the optimization occurs only on the stage of the best found offer selection. First fit slot selection algorithms (backtrack [7] and NorduGrid [8] approaches) assign any job to the first set of slots matching the resource request conditions, while other algorithms use an exhaustive search [2, 9, 10] and some of them are based on a linear integer programming (IP) [2, 9] or mixed-integer programming (MIP) model [10]. Moab scheduler [11] implements the backfilling algorithm and during a slot window search does not take into account any additive constraints such as the minimum required storage volume or the maximum allowed total allocation cost. Moreover, it does not support environments with non-dedicated resources.

Modern distributed and cloud computing simulators GridSim and CloudSim [12, 13] provide tools for jobs execution and co-allocation of simultaneously available computing resources. Base simulator distributions perform First Fit allocation algorithms without any specific optimization. CloudAuction extension [13] of CloudSim implements a double auction to distribute datacenters' resources between a job flow with a fair allocation policy. All these algorithms consider price constraints on individual nodes and not on a total window allocation cost. However, as we showed in [14], algorithms with a total cost constraint are able to perform the search among a wider set of resources and increase the overall scheduling efficiency.

GrAS [15] is a Grid job-flow management system built over Maui scheduler [11]. The resources co-allocation algorithm retrieves a set of simultaneously available slots with the same start and finish times even in heterogeneous environments. However the algorithm stops after finding the first suitable window and, thus, doesn't perform any optimization except for window start time minimization.

Algorithm [16] performs job's response and finish time minimization and doesn't take into account constraint on a total allocation budget. [17] performs window search on a list of slots sorted by their start time, implements algorithms for window shifting and finish time minimization, doesn't support other optimization criteria and the overall job execution cost constraint.

AEP algorithm [18] performs window search with constraint on a total resources allocation cost, implements optimization according to a number of criteria, but doesn't support a general case optimization. Besides AEP doesn't guarantee same finish time for the window slots in heterogeneous environments and, thus, has limited practical applicability.

In this paper, we propose algorithms for effective slot selection based on user de-fined criteria. The novelty of the proposed approach consists in implementing a dynamic programming scheme to allocate a set of simultaneously available slots in heterogeneous HPCS with non-dedicated resources. The paper is organized as follows. Section 3 introduces a general scheme for searching slot sets efficient by the specified criterion. Then several implementations are proposed and considered. Section 4 contains simulation results for comparison of proposed and known algorithms. Section 5 summarizes the paper and describes further research topics.

3 Resource Selection Algorithm

3.1 Problem Statement

We consider a set R of heterogeneous computing nodes with different performance p_i and price c_i characteristics. Each node has a local utilization schedule known in advance for a considered scheduling horizon time L. A node may be turned off or on by the provider, transferred to a maintenance state, reserved to perform computational jobs. Thus, it's convenient to represent all available resources as a set of slots. Each slot corresponds to one computing node on which it's allocated and may be characterized by its performance and price.

In order to execute a parallel job one needs to allocate the specified number of simultaneously idle nodes ensuring user requirements from the resource request. The resource request specifies number n of nodes required simultaneously, their minimum applicable performance p, job's computational volume V and a maximum available resources allocation budget C. The required window length is defined based on a slot with the minimum performance. For example, if a window consists of slots with performances $p \in \{p_i, p_j\}$ and $p_i < p_j$, then we need to allocate all the slots for a time $T = \frac{V}{p_i}$. In this way V really defines a computational volume for each single job subtask. Common start and finish times ensure the possibility of inter-node communications during the whole job execution. The total cost of a window allocation is then calculated as $C_W = \sum_{i=1}^{n} T * c_i$.

These parameters constitute a formal generalization for resource requests common among distributed computing systems and simulators.

Additionally we introduce criterion f as a user preference for the particular job execution during the scheduling horizon L. f can take a form of any additive function and as an example, one may want to allocate suitable resources with the maximum possible total data storage available before the specified deadline.

3.2 General Window Search Procedure

For a general window search procedure for the problem statement presented in Sect. 3.1, we combined core ideas and solutions from algorithm AEP [18] and systems [15, 17]. Both related algorithms perform window search procedure based on a list of slots retrieved from a heterogeneous computing environment.

Following is the general square window search algorithm. It allocates a set of n simultaneously available slots with performance $p_i > p$, for a time, required to compute V instructions on each node, with a restriction C on a total allocation cost and performs optimization according to criterion f. It takes a list of available slots ordered by their non-decreasing start time as input.

1. Initializing variables for the best criterion value and corresponding best window: $f_{max} = 0$, $W_{max} = \{\}$.
2. From the slots available we select different groups by node performance p_i. For example, group P_k contains resources allocated on nodes with performance $p_i \geq P_k$. Thus, one slot may be included in several groups.
3. Next is a cycle for all retrieved groups P_i starting from the max performance P_{max}. All the sub-items represent a cycle body.
   a. The resources reservation time required to compute V instructions on a node with performance P_i is $T_i = \frac{V}{p_i}$.
   b. Initializing variable for a window candidates list $S_W = \{\}$.
   c. Next is a cycle for all slots s_i in group P_i starting from the slot with the minimum start time. The slots of group P_i should be ordered by their non-decreasing start time. All the sub-items represent a cycle body.
      (1) If slot s_i doesn't satisfy user requirements (hardware, software, etc.) then continue to the next slot (3c).
      (2) If slot length $l(s_i) < T_i$ then continue to the next slot (3c).
      (3) Set the new window start time $W_i.start = s_i.start$.
      (4) Add slot s_i to the current window slot list S_W
      (5) Next a cycle to check all slots s_j inside S_W.
         i. If there are no slots in S_W with performance $P(s_j) = P_i$ then continue to the next slot (3c), as current slots combination in S_W was already considered for previous group P_{i-1}.
         ii. If $W_i.start + T_i > s_j.end$ then remove slot s_j from S_W as it can't consist in a window with the new start time $W_i.start$.
      (6) If S_W size is greater or equal to n, then allocate from S_W a window W_i (a subset of n slots with start time $W_i.start$ and length T_i) with a maximum criterion value f_i and a total cost $C_i < C$. If $f_i > f_{max}$ then reassign $f_{max} = f_i$ and $W_{max} = W_i$.
4. End of algorithm. At the output variable W_{max} contains the resulting window with the maximum criterion value f_{max}.

3.3   Optimal Slot Subset Allocation

Let us discuss in more details the procedure which allocates an optimal (according to a criterion f) subset of n slots out of S_W list (algorithm step 3 c (6)).

For some particular criterion functions f a straightforward subset allocation solution may be offered. For example for a window finish time minimization it is reasonable to return at step 3 c (6) the first n cheapest slots of S_W provided that they satisfy the restriction on the total cost. These n slots (as any other n slots from S_W at the current step) will provide $W_i.finish = W_i.start + T_i$, so we need to set $f_i = -(W_i.start + T_i)$ to *minimize* the finish time at the end of the algorithm.

The same logic applies for a number of other important criteria, including window start time, runtime and a total cost minimization.

However in a general case we should consider a subset allocation problem with some additive criterion: $Z = \sum_{i=1}^{n} c_z(s_i)$, where $c_z(s_i) = z_i$ is a target optimization characteristic value provided by a single slot s_i of W_i.

In this way we can state the following problem of an optimal n - size window subset allocation out of m slots stored in S_W:

$$Z = x_1 z_1 + x_2 z_2 + \ldots + x_m z_m, \tag{1}$$

with the following restrictions:

$$x_1 c_1 + x_2 c_2 + \ldots + x_m c_m \leq C,$$
$$x_1 + x_2 + \ldots + x_m = n,$$
$$x_i \in \{0, 1\}, i = 1..m,$$

where z_i is a target characteristic value provided by slot s_i, c_i is total cost required to allocate slot s_i for a time T_i, x_i - is a decision variable determining whether to allocate slot s_i ($x_i = 1$) or not ($x_i = 0$) for the current window.

This problem relates to the class of integer linear programming problems, which imposes obvious limitations on the practical methods to solve it. However we used 0-1 knapsack problem as a base for our implementation. Indeed, the classical 0-1 knapsack problem with a total weight C and items-slots with weights c_i and values z_i have the same formal model (1) except for extra restriction on the number of items required: $x_1 + x_2 + \ldots + x_m = n$. To take this into account we implemented the following dynamic programming recurrent scheme:

$$f_i(C_j, n_k) = \max\{f_{i-1}(C_j, n_k), f_{i-1}(C_j - c_i, n_k - 1) + z_i\} \\ i = 1, .., m, j = 1, .., C, k = 1, .., n, \tag{2}$$

where $f_i(C_j, n_k)$ defines the maximum Z criterion value for n_k-size window allocated out of first i slots from S_W for a budget C_j. After the forward induction procedure (2) is finished the maximum value $Z_{max} = f_m(C, n)$. x_i values are then obtained by a backward induction procedure.

An estimated computational complexity of the presented recurrent scheme is $O(m * n * C)$, which is n times harder compared to the original knapsack problem $(O(m * C))$. On the one hand, in practical job resources allocation cases this overhead doesn't look very large as we may assume that $n < < m$ and $n < < C$. On the other hand, this subset allocation procedure (2) may be called multiple times during the general square window search algorithm (step 3 c (6)).

4 Simulation Study

4.1 Simulation Environment Setup

An experiment was prepared as follows using a custom distributed environment simulator [6, 14, 18]. For our purpose, it implements a heterogeneous resource domain model: nodes have different usage costs and performance levels. A space-shared resources allocation policy simulates a local queuing system (like in GridSim or CloudSim [12]) and, thus, each node can process only one task at any given simulation time. The execution cost of each task depends on its execution time which is proportional to the dedicated node's performance level. The execution of a single job requires parallel execution of all its tasks.

During the experiment series we performed a window search operation for a job requesting $n = 7$ nodes with performance level $p_i \geq 1$, computational volume $V = 800$ and a maximum budget allowed is $C = 644$. During each experiment a new instance for the computing environment was automatically generated with the following properties. The resource pool includes 100 heterogeneous computational nodes. Each node performance level is given as a uniformly distributed random value in the interval [2, 10]. So the required window length may vary from 400 to 80 time units. The scheduling interval length is 1200 time quanta which is enough to run the job on nodes with the minimum performance. However we introduce the initial resource load with advanced reservations and local jobs to complicate conditions for the search operation. This additional load is distributed hyper-geometrically and results in up to 30% utilization for each node.

Additionally an independent value $q_i \in [0; 10]$ is randomly generated for each computing node i to compare algorithms against $Q = \sum_{i=1}^{n} q_i$ window allocation criterion.

4.2 General Algorithms Comparison

We implemented the following window search algorithms based on the general window search procedure introduced in Sect. 3.2.

- *FirstFit* performs a square window allocation in accordance with a general scheme described in Sect. 3.2. Returns first suitable and affordable window found. In fact, performs window start time minimization and represents algorithm from [15, 17].
- *MultipleBest* algorithm searches for multiple non-intersecting alternative windows using *FirstFit* algorithm. When all possible window allocations are retrieved the

algorithm searches among them for alternatives with the maximum Q value. In this way *MultipleBest* is similar to [4] approach.

- *MaxQ* implements a general square window search procedure with an optimal slots subset allocation (2) to return a window with maximum total Q value.
- *MaxQ Lite* follows the general square window search procedure but doesn't implement slots subset allocation (2) procedure. Instead at step 3 c (6) it returns the first n cheapest slots of S_W. The total Q value of these n slots is returned as a target criterion which is then maximized during the search procedure. Thus, *MaxQ Lite* has much less computational complexity compared to *MaxQ* but doesn't guarantee an accurate solution.

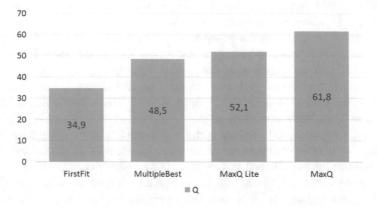

Fig. 1. Simulation results: average window Q value.

Figure 1 shows average $Q = \sum_{i=1}^{n} q_i$ value obtained during the simulation. Parameter q_i was generated randomly and is independent from other node characteristics. Note that as q_i was generated randomly on a $[0;10]$ interval for a single window of 7 slots we have the following practical limit specific for our experiment: $Q \in [0; 70]$.

As can be seen from Fig. 1, *MaxQ* is indeed provided the maximum average criterion value $Q = 61.8$, which is quite close to the practical maximum, especially compared to other algorithms. The advantage over *MultipleBest* and *MaxQ Lite* is almost 20%. *MaxQ Lite* implements a simple heuristic but still is able to provide a better solution compared to the best of 31 different alternative executions retrieved by *MultipleBest*. *First Fit* provided average Q value exactly in the middle of $[0;70]$ which is 44% less compared to *MaxQ*.

4.3   Dependable Resources Allocation

As a practical implementation for a general Q parameter maximization we propose to study a resources allocation placement problem. Figure 2 shows Gantt chart of 4 slots co-allocation (hollow rectangles) in a computing environment with resources pre-utilized with local and high-priority tasks (filled rectangles).

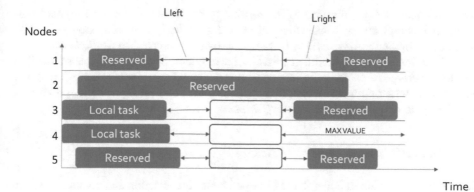

Fig. 2. Dependable window co-allocation metrics.

As can be seen from Fig. 2, even using the same computing nodes (1, 3, 4, 5 on Fig. 2) there are usually multiple window placement options with respect to the slots start time. The window placement generally may affect such job execution properties as cost, finish time, computing energy efficiency, etc. Besides, slots *proximity* to neighboring tasks reserved on the same computing nodes may affect a probability of the job execution delay or failure. For example, a slot reserved too close to the previous task on the same node may be delayed or cancelled by an unexpected delay of the latter. Thus, dependable resources allocation may require reserving resources with some reasonable distance to the neighboring tasks.

As presented in Fig. 2, for each window slot we can estimate times to the previous task finish time: L_{left} and to the next task start time: L_{right}. Using these values we consider the following criteria for the whole window allocation optimization:

$L_{\Sigma} = \frac{1}{n}\sum_{i=1}^{n}(L_{left\ i} \mid L_{right\ i})$ represents average time distance between window and the neighboring tasks reserved on the same nodes; $L_{min\Sigma} = \frac{1}{n}\sum_{i=1}^{n} min(L_{left\ i}, L_{right\ i})$ displays average time distance to the nearest neighboring tasks. For a dependable scheduling we are interested in maximizing L_{Σ} and $L_{min\Sigma}$ values. So, by setting $Q = L_{\Sigma}$ and $Q = L_{min\Sigma}$ we performed scheduling simulation with the same settings described in Sect. 4.1. The results of 1000 independent scheduling cycles are compiled in Table 1.

As a result *MaxQ* provided best values for both L_{Σ} and $L_{min\Sigma}$ criteria (highlighted in bold in Table 1). The advantage over *Multiple Best* is 15% against the total distance L_{Σ}

Table 1. Dependable window co-allocation simulation results

Algorithm	Average L_{Σ}	Average $L_{min\Sigma}$	Average operational time, ms
Multiple Best	719	278	156
First fit	433	99	7
Max Q Lite	713	279	7
Max Q	**842**	**370**	2860

and 25% against $L_{min\Sigma}$ distance to the nearest tasks. However due to a higher computational complexity it took *MaxQ* almost 3 s to find such allocation over 100 available computing nodes, which is 20 times longer compared to *Multiple Best*.

At the same time a simplified *MaxQ Lite* provided almost the same scheduling results as *Multiple Best* for even less operational time: 7 ms. *FirstFit* doesn't perform the target criteria optimization and, thus, provides the smallest L_Σ and $L_{min\Sigma}$ distances and the same operational time as *MaxQ Lite*.

5   Conclusion and Future Work

In this work, we address the problem of dependable slot selection and co-allocation for parallel jobs in distributed computing with non-dedicated resources. For this purpose a general square window allocation algorithm was proposed and considered. A special slots subset allocation procedure is implemented to support a general case optimization problem. Simulation study proved optimization efficiency of the proposed algorithms according to target criteria. A practical implementation was considered for a dependable resources allocation providing up to 25% advantage over traditional scheduling algorithms.

As a drawback, the general case algorithm has a relatively high computational complexity, especially compared to First Fit approach. In our further work, we will refine a general resource co-allocation scheme in order to decrease its computational complexity.

Acknowledgments. This work was partially supported by the Council on Grants of the President of the Russian Federation for State Support of Young Scientists (YPhD-2297.2017.9), RFBR (grants 18-07-00456 and 18-07-00534) and by the Ministry on Education and Science of the Russian Federation (project no. 2.9606.2017/8.9).

References

1. Lee, Y.C., Wang, C., Zomaya, A.Y., Zhou, B.B.: Profit-driven Scheduling for Cloud Services with Data Access Awareness. J. of Parallel Distrib. Comput. **72**(4), 591–602 (2012)
2. Garg, S.K., Konugurthi, P., Buyya, R.: A linear programming-driven genetic algorithm for meta-scheduling on utility grids. Int. J. Parallel Emergent Distrib. Syst. **26**, 493–517 (2011)
3. Buyya, R., Abramson, D., Giddy, J.: Economic models for resource management and scheduling in grid computing. J. Concurr. Comput. Pract. Exp. **5**(14), 1507–1542 (2002)
4. Ernemann, C., Hamscher, V., Yahyapour, R.: Economic scheduling in grid computing. In: Feitelson, D.G., Rudolph, L., Schwiegelshohn, U. (eds.): JSSPP 2002. LNCS, vol. 2537, pp. 128–152. Springer, Heidelberg (2002)
5. Kurowski, K., Nabrzyski, J., Oleksiak, A., Weglarz, J.: Multicriteria Aspects of Grid Re-source Management. In: Nabrzyski, J., Schopf, J.M., Weglarz, J. (eds.) Grid resource management. State of the art and future trends, pp. 271–293. Kluwer Academic Publishers, Norwell (2003)

6. Toporkov, V., Tselishchev, A., Yemelyanov, D., Bobchenkov, A.: Composite scheduling strategies in distributed computing with non-dedicated resources. Procedia Comput. Sci. **9**, 176–185 (2012)
7. Aida, K., Casanova, H.: Scheduling mixed-parallel applications with advance reservations. In: 17th IEEE International Symposium on HPDC, pp. 65–74. IEEE CS Press, New York (2008)
8. Elmroth, E., Tordsson, J.: A standards-based grid resource brokering service supporting advance reservations, co-allocation and cross-grid interoperability. J Concurr. Comput. Pract. Exp. **25**(18), 2298–2335 (2009)
9. Takefusa, A., Nakada, H., Kudoh, T., Tanaka, Y.: An advance reservation-based co-allocation algorithm for distributed computers and network bandwidth on QoS-guaranteed grids. In: Frachtenberg, E., Schwiegelshohn, U. (eds.) JSSPP 2010. LNCS, vol. 6253, pp. 16–34. Springer, Heidelberg (2010)
10. Blanco, H., Guirado, F., Lrida, J.L., Albornoz, V.M.: MIP model scheduling for multi-clusters. In: Euro-Par 2012. LNCS, vol. 7640, pp. 196–206. Springer, Heidelberg (2013)
11. Moab Adaptive Computing Suite. http://www.adaptivecomputing.com
12. Calheiros, R.N., Ranjan, R., Beloglazov, A., De Rose, C.A.F., Buyya, R.: CloudSim: a toolkit for modeling and simulation of cloud computing environments and evaluation of resource provisioning algorithms. J. Softw. Pract. Exp. **41**(1), 23–50 (2011)
13. Samimi, P., Teimouri, Y., Mukhtar, M.: A combinatorial double auction resource allocation model in cloud computing. J. Inf. Sci. **357**(C), 201–216 (2016)
14. Toporkov, V., Toporkova, A., Bobchenkov, A., Yemelyanov, D.: Resource selection algorithms for economic scheduling in distributed systems. In: Proceedings of the International Conference on Computational Science, ICCS 2011, June 1–3, 2011, Singapore, Procedia Computer Science. Elsevier, vol. 4. pp. 2267–2276 (2011)
15. Kovalenko, V.N., Kovalenko, E.I., Koryagin, D.A., et. al.: Parallel Job Management in the Grid with Non-Dedicated Resources, Preprint of Keldysh Institute of Applied Mathematics, Russian Academy of Science, Moscow, no. 63 (2007)
16. Makhlouf, S., Yagoubi, B.: Resources co-allocation strategies in grid computing. In: CIIA, vol. 825, CEUR Workshop Proceedings (2011)
17. Netto, M.A.S., Buyya, R.: A flexible resource co-allocation model based on advance reservations with rescheduling support. In: Technical Report, GRIDSTR-2007-17, Grid Computing and Distributed Systems Laboratory, The University of Melbourne, Australia, 9 October 2007
18. Toporkov, V., Toporkova, A., Tselishchev, A., Yemelyanov, D.: Slot selection algorithms for economic scheduling in distributed computing with high QoS rates. In: New Results in Dependability and Computer Systems. AISC, vol. 224, pp. 459–468. Springer, Heidelberg (2013)

Route Risk Assessment for Road Transport Companies

Agnieszka Tubis[(⊠)]

Faculty of Mechanical Engineering, Wroclaw University of Science
and Technology, 27 Wybrzeze Wyspianskiego St., Wroclaw, Poland
agnieszka.tubis@pwr.edu.pl

Abstract. The transportation process is described by 5 parameters: transport operations and the relations between them, the route, the time of the transportation process and the kind of product that is transported. The direct interviews conducted among carriers indicate that if entered into the TMS (Transport Management System), the route parameters applicable for carriers would speed up cost calculation process and improve accuracy of business decisions. One of those parameters is the level of risk connected with potential disruptions of the transportation process. Therefore, this article aims to present a new approach to risk analysis conducted to facilitate decision process for the management of a transport enterprise. The research presented in this article is part of "Identifying data supporting risk assessment in road transport system in the aspect of safety and meeting logistic standards" program financed as part of MINIATURA 1 program.

Keywords: Road transport · Risk assessment · Route parameters
Decision process

1 Introduction

There are many different views on what risk is and how to define it [6, 19], how to measure/describe it [5, 22], and how to use risk analysis in decision making [3, 4]. The scope of the assessed risk in transport may also depend on the transport system, which is analyzed [23, 24, 30, 34]. Therefore the article is going to be narrowed down to road transport only.

The direct interviews conducted among carriers indicate that if entered into the TMS (Transport Management System), the route parameters applicable for carriers would speed up cost calculation process and improve accuracy of business decisions. One of those parameters is the level of risk connected with potential disruptions of the transportation process. Therefore, this article aims to present a new approach to risk analysis for transport on a particular route, conducted in order to facilitate decision process for the management of a transport enterprise. As a result, Sect. 2 describes the road transport process. Section 3 presents the traditional approach to risk assessment in road transport which focuses mostly on safety aspects. This section also points out the limits of applying the traditional approach to risk analysis performed for road transport enterprises. This line of reasoning has, in turn, become a starting point for presenting in

© Springer International Publishing AG, part of Springer Nature 2019
W. Zamojski et al. (Eds.): DepCoS-RELCOMEX 2018, AISC 761, pp. 492–503, 2019.
https://doi.org/10.1007/978-3-319-91446-6_46

part 4 the new approach to risk assessment that is more applicable in the decision-making process of the managerial staff. The article concludes with a summary and guidelines, including directions for further research.

2 Road Transport Process

According to Polish norm [28], transportation process is a range of interconnected transport operations, performed in a specific order on a particular route in a specific time. The transportation process is described by 5 parameters: transport operations and the relations between them, the route, the time of the transportation process and the kind of product that is transported.

$$PT = (O, Re, Ro, T, P)$$

where: TP – transportation process; O – operations; Re – relation between operations; Ro – route; T – time; P – product.

The starting point for the decision process is the kind of product that is being transported, which for the most part decides about choosing means of transport. The choice of the means of transport will in turn affect the operations connected with it, as well as the choice of route and the estimated time of transportation process. Therefore, there's a need for implementing a classification of product that can be based on different criteria. The most often used criteria are [25]: (1) natural transportability; (2) technical transportability; (3) basic methods of loading; (4) size of cargo and (5) economic transportability. The carrier is supposed to optimize the parameters of route and time for the defined class of cargo and the selected means of transport. Taking into account the scope of this research, further analysis is going to be narrowed down to road transport only.

The route parameter strongly affects the time of completing the transportation process. This interrelation shows clearly in the transport companies approach to serviced routes, which are usually described with two parameters: time (T) and cost (C).

$$R_i = (T_i, C_i)$$

Where: i – number of route.

Assuming that transport is a combination of activities connected with moving single units, comprising moving from place to place and any other operations that might be deemed necessary, such as loading operations (loading, unloading or transshipment) and handling operations [28], the function of the time of carriage can be described as

$$T = t_l + t_c + t_{ul}$$

where: T – transport time; t_l – loading time; t_c – carriage time; t_{ul} – unloading time.

The cost structure for car transport is usually divided into three main kinds of costs [11]: (1) the cost of amortization of the means of transport; (2) the cost of wear of operational and maintenance materials; (3) the cost of remuneration of staff. However, when considering transport route, the main approach to cost calculation that the companies use is summing up:

- remuneration of the driver (salary plus expenses) (C_s),
- cost of fuel (C_f),
- insurance cost (C_i),
- road toll cost (C_R) and
- parking costs (C_p).

$$C = C_s + C_f + C_i + C_R + C_p$$

The market for transport companies is at the moment highly competitive and turbulent. Therefore, the description of the route that relies solely on the two factors of time and cost seems to be insufficient, especially if taking into consideration the management of transportation processes.

The interviews conducted among managers in the company that was researched indicate that the right description of routes, at least those serviced regularly, is indispensable for the planning process. If entered into the TMS system, the route parameters meeting information needs of the carriers would greatly improve both the cost estimating process and the accuracy of business decisions.

Therefore, the description of each route (R_i, where i – number of route) should be characterized by 3 parameters: time of completing transportation (T_i), the cost of transport (C_i) and the set of risk (RS_i).

$$R_i = (T_i, C_i, RS_i)$$

The set of risk priority number informs the decision maker about potential hazards occurring on the specific route and their possible consequences. It relies on the risk analysis conducted specifically for the transportation process on the selected route.

3   Risk Analysis in Road Transport

Currently, there is no one, unified and commonly used definition of risk term [6]. We can even state that the underlying concepts of risk are hard to define and even harder to assess [20]. In one of his works Aven presented a proposal for a risk classification system that comprises 9 groups of defining it [6]. This system was used i.a. to asses risk in maritime transportation systems by Montewka [18].

For transport management purposes, the most applicable qualitative definition describes risk as the potential for realization of unwanted, negative consequences of an event [8]. This approach is in line with the ISO norm [29] which defines risk as the effect of uncertainty on objective. Therefore, we can assume that all the events that have negative impact on realizing the transport process should be treated as hazards and analyzed as part of the risk assessment.

Meanwhile, a review of the literature from the risk assessment in road transport field shows that risk analysis is treated as an important tool that can be useful for improving and/or optimizing the safety level on the roads [13]. Therefore, for the purpose of this process there is a need to gather data on:

- frequency of accidents [1, 2, 10, 12, 14, 15],
- number of deaths and injuries [10, 12, 14],
- average traffic [2, 21, 26, 33],
- recorded vehicles speed [16, 21, 26, 33],
- class of road [16, 26],
- weather conditions [17, 26].

This data supports risk analysis, but mainly when it comes to road safety. The scenarios that are created concern mostly the hazards of the transport system and its functioning on the macro scale.

However, when assessing risk of a transport company it is impossible to narrow down the analysis only to the safety aspect. This is because of the objective that the managers responsible for the transportation process have to achieve – namely, efficient delivery of the load in accordance with the declared logistic and quality parameters. This, in turn, stems from the fact that the main purpose of management in transport is not only safety, but also efficient and economically justified realization of transport processes with existing technological, organizational and economic conditions [31]. The transport process is therefore realized in specific circumstances (environment) using the resources available to the company. All these elements comprise a potential source of risk. They affect the course of transport operations and the transformation from input states to output states, aiming to achieve a measurable result which is of considerable value from the point of view of a client (according to the definition of a process [27]).

For the purpose of risk assessment conducted for the decision making managers it is important to gather relevant information connected with all the resources used in the transportation process. This data will form a base to estimate risk priority number (RPN) which should constitute a component of the transport route description.

In the conducted risk analysis it will be possible to identify and evaluate all the events that may cause any deviation from the scheduled time of starting and completing the process, as well as those that may generate additional costs connected with the transportation process. It is also worth noting that the RPN will also directly influence the remaining two factors describing the route.

4 Risk Assessment for Transport Route for the Selected Road Transportation Company

As part of the research on identifying sets of information improving the system for risk assessment in road transport, a new approach has been introduced, which assesses the risk for transport routes. The procedure has been verified through research conducted in the selected transport company. The analyzed company provides transport-forwarding services, mainly in road transport. It operates the cargo loads transported both in the domestic and international distributions. The company has 9 branches in Poland and

3 branches abroad. Currently, the company has 70 own vehicles, which are used primarily to serve regular customers.

The proposed procedure is in accordance with ISO 31000 guidelines and is divided into two stages of the risk assessment process (Fig. 1).

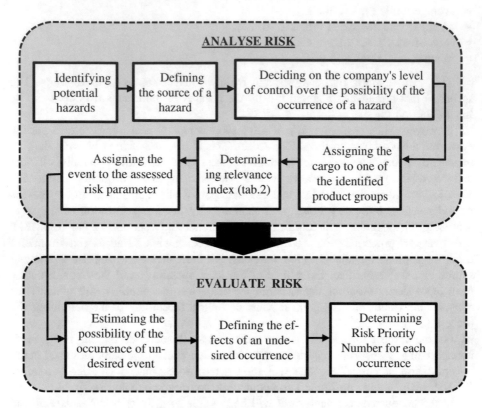

Fig. 1. The risk assessment process

At the stage of risk analysis the managers should identify possible hazards connected with the transportation process on a given route. By hazards, it is meant here any undesired occurrence which might affect the correctness of realizing the transport process. In order to improve the process of risk evaluation, it has been proposed to group the hazards according to their potential sources, basing on process analysis. The traditional division of hazards refers to the damages to the elements of the transport system structure (people, infrastructure, means of transport) and to the surroundings (environment) [31]. Taking into account the resources at the input of the transport system, the classification should be expanded by adding an extra parameter of the available information. Therefore, the proposed classification of the sources of hazards comprises the following:

- S1 – people – company employees (including drivers)
- S2 – people – third parties

- S3 – vehicle
- S4 – infrastructure
- S5 – environment
- S6 – information.

Grouping hazards according to the proposed classification of sources will allow the managers to quickly set the level of control over a given event. However, in order for it to be possible, each source should be assigned the indicator of control estimated during the direct interviews conducted with the managers of the company being researched. The research procedure assumed 4 steps.

Step 1. The level of control over a given element has been described using cross comparison by managers of pairs of elements S_i and S_j, for ($i = 1, 2...6; j = 1, 2....6$). It has allowed to create a pairwise comparison matrix of control over the particular components of the process.

$$S = \begin{bmatrix} X_{11} & X_{12} & \cdots & X_{16} \\ X_{21} & X_{22} & \cdots & X_{26} \\ \cdots & \cdots & \cdots & \cdots \\ X_{61} & X_{62} & \cdots & X_{66} \end{bmatrix}$$

The managers participating in the research described subjectively the value of particular parameters X_{ij} ($i = 1, 2...6; j = 1, 2....6$), which were the result of the conducted comparison:

- if the level of control $S_i > S_j$, then $X_{ij} = i$;
- if the level of control $S_i < S_j$, then $X_{ij} = j$.

Step 2. The frequency of indicating particular element across the matrix S_i has been calculated - how many times in cross-comparison the particular element was indicated.

Step 3. The importance for each S_i has been specified, basing on the number of times each element was indicated against the total of all indications.

Step 4. The level of exercised control (M_i) has been established relying on the importance assigned to particular components, with 1 being the highest level of control and 6 – the lowest.

The answers was the same for all the managers participating in the research. Therefore, the assigned level of control for particular hazards was assumed as a model and applicable for further processing. The results are shown in Table 1.

Table 1. The matrix described the level of control

	S1	S2	S3	S4	S5	S6	$\sum S_i$	M_i
S1	1	1	3	1	1	6	7	3
S2	1	2	3	4	2	6	3	5
S3	3	3	3	3	3	3	11	1
S4	1	4	3	4	4	6	5	4
S5	1	2	3	4	5	6	1	6
S6	6	6	3	6	6	6	9	2

As pointed out at the beginning of the article, the point of departure for the process of transport planning is the kind of a load to be carried. This is also of importance when conducting risk analysis. The particular kinds of loads are connected with the possibility of the occurrence or escalation of undesired events. Therefore, it is important to assign the importance to each feature characterizing the transport process of the particular product group. Taking into account the goals for transport management, two basic parameters characterizing the process have been indicated, namely (1) safety; (2) logistic efficiency. These two parameters are guidelines for grouping the events at the initial stage of risk assessment, but also for assigning relevance index.

The researched company services 4 basic groups of products: (1) dangerous goods; (2) perishable goods; (3) sensitive/valuable goods (especially prone to e.g. theft); (4) others. Each of these product groups was considered during a brainstorming session and for each of the two basic parameters characterizing transport process importance has been assigned, expressed by means of a relevance index. The results are presented in Table 2.

Table 2. The relevance index for basic group of products

	Dangerous goods	Perishable goods	Sensitive goods	Others
SAFETY	0,8	0,5	0,4	0,2
LOGISTICS	0,2	0,5	0,6	0,8

The importance assigned to a given feature constitutes relevance index, a crucial component when assessing risk for the specific event.

A significant challenge when evaluating the risk of transporting a product on a given route is the assessment of the possibility of the occurrence of such an event. When assessing it for events in the safety group, the managers can use risk maps published by an international organization EuroRAP. However, in case of the other two groups the historical data which would facilitate estimating the likelihood of the event occurrence is not always available. Therefore, it has been proposed that the possibility and likelihood of a given occurrence can be estimated basing on the knowledge and experience of managers responsible for the transport on that part of the route. Since when human experts are asked to evaluate a variable, they feel more comfortable in giving the answer in words [9], it has been proposed to use linguistic assessment to describe likelihood of event occurrence. The possibility of applying the linguistic variable to describe likelihood of occurrence are presented in the example below.

When assessing the consequences of the occurrence we can assign them to different classes. In the researched company, 5 groups of recorded consequences of disruption in the transport process were identified: (1) delay/extending the time for completing the transport; (2) damage/loss of the load; (3) damage to the vehicle; (4) decline of the driver's health; (5) damage to the environment. The consequences can be assessed in the category occurred (1)/did not occur (0). The value of the consequences of the event is the total sum of points of its consequences.

When establishing likelihood and consequences the database is becoming of particular importance. It contains information on resources used in the transport process on a given route. The proposed range of data reported in this database for the purpose of risk assessment has been presented in [32]. Because of the assigned control index, it is crucial to have a full range of up-to-date information on the company employees, including drivers, and the technical condition of the vehicles used. The conducted analytical process would enable us to evaluate risk as a product of 4 factors for identified hazards (h).

$$RPN_h = L_h \times C_h \times M_h \times RI_h$$

where: L_h – likelihood; C_h – consequences; M_h – level of control; RI_h – relevance index.

Based on the set of RPN_h, we can determine a set of risks for the described route (RS_i) Using the risk index calculated in this way it is possible to assess the acceptable level of risk for a given route and identify area for improvement within the risk management process.

5   Case Study for Wrocław - Magdeburg Route Assessment

The company being researched offers transport services for two clients along the Wrocław - Magdeburg route. The direction from Wrocław to Magdeburg concerns transporting products from the dangerous goods group, while from Magdeburg to Wrocław it concerns sensitive goods. With the application of the method described above, the analysis and risk assessment were conducted for both projects.

Step 1
Specialists identified 9 threats and then they indicated the sources of hazards. M_i – index was assigned in accordance with the Table 1. The results are presented in Table 3.

Table 3. The source of the identified hazards

Hazard symbol	Description of the threat	Source of hazard	M_i
H1	Congestion on the road as a result of an incident of other road users	S2	5
H2	An accident involving our vehicle	S1	3
H3	A collision involving our vehicle	S1	3
H4	Theft of the cargo	S2	5
H5	Breakdown of the vehicle	S3	1
H6	Adverse weather conditions that affect driving	S5	6
H7	Road works affecting driving time	S4	4
H8	Lack of communication with the driver while servicing the route	S6	2
H9	Dangerous/fast driving	S1	3

Step 2
The possibility of occurrence of particular hazards was described by carriers servicing the selected route. The carriers evaluating it are persons employed in the company for more than 3 years, who have the knowledge and experience in servicing this route. The scale for evaluating the possibility of the occurrence has been presented in Table 4:

Table 4. Linguistic variable and its defuzzified values for the possibility of a particular adverse event occurrence

Possibility of event occurrence	Characteristics	L [rank]
Low	Not occurred in the last year	1
High	Occurred at least once in the last year	2–3
Very high	Occurred at least once in the last quarter	4–5

Step 3
An evaluation of consequences of the occurrence has been conducted, with the additional participation of manager of transport department. The assessment of the consequences of the event is conducted in accordance with scoring presented in Table 5:

Table 5. Values of the consequences for dangerous goods (DG) and sensitive goods (SG)

Hazard symbol	(1) DG/SG	(2) DG/SG	(3) DG/SG	(4) DG/SG	(5) DG/SG	SUM DG/SG
H1	1/1	0/0	0/0	0/0	0/0	1/1
H2	1/1	1/1	1/1	1/1	1/0	5/4
H3	1/1	1/1	1/1	1/0	1/0	5/3
H4	0/0	1/1	1/1	0/0	1/0	3/2
H5	1/1	0/0	1/1	0/0	1/0	3/2
H6	1/1	0/0	1/1	0/0	0/0	2/2
H7	1/1	0/0	0/0	0/0	0/0	1/1
H8	1/1	0/0	0/0	0/0	0/0	1/1
H9	0/0	1/1	1/1	1/0	1/0	4/2

The value of the consequences of the occurrence is the total sum of points of its consequences, which were identified in 5 groups of consequences of disruption in the transport process - (1) delay/extending the time for completing the transport; (2) damage/loss of the load; (3) damage to the vehicle; (4) decline of the driver's health; (5) damage to the environment.

Step 4
The evaluation of the hazards identified for transporting cargo on the selected route for both groups of products has been presented in Table 6 (dangerous goods) and in Table 7 (sensitive goods).

Table 6. RPN indicator for the identified hazards for the dangerous goods

Hazard symbol	L_h	C_h	M_h	RI_h	RPN_h
H1	5	1	5	0,2	5,0
H2	1	5	3	0,8	12,0
H3	2	5	3	0,8	24,0
H4	1	3	5	0,2	3,0
H5	1	3	1	0,2	0,6
H6	3	2	6	0,8	28,8
H7	3	1	4	0,2	2,4
H8	1	1	2	0,2	0,4
H9	1	4	3	0,8	9,6

Table 7. RPN indicator for the identified hazards for the sensitive goods

Hazard symbol	L_h	C_h	M_h	RI_h	RPN_h
H1	5	1	5	0,6	15,0
H2	2	4	3	0,4	9,6
H3	3	3	3	0,4	10,8
H4	2	2	5	0,6	12,0
H5	3	2	1	0,6	3,6
H6	3	2	6	0,4	14,4
H7	3	1	4	0,6	7,2
H8	2	1	2	0,6	2,4
H9	3	2	3	0,4	7,2

The analysis of the results proves that the RPN risk indicator takes on different values, depending on the type of load being transported. The parameters that differentiate the value are the probability of the occurrence of a particular incident, which depends on the type of cargo, as well as the priority assigned to the aspects of safety and logistics.

6 Summary

The results presented in the article are part of a research conducted by the author and connected with the development of risk management model dedicated to road transport companies. The interviews conducted among managerial staff indicate that transport companies are looking for a risk assessment model for their operations that would correspond to their information needs. The approach to risk assessment presented here bases on the study of literature on the subject. The preliminary validation in the actual environment proves the assumed scheme of analysis to be correct. In order to verify its correctness fully, it is necessary to implement this method in combination with a correctly prepared database which will support the analysis process. Therefore, further research by the author will aim at creating for the selected transport company a complex information system which will support the procedure of risk assessment for the services provided.

The research presented in this article is part of "Identifying data supporting risk assessment in road transport system in the aspect of safety and meeting logistic standards" program financed as part of MINIATURA 1 program.

References

1. Ambituuni, A., Amezaga, J.M., Werner, D.: Risk assessment of petroleum product transportation by road: a framework for regulatory improvement. Saf. Sci. **79**, 324–335 (2015)
2. Ambros, J., Havránek, P., Valentová, V., Krivankova, Z., Striegler, R.: Identification of hazardous locations in regional road network – comparison of reactive and proactive approaches. Transp. Res. Procedia **14**, 4209–4217 (2016)
3. Apostolakis, G.E.: How useful is quantitative risk assessment? Risk Anal. **24**, 515–520 (2004)
4. Aven, T.: Perspectives on risk in a decision-making context – review and discussion. Saf. Sci. **47**, 798–806 (2009)
5. Aven, T.: On how to define, understand and describe risk. Reliab. Eng. Syst. Saf. **95**, 623–631 (2010)
6. Aven, T.: The risk concept—historical and recent development trends. Reliab. Eng. Syst. Saf. **99**, 33–44 (2012)
7. Aven, T., Zio, E.: Foundational issues in risk analysis. Risk Anal. **34**(7), 1164–1172 (2014)
8. Aven, T.: Risk assessment and risk management: review of recent advances on their foundation. Eur. J. Oper. Res. **253**, 1–13 (2016)
9. Bajpai, S., Sachdeva, A., Gupta, J.P.: Security risk assessment: applying the concepts of fuzzy logic. J. Hazard. Mater. **173**, 258–264 (2010)
10. Benekos, I., Diamantidis, D.: On risk assessment and risk acceptance of dangerous goods transportation through road tunnels in Greece. Saf. Sci. **91**, 1–10 (2017)
11. Bronk, H.: Features and costs structure in transport allowing for decision making. Zesz. Nauk. Uniw. Szczec. **813**, 21–39 (2014)
12. Caliendo, C., De Guglielmo, M.L.: Accident rates in road tunnels and social cost evaluation. Procedia Soc. Behav. Sci. **53**, 166–177 (2012)
13. Caliendo, C., De Guglielmo, M.L.: Quantitative risk analysis on the transport of dangerous goods through a bi-directional road tunnel. Risk Anal. **37**(1), 116–129 (2017)
14. Chakrabarti, U.K., Parikh, J.K.: Applying HAZAN methodology to hazmat transportation risk assessment. Process Saf. Environ. Prot. **90**, 368–375 (2012)
15. Conca, A., Ridella, Ch., Sapori, E.: A risk assessment for road transportation of dangerous goods: a routing solution. Transp. Res. Procedia **14**, 2890–2899 (2016)
16. Douglas, J., Serrano, J.-J., Coraboeuf, D., Bouc, O., Arnal, C., Robida, F., Modaressi, H., Atkinson, M., Vowles, G., Holt, I.: Risk assessment for the road network in the French-Italian border region using web services. In: First European Conference on Earthquake Engineering and Seismology (a joint event of the 13th ECEE & 30th General Assembly of the ESC), September 2006, Geneve, Switzerland, Paper number 827 (2006)
17. East, A., Smale, N., Kang, S.: A method for quantitative risk assessment of temperature control in insulated boxes. Int. J. Refrig. **32**, 1505–1513 (2009)
18. Goerlandt, F., Montewka, J.: Maritime transportation risk analysis: review and analysis in light of some foundational issues. Reliab. Eng. Syst. Saf. **138**, 115–134 (2015)

19. Hampel, J.: Different concepts of risk – a challenge for risk communication. Int. J. Med. Microbiol. **296**, 5–10 (2006)
20. Heckmann, I., Comes, T., Nickel, S.: A critical review on supply chain risk – definition, measure and modelling. Omega **52**, 119–132 (2015)
21. Jurewicz, Ch., Excel, R.: Application of a crash-predictive risk assessment model to prioritise road safety investment in Australia. Transp. Res. Procedia **14**, 2101–2110 (2016)
22. Kaplan, S.: The words of risk analysis. Risk Anal. **17**, 407–417 (1997)
23. Kierzkowski, A., Kisiel, T.: Simulation model of security control system functioning: a case study of the Wroclaw Airport terminal. J. Air Transp. Manag. (2016)
24. Młyńczak, M., Nowakowski, T., Valis, D.: How to manage risks? The normative approach. Probl. Eksploat. **1**, 137–147 (2011). (in Polish)
25. Mokrzyszczak, H.: Ładunkoznawstwo. Technologia zabezpieczania ładunków w transporcie. WKiŁ, Warszawa (1974)
26. Norros, I., Kuusela, P., Innamaa, S., Pilli-Sihvola, E., Rajamäki, R.: The Palm distribution of traffic conditions and its application to accident risk assessment. Anal. Methods Accid. Res. **12**, 48–65 (2016)
27. Nowosielski, S. (ed.): Logistic Processes and Projects. Wydawnictwo UE we Wrocławiu, Wroclaw (2008). (in Polish)
28. PN-72/M-78000: Transport – określenia podstawowe i podział. PKN, Warsaw (1972)
29. PN-ISO 31000:2010: Risk management – principles and guide-lines. PKN, Warsaw (2010)
30. Restel, F.J.: The Markov reliability and safety model of the railway transportation system. In: Safety and Reliability: Methodology and Applications - Proceedings of the European Safety and Reliability Conference, ESREL 2014, pp. 303–311. Taylor and Francis (2015)
31. Szymanek, A.: Conception of "4 Goals and 3 Levels" in risk management in road transport systems. Arch. Transp. **22**(3), 359–375 (2010)
32. Tubis, A., Werbińska-Wojciechowska, S.: The scope of the collected data for a holistic risk assessment performance in the road freight transport companies. In: Zamojski, W. (ed.) Advances in Dependability Engineering of Complex Systems: Proceedings of the Twelfth International Conference on Dependability and Complex Systems DepCoS-RELCOMEX, 2–6 July 2017, Brunów, Poland, pp. 450–463 (2018)
33. Tu, H., Li, H., van Lint, H., van Zuylen, H.: Modeling travel time reliability of freeways using risk assessment techniques. Transp. Res. Part A **46**, 1528–1540 (2012)
34. Zajac, M., Swieboda, J.: Process hazard analysis of the selected process in intermodal transport. In: 2015 International Conference on Military Technologies (ICMT), pp. 1–7. IEEE (2015)

User Estimates Inaccuracy Study in HPC Scheduler

Mariusz Uchroński[1]([✉]), Wojciech Bożejko[2], Zdzisław Krajewski[2],
Mateusz Tykierko[1], and Mieczysław Wodecki[3]

[1] Wroclaw Centre for Networking and Supercomputing, Wyb. Wyspiańskiego 27,
50-370 Wrocław, Poland
{mariusz.uchronski,mateusz.tykierko}@pwr.edu.pl

[2] Department of Control Systems and Mechatronics, Faculty of Electronics, Wrocław
University of Scicnce and Technology, Janiszewskiego 11/17, 50-372 Wrocław, Poland
{wojciech.bozejko,zdzislaw.krajewski}@pwr.edu.pl

[3] Telecommunications and Teleinformatics Department, Faculty of Electronics,
Wrocław University of Technology, Janiszewskiego 11-17, 50-372 Wrocław, Poland
mieczyslaw.wodecki@pwr.edu.pl

Abstract. The article is focused on user estimates inaccuracy study
in high performance computing scheduler. The main goal of this survey
is to study historical job submission data from the real supercomputer
cenre and investigate accuracy of resources estimated by user on job
submission. Based on this study 68.94% of jobs used less than 25% of
their requested walltime, which indicates a high estimation inaccuracy.
In case of estimation of amount of memory only 31.46% of jobs uses
more than 75% of requested memory. Analysis of historical jobs from
the further prediction of resources accuracy has been made. Similarities
of the new submitted jobs with historical data were described by keys
based on user names, software name, walltime and number of processors.
Also mathematical model for jobs scheduling in HPC system, with a goal
functions, has been proposed and verified.

Keywords: Scheduling · High performance computing
Resources accuracy

1 Introduction

Modern supercomputers offers computation power on the level of thousands of
TFLOPS. Researchers submit highly parallel jobs and perform scientific simulations in chemistry, physics, life science, big data and in many other areas of
science. Increase usage of supercomputers makes jobs scheduling critical task.
Small changes in scheduling policies may cause a great changes in performance,
optimal usage of resources and user waiting time for start job execution. Accurate resources estimation made by user on job submission has great influence on
efficiency of scheduling policies.

© Springer International Publishing AG, part of Springer Nature 2019
W. Zamojski et al. (Eds.): DepCoS-RELCOMEX 2018, AISC 761, pp. 504–514, 2019.
https://doi.org/10.1007/978-3-319-91446-6_47

Schedulers optimize the HPC system execution scheduling taking into account the highest system utilization that can be achieved in a reasonable amount of time taking into account the priorities of the tasks. The most common technique are varieties of FCFS (First Come First Served) in which tasks are carried out in the order of arrival with regard to the priorities and walltimes declared by users, which are usually exaggerated [9]. This paper is focused on the problem of user estimates inaccuracy related to the resources definition by users in jobs submission to the HPC scheduler. Analysis is focused not only on walltime inaccuracy but also on amount of memory.

2   BEM Cluster

In April 2015 Wroclaw Center for Networking and Supercomputing (WCSS) started operation of its latest supercomputer – BEM, named after founder of WCSS: prof. Daniel J. Bem.

BEM is a cluster of almost 1000 worker nodes, connected with dual networks: 1+10 G Ethernet and InfiniBand FDR. Thanks to enormous speed and ultra low latency of InfiniBand BEM can efficiently calculate very large computing models. Network topology is fat-tree with 3:1 blocking factor. This allows to reduce network cost and is sufficient for most of users.

Each compute node of BEM is equipped for 24 or 28 compute cores of latest Intel Haswell architecture. Clock speed is 2.3 GHz. Nodes have 64, 128 or 512 GB of operating memory for most demanding users. Total number of cores is 22 656 with total operating memory 76.5 TB. Nodes run Linux CentOS operating system.

Theoretical computing performance is 860 TFLOPS. Measured peak performance is 700 TFLOPS, placing BEM on Top500 list as 3rd supercomputer in Poland and 128th in the world in November 2015.

Users would not be able to use BEM without high performance storage. All nodes boot from NFS NAS heads, which also provide home directories and all other utility storage with speed of 4 GB/s and total capacity of about 100 TB. Each node is equipped with two SATA hard drives providing local scratch space for compute jobs that not require a shared filesystem. Each nodes has also access to Lustre shared filesystem with capacity of 1.1 PB and read/write speed of 80GB/s using InfiniBand network.

3   Workload Manager and Job Scheduler

Cluster BEM at WCSS runs the PBS Pro batch server and scheduler to satisfy users' needs. To run his calculations a user has to decide (Tables 1 and 2):

1. how much and which resources does he needs,
2. how long will the calculation run,
3. when should the calculation start.

Table 1. BEM cluster specification

Parameter	Value
Peak performance	700 TFLOPS
Operating system	CentOs
Number of nodes	1000
Memory per node	64/128/512 GB
Total memory	76.5 TB
Processor clock rate	2.3 GHz
Cores per node	24/28
Total CPU cores	22656
Network architecture	10 G Ethernet and InfiniBand FDR
Storage system	NFS NAS and Lustre
Storage capacity	1.1 PB

The first point is essential for the user, the second one – for the system. A job duration is used by the scheduler to plan an optimal resources utilization. The third point in typical cases is not very important (usually jobs are expected to start as soon as possible), but sometimes a job has to wait for some time or for a specific event (for example another job end). If an user need specific time slot for his job – he has to use a resource reservation.

Resources that can be requested on cluster BEM are following:

- tokens for scientific software (per server) for example: Lumerical FDTD, MODE, Matlab, ADF etc.
- hardware (per node),
- hardware model,
- number of cpus,
- amount of memory,
- local scratch space size (if available).

Hardware resources are specified in allocation units called "chunk". Each chunk will be allocated on one node, but more of them can also be allocated on one node. Example:

a user need 2 nodes, witch 4 cores and 16 GB memory on each. PBS Pro specification:

```
select=2:ncpus=4:mem=16GB
```

Software resources are specified in simple way. If a user need 4 ADF license tokens, he should type (after resource option switch):

```
adf=4
```

Example jobs submission:

```
sub -l select=1:ncpus=8:mem=32GB -l walltime=24:00:00 script.sh
```

script.sh will be run on 8 cores and 32 GB of RAM with the overall time limit set to 24 h.

On the BEM the server is configured in following way – users has to submit jobs to a main (routing) queue. Then the jobs are passing to the executing queues based on their walltime limit. It allows system administrator to configure appropriate limits for short and long jobs. Jobs are considered to run during scheduling cycle. The cycle is triggered by some events (a job submission or a job end) and periodically. During the cycle scheduler is matching needed resources to available resources – if they are met, the job is run. If the job cannot be started in specific (configurable) amount of time – it is set as a "top-job" to avoid starvation. Smaller jobs will not start before the top-job unless they are expected to finish before the job starts.

To provide fair resources sharing – pbs_mom (process running on a worker node) checks periodically resource usage of a job. This usage is reported to the server and if the job uses more resources than declared – it is interrupted by PBS. A user can query the server to get job's maximum usage (with "qstat -f JOBID" command) to plan the next calculations requirements. There is also a convenient way to check it – a user may request mail notification after job end with resource usage information.

4   Inaccuracy Study

The users frequently execute complex jobs, especially for problems with exponential size of the solution space [1–4] without exact knowledge about their runtime. Resources (walltime, amount of memory, CPU time) estimation provided to the scheduler by user in most cases are inaccurate [5–7,12]. Many results [11] show that inaccurate estimation of job walltime has impact on computing system performance. On the other hand studies from [8,10] shows that more accurate requested walltime has only minimal impact on computing system performance.

We collect jobs executed on the BEM cluster from 01.04.2015 to 10.03.2016. Each job is characterized by attributes such as:

- start time,
- end time,
- user requested/used wall time,
- user requested/used amount of memory,
- user requested/used number of processors, etc.

The accuracy of user resources estimates is defined with equation:

$$A = \frac{r_{act}}{r_{req}} \qquad (1)$$

Table 2. Basic information about analyzed data set

Attribute	Value
Number of jobs	844211
Avg. job running time	12311 s
Avg. job requested wall time	484072 s
Avg. job waiting time	45341 s
Avg. memory used	5 GB
Avg. memory requested	13 GB
Number of software types	105
Number of users	316

as a ratio of the job's actual resource usage (r_{act}) to the user requested resources (r_{req}). Higher value of A corresponds to the higher resources estimation accuracy. Equation 1 can be used for calculating inaccuracy of walltime, amount of memory or CPU usage. The reason why users overestimate resources is order to avoid having the job killed at the requested resources expiration. Values of A can be between 0.0 and 1.0. The bigger value of accuracy means that user better estimates resources.

Table 3 shows distribution of A values for analyzed data set. Regarding walltime most of the jobs are underestimated – 68.94% of jobs used less than 25% of their requested walltime. Only 0.32% of jobs uses more than 75% of estimated walltime. In case of estimation of amount of memory 31.46% of jobs uses more than 75% of requested memory.

Table 3. Distribution of the A values for walltime and memory

A range	% of jobs (walltime)	% of jobs (memory)
(0:0,25]	68.94	44.07
(0,25:0,5]	29.92	15.59
(0,5:0,75]	0.82	8.88
(0,75:1]	0.32	31.46

Resources related to the submitted jobs can be adjusted using dedicated schemes. In [7] was presented basic methodology of predicting accuracy of walltime estimation instead of directly predicting values of walltime. Using predicted value of accuracy denoted by A' value of resources can be estimated:

$$r_{sched} = r_{req} \cdot A'. \tag{2}$$

Value of A' can be predicted at the job submission based on the A values of the historical jobs. Historical jobs used for the prediction of A' value have some

Table 4. Average standard deviation of A for job groups by keys

Key	Jobs in group	Avg. std. of A (walltime)	Avg. std. of A (memory)
All jobs	1	0.192	0.267
User	316	0.085	0.123
Software	105	0.091	0.122
User/software	730	0.085	0.131
User/software/walltime	2697	0.089	0.128
User/software/ncpus	2068	0.087	0.127

common attributes with newly submitted job. This similarities can be described with some keys, such as:

- user name,
- software name/type,
- user and software name/type,
- user and software name/type and number of processors,
- user and software name/type and walltime.

Table 4 shows average standard deviation of A for job groups by keys for walltime and memory.

Values of average standard deviation of A for job groups are smaller in all cases than value for all jobs. Smaller values in groups means that the A values within group are less scattered and can provide more precise prediction value of job related resources.

5 HPC System as Scheduling Problem

Problem description. The HPC system can be described with mathematical model as follows: there is a set of n tasks $\mathcal{J}$, which are to be executed on m parallel machines from set $\mathcal{M}$. The problem consist in the allocation of jobs to machines and the schedule of jobs determination on each machine. For each job following restrictions must be fulfilled:

(a) the execution of any job cannot be interrupted before its completion,
(b) each job can be executed on one ore more machines depends on requested resources,
(c) machine can execute more than one job in each moment of time.

Every job can be described by parameters:

- p_i – job execution time (*walltime*),
- r_i – job delivery time,
- u_i – number of machines,
- v_i – number of processors,
- y_i – amount of memory.

Every machine (computing node) can be described by parameters:

- w_j – total number of processors,
- z_j – total amount of memory.

Job allocation and schedule on each machine can be made for one of goal functions:

- minimization average jobs waiting time,
- maximize machines workload,
- minimization number of used machines,
- improve accuracy of requested resources.

Mathematical model. Set of jobs $\mathcal{J}$ can be divided into disjoint sets:

$$\bigcup_{j=1}^{m} \mathcal{J}_j = \mathcal{J}, \quad \bigcap_{j=1}^{m} \mathcal{J}_j = \emptyset, \tag{3}$$

which reflects to the jobs allocation to the machines from set $\mathcal{M}$. For every set $\mathcal{J}_j$ ($j = 1, 2, \ldots, m$) sum of requested processors must be equal or less than total amount of processors w_j available on j-th machine:

$$\sum_{k \in \mathcal{J}_j} v_k \leq w_j. \tag{4}$$

Similar for every set $\mathcal{J}_j$ ($j = 1, 2, \ldots, m$) sum of requested amount of memory must be equal or less than total amount of memory z_j available on j-th machine:

$$\sum_{k \in \mathcal{J}_j} y_k \leq z_j. \tag{5}$$

Solution of considered problem is represented by vector of times of the jobs execution beginning $S = (S_1, S_2, \ldots, S_n)$ and vector of jobs assignment to the machines $K = (K_1, K_2, \ldots, K_n)$. For every job we can define job waiting time for execution:

$$d_i = S_i - r_i \; i = 1, 2, \ldots, n. \tag{6}$$

Based on definition of delay time goal function for minimization average jobs waiting time can be defines as:

$$F(S) = \min \frac{1}{n} \sum_{i=1}^{n} d_i. \tag{7}$$

For every machine variable x_j denotes if machine is used:

$$x_j = \begin{cases} 1 & \text{if } j \text{ machine is used,} \\ 0 & \text{otherwise.} \end{cases} \tag{8}$$

Some machine can be turned off in order to save energy. Goal function for minimization number of used machines can be defined as:

$$\min \sum_{j=1}^{m} x_j. \tag{9}$$

6 Case Study

A solution for dealing with inaccuracy of declaring resources is to allow machine to schedule more jobs by virtually expanding its available memory from physical z_j to z_j':

$$z_j' = z_j(1 + \varepsilon) \tag{10}$$

where ε is expanding factor (e.g. from physical 64 GB to theoretical 70 GB – as 'overbooking'). In such a case scheduler can take risk and assign more jobs in a node. If the machine memory limit z_j is reached during the node work, one of jobs which try to use 'overbooked' memory is being killed and rescheduled, so that finally constraint defined by Eq. (5) is fulfilled at any time.

In this paper we present results of simulation of such a machine, which executes jobs basing on provided set of real users data collected from BEM supercomputer. The data were anonymized before the experiment. The simulator schedules jobs basing on declared memory and real walltime what corresponds to ideal on-line job scheduling – there are no unused time slots between scheduled jobs. Such a virtual machine was fed with set of data which has been limited to jobs which declared memory was not exceeding 10 GB, which is shown in Table 5.

Table 5. Basic informations about used data set

Parameter	Value
Number of jobs	5000
Declared memory mean/median	3.898/4 GB
Real memory mean/median	0.514/0.507 GB
Maximum declared memory	9.766 GB
Maximum real memory	8.02 GB

Simulation has been performed for different number of jobs $m = (500, 2000, 5000)$, different amount of memory $z + j = (32, 128)\, GB$ and for different values of $\varepsilon = (0, 0.5, 1, \ldots, 4.5)$. Table 6 shows reference values of total execution time for $\varepsilon = 0$.

Table 6. Reference values of total execution time for $\varepsilon = 0$

Number of jobs m	Machine memory z_j	Execution time [s]
500	32	239570
2000	23	1214085
2000	128	483982
5000	128	1018574

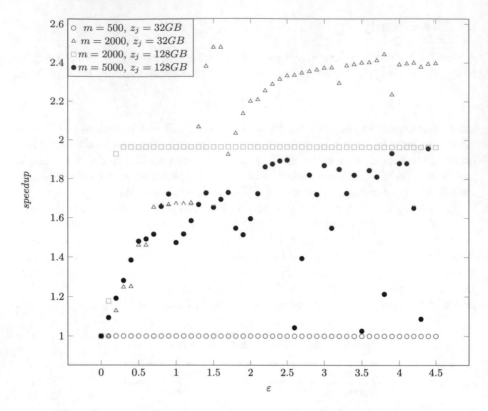

Fig. 1. Speedup values for different values of ε and machine parameters

Figure 1 presents speedup values for different ε values. Analyzing it one can observe that for simulations with $m = 500$, $z_j = 32$ GB and $m = 2000$, $z_j = 128$ GB speedup reaches constant level. This happens because the number of provided jobs is too small for machine with such parameters. One job has longer walltime than others and remaining jobs finish their execution (are killed) and must be run once again, which makes the speedup quiet week. Results from simulations with $m = 2000$, $z_j = 32$ and $m = 5000$, $z_j = 128$ show that the speedup is increasing with ε but reaches a limit.

7   User Inaccuracy

In order to give user chance to improve accuracy of jobs submitted to the BEM cluster two command are available. With command `resused` user can obtain information about memory and walltime requested and used for given number of last jobs. Figure 2 shows resource usage of the last 10 jobs obtained by calling command `resused -c 10`. Command `resstat` provides output with total resource usage (Fig. 3). Using this tools user can monitor resource usage and then improve accuracy of submitted jobs.

```
Resource usage of the last 10 jobs

Job Id Job Name Memory    Used/Req [GB] Walltime Used/Req [h]
------- -------- -------- ------------- -------- ------------
5787184 STDIN    WARNING  0.01/0.02     OK       0/0
5787113 STDIN    OK       0.11/0.02     OK       0/0
5773694 test.sh  CRITICAL 0.00/0.02     WARNING  0/0
5773693 STDIN    CRITICAL 0.00/0.02     CRITICAL 0/0
5773692 STDIN    CRITICAL 0.00/0.02     CRITICAL 0/0
5773690 STDIN    CRITICAL 0.00/0.02     CRITICAL 0/0
5723754 STDIN    CRITICAL 0.00/0.02     CRITICAL 0/0
5700033 STDIN    WARNING  0.01/0.02     OK       0/0
5624417 STDIN    OK       0.07/0.02     OK       0/0
5618655 STDIN    OK       0.04/0.02     OK       1/1
```

Fig. 2. Results of resused -c 10 command

```
Total resources usage of the last:

          Memory used/req [GB]  [%]  Walltime used/req [h]  [%]
--------- -------------------- ---- --------------------- ----
  10 jobs                  0/0   42                    3/4   56
 100 jobs              211/370   78                354/471   87
1000 jobs              228/387   51             9467/12377   72
```

Fig. 3. Results of resstat command

8   Conclusions

In this work there was presented resources inaccuracy study for BEM super-computer. Based on the study there is a place to research and improvement scheduling policies. Also mathematical model for jobs scheduling in HPC system (with goal functions) has been proposed and can be used for further research, especially with the use of potential 'overbooking' of the node memory.

Acknowledgement. The paper was partially supported by the National Science Centre of Poland, grant OPUS no. DEC 2017/25/B/ST7/02181.

References

1. Bożejko, W.: Solving the flow shop problem by parallel programming. J. Parallel Distrib. Comput. **69**(5), 470–481 (2009)
2. Bożejko, W.: On single-walk parallelization of the job shop problem solving algorithms. Comput. Oper. Res. **39**(9), 2258–2264 (2012)
3. Bożejko, W., Uchroński, M., Wodecki, M.: Block approach to the cyclic flow shop scheduling. Comput. Ind. Eng. **81**, 158–166 (2015)
4. Bożejko, W., Wodecki, M.: Parallel genetic algorithm for minimizing total weighted completion time. LNAI (L NCS), vol. 3070. Springer, Heidelberg (2004)

5. Chiang, S.H., Arpaci-Dusseau, A., Vernon, M.K.: The Impact of More Accurate Requested Runtimes on Production Job Scheduling Performance. In: Job scheduling strategies for parallel processing, 8th International Workshop, JSSPP 2002, Edinburgh, Scotland, UK, 24 July 2002 Revised Papers, pp. 103–127. Springer, Heidelberg (2002)
6. Cirne, W., Berman, F.: A comprehensive model of the supercomputer workload. In: 2001 IEEE International Workshop on Workload Characterization 2001, WWC-4, pp. 140–148, December 2001
7. Lan, Z., Buettner, D., Desai, N., Tang, W.: Job scheduling with adjusted runtime estimates on production supercomputers. J. Parallel Distrib. Comput. **73**, 926–938 (2013)
8. Mu'alem, A.W., Feitelson, D.G.: Utilization, predictability, workloads, and user runtime estimates in scheduling the IBM SP2 with backfilling. IEEE Trans. Parallel Distrib. Syst. **12**(6), 529–543 (2001)
9. Rodrigo, G.P., Östberg, P.O., Elmroth, E., Antypas, K., Gerber, R., Ramakrishnan, L.: Towards understanding HPC users and systems: a NERSC case study. J. Parallel Distrib. Comput. **111**, 206–221 (2018)
10. Smith, W., Taylor, V., Foster, I.: Using run-time predictions to estimate queue wait times and improve scheduler performance. In: Job Scheduling Strategies for Parallel Processing: IPPS/SPDP'99 Workshop, JSSPP 1999, San Juan, Puerto Rico, 16 April 1999 Proceedings, pp. 202–219. Springer, Heidelberg (1999)
11. Srinivasan, S., Kettimuthu, R., Subramani, V., Sadayappan, P.: Characterization of backfilling strategies for parallel job scheduling. In: International Conference on Parallel Processing Workshops 2002 Proceedings, pp. 514–519 (2002)
12. Ward, W.A., Mahood, C.L., West, J.E.: Scheduling jobs on parallel systems using a relaxed backfill strategy. In: Job Scheduling Strategies for Parallel Processing: 8th International Workshop, JSSPP 2002, Edinburgh, Scotland, UK, 24 July 2002, Revised Papers, pp. 88–102. Springer, Heidelberg (2002)

Software Package Development for the Active Traffic Management Module Self-oscillation Regime Investigation

Tatyana R. Velieva[1,3]([✉]), Anna V. Korolkova[1,3], Anastasiya V. Demidova[1,3], and Dmitry S. Kulyabov[1,2,3]

[1] Department of Applied Probability and Informatics,
Peoples' Friendship University of Russia (RUDN University), 6 Miklukho-Maklaya Street, 117198 Moscow, Russian Federation
{velieva_tr,korolkova_av,demidova_av,kulyabov_ds}@rudn.university
[2] Laboratory of Information Technologies, Joint Institute for Nuclear Research, 6 Joliot-Curie Street, Dubna, 141980 Moscow Region, Russian Federation
[3] Bogoliubov Laboratory of Theoretical Physics, Joint Institute for Nuclear Research, 6 Joliot-Curie Street, Dubna, 141980 Moscow Region, Russian Federation

Abstract. Self-oscillating modes in control systems of computer networks quite negatively affect the characteristics of these networks. The problem of finding the areas of self-oscillations is actual and important as the study of parameters of self-oscillations. These studies are extremely labor-intensive because of the substantial non-linear nature of the mathematical model. To investigate the self-oscillation parameters, the authors used the method of harmonic linearization. However, the previously used model with symmetric oscillations showed its limitations. In this paper, a more complex model with asymmetric oscillations is applied. For this purpose, the authors developed the program complex with both numerical analysis elements and elements of symbolic calculations.

Keywords: Traffic active management · Control theory
Self-oscillating mode

1 Introduction

While modeling technical systems with control it is often required to study not only characteristics of these systems, but also the influence of system parameters on these characteristics.

In systems with control there is a parasitic phenomenon as the self-oscillating mode. Earlier, we conducted the study to determine the regions of occurrence of self-oscillations in the system with the RED Active Queue Management algorithm [8]. To do this, we applied the method of harmonic linearization. We used a simplified model with unbiased oscillations. However, the asymmetry of the

© Springer International Publishing AG, part of Springer Nature 2019
W. Zamojski et al. (Eds.): DepCoS-RELCOMEX 2018, AISC 761, pp. 515–525, 2019.
https://doi.org/10.1007/978-3-319-91446-6_48

drop function generates a constant bias of self-oscillations. Thus this case has limited use.

In this paper, the method of harmonic linearization with asymmetric oscillations is applied. For this more labor-intensive method (than the one considered earlier) the authors have developed the software package for symbolic formulas obtaining and programs for numerical computation generating.

The structure of the paper is as follows. In the Sect. 2, we provide a brief description of the RED algorithm. In the Sect. 3, we describe the method of harmonic linearization for nonsymmetric oscillations. In the Sect. 4, we calculate the harmonic linearization coefficients for nonsymmetric oscillations. In the Sect. 5 we describe the main elements of the developed software package. In Sect. 6 a concrete example of calculation of self-oscillation parameters is given.

2   The RED Congestion Adaptive Control Mechanism

The RED algorithm uses a weighted queue length as a factor determining the probability of packets drop. As the average queue length grows, the probability of packets drop also increases. The algorithm uses two threshold values of the average queue length to control the drop function.

The RED algorithm is quite effective due to simplicity of its implementation in the network hardware, but it has a number of drawbacks. In particular, for some parameters values there is a steady oscillatory mode in the system, which negatively affects the quality of service (QoS) indicators [4, 10, 17]. Unfortunately there are no clear selection criteria for RED parameters values, at which the system does not enter in self-oscillating mode.

To describe the RED algorithm we will use the continuous model (see [1, 2, 6, 11, 12, 18]) with some simplifying assumptions: the model is written in the moments; the model describes only the phase of congestion avoidance for TCP Reno protocol; in the model the drop is considered only after reception of 3 consistent ACK confirmations.

3   Harmonic Linearization Method

The method of harmonic linearization was proposed by Bogolyubov, Krylov [7] and Nyquist [13]. The content of this method is reduced to separating the 'slow' variables from the 'fast' variables. The harmonic-linearized system depends on the amplitudes and frequencies of the periodic processes. This is an essential difference between harmonic linearization and the usual linearization method, leading to purely linear expressions, which allows us to investigate the basic properties of nonlinear systems.

The method of harmonic linearization is used for systems of a certain structure (see Fig. 1). The system consists of a linear link H_l and a nonlinear link H_{nl}, given by the function $f(x)$. A static nonlinear element is usually considered.

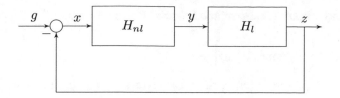

Fig. 1. Block structure of the system for the harmonic linearization method

The free harmonic oscillations are applied to the input of the nonlinear element:

$$x(t) = x_0 + \tilde{x} := x_0 + A\sin(\omega t).$$

On the output of the nonlinear element $f(x)$ we get a periodic signal. Let's expand it in a Fourier series:

$$y = \frac{a_0}{2} + \sum_{k=1}^{\infty}(a_k\sin(k\omega t) + b_k\cos(k\omega t)).$$

where the coefficients of the Fourier series have the following form:

$$a_0 = \frac{1}{\pi}\int_0^{2\pi} f(x_0 + A\sin(\omega t))\,\mathrm{d}(\omega t)\,;$$

$$a_k = \frac{1}{\pi}\int_0^{2\pi} f(x_0 + A\sin(\omega t))\sin(k\omega t)\,\mathrm{d}(\omega t)\,;$$

$$b_k = \frac{1}{\pi}\int_0^{2\pi} f(x_0 + A\sin(\omega t))\cos(k\omega t)\,\mathrm{d}(\omega t)\,;\quad k = \overline{1,\infty}.$$

The linear element is a low-pass filter, that is, when k is increasing the linear elements suppress higher harmonics.

Let's write the signal after the non-linear element:

$$y = y_0 + \tilde{y} \approx \varkappa_0(A,\omega,x_0) + [\varkappa(A,\omega,x_0) + i\varkappa'(A,\omega,x_0)]\tilde{x}, \tag{1}$$

$\varkappa_0$ is a constant shift, $\varkappa$ and $\varkappa'$ are the harmonic linearization coefficients:

$$\varkappa_0(A,\omega,x_0) = \frac{1}{2\pi}\int_0^{2\pi} f(x_0 + A\sin(\omega t))\,\mathrm{d}(\omega t)\,;$$

$$\varkappa(A,\omega,x_0) = \frac{a_1}{A} = \frac{1}{A\pi}\int_0^{2\pi} f(x_0 + A\sin(\omega t))\sin(\omega t)\,\mathrm{d}(\omega t)\,; \tag{2}$$

$$\varkappa'(A,\omega,x_0) = \frac{b_1}{A} = \frac{1}{A\pi}\int_0^{2\pi} f(x_0 + A\sin(\omega t))\cos(\omega t)\,\mathrm{d}(\omega t)\,.$$

In addition to (1), we will write

$$z = z_0 + \tilde{z} = (y_0 + \tilde{y})H_l(\omega),$$
$$x = x_0 + \tilde{x} = g(\omega) - (z_0 + \tilde{z}).$$

Then we may obtain the harmonic linearization equation:

$$\left[x_0 + H_l(\omega)\Big|_{\omega=0} \varkappa_0(A,\omega,x_0)\right] +$$
$$+ [1 + H_l(\varkappa(A,\omega,x_0) + i\varkappa'(A,\omega,x_0))]\tilde{x} = g(\omega) := g_0(\omega) + \tilde{g}(\omega).$$

Separating for constant and harmonic components, we may write:

$$\left[x_0 + H_l(\omega)\Big|_{\omega=0} \varkappa_0(A,\omega,x_0)\right] = g_0(\omega),$$
$$[1 + H_l(\varkappa(A,\omega,x_0) + i\varkappa'(A,\omega,x_0))]\tilde{x} = \tilde{g}(\omega).$$

In the study of self-oscillatory mode it is assumed that there is no external signal $(g = 0)$.

4   RED Model Harmonic Linearization

The linearization of the RED model and the derivation of the function H_l are described in detail in the article [8]:

$$H_l(\omega) = -\frac{C^4 T_f^3 w_q}{2N(Cw_q + i\omega)(iT_f\omega + 1)\left(iCT_f^2\omega - iNT_f\omega + 2N\right)}.$$

The linearized function P_{RED} has the form shown in Fig. 2

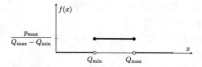

Fig. 2. The function P_{RED}

Let us compute the coefficients of harmonic linearization $\varkappa_0(A,x_0)$, $\varkappa(A,\omega,x_0)$ and $\varkappa'(A,\omega,x_0)$ (2) for the static nonlinearity P_{RED}:

$$\varkappa_0(A,\omega,x_0) = \frac{1}{2\pi}\int_0^{2\pi} P_{\text{RED}}(x_0 + A\sin(\omega t))\,\mathrm{d}(\omega t)\,;$$

$$\varkappa(A,\omega,x_0) = \frac{1}{A\pi}\int_0^{2\pi} P_{\text{RED}}(x_0 + A\sin(\omega t))\sin(\omega t)\,\mathrm{d}(\omega t)\,;$$

$$\varkappa'(A,\omega,x_0) = \frac{1}{A\pi}\int_0^{2\pi} P_{\text{RED}}(x_0 + A\sin(\omega t))\cos(\omega t)\,\mathrm{d}(\omega t)\,.$$

Depending on the relations between the thresholds $Q_{\min}$, $Q_{\max}$, the shift x_0 and the amplitude A, it is possible to obtain different limits of integration. For example, let's consider the case when

$$Q_{\min} < x_0 < Q_{\max}, \quad x_0 - A > Q_{\min}, \quad x_0 + A > Q_{\max}.$$

Then we will obtain:

$$\varkappa_0(A, x_0) = \frac{1}{2\pi} \frac{p_{\max}}{Q_{\max} - Q_{\min}} \int_0^{\alpha_{\max}} \mathrm{d}(\omega t) +$$

$$+ \int_{\pi - \alpha_{\max}}^{2\pi} \mathrm{d}(\omega t) = \frac{1}{2\pi} \frac{p_{\max}}{Q_{\max} - Q_{\min}} [2\alpha_{\max} + \pi];$$

$$\varkappa(A, x_0) = \frac{1}{A\pi} \frac{p_{\max}}{Q_{\max} - Q_{\min}} \left[\left. (-\cos(\omega t)) \right|_0^{\alpha_{\max}} + \left. (-\cos(\omega t)) \right|_{\pi - \alpha_{\max}}^{2\pi} \right]; \tag{3}$$

$$\varkappa'(A, x_0) = \frac{1}{A\pi} \frac{p_{\max}}{Q_{\max} - Q_{\min}} \left[\left. \sin(\omega t) \right|_0^{\alpha_{\max}} + \left. \sin(\omega t) \right|_{\pi - \alpha_{\max}}^{2\pi} \right].$$

The values of the integration limits are follows:

$$\sin \alpha_{\max} = \frac{Q_{\max} - x_0}{A}; \quad \cos \alpha_{\max} = \sqrt{1 - \frac{(Q_{\max} - x_0)^2}{A^2}}. \tag{4}$$

Thus, from (3) with the help of (4) we may get:

$$\varkappa_0(A, x_0) = \frac{p_{max}}{2\pi \left(Q_{max} - Q_{min} \right)} \left[2 \operatorname{asin} \left(\frac{Q_{max} - x_0}{A} \right) + \pi \right];$$

$$\varkappa(A, x_0) = -\frac{2p_{max}}{\pi A \left(Q_{max} - Q_{min} \right)} \sqrt{1 - \frac{(Q_{max} - x_0)^2}{A^2}};$$

$$\varkappa'(A, x_0) = 0.$$

The resulting harmonic linearization coefficients are used to generate the program by means of a computer algebra system.

5   Software Implementation of the RED Model

The software implementation of the RED model was carried out in two stages. At the first stage, a computer algebra system was employed. With the help of this system the whole time-consuming processing of the formulas is carried out. The resulting expressions are used both in the generation of numerical programs and in the transfer of formulas to the text of articles. Then the resulting formulas are used to generate computational programs. We suggest to use the Julia language [5] as a numerical programming language. It is unlikely that this language is really a silver bullet. However, it has a number of interesting features. This language is positioned as a modern reincarnation of the FORTRAN language. It supports both the stage of prototyping and writing the final version of the program. This language is intensively developing. All this attracted our attention to this language.

The most interesting system of symbolic calculations for us is the SymPy system [9]. This system appeared as a library of symbolic calculations for the Python language. But the Python language has become a universal glue language. Its application in a variety of projects led to explosive growth of related tools and libraries. Therefore, SymPy developed along with it. Now this is a fairly powerful system of computer algebra. SymPy suits us for the following reasons:

- As an interactive shell, it is convenient to use the Jupyter notepad, which is a component of the system iPython [16], with the REPL ideology.
- Python is actually used as a glue language, that allows you to integrate different software products. In addition, within the SciPy library [14] is supported a large number of output formats.
- The output of SymPy can be naturally passed for further numerical calculations to the NumPy [15] library and various programming languages.

Consider the fragments of the developed software. First, the SymPy library is included.

```
1  from sympy import *
2  init_printing(pretty_print=True, use_latex=True,    ↩
       use_unicode=True, fontsize='14pt')
```

In the following fragment the calculations for the nonlinear transfer function H_{nl} are performed.

```
1   Q_min = symbols('Q_min',real=True)
2   Q_max = symbols('Q_max',real=True)
3   p_max = symbols('p_max',real=True)
4   A = symbols('A',real=True)
5   x_0 = symbols('x_0',real=True)
6
7   sQ_min,sQ_max = symbols('sQ_min,sQ_max',real=True)
8   sQ_subst = [(sQ_min,sqrt(1- (x_0 -   ↩
        Q_min)**2/A**2)),(sQ_max,sqrt(1- (Q_max -x_0)**2/A**2))]
9
10  k0 = 1/(2*pi) *p_max/(Q_max - Q_min) *(2 * asin((Q_max-x_0)/A) + pi)
11  print(latex(k0))
12  k1 = 1 /(A * pi) * p_max/(Q_max - Q_min) * (- 2 * sQ_max)
13  print(latex(k1.subs(sQ_subst)))
14  k2 = 1 /(A * pi) * p_max/(Q_max - Q_min) * 2 * (Q_max - x_0)/A
15  print(latex(k2.subs(sQ_subst)))
16
17  Hnl = k1 + I * k2
```

For further calculations, the numerator and denominator of the nonlinear transfer function H_{nl} are obtained, as well as the real and imaginary parts.

```
1   reHnl = re(Hnl)
2   imHnl = im(Hnl)
3   nreHnl,dreHnl=fraction(reHnl)
4   nimHnl,dimHnl=fraction(imHnl)
```

In the following fragment, the initial model is linearized and the transfer function H_l is obtained. This operation is quite labor-intensive.

```
1   T_p, N, C, w_q = symbols('T_p, N, C, w_q',real=True)
2   t, p, W, W_T, Q, Qh  = symbols('t, p, W, W_T, Q, \hat{Q}',real=True)
3   T_f, W_f, p_f, Q_f = symbols('T_f, W_f, p_f, Q_f',real=True)
4   s, omega = symbols('s, \omega',real=True)
5   dW,dW_T,dQ,dQh,dp =    ←
          symbols('\delta{W},\delta{W_T},\delta{Q},\delta{\hat{Q}},    ←
          \delta{p}', real=True)
6
7   T = T_p + Q / C
8   L_W = 1 / T - W * W_T * p / (2 * T)
9   L_Q = W * N / T - C
10  L_Qh = -w_q * C * Qh + w_q * C * Q
11
12  W_f = C * T_f / N
13  p_f = 2 / W_f**2
14  Qh_f = Q_f
15
16  f_subst = [(T, T_f), (2*T, 2*T_f), (W_T, W_f), (W, W_f), (p, p_f)]
17  s_subst = [(dW_T,dW*(1-s*T_f))]
```

We calculate the variations with automatic substitution of the coupling equations.

```
1   dL_WdW = L_W.diff(W).subs(f_subst)
2   dL_WdW_T = L_W.diff(W_T).subs(f_subst)
3   dLWdQ = L_W.diff(Q).subs(f_subst)
4   dL_Wdp = L_W.diff(p).subs(f_subst)
5   dL_QdW = L_Q.diff(W).subs(f_subst)
6   dL_QdQ = L_Q.diff(Q).subs(f_subst)
7   d_QhdQh = L_Qh.diff(Qh).subs(f_subst)
8   dL_QhdQ = L_Qh.diff(Q).subs(f_subst)
9   dDotW = dL_WdW * dW + dL_WdW_T * dW_T + dLWdQ * dQ + dL_Wdp *dp
10  dDotQ = dL_QdW * dW + dL_QdQ * dQ
11  dDotQh = d_QhdQh * dQh + dL_QhdQ * dQ
12  dWfin = solve((dDotW - dW*s).subs(s_subst),dW)
13  dQfin = solve(dDotQ - dQ*s,dQ)
14  dQhfin = solve(dDotQh - dQh*s,dQh)
15  dW_subst = [(dW,dWfin[0])]
16  dQfin2 = dQfin[0].subs(dW_subst)
17  dQ_subst = [(dQ,dQfin2),(dp,1)]
```

We extract the real and imaginary parts.

```
1  nH1,dH1=fraction(H1)
```

The export of the resulting expressions to the Julia system.

```
1   x = symbols('x')
2   julia_x_subst = [(A,x)]
3   julia_code(eq_re_fin_omega2.subs(julia_x_subst),assign_to='f(x)')
4
5   x1,x2,x0 = symbols('x[1],x[2],x[3]')
6   julia_subst = [(A,x1),(omega,x2),(x_0,x0)]
7
8   print(julia_code(eq_re_fin.subs(julia_subst),assign_to='F[1]'))
9   print(julia_code(eq_im_fin.subs(julia_subst),assign_to='F[2]'))
10
11  omega0_subst = [(omega,0)]
12  H10 = H1.subs(omega0_subst)
13  H0 = x_0 + H10 * k0
14  print(julia_code(H0.subs(julia_subst),assign_to='F[3]'))
```

For numerical calculations we use the Julia language. Here is the fragment of the program code, most of which was obtained by using a computer algebra system.

```
1   using NLsolve
2
3   include("parameters.jl")
4
5   function f!(F, x::Vector{Float64})
6       F = Array{Float64}(length(x))
7       F[1] = 2*C.^4.*T_f.^3.*p_max.*w_q.*sqrt(1 - (Q_max -    ←
           x[3]).^2./x[1].^2) + 2*pi*N.*x[1].*(Q_max -    ←
           Q_min).*(-C.^2.*T_f.^3.*w_q.*x[2].^2 +    ←
           C.*N.*T_f.^2.*w_q.*x[2].^2 + 2*C.*N.*w_q -    ←
           C.*T_f.^2.*x[2].^2 - N.*T_f.*x[2].^2)
8       F[2] = sqrt((C.^2.*T_f.^2.*w_q + C.*N.*T_f.*w_q +    ←
           2*N)./(C.*T_f - N))./T_f
9       F[3] = -C.^3.*T_f.^3.*p_max.*(2*asin((Q_max - x[3])./x[1]) -    ←
           pi)./(8*pi*N.^2.*(Q_max - Q_min)) + x[3]
10  end
11
12  res = nlsolve(f!, x_init).zero
```

This program uses the library for solving systems of nonlinear equations NLsolve. The parameters are set in a separate file. The results are output in the form of a text file.

6 Influence of Parameters on Occurrence of Self-oscillations

Based on the developed algorithm, one can investigate the dependence of self-oscillation regions on the parameters of the RED algorithm. Naturally, we can consider other variants of RED-like algorithms.

As an illustration, we give a concrete example. Let's set the following parameters of the RED algorithm: the number of sessions $N = 60$, round-trip time $T_p = 0.5$ s, thresholds $Q_{min} = 75$ packages and $Q_{max} = 150$ packets, drop probability $p = 0.1$, parameter $w_q = 0.002$. Let us investigate the dependence of self-oscillation on the link capacity C. We obtain that the transition to the self-oscillatory regime occurs at $C_a = 15$ Mbps. That is, for $C \geqslant C_a$ the system will be in self-oscillation mode. This can be demonstrated using the network simulation tool NS-2 [3] (see Figs. 3 and 4). For the transition point to the self-oscillating regime, we obtain the following self-oscillation parameters: the frequency of self-oscillations $\nu = \omega/2\pi = 0.64$ Hz, the amplitude of the oscillations $A = 145$ packets.

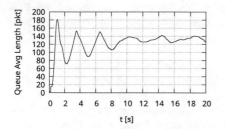

Fig. 3. Average queue length at link capacity $C = 5$ Mbps

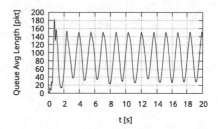

Fig. 4. Average queue length at link capacity $C = 15$ Mbps

7 Conclusion

The authors developed the software package for RED-like Active Queue Management algorithms investigation. The software package includes components for symbolic and numerical calculations. In constructing the calculation technique, the asymmetric nature of self-oscillations arising in the system was taken into account.

Acknowledgments. The publication has been prepared with the support of the "RUDN University Program 5–100" and funded by Russian Foundation for Basic Research (RFBR) according to the research project No 16-07-00556.

References

1. Brockett, R.: Stochastic analysis for fluid queueing systems. In: Proceedings of the 38th IEEE Conference on Decision and Control (Cat. No.99CH36304), vol. 3, pp. 3077–3082. IEEE (1999). https://doi.org/10.1109/CDC.1999.831407
2. Hollot, C.V.V., Misra, V., Towsley, D., Gong, W.-B.: On designing improved controllers for AQM routers supporting TCP flows. In: Proceedings of the IEEE INFOCOM 2001 Conference on Computer Communications, Twentieth Annual Joint Conference of the IEEE Computer and Communications Society (Cat. No.01CH37213), vol. 3, pp. 1726–1734. IEEE (2001). https://doi.org/10.1109/INFCOM.2001.916670
3. Issariyakul, T., Hossain, E.: Introduction to Network Simulator NS2. Springer, Boston (2012)
4. Jenkins, A.: Self-oscillation. Phys. Rep. **525**(2), 167–222 (2013). https://doi.org/10.1016/j.physrep.2012.10.007
5. Joshi, A., Lakhanpal, R.: Learning Julia. Packt Publishing, Birmingham (2017)
6. Korolkova, A.V., Velieva, T.R., Abaev, P.A., Sevastianov, L.A., Kulyabov, D.S.: Hybrid simulation of active traffic management. In: Proceedings of the 30th European Conference on Modelling and Simulation, pp. 685–691 (2016). https://doi.org/10.7148/2016-0685
7. Kryloff, N., Bogoliuboff, N.: Les methodes symboliques d l Mecanique non Lineaire dans leur application a l'etude de la resonance dans l'oscillateur. Bulletin del'Academie des Sciences de l'URSS. Classe des sciences mathematiques et na (1), 7–34 (1934)
8. Kulyabov, D.S., Korolkova, A.V., Velieva, T.R., Eferina, E.G., Sevastianov, L.A.: The methodology of studying of active traffic management module self-oscillation regime. In: Zamojski, W., Mazurkiewicz, J., Sugier, J., Walkowiak, T., Kacprzyk, J. (eds.) DepCoS-RELCOMEX 2017. Advances in Intelligent Systems and Computing, vol. 582, pp. 215–224. Springer International Publishing, Cham (2018). https://doi.org/10.1007/978-3-319-59415-621
9. Lamy, R.: Instant SymPy Starter. Packt Publishing, Birmingham (2013)
10. Lautenschlaeger, W., Francini, A.: Global synchronization protection for bandwidth sharing TCP flows in high-speed links. In: Proceedings of the 16-th International Conference on High Performance Switching and Routing, IEEE HPSR 2015, Budapest, Hungary (2015)
11. Misra, V., Gong, W.B., Towsley, D.: Stochastic differential equation modeling and analysis of TCP-windowsize behavior. In: Proceedings of PERFORMANCE, vol. 99 (1999)
12. Misra, V., Gong, W.B., Towsley, D.: Fluid-based analysis of a network of AQM routers supporting TCP flows with an application to RED. ACM SIGCOMM Comput. Commun. Rev. **30**(4), 151–160 (2000). https://doi.org/10.1145/347057.347421
13. Nyquist, H.: Regeneration theory. Bell Syst. Tech. J. **11**(1), 126–147 (1932). https://doi.org/10.1002/j.1538-7305.1932.tb02344.x
14. Oliphant, T.E.: Python for scientific computing. Comput. Sci. Eng. **9**(3), 10–20 (2007). https://doi.org/10.1109/MCSE.2007.58
15. Oliphant, T.E.: Guide to NumPy, 2nd edn. CreateSpace Independent Publishing Platform, Santa Monica (2015)
16. Perez, F., Granger, B.E.: IPython: a system for interactive scientific computing. Comput. Sci. Eng. **9**(3), 21–29 (2007). https://doi.org/10.1109/MCSE.2007.53

17. Ren, F., Lin, C., Wei, B.: A nonlinear control theoretic analysis to TCP-RED system. Comput. Netw. **49**(4), 580–592 (2005). https://doi.org/10.1016/j.comnet.2005.01.016
18. Velieva, T.R., Korolkova, A.V., Kulyabov, D.S.: Designing installations for verification of the model of active queue management discipline RED in the GNS3. In: 6th International Congress on Ultra Modern Telecommunications and Control Systems and Workshops (ICUMT), pp. 570–577. IEEE Computer Society (2015). https://doi.org/10.1109/ICUMT.2014.7002164

Bag-of-Words, Bag-of-Topics and Word-to-Vec Based Subject Classification of Text Documents in Polish - A Comparative Study

Tomasz Walkowiak[✉], Szymon Datko, and Henryk Maciejewski

Faculty of Electronics, Wrocław University of Science and Technology,
Wrocław, Poland
{tomasz.walkowiak,szymon.datko,henryk.maciejewski}@pwr.edu.pl

Abstract. This paper deals with the problem of classification of Polish language documents in terms of a subject category. We compare four state-of-the-art approaches to this task which differ primarily in the way the documents are represented by feature vectors. Two methods considered in the study use frequency-of-words or frequency-of-topics representation of the documents and rely on the Natural Language Processing (NLP) technology to pre-process the raw text. Two alternative methods do not involve the NLP technology. They construct feature vectors using vector representation of words (Word2Vec method) or using a frequency of topics derived from the raw text. These four approaches are evaluated using 3 corpora with 5, 34 and 25 subject categories respectively and with a different level of class discrimination. Results suggest that no single method outperforms other method in all tests, however tests with large number of training observations seem to favour the NLP-free Word2Vec methods.

Keywords: Text mining · Subject classification · Bag of words
Bag of topics · Latent Dirichlet Allocation · Word embedding · fastText

1 Introduction

Subject classification of text documents is a task aiming for automatic categorization of given document into one of known/defined groups. Such assignment can be important for press agencies, editors or even just for readers. However, due to complex nature of Polish and other Slavic languages (e.g. rich inflection), classification of text is often a challenging problem.

Several approaches to representing documents by feature vectors are available with no clear advice as to which of the methods is expected to be most successful in a particular study. While standard approaches (e.g., bag-of-words or bag-of-topics) strongly rely on NLP tools used to prepare feature vectors for classification, recent developments (e.g. [8]) seem to suggest that NLP-free methods

© Springer International Publishing AG, part of Springer Nature 2019
W. Zamojski et al. (Eds.): DepCoS-RELCOMEX 2018, AISC 761, pp. 526–535, 2019.
https://doi.org/10.1007/978-3-319-91446-6_49

operating directly on raw text may be a promising alternative. The purpose of this work is to empirically compare these leading state-of-the-art methods using text corpora with different characteristics in terms of size and class separability. We attempt to find characteristics of data which seem to favour NLP-based on NLP-free methods. In the following section we outline the current research in classification of documents in Polish. In Sect. 2 we provide technicalities about the four methods we compare; in Sects. 3 and 4 we present test corpora used in the study and results of the comparative study.

1.1 State of the Art – Polish Text Classification

The paper [1] provides an overview of the categorization methods for Polish language. It compares the Naïve Bayes classifier [6] and Roccio algorithm [7] in the context of job advertisements processing system. The Naïve Bayes classifier reached 2% better result than tf-idf based Roccio algorithm.

In [12] the authors analyzed several linguistic features in classification of Polish text documents. The classical centroid-based classification, utilizing the tf-idf model, was used in a series of experiments on a corpus of press news. About 12.000 articles were divided into 8 different subject categories. The best achieved result (accuracy equal to 86.3%) was reached when using lemmatized terms.

Zadrożny et al. published [21] a comprehensive analysis of the categorization problem, focusing on applications for Polish language. Later, in 2013 the same authors presented a concept of in-depth classification [20].

The paper [13] compares Natural Language Processing tools suitable for English language. The author provides also a list of corresponding tools relevant for Polish language.

In [3] a new method of Polish text classification with application of distant supervision is presented. The research was conducted on collection of Wikipedia resources.

Authors of [19] compare different techniques of text representation and several classifiers on a corpus of press news in Polish (the same corpus is used in this paper, see section 3). They achieve classification accuracy of 95%. They recommend to use frequencies of the most frequent 1000 base forms of nouns and verbs as feature vectors.

In [18] authors compared the bag of words method and the fastText [8] in subject classification of text from Polish Wikipedia achieving accuracy of 89%.

It is worth to notice that in this work we are not focused on multicategory classification problem, where each document can be labeled by more than one category. We follow the standard approach, so each document is assigned to exactly one subject group.

2 Text Representation Methods

2.1 Bag of Words

The bag of words [5] is a common method used for building a vector representation of text documents. The key assumption in this approach is that the text sentence can be expressed using unordered set of frequencies of selected words (or dictionary) [15]. Currently a number of task-specific modifications of this method are available. For example, in the area of subject classification, the selected dictionary is typically filtered off the most and least common phrases, as well as stop words [16].

In addition, the number of selected features (words) can be often reduced by transforming the words to their generic form (stemming, lemmatization). The literature commonly suggests to weight the raw counts of word occurrences, e.g. in relation to the total length of each document and uniqueness of a word for single category.

However, it is worth to notice that recent experiments [19], as well as research conducted by us, suggest that selecting any particular weighing schema may be not really important. It is most likely because the weighing process is often just a linear transformation of the feature vectors, while many classification algorithms (e.g. multilayer perceptron, logistic regression, SVM) perform similar transformations during the learning process.

For our research, we followed the BoW schema featured in [19]. At first, all text documents were processed using the WCRFT tagger [14]. It is a morphosyntactic tagger designed for Polish language, utilizing the Conditional Random Fields (CRF) and tiered tagging of plain text for part of speech tagging and lemmatization. Next, we chose only words that were nouns, additionally selecting only a certain number (1000) of the most frequent ones (nouns/lemmas) from the training set. Then, each document in corpora was expressed as a vector of counts of occurrences of selected lemmas.

All the processing (feature generation) was performed using Clarin-PL infrastructure [17].

2.2   fastText

The fastText [8] is a recently proposed, deep learning method for classification of text documents. It utilizes a linear soft-max classifier [4] and is based on representation of text using so called word embeddings, often called word-to-vec. The main concept consists in simultaneous generation of word representation and classifier learning process. The resulting model is linear and appears highly effective in training, reaching the solution faster than any other method [8]. Moreover, in many text mining tasks it outperforms approaches the BoW-involving approaches.

FastText constructs the word embedding model using the training data. Is is a look-up table-like mapping for each word to a real number vector of fixed dimension. Finally, each document in corpora is represented as an average of sum of the word embeddings existing within the document. If some words are unknown (they did not exist in training set, so no word embeddings are available for them), then they are ignored during averaging. The dimension of word embedding and hence the dimension of feature vectors (doc embeddings) is equal 100 in our study. Preliminary tests with different size of the word embedding vector size do not improve the results.

Similarly to the BoW model, fastText ignores (by default) the word order. It is possible to take word n-grams into considerations, so the local word ordering will be taken into account - this approach, however, was not used in our experiments.

2.3 Latent Dirchlet Allocation

The Latent Dirichlet Allocation (LDA) is a popular model used in various topic modeling algorithms [2]. The aim of such algorithms is to generate words collections (interpreted as topics) that are significant and specific to a given corpus on the basis of the words frequency and the scope of their presence in the samples [2]. Similarly to previous methods, it ignores the word order.

We filtered out terms that appeared in more than 70% of the documents (a fraction of the total corpus size). This threshold is often referred as the maximum document frequency max_df (for example in the scikit library[1]) and is used to remove the terms that appear too frequently. Such words are often also qualified as corpus-specific stop words.

The package Mallet [9] (popular in Digital Humanities) was used to perform the LDA to generate topics and feature vectors for documents. The topics (probabilities of words in given topics) were generated from the training corpora. They were used to get the document vector representation (i.e. the topic probabilities in that document) for training and testing corpora. The feature vector forms a bag of topics. In contrast to the BoW, here we do not count the word occurrences but probabilities of each topics.

We used 100 topics in our experiments. This value proved to maximize classification results (as compared to other values tested: 50, 100, 300 and 500). Therefore, the dimension of the feature vector from LDA is 100.

Additionally, in our research we used a modified LDA method. The modification consists in using previously lemmatized nouns, instead of raw words from text documents. For this purpose, again the WCRFT tagger [14] was used for lemmatization and selection of nouns.

2.4 Methods Comparison

Finally, all feature vectors (regardless of the method) were normalized and standardized using the mean and variance of the data from the training set - to match the requirement of normal data distribution, needed by many classifiers.

Summarizing - we used 4 methods of features extraction from text: BoW, fastText, LDA (based on raw text) and LDA (based on lemmatized nouns only). Two of these methods (fastText, LDA (raw)) do not involve any NLP knowledge, while the two other methods (BoW, LDA (nouns)) depend on such knowledge. In the NLP-based experiments, all text documents were pre-processed using a morphosyntactic WCRFT tagger [14] for Polish.

---

[1] http://scikit-learn.org/0.18/modules/generated/sklearn.feature_extraction.text.
CountVectorizer.html.

The Table 1 provides a brief overview of the methods examined in our research, along with summary of their properties.

Table 1. Methods comparison

Method	BoW	fastText	LDA	LDA (nouns)
Number of features	1000	100	100	100
Usage of tagger	Yes	No	No	Yes

3 Data Sets

We evaluated the four algorithms, described in Sect. 2, using three collections of text documents: *Press*, *Wiki* and *Web*. Table 2 summarizes properties of these collections.

Table 2. Corpora features

Feature n Corpora	*Press*	*Wiki*	*Web*
Number of classes	5	34	25
Documents per class (ca.)	1300	300	80
Subject quality	Very good	Good	Satisfactory

The first corpus (*Press*) consists of Polish press news and it was also used previously for experiments reported in [19]. It is a good example of a complete, high quality and well defined data set. The texts were assigned by press agency to 5 subject categories. All the subject groups are very well separable from each other and each group contains reasonably large number of members. There are 6564 documents in total in this corpus, which gives an average of ca. 1300 documents per class. In the study [19], the authors reported 95% accuracy achieved on this data set in cross-validation tests with 4 folds [6].

The second corpus (*Wiki*) consists of articles extracted from the Polish language Wikipedia. This corpus has also good quality, however some of the class assignments may be doubtful. It is characterized by a significant number of 34 subject categories. This data set was created by merging two publicly available collections [11] and [10]. It includes 9,837 articles in total, which translates into about 300 articles per class (however one class is slightly underrepresented). This corpus is also known in literature [18], where the classification accuracy reported for this set reached 88.7%.

The third corpus (*Web*) contains articles from one of Polish web portals. Articles were assigned to subject groups by the portal administrator. However, due to overlapping of subject areas (e.g. philosophy, religion, worldview), it can be considered as the worst of these three corpora for subject classification. To achieve the data set as balanced as possible, only a subset of articles was included in the corpora. Among all available categories, only those with the number of articles larger then 50 and smaller then 150 were used. This resulted in 25 classes and 2100 documents in total, 84 in average per class.

4   Evaluation

Each experiment was organized as follows. First, the corpus was divided into training set and testing set. Then, each of the 4 methods was used to generate feature vectors for all the documents. Finally, a classifier was trained and tested on generated sets of feature vectors. For every setting the experiment was performed 20 times and average accuracy was calculated for the report. After preliminary experiments, we selected the multilayer perceptron (MLP) [6] with Broyden-Fletcher-Goldfarb-Shanno (BFGS) nonlinear optimization learning algorithm for the classification using BoW and LDA feature vectors as it gave the best or almost the best results for all analyzed cases. In case of fast-Text methods we used the built-in classifier (linear soft-max) [8].

Results of the experiments are presented in Figs. 1, 2 and 3.

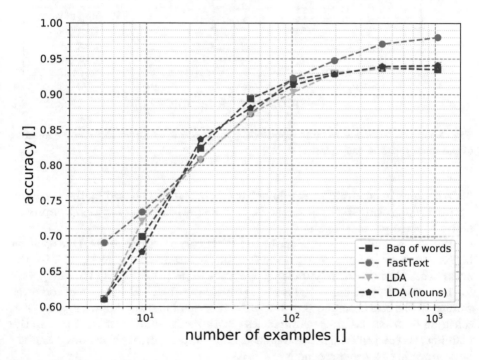

Fig. 1. Classification accuracy for the *Press* corpus as a function of the average number of per-class training examples.

Figure 1 presents the classification accuracy achieved for the *Press* corpus. Little number of classes and high diversity between them resulted in a high precision (69% for fastText, 61% for BoW, LDA and LDA(nouns)) even though the number of training examples for each class was small. With growing number of examples the difference between methods becomes less significant (3% at most) and furthermore the NLP-based approaches (BoW, LDA (nouns)) performe slightly better that other methods. However, at some point bigger number

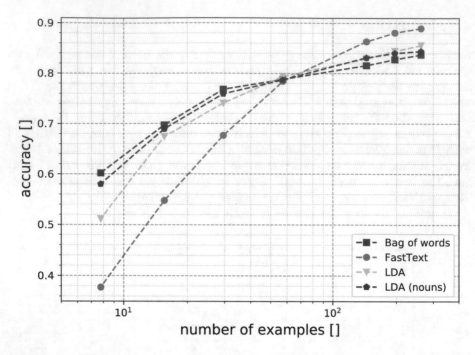

Fig. 2. Classification accuracy for the *Wiki* corpus as a function of the average number of per-class training examples.

of documents in the training set do not result in the NLP-based methods clearly outperforming the NLP-free methods, and the Word2Vec (fastText) approach starts to have the edge again.

In Fig. 2 a similar observations for the *Wiki* corpus can be made. Due to much bigger number of subject groups, the starting accuracy was much smaller for the approaches that do not rely on language knowledge (51% for LDA(nouns), 45% for fastText). Surprisingly, the accuracy observed for NLP-based methods was similar to the one observed previously. Clearly the language knowledge helps in selection of better features. Nevertheless, with rising number of examples in the training set, the fastText method gained much improvement in accuracy, rapidly outperforming all other methods.

Figure 3 shows results obtained for the third i.e. the *Web* corpus. The presence of a number of similar (tough to distinguish) categories clearly has an impact on the overall accuracy - the best results achieved are still worse than the lowest results reached for *Press* and *Wiki* corpora. In this case, for almost all numbers of examples tested the achieved accuracy of fastText approach was worse than accuracy of the other methods. We can observe however that with increasing number of examples improvement in accuracy for fastText algorithm grows fastest. Unfortunately this corpus was too small to verify whether this trend will be preserved and will finally allow fastText to outperform other methods. Another interesting effect is the visible fall in accuracy of the NLP-based

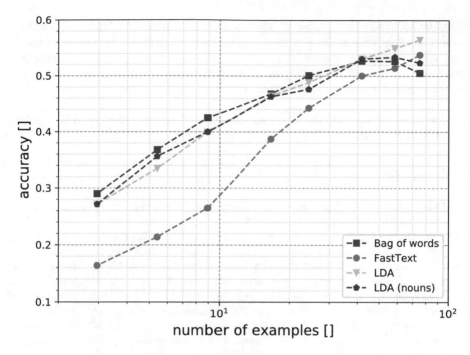

Fig. 3. Classification accuracy for the *Web* corpus as a function of the average number of per-class training examples.

approaches (BoW, LDA(nouns)) for the largest numbers of examples considered in the study.

Although the language knowledge seems important and provides additional advantage when number of examples is small, it is clearly visible that the NLP-free Word-to-Vec method starts to outperform other methods when the training set is big enough. Moreover, the higher accuracy of prediction with NLP-free approaches is obtained with the lower dimensionality of feature vectors.

5   Conclusion

This paper presented a comparison of 4 approaches to represent text documents using feature vectors. We studied two methods that involve the Natural Language Processing (NLP) technologies and two methods that did not require any language knowledge. Our study involved 3 completely different sets of text documents.

Our research showed that the language knowledge, clearly brings important advantage during subject classification with many categories and few examples in training set. However, the language knowledge is no longer so crucial when the training set contains sufficiently many examples. For such datasets the NLP-free approaches can perform as good as the NLP-based methods or even better. It is worth to notice that such results were observed although NLP-free methods used much smaller feature vectors as compared to e.g. BoW. However, when the number of training examples is very small and there is a high number categories which are of difficult to distinguish, the classical, NLP-based approaches can promise better result of classification.

Acknowledgment. This work was sponsored by National Science Centre, Poland (grant 2016/21/B/ST6/02159).

References

1. Żak, I., Ciura, M.: Automatic text categorization. Information Systems Architecture and Technology, September 2005
2. Blei, D.M., Ng, A.Y., Jordan, M.I.: Latent dirichlet allocation. J. Mach. Learn. Res. **3**, 993–1022 (2003). http://dl.acm.org/citation.cfm?id=944919.944937
3. Ciesielski, K., Borkowski, P., Kłopotek, M.A., Trojanowski, K., Wysocki, K.: Wikipedia-based document categorization. In: Bouvry, P., Kłopotek, M.A., Leprévost, F., Marciniak, M., Mykowiecka, A., Rybiński, H. (eds.) Security and Intelligent Information Systems, pp. 265–278. Springer, Heidelberg (2012)
4. Goodman, J.: Classes for fast maximum entropy training. In: 2001 IEEE International Conference on Acoustics, Speech, and Signal Processing. Proceedings (Cat. No. 01CH37221). vol. 1, pp. 561–564 (2001)
5. Harris, Z.: Distributional structure. Word (1954)
6. Hastie, T.J., Tibshirani, R.J., Friedman, J.H.: The Elements of Statistical Learning: Data Mining, Inference, and Prediction. Springer series in statistics. Springer, New York (2009). autres impressions: 2011 (corr.), 2013 (7e corr.)
7. Joachims, T.: A probabilistic analysis of the Rocchio algorithm with TFIDF for text categorization. In: Fisher, D.H. (ed.) Proceedings of the ICML-97, 14th International Conference on Machine Learning, pp. 143–151. Morgan Kaufmann Publishers, San Francisco, Nashville, US (1997). http://citeseerx.ist.psu.edu/viewdoc/summary?doi=10.1.1.21.7950
8. Joulin, A., Grave, E., Bojanowski, P., Mikolov, T.: Bag of tricks for efficient text classification. In: Proceedings of the 15th Conference of the European Chapter of the Association for Computational Linguistics, vol. 2, Short Papers, pp. 427–431. Association for Computational Linguistics (2017). http://aclweb.org/anthology/E17-2068
9. McCallum, A.K.: Mallet: A Machine Learning For Language Toolkit (2002). http://mallet.cs.umass.edu
10. Młynarczyk, K., Piasecki, M.: Wiki Test - 34 Categories (2015), CLARIN-PL digital repository. http://hdl.handle.net/11321/217
11. Młynarczyk, K., Piasecki, M.: Wiki Train - 34 Categories (2015). CLARIN-PL digital repository http://hdl.handle.net/11321/222
12. Piskorski, J., Sydow, M.: Experiments on classification of polish newspaper. Arch. Control Sci. **15**, 613–625 (2005)
13. Przybyła, P.: Issues of polish question answering. In: Hryniewicz, O., Mielniczuk, J., Penczek, W., Waniewski, J. (eds.) Proceedings of the First Conference 'Information Technologies: Research and their Interdisciplinary Applications' (ITRIA 2012), pp. 122–139. Institute of Computer Science, Polish Academy of Sciences (2012)

14. Radziszewski, A.: A tiered CRF tagger for Polish. Intelligent Tools for Building a Scientific Information Platform. Studies in Computational Intelligence, vol. 467, pp. 215–230. Springer, Heidelberg (2013)
15. Salton, G., Buckley, C.: Term-weighting approaches in automatic text retrieval. Inf. Process. Manag. **24**(5), 513–523 (1988)
16. Torkkola, K.: Discriminative features for text document classification. Form. Pattern Anal. Appl. **6**(4), 301–308 (2004). https://doi.org/10.1007/s10044-003-0196-8
17. Walkowiak, T.: Language processing modelling notation - orchestration of nlp microservices. In: Zamojski, W., Mazurkiewicz, J., Sugier, J., Walkowiak, T., Kacprzyk, J. (eds.) Advances in Dependability Engineering of Complex Systems, pp. 464–473. Springer, Cham (2018)
18. Walkowiak, T., Datko, S., Maciejewski, H.: Feature extraction in subject classification of text documents in polish. Artificial Intelligence and Soft Computing. Springer, Cham (2018)
19. Walkowiak, T., Malak, P.: Polish texts topic classification evaluation. In: Proceedings of the 10th International Conference on Agents and Artificial Intelligence, vol. 2, ICAART, pp. 515–522. INSTICC, SciTePress (2018)
20. Zadrożny, S., Kacprzyk, J., Gajewski, M., Wysocki, M.: A novel text classification problem and its solution. Czasopismo Techniczne. Automatyka R. 110, z. 4-AC, 7–16 (2013)
21. Zadrożny, S., Kacprzyk, J.: Computing with words for text processing: an approach to the text categorization. Inf. Sci. **176**(4), 415–437 (2006). http://dx.doi.org/10.1016/j.ins.2005.07.017

Use of Distributed Machine Learning Toolkit for Searching Content Promoting Hate Speech on the Web

Marek Woda[✉] [ID] and Mateusz Torbiarczyk

Department of Computer Engineering, Wroclaw University of Technology,
Janiszewskiego 11-17, 50-372 Wrocław, Poland
marek.woda@pwr.edu.pl

Abstract. The paper describes results of research on applicability of a new tool called *Distributed Machine Learning Toolkit* (DMTK) to detect hate speech on the Internet. For this purpose, the *Word Embedding* module was used, which uses the *word2vec* method to create a vector representation of the word. These representations were used for vector recording of entries posted on twitter and then they were subjected to classification using *LightGBM*, a classifier using *gradient boosting* methods. As a reference, in order to compare results provided by *DMTK*, two free of charge machine learning algorithms *Gensim* and *GloVe* were scrutinized.

Keywords: DMTK · Hate speech · *word2vec* · *GloVe* · *Gensim*

1 Introduction

Nowadays, hate speech, which is posted on websites, is becoming a bigger problem [18]. People hide behind the invented nicknames and feel anonymous, and thus go unpunished. They offend people of other faith, color, gender, or orientation. They do not realize what harm they can cause in the psyche of other people. Site administrators try to block this type of content on their internet domains, however, the number of entries to check is too large and the harmful comment can reach many people before it is blocked.

The result is that a new solution is sought that could recognize hate speech automatically [12, 17]. Thanks to this, such entries will not be published, and the person trying to place them will be excluded from the given online community.

The methods of artificial intelligence and machine learning come with help. For many years, scientists have been using them to analyze the text and extract interesting information from it, such as the theme of the text or the sentiment that it brings. They use different approaches. A number of programs and libraries adapted to the needs of text analysis have also been created. Such tools facilitate research on finding the best method of analysis. Main aim of authors was to provide the answer how efficient in terms of hate speech classification is a new framework called *Distributed Machine Learning Toolkit* in comparison to following NLP (Natural Language Processing) algorithms: *GloVe*, *Gensim* with already established reputation. All three utilize

© Springer International Publishing AG, part of Springer Nature 2019
W. Zamojski et al. (Eds.): DepCoS-RELCOMEX 2018, AISC 761, pp. 536–544, 2019.
https://doi.org/10.1007/978-3-319-91446-6_50

Word2vec which is a group of related models that are used to produce *word embeddings* (a method of representing words using vectors).

2   Related Work

Recently, *Word embeddings* is gaining its momentum. It is used for various tasks in NLP where words or phrases from the vocabulary are mapped to vectors of real numbers. This approach provides significantly better results than other methods [13]. Simply put, this method converts text into numbers and there may be different numerical representations of the same text. Generally, *Word embeddings* could be classified into categories: Frequency- and Prediction- based Embedding.

Vector representation tries to encode the meaning of a word. Similar words should have similar vectors, thanks to which it is possible to find connections between different words denoting the same or similar things. This can be achieved by processing a large set of texts in various ways, e.g. each one-way neural network that takes a set of words for input, and presents them with vectors, then tunes them through backward propagation return *Word Embeddings* as the weight of the first layer, called *Embedding Layer*.

Generating *Word Embeddings* using deep learning is too complex and takes too much time for a large dictionary [3]. This is the main reason why only in 2013 *Word Embeddings* began to be used more widely for NLP purposes.

The individual models of *Word Embeddings* differ [1, 7], among others the purpose of the training, for example, *word2vec* is set to produce *Word Embedding*, which encodes the syntactic relation of the sentence. This is helpful in many tasks, but it does not work in cases where these relationships do not matter.

Word Embedding models are associated with language models. The quality of the model is measured on the basis of its ability to recognize the probability distribution of words in the sequence of words. Many models try to pinpoint the next word in a sequence of words. In addition, these models are being developed using a measure of indeterminacy or cross-entropy based on a measure taken from language models.

The classic language model presented in [2] is based on classic neural model, and consists of one hidden layer of a one-way neural network that predicts the next word in the sequence.

In the work [2] the first technique how to obtain *Word embeddings* is shown, presenting the words as vectors of real numbers. Recent methods have only improved this model, but the core elements have remained the same:

- *Embedding Layer* - generates representations by multiplying index vectors and word embedding matrix.
- *Intermediate Layer* - one or more layers that produce intermediate input representations, e.g. a combined layer that adds non-linearity to a combination of processed words with n-previous words.
- *Softmax Layer* - the last layer providing the probability of words distribution in a dictionary.

Bengio, in [2] pointed out two problems that underlie today's models:

- a possibility to replace the middle layer by Long short-term memory (LSTM) as it was done in [9, 12].
- the bottleneck of *softmax layer*, which increases the calculation time in proportion to the size of dictionary.

Finding a way to lower computational cost (imposed by *Softmax Layer*) for large datasets is the main challenge both in the neural and *Word Embeddings* models.

After pioneered research on neural language models [2], next were Collobert and Weston [4] that presented *Word Embeddings* application. In [5] they proved that trained on a sufficiently large learning dataset brings meaningful results in various areas.

Collobert and Weston model introduced a changed *objective function*, thanks to that exhaustive calculations in *Softmax Layer* could be avoided. Instead of the cross-entropy used by Bengio, the *pairwise ranking criterion* network was used. Whereas *objective ranking* eliminates the computational complexity of the *Softmax Layer*, the hidden middle layer of the Bengio's model is still a source of costly calculations.

Word2Vec is the most popular model of *Word Embeddings*. It was first described in [10, 11]. *Word2Vec* is often confused as learning deep of NLP models, but this is not true. *Word2Vec* is not part of deep learning because its architecture is neither deep nor linear.

In [10], Mikolow proposed two architectures of *word embeddings*, which are less computationally expensive than the methods described in previously cited works. In [11], proposed models are further improved in terms of faster calculation as well as greater accuracy. Its architecture differs from the Bengio's and Collobert and Weston models by the lack of a hidden *Intermediate Layer* and in addition it allows the model to use additional context.

Previous models learn to improve their ability to predict the next word or words that may appear next to a word, which is why they are named *predictive models*. In contrast, *Glove - Global Vectors for Word Representation* [14] is a *computational model* because it is based on reductions in the dimensionality of the number of words co-occurrences matrix.

At the beginning, a large matrix of word co-occurrences is built. For every word (matrix lines) it is counted how often it occurred in a sentence or a context (columns). The number of sentences is (usually very) large, depending on the size of the text being processed, therefore the factorization is carried out to maintain a matrix with fewer dimensions.

The main disadvantage of the model obtained is that each word has the same weight. This is problematic in the case of words that have been rare and introduces distortions to the final results. To solve this problem, *GloVe* authors, created a new regression model of weighted least squares.

3   Tools Used in the Research

3.1   Distributed Machine Learning Toolkit - DMTK

Distributed Machine Learning Toolkit (called *DMTK*) is a framework created by Microsoft containing various types of algorithms and methods of *machine learning* and *artificial intelligence*. It consists of *LightLDA* (a scalable topic model algorithm), D(M) WE (a distributed version of (multi-sense) word embedding algorithm) and *LightGBM* (implementation of a very high-performance gradient boosting method). DMTK is an open-source project, everyone can contribute to its development as well as use it. What distinguishes *DMTK* from other tools of this type is the ability to perform calculations in parallel on many computers, thanks to which the task is completed faster and it is possible to load more data into shared operating memory. The reduction of the operation time is not at the expense of the quality of the results obtained.

A built-in *DMTK* tool *Word Embedding* was used for creating *Word Embeddings*. The algorithm *word2vec* described for the first time in [10, 11] was implemented for this reason. The algorithm has been implemented in such a way that it can be run on a distributed system. This was achieved thanks to *Multiverso* tool (also a part of *DMTK*).

3.2   LightGBM

LightGBM is a framework using *gradient boosting methods* [8]. It supports algorithms such as *GBDT*, *GBRT* and *GBM*. The official webpage of *DMTK* (www.DMTK.io) claims *LightGBM* is several times faster than other similar implementations. It owes this to the histogram memory and optimization of calculations. It also allows for distributed learning based on *DMTK*. The authors boast that one to two hours is sufficient to have a predictor trained based on a database with 1.7 billion records with 67 features on a cluster with only 16 machines. Another advantage of *LightGBM* is that it is available as a library for the Python programming language.

3.3   Gensim

Gensim [15] is an abbreviation of Generate similar. This Python library is intended for the implementation of unattended semantic modeling based on a given text. Developers paid a lot of attention to optimizing the execution time of algorithms [21].

One of the methods available in it is *word2vec*, based on [10]. However, this is not the only method for representing words in the vector space. The library also includes such methods as *Fasttext* (an extension of *word2vec* model), *VarEmbed* and *WordRank*.

3.4   Global Vectors for Word Representation

GloVe, written in C [14], is an unsupervised learning algorithm for obtaining vector representations for words. Training is performed on aggregated global word-word co-occurrence statistics from a corpus, and the resulting representations showcase interesting linear substructures of the word vector space.

4 Results

4.1 Research Methodology

During the research in order to create *Word Embeddings*, various tools were tested. First, we focused on the *DMTK* tool. Based on the results achieved by this program, it was determined what model, architecture and configuration achieves the highest accuracy and in what time the process of creating "*word embedded*" model is performed. The program was run five times for each configuration, and the model that achieved the best results was used for further analysis.

In order to be able to compare the tools, each of them was fired in the same configuration, determined during the *DMTK* research. It was checked at what time (the number of iterations was changed during model learning) the best results are achieved.

While determining the *accuracy* of *word embedded*, a Python library called *Gensim* was used. It allows loading an existing *word embedded* model. Using the *accuracy* function and a special file containing sets of words related to each other, it can be checked if the relationship between the words in the model is correct. A set of words related to each other is e.g.

brother he she sister

the "*accuracy*" function, substitutes the appropriate vector for the words, and then checks the dictionary whether the nearest vector for operation:

brother - he + she

is a vector representing *sister*. The nearest vector is determined based on one of the most popular measures of semantic similarity – a *cosine distance* given by the dot product between two normalized vectors. Denoting the cosine of the angle between the two vectors, the cosine similarity can take values in the interval $[-1, 1]$.

The file used to check the *accuracy* is a file provided by the creators of the program "*word2vec*" [10]. It contains collections of words grouped according to their relationship. The given example comes from the group of family relationships; other available examples: the capitals of states, cities, adjectives, adverbs and antonyms.

The recognition of hate speech consisted in checking whether the generated vector representation could be used to recognize hate speech.

For this purpose, the *LightGBM* tool was used, an add-on for *Python - train* function. Before each test, the base was subjected to cross validation. The input embedding size was cross validated (we performed multiple evaluations on 3 different test sets, then to the scores from those evaluations were combined).

The database used to create *word embeddings* models is a collection of tweets in English. This collection was created by combining two datasets made available on the internet in the form of a *csv* file [6, 19]. 1 603 399 entries are available after the merger. These datasets were previously used for various tasks of text classification [6].

The training set consisted of 70% of all entries, while the remaining data was used as a test base. Before the data from the files could have been used for *Word Embeddings*, entries had to be transformed, namely, special characters such as periods and commas were deleted, links to web pages, hashtags removed and all words were changed to lowercase letters. As a result, a text containing 403 072 unique words was obtained. However, not all of them were used, because the assumption was made that a word must appear at least 2 times in a dataset.

The *hate-speech-identification* file [6] was used to study classifiers. It consists of 24784 marked entries. Each entry was given one of three classes - hate speech, offensive and neutral. This data has been processed so that the entries are divided into two groups: neutral and offensive. The offensive group included entries tagged as hate speech and offensive. Thanks to this, classifiers can immediately identify which comments should be blocked or checked, and which do not require any intervention.

Another file [19] has marked data, for the purpose of whether the entry contains positive or negative emotions, and does not contain information about hate speech. Therefore, only some of entries from this file were used in the training database and marked as neutral.

4.2 Comparison of *Word Embedding* Methods

Charts in Fig. 1 show how the *word2vec* method of *DMTK* performs in comparison to *GloVe* and *Gensim*. Despite the fact that *DMTK* uses a similar algorithm to that used in *Gensim*, it provides much worse results in terms of *precision* (relation between true positives and the total number of true positives and false positives), even after the same number of iterations. Additionally, iterations (epochs) in *Gensim* are performed 5 times faster than those in *DMTK*.

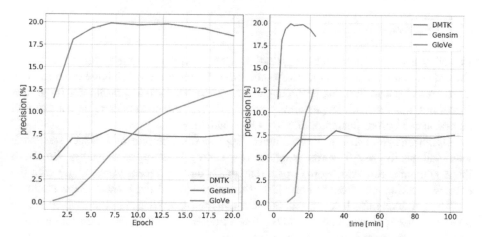

Fig. 1. Precision vs. # of iterations vs. / precision vs. time charts

GloVe also performs better than *DMTK*, but not as well as *Gensim*, which achieves better *precision* much faster than *GloVe*.

A python library *Gensim*, which uses the *word2vec* method, proved to be the best of the solutions to create *word embedded* with highest *precision* of three solutions.

4.3   Hate Speech Recognition – Results

The results of the research are presented in Table 1. One can notice very good results reaching up to 95%, which is quite an achievement. One can also observe the influence of the size of the learning set, i.e. number of entries on accuracy of the classifier. The more entries, the better results.

Table 1. Accuracy of the classifier vs. learning dataset size

Total entries	Hate speech entries	Accuracy [%]		
		DMTK	Gensim	GloVe
34676	20609	89	**90**	88
54582	20609	90	**92**	91
124225	20609	94	**95**	**95**

The best results are achieved for the vectors generated by the *Gensim* tool. For each size of the database, the classifier achieves the best results. Only for the largest data set *GloVe* results are matched by *Gensim*. Nevertheless, *DMTK* is also performing well, achieving the accuracy of the classifier at 90%, a few percent less than *Gensim*.

Total entries in Table 1 means total number of entries in a dataset used for learning. Test dataset was enriched by file [18] which had no (direct) information about hate speech and its entries were considered as neutral, hence *Hate speech* column has always constant value of hate phrases.

5   Conclusion

Detecting hate speech on websites is not an easy task, mainly due to large number of phrasemes and collocations [1, 2]. In recent years, many solutions have emerged that allow us to believe that this problem can be eliminated by using computer programs. Hopefully, this task will be performed automatically, and thus undesirable content will be eradicated before posted online for wider audience.

During the research the *Word Embedding* module (the part of *DMTK*) was used, which allows to present words using vectors, coding in them the meaning. Thanks to this a computer can learn words of similar importance. In recent years, along with the extraction of features from the text [20], this is considered the most popular method.

Then, to recognize hate speech, one can use the *LightGBM* classifier, which is also a ready-to-use tool provided by *DMTK*. It has implemented machine learning methods, specifically *gradient boosting* algorithms.

To generate a vector representation of words, one can also use tools such as *GloVe* and *Gensim*. Direct comparison demonstrated that they get better results (*precision*), and the process of creating vectors runs faster than in the case of *DMTK*. The final effect, that is, the classification of comments, using these representations, speaks in favor of *GloVe* and *Gensim*. However, the difference in *accuracy* of the classification does not differ significantly, only by 1–2%.

The *accuracy* of the *LightGBM* classifier, using the vector representation provided enormously good results. *Accuracy* reaches even 95% for the representation created by *Gensim* and *GloVe*, for a database of 124 225 entries. *DMTK* falls behind by only 1%.

The research also shows which vocabulary models learned best. These are words related to the family and various forms of verbs. This is due to the nature of entries posted on Twitter. These are usually short entries about who, with whom and what did.

Differences in *precision* of scrutinized tools originates in implementation of built-in algorithms *DMTK* uses traditional *word2vec*, whereas *Gensim* uses *Fasttext* (which is essentially an extension of *word2vec* model [22]) that has following advantages over *Glove* and *DMTK*: generate better word embeddings for rare words and can construct the vector for a word even if word doesn't appear in training corpus. Main differences between Glove and word2vec can be [23].

In summary, *DMTK* is a good tool providing ready-to-use methods that allowing for fast hate speech recognition even for fairly inexperienced researchers. Recognition with the accuracy of 94% is a great result, giving the opportunity to create an application to block inappropriate content on websites. The only annoyance is that training consumes disproportionately more time than in case of competitors. *DMTK* remains a good worthy alternative, which is far more versatile than other free-of-charge tools.

References

1. Badjatiya, P., Gupta, S., Gupta, M., Varma, V.: Deep learning for hate speech detection in tweets. In: Proceedings of the 26th International Conference on World Wide Web Companion, pp. 759–760. International WWW Conferences Steering Committee (2017)
2. Bengio, Y., Ducharme, R., Vincent, P., Jauvin, C.: A neural probabilistic language model. J. Mach. Learn. Res. **3**(6), 1137–1155 (2003)
3. Bian, J., Gao, B., Liu, T.Y.: Knowledge-powered deep learning for word embedding. In: Joint European Conference on Machine Learning and Knowledge Discovery in Databases, pp. 132–148. Springer, Heidelberg (2014)
4. Collobert, R., Weston, J.: A unified architecture for natural language processing: deep neural networks with multitask learning. In: Proceedings of the 25th International Conference on Machine Learning, pp. 160–167. ACM (2008)
5. Collobert, R., Weston, J., Bottou, L., Karlen, M., Kavukcuoglu, K., Kuksa, P.: Natural language processing (almost) from scratch. J. Mach. Learn. Res. **12**, 2493–2537 (2011)
6. Davidson, T., Warmsley, D., Macy, M., Weber, I.: Automated hate speech detection and the problem of offensive language. In: Proceedings of the 11th International Conference on Web and Social Media (ICWSM) (2017). https://data.world/crowdflower/hate-speech-identification. Accessed 21 Jan 2018
7. Djuric, N., Zhou, J., Morris, R., Grbovic, M., Radosavljevic, V., Bhamidipati, N.: Hate speech detection with comment embeddings. In: Proceedings of the 24th International Conference on World Wide Web, pp. 29–30. ACM, May 2015
8. Hastie, T., Tibshirani, R., Friedman, J.: The Elements of Statistical Learning. Springer Series in Statistics (2001)
9. Jozefowicz, R., Vinyals, O., Schuster, M., Shazeer, N., Wu, Y.: Exploring the limits of language modeling (2016). arXiv preprint arXiv:1602.02410
10. Mikolov, T., Chen, K., Corrado, G., Dean, J.: Efficient estimation of word representations in vector space (2013). arXiv preprint arXiv:1301.3781

11. Mikolov, T., Sutskever, I., Chen, K., Corrado, G.S., Dean, J.: Distributed representations of words and phrases and their compositionality. In: Advances in Neural Information Processing Systems, pp. 3111–3119 (2013)
12. Kim, Y., Jernite, Y., Sontag, D., Rush, A.M.: Character-aware neural language models. In: AAAI, pp. 2741–2749, February 2016
13. Li, Y., Xu, L., Tian, F., Jiang, L., Zhong, X., Chen, E.: Word embedding revisited: a new representation learning and explicit matrix factorization perspective. In: IJCAI, pp. 3650–3656, July 2015
14. Pennington, J., Socher, R., Manning, C.: Glove: global vectors for word representation. In: Proceedings of the 2014 Conference on Empirical Methods in Natural Language Processing (EMNLP), pp. 1532–1543 (2014)
15. Ross, B., Rist, M., Carbonell, G., Cabrera, B., Kurowsky, N., Wojatzki, M.: Measuring the reliability of hate speech annotations: the case of the European refugee crisis (2017). arXiv preprint arXiv:1701.08118
16. Rehurek, R., Sojka, P.: Software framework for topic modelling with large corpora. In: Proceedings of the LREC 2010 Workshop on New Challenges for NLP Frameworks (2010)
17. Silva, L.A., Mondal, M., Correa, D., Benevenuto, F., Weber, I.: Analyzing the targets of hate in online social media. In: ICWSM, pp. 687–690, March 2016
18. Schmidt, A., Wiegand, M.: A survey on hate speech detection using natural language processing. In: Proceedings of the Fifth International Workshop on Natural Language Processing for Social Media, pp. 1–10 (2017)
19. Twitter Sentiment Analysis Training Corpus (Dataset). http://thinknook.com/twitter-sentiment-analysis-training-corpus-dataset-2012-09-22/. Accessed 21 Jan 2018
20. Waseem, Z., Hovy, D.: Hateful symbols or hateful people? Predictive features for hate speech detection on twitter. In: Proceedings of the NAACL Student Research Workshop, pp. 88–93 (2016)
21. Word2Vec algorithm. https://code.google.com/archive/p/word2vec/. Accessed 21 Jan 2018
22. https://www.quora.com/What-is-the-main-difference-between-word2vec-and-fastText. Accessed 15 Mar 2018
23. Baroni, M., Dinu, G., Kruszewski, G.: Don't count, predict! A systematic comparison of context-counting vs. context-predicting semantic vectors. In: Proceedings of the 52nd Annual Meeting of the Association for Computational Linguistics (Volume 1: Long Papers), pp. 238–247 (2014)

Optimization Techniques for Modelling Energy Generation Portfolios in Ukraine and the EU: Comparative Analysis

Volodymyr Zaslavskyi[✉] and Maya Pasichna

Taras Shevchenko National University of Kyiv,
Volodymyrska Street 64, Kyiv 01033, Ukraine
zas@unicyb.kiev.ua, maypas@gmail.com

Abstract. The article explores application of LEAP (Long-range Energy Alternatives Planning System) optimization software for developing optimal structure (portfolio) of the electricity generation for Ukraine and seven European Union (EU) member countries. Given the decline of energy demand in Europe, the newly-built power plants are being put on hold, thus resulting in considerable financial losses for energy companies and loss of competitiveness for some energy generation technologies. To find a solution for this situation the developed scenarios for energy transformation pathways have to be evaluated in more detail. The research undertakes bottom-up approach for the analyses of different sectors of the considered economies, which are directly connected with electricity generation industry within two previously developed scenarios – Reference and European. Optimization simulation delivered by the model results in a third Optimum scenario is called to define the optimal expansion and dispatch pathway for the electricity generation system in the upcoming decades.

Keywords: Optimization · Energy planning · Linear programming

1 Introduction

Modern versions of software tools for energy planning, which are supplemented by the possibilities of using integer variables, nonlinear functions, multi-criteria optimization and stochastic programming, have, as a rule, macroeconomic approach. Given that, the representation of the energy generation sector can be somewhat simplified.

Thus, there is a need to use energy modelling that could allow the formation and prompt change of input information (scenarios), criteria and indicators of the decisions taken, make it possible to present the country's energy generation sector in more detail. One solution is the approach embodied in reporting models such as LEAP, the Long-range Energy Alternatives Planning System, a software tool for energy policy analysis. In LEAP the user indicates the expected results, and the tool manages the data and analyzes the effect of the result on a certain "what if" energy scenario.

A bottom-up modelling approach, characteristic for LEAP, allows addressing sectoral and sub-sectoral levels of the entire energy system (consisting of energy exploration, generation, transmission and distribution), thus contributing to Types

© Springer International Publishing AG, part of Springer Nature 2019
W. Zamojski et al. (Eds.): DepCoS-RELCOMEX 2018, AISC 761, pp. 545–555, 2019.
https://doi.org/10.1007/978-3-319-91446-6_51

Variety Principle (TVP) [1, 2]. Within LEAP modelling TVP can be seen as a combination of different areas influencing energy generation sector, e.g. economy trends, final energy demand, energy transformation and other. When combined in a system the analysis of all these areas could improve the understanding of how to deliver reliable, long-term performance of such critical system as energy generation sector.

Widely-used approaches for the optimization of energy generation structure (that is a distribution of primary energy sources consumed to produce various types of energy), such as Markowitz Mean-Variance Portfolio (MVP) analysis, are based on quadratic programming and are mainly a static methodology, strongly relying on past and cost/revenue data [3]. Linear programming is characterized for a fuzzy portfolio optimization, which is based on the mean-absolute deviation (MAD); however it is characterized by a large set of possible choices thus making the decision-making more difficult [4]. Exploration and production portfolio optimization model (EPPO) approach is also based on Monte Carlo simulation and linear programming, but is criticized for deterministic NPV analysis and is intended for petroleum industry [5]. There are studies on using AHP (Analytic Hierarchy Process) for defining priorities and ranking of different electricity generation sources. Given a possibility to combine qualitative along quantitative assessment, AHP cannot account for all the necessary factors influencing a studied object [6, 7]. By contrast, LEAP is characterized by low initial data requirements, detailed energy sector coverage, ability to account for significant technological changes.

LEAP has been applied to explore how global energy generation structures can be reconfigured all over the world. For the EU, LEAP was utilized in 2009 to create a detailed sector-by-sector emission mitigation scenario for 27 EU countries [6]. In Ukraine for the purposes of assessment of the energy transformation pathways "TIMES-Ukraine" modelling was applied [9]. Although the use of LEAP is not new, there is no information about LEAP application for Ukraine's energy generation sector; moreover, comparable analysis of energy generation structure optimization for Ukraine and the EU has not been widely covered before.

This study applies LEAP within the strategic planning of an optimal diversified electricity generation structure. It presents analysis of the development and optimization of the current energy generation portfolios for Ukraine and seven EU member states (Germany, France, Belgium, Italy, Spain, Sweden and Finland). The selection of EU countries is justified by the fact that they are headquarters to the leading EU energy companies and their energy policies have impact on the global energy market.

The findings of the research can be used by electricity generating companies in general and in other countries and/or regions while developing the energy generation portfolio, by the governments while delivering energy and environmental strategies.

The paper is organized into three sections: (1) In "Materials and Methods" the methodology as well as methods of collecting input data and delivering calculations are described; "Results" elaborates on the outcomes of the LEAP software calculations; Discussions on the findings of the research are covered in section "Conclusions".

2 Materials and Methods

A General Description of the Model. The model analysis in LEAP has been built around a series of integrated modules, namely Key Assumptions, Demand, Transformation and Resources (Fig. 1).

Key assumptions cover levels of GDP (Gross Domestic Product), income and population, number and size of private households (PHHs). Demand module covers four energy end-use sectors: PHHs, industry, transport and services. The PHHs sector takes into account electrified buildings; industry includes chemical and petrochemical, metallurgical and pulp and paper sectors; transport covers passenger and freight, both automobile (including electric vehicles) and railway; the service sector covers commercial buildings (non-residential). The transformation module includes natural gas losses during transmission and distribution of electricity. Coal, natural gas, hydropower, wind energy, solar energy, biomass and nuclear energy are taken for primary energy resources, diesel, gasoline and electricity - for secondary energy.

The timeline of the analysis is 2015 (base year) to 2030. Forecast includes the "Reference scenario" (RS), the "European" and "Optimum" scenarios.

The RS provides for such an economic development that would follow the past trends. In demand area, on the side of PHHs there is a slow decline in the consumption of solid fuels and electricity, in the industry there is a decrease in final energy demand by 1.3% per annum on average with a reduced use of coal and electricity and increased use of natural gas and RES (renewable energy sources). There is an increase in consumption of diesel fuel and electricity and decrease in consumption of gasoline fuel in transport, decline in energy use by the commercial sector. In the area of energy transformation, natural gas losses make up 20% for Ukraine and an average of 3.9% for the EU countries. Installed capacity of coal, gas and nuclear generation in the EU is being reduced compared to 2015; while Germany and Belgium expect phase-out of the nuclear generation facilities. By contrast, wind and solar generation show a moderate capacity expansion. In Ukraine there is a moderate growth of almost all types of electricity generation, except for nuclear.

The European scenario shows more ambitious targets in terms of economic development, increased energy efficiency, accelerated retirement of fossil- and nuclear-based generation, transformation towards the use of RES, etc. The European scenario differs from the RS in terms of higher GDP growth (1.8% against 1.2% per annum in RS for EU); In Ukraine, there is no further decline in population and GDP. On demand side for all the countries there is a steep drop in the use of electricity in PHHs (−60% for lighting, −50% for refrigeration, −42% for cooking, −30% for heating and cooling), coal (increase for Ukraine) and natural gas. In industry, for all the countries, there is a decline in the use of coal (1.1% per year) and growth (0.5% per year) in the consumption of natural gas, solar energy, biomass and electricity. In transport (passenger cars), diesel and gasoline consumption is reduced, share of electrical vehicles grows to 10% in 2030 (5% for Ukraine). In the service sector, consumption of natural gas and electricity in the period 2015–2030 falls by 33% and 50% respectively.

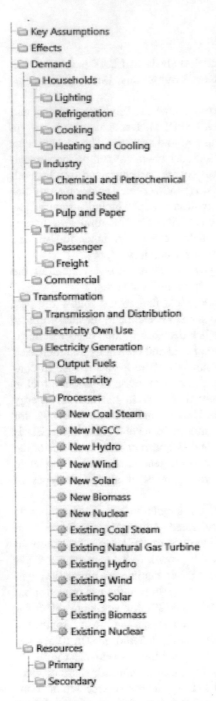

- Key Assumptions
- Effects
- Demand
  - Households
    - Lighting
    - Refrigeration
    - Cooking
    - Heating and Cooling
  - Industry
    - Chemical and Petrochemical
    - Iron and Steel
    - Pulp and Paper
  - Transport
    - Passenger
    - Freight
  - Commercial
- Transformation
  - Transmission and Distribution
  - Electricity Own Use
  - Electricity Generation
    - Output Fuels
      - Electricity
    - Processes
      - New Coal Steam
      - New NGCC
      - New Hydro
      - New Wind
      - New Solar
      - New Biomass
      - New Nuclear
      - Existing Coal Steam
      - Existing Natural Gas Turbine
      - Existing Hydro
      - Existing Wind
      - Existing Solar
      - Existing Biomass
      - Existing Nuclear
- Resources
  - Primary
  - Secondary

Fig. 1. LEAP hierarchical tree with modules and corresponding branches

Natural gas losses are reduced by 50%. Installed capacity in Ukraine is growing; in the EU the decline in fossil-based and nuclear capacity is expected against the increase of the RES-based generation.

In order to find the Optimum scenario for the development of energy generating capacities, that takes into account generation costs, necessity to meet the required demand, fluctuations in the daily loadings of the grid, and the GHG emissions, the linear programming of the LEAP OSeMOSYS (Open Source Energy Modelling System) was applied [8].

Data Collection. The data for RS were borrowed from the developed RS scenario using TIMES-Ukraine for Ukraine [9] and existing RS based on PRIMES modelling for the EU countries [10]. In the absence of the necessary data, a linear forecast based on trends for the period 2010–2015 was calculated. The data for the European scenario were borrowed from the developed scenarios for the countries which are not reference scenarios, but are more progressive in terms of the targets for economy development, fuel consumption and the transition to RES [11–18].

The environmental emissions and impacts associated with fossil-based technologies are calculated by LEAP based on the IPCC (Intergovernmental Panel on Climate Change) emission factors.

Main data sources include the database of Eurostat [19], the State Statistics Service of Ukraine [20] and ENTSO-E [21].

Calculations. The calculation algorithm in LEAP is constructed around the calculation of variables such as:

Demand calculations:

$$D_{t,s,y} = AL_{t,s,y}EI_{t,s,y}, \qquad (1)$$

where D - energy demand which is the product of the total activity level (AL) and energy intensity (EI) at each given technology branch (fuel consumed); AL - measure of social or economic activity for which fuel is consumed (e.g. share of households consuming biomass); EI - annual energy consumption per unit of activity (e.g. MWh per household); t - type of fuel consumed; s - scenario; y - year in scenario.

Transformation calculations:

Each module (electricity generation, electricity own use, transmission and distribution (power losses) (Fig. 1)) is operated to meet the demands that arise from domestic requirements and export demands. After each module is calculated, domestic requirements are decremented by the outputs produced from the module, but incremented by the input fuels required by the module.

Emission calculations:

$$DCE_{t,s,y} = \sum_t \sum_s \sum_y AL_{t,s,y} EI_{t,s,y} EF_{t,s,y}, \tag{2}$$

where DCE - carbon emissions from energy demand, $EF_{t,s,y}$ - carbon emission factor for fuel type t in scenario s in year y.

The optimization function in LEAP is performed through integration with OSeMOSYS which is based on the GNU Linear Programming Kit (GLPK), for solving large linear programming tasks.

The objective function of the OSeMOSYS is called to evaluate such an electricity generation portfolio which is to meet energy demand by minimizing the total discounted cost (5):

$$Min \sum_t \sum_s \sum_y TC_{t,s,y} = OC_{t,s,y} + CC_{t,s,y} + EP_{t,s,y} - SV_{t,s,y} \forall t, s, y \tag{5}$$

where TC, OC, CC, EP, SV represents discounted (5%) total cost, operating cost, capital cost, technology emission penalty and salvage value respectively. The costs of transmission and distribution grids development are not included and has to be accounted for in the future researches.

Constraints are as the following [22]:

Total capacity constraints ensure that the total capacity of each technology in each year is greater than and less than the user-defined parameters:

$$\forall_{y,t,r} TotalCapacityAnnual_{y,t,r} \leq TotalAnnualMaximumCapacity_{y,t,r} \tag{6}$$

$$\forall_{y,t,r} TotalCapacityAnnual_{y,t,r} \geq TotalAnnualMinimumCapacity_{y,t,r} \tag{7}$$

where r - region

New capacity constraints ensure that the new capacity of each technology installed in each year is greater than and less than the user-defined parameters:

$$\forall_{y,t,r} NewCapacity_{y,t,r} \leq TotalAnnualMaximumCapacityInvestment_{y,t,r} \tag{8}$$

$$\forall_{y,t,r} NewCapacity_{y,t,r} \geq TotalAnnualMinimumCapacityInvestment_{y,t,r} \tag{9}$$

Annual activity constraints ensure that the total activity of each technology over each year is greater than and less than the user-defined parameters:

$$\forall_{y,t,r} TotalTechnologyAnnualActivity_{y,t,r} =$$
$$\sum_i RateOfTotalActivity_{y,i,t,r} * YearSplit_{y,i} \tag{10}$$

where i - intra-annual time step within a year

$$\forall_{y,t,r} TotalTechnologyAnnualActivity_{y,t,r}$$
$$\leq TotalTechnologyAnnualActivityUpperLimit_{y,t,r} \tag{11}$$

$$\forall_{y,t,r} TotalTechnologyAnnualActivity_{y,t,r}$$
$$\leq TotalTechnologyAnnualActivityLowerLimit_{y,t,r} \tag{12}$$

Total activity constraints ensure that the total activity of each technology over the entire model period is greater than and less than the user-defined parameters:

$$\forall_{t,r} TotalTechnologyModelPeriodActivity_{t,r} =$$
$$\sum_y TotalTechnologyAnnualActivity_{y,t,r} \tag{13}$$

$$\forall_{t,r} TotalTechnologyModelPeriodActivity_{t,r}$$
$$\leq TotalTechnologyModelPeriodActivityUpperLimit_{t,r} \tag{14}$$

$$\forall_{t,r} TotalTechnologyModelPeriodActivity_{t,r}$$
$$\leq TotalTechnologyModelperiodActivityLowerLimit_{t,r} \tag{15}$$

Reserve Margin constraints ensure that sufficient reserve capacity of specific technologies (*ReserveMarginTagTechnology*) is installed to maintain reserve margin defined by user:

$$\forall_{y,r} TotalCapacityReserveMargin_{y,r} =$$
$$\sum_t TotalCapacityAnnual_{y,t,r} * ReserveMarginTagTechnology_{y,t,r} *$$
$$CapacityToActivityUnit_{t,r} \tag{16}$$

$$\forall_{y,i,r} DemandNeedingReserveMargin_{y,i,r} =$$
$$\sum_t RateOfDemand_{y,i,t,r} * ReserveMarginTagFuel_{y,t,r} \tag{17}$$

$$\forall_{y,i,r} TotalCapacityInReserveMargin_{y,r} \geq$$
$$DemandNeedingReserveMargin_{y,i,r} * ReserveMargin_{y,r} \tag{18}$$

3   Results

Demand. As can be seen from Fig. 2 the electricity demand projections for both Ukraine and selected EU countries is mainly falling in Reference and European scenarios.

Whereas in Ukraine the fall is expected in almost all end-use sectors, in the EU countries the fall is associated with all the sectors, except for transport.

The forecast reveals decreasing consumption trends for Ukraine for all the fuels, especially for natural gas and electricity (Fig. 3). For the selected EU countries the decrease in fuel consumption is associated with the use of natural gas mainly.

Observed and forecasted decrease in energy demand in Ukraine can be explained by reduction of own coal production (approximately by 40%) due to the situation in the

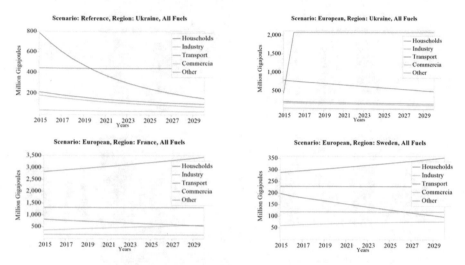

Fig. 2. Eletricity demand forecasts for Ukraine and two of the EU member states (France and Sweden) – split by sectors of final energy use.

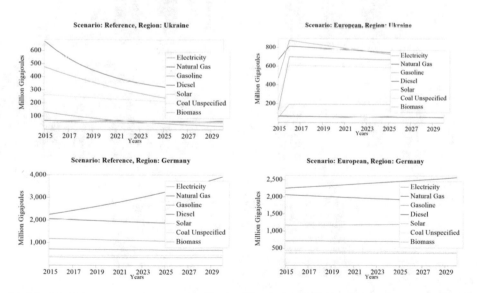

Fig. 3. Eletricity demand forecasts for Ukraine and one of the EU member states (Germany) – split by fuel types.

East of the country, falling demand for electricity among the population and budget institutions due to higher tariffs, as well as in the industry due to the stagnation of the metallurgical complex and machine building, insufficient increase in capacity of RES, in particular, by reducing the value of the "green" tariff (for solar energy producers by 55%, for other producers of electricity from RES - by 50%) [19].

The fall in energy demand in the EU can result from the financial crisis, expansion of the alternative energy carriers (coal) and carbon prices' collapse.

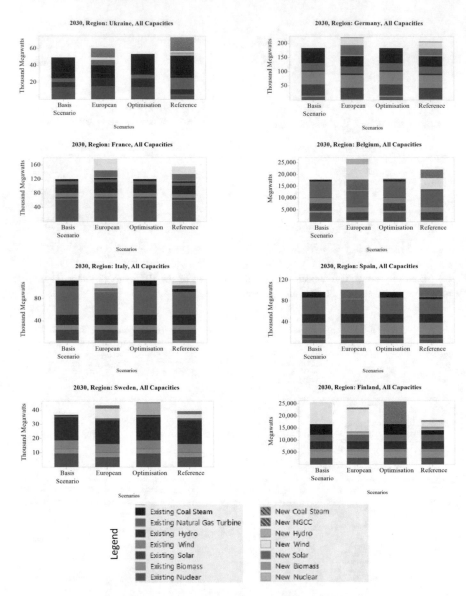

Fig. 4. Forecasts of installed capacity for Ukraine and EU member states for 2030; Basis scenario – situation for the year 2015.

Installed Capacity. Given the projected demand a LEAP model predicts the following optimum structure of the electricity generation (Fig. 4) for the year 2030.

As a result, the Optimum scenario in the EU differs considerably from both other scenarios and shows very moderate increase of capacities; the increase, if any, concerns mainly natural gas and hydro, the Optimum scenario also favors nuclear energy. For Ukraine the increase in capacities results in expansion of hydro facilities.

Thus, the fall in demand for energy and unstable economy contribute to over-capacity in generation (with reserve capacity of more than 40% (for both EU countries and Ukraine) during the highest peak demand). Wind and solar generation appear to be more relevant for EU energy generation portfolio, rather than Ukraine as dependent to a large degree on a green tariff scheme. Biomass energy carrier is as well more characteristic for the EU energy portfolio, especially in Sweden and Finland, but remain capital-intensive fuel. In Sweden and Finland Optimum scenario "prefers" hydro and natural gas capacities for expansion rather than planned wind capacities as those options are cheaper. Nuclear potential is also supported by the Optimum scenario, but can be disadvantaged by the integration of wind and solar generation, and as a consequence, the reduction of load factors (provided that RES is subsidized).

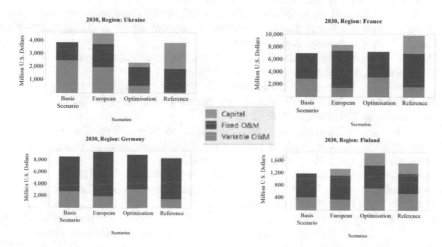

Fig. 5. Cumulative discounted costs by costs categories for Ukraine and some of the EU member states

Costs Associated with Generation. The Optimum scenario appears to be a least cost scenario if compared to RS and European scenario for Ukraine, but not for all the EU countries (Fig. 5), that can be explained by targets on installed capacity. The result for Ukraine has to be taken carefully as there is a lack of data on the costs associated with the reconstruction of coal and the expansion of the hydro power plants.

4 Conclusions

The presented model delivers an assessment of different scenarios for energy planning. Detailed insight into final energy demand is an important determinant for the forecasting. Given the trends, demand management (especially smart houses) and energy storage deserve more attention in terms of its potential for further reductions in energy consumption and peak loading.

References

1. Zaslavskyi, V.: Type variety principle and peculiarities of study of complex systems with a high cost of failure. Bull. Kyiv Natl. Taras Shevchenko Univ. **1**, 136–147 (2006)
2. Zaslavskyi, V., Ievgiienko, Y.: Risk analyses and redundancy for protection of critical infrastructure. Monographs of System Dependability. Oficyna Widawnicza Politechniki Wroclawskiej, pp. 161–173 (2010)
3. Lang, J., Madlener, R.: Portfolio optimization for power plants: the impact of credit risk mitigation and margining. Institute for Future Energy Consumer Needs and Behavior (FCN), E.ON ERC, RWTH Aachen University, Aachen (2010)
4. Madlener, R., Glensk, B., Weber, V.: Fuzzy portfolio optimization of onshore wind power plants. Institute for Future Energy Consumer Needs and Behavior (FCN), E.ON ERC, RWTH Aachen University, Aachen (2014)
5. Qing, X., Zhen, W., Sijing, L.: An improved portfolio optimization model for oil and gas investment selection. Pet. Sci. **11**(1), 181–188 (2014)
6. Zaslavskiy, V., Krasovska, K., Pasichna, M.: Towards the creation of a competitive, diversified energy portfolio for electricity generating companies in EU energy market conditions. Eur. Data Q. **1**(1), 3–20 (2017)
7. Zaslavskyi, V., Pasichna, M.: Type variety principle and the algorithm of strategic planning of diversified portfolio of electricity generation sources. In: Advances in Intelligent Systems and Computing, vol. 582, pp. 474–485. Springer (2017)
8. LEAP Homepage. https://www.energycommunity.org. Accessed 31 Jan 2018
9. Diachuk, O., Chepelev, M., Podoloets, R.: Transformation of Ukraine Towards Renewable Energy in 2050. Heinrich Boll Stiftung, Kiev (2017)
10. Capros, P., Höglund-Isaksson, L., Witzke, H.: EU Reference Scenario 2016. Energy, Transport and GHG Emissions Trends to 2050. The European Commission (2016)
11. REmap 2030: Perspectives of the Development of Renewable Energy in Ukraine. IRENA, Abu-Dhabi (2015)
12. Hasbani, K.: UNECE Renewable Energy Status Report. REN21 Secretariat, Paris (2015)
13. Meunier, L., Vidalenc, E., Vincent, I.: Environmental impacts of French households' final consumption. In: ECEEE Summer Study Proceedings, pp. 247–254 (2015)
14. Energiewende Homepage. https://www.agora-energiewende.de. Accessed 20 Jan 2018
15. Edelenbosch, O.: Comparing projections of industrial energy demand and greenhouse gas emissions in long-term energy models. Energy **122**, 701–710 (2017)
16. Global Transportation Energy Consumption: Examination of Scenarios to 2040 using ITEDD. U.S. Department of Energy, Washington (2017)
17. Conti, J., Holtberg, P., Beamon, J.: International Energy Outlook 2013. With Projections to 2040. U.S. Energy Information Administration, Washington (2013)
18. Transformation of Europe's Power System until 2050, including specific considerations for Germany. McKinsey & Company, Dusseldorf (2010)

19. Eurostat Homepage. http://ec.europa.eu/eurostat/data/database. Accessed 31 Jan 2018
20. State Statistics Service of Ukraine Homepage. http://www.ukrstat.gov.ua/. Accessed 31 Jan 2018
21. ENTSO-E Homepage. https://www.entsoe.eu/data/statistics/. Accessed 31 Jan 2018
22. Howells, M., Rogner, H., Strachan, N.: OSeMOSYS: the open source energy modeling system: an introduction to its ethos, structure and development. Energy Policy **39**(10), 5850–5870 (2011)

An Information Rate of Key Predistribution Schemes with Mutually Complementary Correctness Conditions

Alexander Zatey[(✉)]

National Research University "Moscow Power Engineering Institute",
Moscow, Russian Federation
zateyav@mpei.ru

Abstract. In this paper we study the key predistribution schemes. Such schemes are used to reduce the total amount of secret information distributed to users of the network through secure channels, through which the secret key is calculated to establish a secure network connection between the participants. It is shown as a theoretical and practical advantage of the combined scheme in the scheme efficiency parameters with the help of computer experiments, as well as lower and sufficient estimates of the probabilistic algorithm synthesis of a key distribution patterns with mutually complementary correctness conditions of the two known schemes (KDP, Key Distribution Pattern and HARPS, Hashed Random Preloaded Subset Key Distribution). The concept of the information rate of the scheme is considered as a criterion for the efficiency of the key distribution scheme in the task of minimizing the distributed volume of secret information stored by the network participants. Calculated lower estimates shows potential advantage of combined correcntess conditions in addition to efficiency discovered previously.

Keywords: Computer networks · Information rate · Key predistribution
Probabilistic method of synthesis · Lower estimate · Sufficient estimate
Hashed key distribution scheme · Crypto has-function · Correctness condition

1 Introduction

In this paper, it is proposed to consider the effectiveness of predistribution schemes for keys with mutually complementary conditions of correctness with respect to other known key predistribution schemes.

To connect in a secure mode (in particular, for creating and updating the shared secret information on the server), the parties must have secret keys. These keys must be delivered through secure channels. To reduce the total amount of this secret information, there are previously known predistribution methods, when the parties compute secret keys using the previously received key information and public information from the server.

To reduce the probability of error in the transmission of sensitive information over secure channels, it is desirable to minimize the amount of secret information stored by network participants, as a consequence, by reducing the amount of transmission.

© Springer International Publishing AG, part of Springer Nature 2019
W. Zamojski et al. (Eds.): DepCoS-RELCOMEX 2018, AISC 761, pp. 556–564, 2019.
https://doi.org/10.1007/978-3-319-91446-6_52

This approach is characterized by the concept of information rate scheme [1]. Briefly – key predistribution scheme the more effective the secret information is transmitted over a secure channel.

Key predistribution schemes in the computer network provide for the formation, by a trusted center on the basis of the original secret system key information of packages that are identical in volume, of the secret key information units for each network participant and the transmitting of these packets to the relevant participants. At the same time, the composition of these packages and, possibly, some additional unclassified information about them is published on a public server. The secret key information received by each participant should be sufficient to calculate each of them working keys to communicate with the participants of a certain group from among the groups to which it belongs and the composition of the secret information packs to which it is known. The composition of the groups themselves is also well known and published. Such groups are called privileged. On the other hand, there are so-called forbidden groups of participants. In a correctly constructed scheme, members of such a group on the basis of combining secret information packets received by each of its participants should not be able to compute the working key of any privileged group. This is guaranteed by the condition of correctness of the scheme.

2 Hashed Key Distribution Schemes

In this paper we give a brief description of the properties of predistribution schemes for keys with hashing, a more detailed description can be found in [2].

Let's describe the scheme of preliminary distribution q of system keys with hashing HAKDP (P, F, L) (n, q) (Hashed Key Distribution Pattern) in a computer network of n users, in which coalitions F of participants from the set of F "forbidden" coalitions and groups P participants from the set of P privileged groups of participants. Below, we will interpret such coalitions and groups as the sets of numbers of the subscribers of the network included in the subset of the set $U = \{1, 2, .., n\}$. To obtain such a scheme, n subsets $Ki, i = 1, \ldots, n,$, of system keys assigned to the i-th participant are formed from the initial set $K, |K| = q$ system keys (binary sets of fixed length). For each participant, a pair of numerical sets (Si, Di) is defined and published on the server. The Si sets contain the numbers of the system keys from subsets i, and the Di sets contain the numbers $Di(s), 0 \leq Di \leq L$, applications to these keys of the cryptographic keyless hash function $h : \{0, 1\}^k \rightarrow \{0, 1\}^k$. The images of the system keys obtained as a result of possible repeated application of the hash function to the system key are transferred to the i-th participant via a closed channel.

To calculate the common key of the privileged group P, each participant (i-th network subscriber) must apply the hash function to the received image of the s-th system key $max_{j\in P}D_j(s) - D_i(s)$ times.

The working key computed by each participant of group P from the images of system keys with the numbers from the set $\cap_{i\in P}S_i$, available for each such participant should not be calculated by the participants of the forbidden group on the basis of combining the images of system keys obtained by them, that is, by the images of keys from the set $\cup_{j \in F} S_i$.

Then, if a cryptographic hash function is used, the participants of any "forbidden" coalition can not calculate the common key of participants of any privileged coalition.

Studied in this paper HAKDP(P,F,L)(n,q)-schemes, on the one hand, a generalization KDP(P,F)(n,q)-schemes (Key Distribution Pattern) [1, 3], in which hashing is not applied, and which are described by sets of sets S_i of system key numbers, such schemes were first described in [6]. In this paper, as in [1], they are called systems of intersection systems. On the other hand, they are a special subclass of the so-called HARPS(n,q)-schemes (HARPS—Hashed Random Preloaded Subset Key Distribution) [8,11], in which each participant is delivered all the system keys from the set K ($\forall i \in P \cup F : S_i = \{1, \ldots, q\}$).

The advantage of the KDP(P,F)(n,q)-schemes and their particular version of theKDP(n,q)-schemes (in them the set P includes all two-element subsets of P, and the set F is all one-element subsets of F) is their unconditional secrecy. The HAKDP(P,F, L)(n,q)-schemes and HARPS(n,q)- schemes are related to the limitation of the computing capabilities of network participants, because they use hash functions.

The concept of HAKDP(P,F,L)(n,q)-schemes was introduced in [6].

An informal explanation of the notion of the HAKDP(P,F,L)(n,q)-scheme is formalized by the following definition.

Definition [2]. HAKDP(P,F,L)(n,q)-scheme, where P and F – are families of subsets of the set $U = \{1, \ldots, n\}$, is called a pair $(\tilde{K}, D)$ of families $\tilde{K} = \{K1, \ldots, Kn\}$ subsets of a finite set K of q elements (system keys) and $D = \{D_1, \ldots, D_n\}$ subsets of the set $\{0, 1, \ldots, L\}$, $|D_i| = |K_i|$ and the elements of the sets D_i correspond in one-to-one correspondence to the elements of the sets $Ki, i = 1, \ldots, n$, satisfying the predicate condition:

$$\forall P \in \mathrm{P},\ F \in \mathrm{F},\ P \cap F = \varnothing :\ \cap_{i \in P} S_i \neq \varnothing$$
$$\wedge \left\{ \left[\cap_{i \in P} S_i \not\subset \cup_{j \in F} S_j \right] \vee \left[\exists s \in \cap_{j \in P} S_j : max_{j \in P} D_j(s) < min_{i \in F} D_i(s) \right] \right\}$$

$$(1)$$

where Si (or Sj)—the set of numbers of elements of the set K, that form the set Ki (or Kj).

Then if there is used a cryptographic hash function, members of any forbidden coalition cannot compute a key shared by participants of any preferred coalition. On the one hand, described HAKDP(P,F,L)(n,q)-schemes, are a generalization of KDP(P,F)(n, q)-schemes (Key Distribution Pattern) that do not use hashing, and which are described by sets of sets Si satisfying the predicate

$$\forall P \in \mathrm{P},\ F \in \mathrm{F},\ P \cap F = \varnothing :\ \cap_{i \in P} S_i \neq \varnothing \wedge \left[\cap_{i \in P} S_i \not\subset \cup_{j \in F} S_j \right] \qquad (2)$$

On the other hand, they are a special subclass of the so-called HARPS(P,F,L)(n,q)-schemes (HARPS, Hashed Random Preloaded Subset Key Distribution) [6] in which each participant recieves all system keys from the set K ($\forall i \in P \cup F : S_i = \{1, \cdots, q\}$) and has a matching predicate of the simple form:

$$\forall P \in \mathrm{P}, \ F \in \mathrm{F}, \ P \cap F = \varnothing : \cap_{i \in P} S_i \neq \varnothing$$
$$\wedge \left[\exists s \in \cap_{j \in P} S_j : max_{j \in P} D_j(s) < min_{i \in P} D_i(s) \right] \tag{3}$$

Predicates (2) and (3) are mutually complementary: for the correctness of the scheme (correspondence to the predicate (1)): it is sufficient to perform at least one of them. Below we shall call the first of them KDP-condition, and the second—HA-condition.

This makes it possible to reduce both the number q of distributed system keys and the number of system keys sent from the trusted center to participants in the network via secure channels, i.e., increasing the information speed of the key distribution system.

3 Estimates of the Volume of Distributed Key Information for the Successful Synthesis of the Key Distribution Scheme with Hashing with the Help of a Probability Algorithm

In [6], the advantages of a probabilistic algorithm over deterministic algorithms are substantiated, namely, the ability to substantially reduce the number of system keys in such a scheme. In work [2] the calculation of the number of keys q, sufficient for successful synthesis of the scheme, was given. In this paper, we describe the method of calculating the minimum number of keys q_{min}, at which a scheme can be constructed. Such a number is practically unattainable in experiments on the synthesis of the circuit; nevertheless, it explains the possibility of constructing schemes with a smaller volume of the key information being distributed relative to a sufficient estimate, and also will allow studying the information speed of the scheme under ideal conditions and finding its maximum value.

Let's describe the calculation of a sufficient estimate:

By definition, the pair $(\tilde{K}, D)$ corresponds in a one-to-one correspondence to a pair of families (S,D), where $S = \{S_1, \ldots, S_n\}$. The probabilistic method for synthesizing a HAKDP(P,F,L)(n,q)-scheme consists in randomly choosing a pair of families (S,D) ((here the numbers s are included in the sets Si with probability p, and the numbers $Di(s)$ in the sets Di — with probability $1/(L+1)$) and subsequent verification of its correspondence to the condition (1).

If the powers of the elements of the set P are equal to g, and the powers of the elements of the set F are w, then we denote the HAKDP(P,F,L)(n,k)-scheme by HAKDP(g,w,L)(n,k).

Let P be the family of all subsets of the set U having power g, F – the family of all subsets of the set U having power w, and $g + w \leq n$.

We estimate the probability $P_{L,g,w}$ that the inequality

$$\max_{i \in P} D_i(s) < \min_{i \in P} D_i(s)$$

in the predicate condition (1) holds for all $S \in \cap_{i \in P} S_i$.

The probability of $P_{L,g,w}$ that these events occur for some t and the specified concrete value $D_{i \in F}(s)$ is minimal (since there are w pieces of such sets) is not less than

$$P'_{L,g,w} = \sum_{i=0}^{L} \frac{1}{L+1} \left(\frac{L-t}{L+1}\right)^g$$

The mathematical expectation of the number of pairs X of sets (P,F), for which the inequality $max_{j \in P} D_j(s) < min_{i \in F} D_i(s)$ is violated in the predicate condition (1) for each s

$$E'[X](q,p,g,w) = \sum_{p \in P} \sum_{\substack{F \in F \\ P \cap F = \varnothing}} (1 - p^g(1-p)^w - p^g(1-(1-p)^w)P'_{L,g,w})^q$$
$$= C_n^g C_{n-g}^w (1 - p^g(1-p)^w - p^g(1-(1-p)^w)P'_{L,g,w})^q \tag{4}$$

In this formula $1 - p^g(1-p)^w - p^g(1-(1-p)^w)P'_{L,g,w}$ is a probability of collision—violation of the above inequality for a particular pair of sets (P,F).

The number q of keys obtained by logarithm of the inequality

$$E'(q,p,g,w) < (1 - E) \tag{5}$$

for E→1 is sufficient for a successful synthesis of the scheme [2]:

$$q < \frac{\log(\frac{1-E}{C_n^g C_{n-g}^w})}{\log(1 - p^g((1-p)^w + p^w P_{L,g,w}))} = \frac{\log((1-E)\frac{g!w!}{(n-g-w+1)\ldots n})}{\log(1 - (p^g((1-p)^w + p^w P_{L,g,w})))} \tag{6}$$

The main difference in the calculation of the minimum estimate is the avoidance of the strict condition of the condition when the specified specific value $D_{i \in F}(s)$ is minimal (since there are w pieces of such a set). In this case, several subscribers of the forbidden group may have a minimum value $D_{i \in F}$, and subscribers of the privileged group, in turn, may have the maximum value $D_{i \in P}$, while the inequality $\max_{i \in P} D_i(s) < \min_{i \in F} D_i(s)$ in the predicate condition (1) is still satisfied.

We present the calculation of the estimate of the minimum distributed volume of key information for the key predistribution scheme with mutually complementary correctness conditions.

We denote the number q_{min} of keys - the minimum, in which the predistribution scheme of keys with mutually complementary conditions can be synthesized by a probabilistic algorithm, for an unlimited number of iterations, i.e. $\lfloor q_{min} - 1 \rfloor$ - is the maximum number of keys at which the distributed key material will not be sufficient to synthesize the circuit for arbitrarily many iterations of the probability algorithm.

We estimate the probability $P_{L,g,w}$ that the inequality

$$\max_{i \in P} D_i(s) < \min_{i \in F} D_i(s),$$

holds for all $S \in \cap_{i \in P} S_i$.

It is suggested immediately to calculate the probability that the selected value $D_{i \in P}$ (s) is greater than all the others in the set, therefore, the probability of this event will be equal to

$$g^{-1}$$

The probability $P_{L,g,w}$ for these events to occur for some i, the specified concrete value $D_{i \in F}$ (s) is minimal (since there are w pieces of such sets $D_{i \in F}$) and the value $D_{i \in P}$ (s) is less than $D_{i \in F}$ (s) is defined as

$$P'_{L,g,w} = \sum_{i=0}^{L} w^{-1} g^{-1} \left(\frac{L-t}{L+1} \right)$$

Thus, when calculating the minimum possible value of q_{min} strict conditions for the fulfillment of inequalities are eliminated, and the most favorable case for the synthesis of a preliminary key distribution scheme with mutually complementary conditions is chosen, when a simple probability of meeting these conditions is calculated.

Further calculations of the value of q_0 made similarly to the above, setting the probability synthesis scheme E = 0 in one iteration of the probabilistic algorithm.

Let us consider the results of calculation of estimates and computer experiments for a series HAKDP(3,w,20)(16,q), w = 2,3,4 and with variable parameters p of the probabilistic algorithm. For p = 1 the series schemes correspond to the correctness condition (3), for p < 1 they correspond to either the correctness condition (2) or (3), that is, the aggregate—the predicate condition (1).

In Table 1, for a series of schemes, the values of q obtained by formula (6), q', values, at which it was possible to synthesize the correct schemes on the computer (Intel (R) Core (TM) i7-4790 CPU @ 3.6 GHz) < 100 s, q_{min} – minimum, at which the predistribution scheme of keys with mutually complementary conditions can be synthesized by a probabilistic algorithm, for an unlimited number of iterations. In [2], a series of schemes was also considered that satisfied the correctness condition (2). Key predistribution schemes that satisfy only the condition of correctness (2) are much inferior in terms of the volume of the key information being distributed both to schemes that satisfy the condition of correctness (3) and to schemes with mutually complementary conditions [3].

Obviously, the minimum value of the number of units of the key information being distributed is significantly less than the sufficient estimate, as well as the value attained during the experiments on the synthesis of schemes. Hence, we can conclude that we should adhere to a sufficient estimate when choosing the parameters of the circuit. It should also be noted that as the parameter p, increases, as the size of privileged and forbidden groups increases, it becomes more difficult to achieve smaller values of the

Table 1. Table of values of volume of distributed key information

p	w								
	2			3			4		
	q	q_{min}	q'	q	q_{min}	q'	q	q_{min}	q'
0,5	466	58	145	603	88	290	1194	124	500
0,6	380	33	110	534	51	230	1142	73	450
0,7	255	21	94	497	32	210	1041	47	430
0,8	180	14	80	441	22	205	835	32	410
0,9	130	10	76	351	15	205	613	23	410
0,95	112	9	75	303	13	205	522	20	410
0,99	99	8	75	268	12	205	461	17	410
1	96	8	75	260	11	205	447	17	410

parameter q. Thus, with the growth of the scheme and groups of subscribers, a problem arises with the increasing volume of necessary computations for the synthesis of the new scheme. The correctness condition (2), corresponding to the KDP(P,F) scheme, allows to reduce the amount of calculations, which, in turn, is of great practical value in the synthesis of the scheme.

4 Information Rate of Key Predistribution Schemes

In [1] D.R. Stinson introduces the concept of the information speed of the schemes. This concept describes the effectiveness of the scheme, which directly depends on the amount of secret information sent over a secure channel from the trusted center to network participants. Minimization of this volume leads to a lower probability of error in the transmission, and to less computing load to the users of the network.

There are two definitions in [1]:

The information rate - minimum value based on distributed secret information for each user:

$$\rho = \min\left\{\frac{1}{S_i}; 1 \leq i \leq n\right\}$$

And the *total information rate* distributed to all users. For the key predistribution schemes, the information rate is defined as the inverse of the volume of the key information being distributed:

$$\rho = \frac{1}{k_p}, k_p \sum_{\{i:P\in P\}} S_i$$

In Table 2 calculate the information rate for the key distribution schemes with mutually complementary correctness conditions. For ease of reading of these values,

Table 2. The values of the information rate parameter $HAKDP(\textbf{P},\textbf{F},20)(16,q)$ - the key predistribution scheme.

p	w								
	2			**3**			**4**		
	ρ	ρ_{max}	ρ'	ρ	ρ_{max}	ρ'	ρ	ρ_{max}	ρ'
0,5	0.268	2.155	0.862	0.207	1.420	0.431	0.105	1.008	*0.250*
0,6	0.274	3.157	0.947	0.195	2.042	*0.453*	0.091	1.427	0.231
0,7	0.350	4.252	0.950	0.180	2.790	0.425	0.086	1.900	0.208
0,8	0.434	5.580	*0.977*	0.177	3.551	0.381	0.094	2.441	0.191
0,9	0.534	6.944	0.914	0.198	4.630	0.339	0.113	3.019	0.169
0,95	0.587	7.310	0.877	0.217	5.061	0.321	0.126	3.289	0.160
0,99	**0.638**	**7.891**	0.842	**0.236**	**5.261**	0.308	**0.137**	**3.714**	0.154
1	0.651	7.813	0.833	0.240	5.208	0.305	0.140	3.676	0.152

each of them is multiplied by 10^{-3}. Also, for q_{min}, the value of the information rate will be ρ_{max}, because of the inverse relationship.

As you can see in this table, the maximum value of the information speed of the circuit, expectedly, is achieved in the column of minimum values q_{min}. In the calculated estimates, both in the sufficient and in the lower, the maximum value of the information rate is achieved not in the particular case (p = 1), when the synthesized circuit corresponds to the HARPS(n,q)-scheme, which confirms the influence of the condition corresponding to KDP(P,F)-schemes.

However, in practical experiments, the increasing demands on the volume of calculations in the synthesis of the scheme exert a strong influence: the larger the volume of privileged and forbidden groups, the longer the scheme is synthesized. Thus, in large schemes, the positive effect of the correctness condition (2), corresponding to the KDP (P,F) scheme, increases, so that smaller values of the parameter p should be chosen. Also, it is obvious that the synthesis of the scheme with the value $q' < q$, gives a positive result with respect to the efficiency of the scheme (the maximum values of the information rate in the corresponding columns); nevertheless, when choosing this parameter one should choose a value close to q', and not to q_{min}.

This study of the information speed and efficiency of the key distribution scheme allowed to substantiate the advantages of $HAKDP(\textbf{P},\textbf{F},L)(n,q)$-schemes of key predistribution with mutually complementary correctness conditions over particular cases of this scheme. Also, it became possible to refine the effective input parameters chosen for the probability algorithm of scheme synthesis.

5 Conclusion

The positive effect of combination key predistribtuion schemes correctness conditions was found on the basis possibility to compute scheme with a lesser volume of secret information, than two separate schemes needed.

In addition to the found positive effect it is became possible to substantiate this effect with an efficiency parameter of such schemes – the information rate of key predistribuition schemes. Also the lower estimate allows us to find the lower bound of secret key information volume have to be transmitted to each participant. The difference between lower and sufficient estimate to scheme synthesis allows us to make fine-selected parameters, which can be extremely useful in the network with a big number of participants. The balance between the lower secret information volume and computing powers will help to synthesize the key predistribution scheme with mutually complementary correctness conditions, which will help to provide to establish a secure connection between coalitions with volume of a chosen number of network participants up to P scheme parameter. This connection can not to be hacked by any coalition of participants up to a specified number in F scheme parameter, by their own secret information obtained from the trusted center.

Acknowledgment. The work was financially supported by the Russian Foundation for Basic Research, project No.17-01-00485a.

References

1. Stinson, D.R.: On some methods for unconditionally secure key distribution and broadcast encryption. Des. Codes Cryptogr. **12**, 215–243 (1997)
2. Frolov, A., Zatey, A.: Probabilistic synthesis of KDP satisfying mutually complementary correctness conditions. In: 2014 Proceedings of International Conference on Advances in Computing, Communication and Information Technology, 16–17 November 2014, Birmingham, UK
3. Dyer, M., Fenner, T., Frieze, A., Thomason, A.: On key storage in secure networks. J. Cryptol. **8**, 189–200 (1995)
4. Mitchell, C., Piper, F.: Key storage in secure networks. Discret. Appl. Math. **21**, 215–228 (1988)
5. Ramkumar, M., Memon, N.: An efficient key predistribution scheme for ad hoc network security. IEEE J. Sel. Areas Commun. **23**(3), 611–621 (2005)
6. Frolov, A., Shchurov, I.: Non-centralized key predistribution in computer networks. In: IEEE Proceedings of International Conference on Dependability of computer Systems DepCos-RELCOMEX 2008, Szklarska Poreba, Poland. Computer Society Conference Publishing Services, Los Alamitos, California, Washington, Tokyo, pp. 179–188 (2008)
7. Ramkumar, M.: Symmetric Cryptographic Protocols. Springer, Berlin (2014)

Author Index

© Springer International Publishing AG, part of Springer Nature 2019
W. Zamojski et al. (Eds.): DepCoS-RELCOMEX 2018, AISC 761, pp. 565–566, 2019.
https://doi.org/10.1007/978-3-319-91446-6

Printed in the United States
By Bookmasters